Gerhard Krauss
Biochemistry of
Signal Transduction and
Regulation

Related Titles

Finkel, T., Gutkind, J. S. (eds.)

Signal Transduction and Human Disease

2003
ISBN 978-0-471-02011-0

Helms, V.

Computational Cell Biology

A Textbook

2008
ISBN 978-3-527-31555-0

Klipp, E., Herwig, R., Kowald, A., Wierling, C., Lehrach, H.

Systems Biology in Practice

Concepts, Implementation and Application

2005
ISBN 978-3-527-31078-4

von Bohlen und Halbach, O., Dermietzel, R.

Neurotransmitters and Neuromodulators, 2nd Edition

Handbook of Receptors and Biological Effects

2006
ISBN 978-3-527-31307-5

Devlin, T. M. (ed.)

Textbook of Biochemistry with Clinical Correlations, 6th Edition

2006
ISBN 978-0-471-67808-3

Schepers, U.

RNA Interference in Practice

Principles, Basics, and Methods for Gene Silencing in C. elegans, Drosophila, and Mammals

2005
ISBN 978-3-527-31020-3

Gerhard Krauss

Biochemistry of Signal Transduction and Regulation

4th, Enlarged and Improved Edition

WILEY-VCH

WILEY-VCH Verlag GmbH & Co. KGaA

The Author

Prof. Dr. Gerhard Krauss
Laboratorium für Biochemie
Universität Bayreuth
95440 Bayreuth
Germany

1st edition 1999
2nd edition 2001
3rd edition 2003
1st corrected reprint 2005
4th edition 2008

Cover
Illustration by Hannes Krauss, Bayreuth

■ All books published by Wiley-VCH are carefully produced. Nevertheless, authors, editors, and publisher do not warrant the information contained in these books, including this book, to be free of errors. Readers are advised to keep in mind that statements, data, illustrations, procedural details or other items may inadvertently be inaccurate.

Library of Congress Card No.: Applied for

British Library Cataloguing-in-Publication Data.
A catalogue record for this book is available from the British Library.

Bibliographic information published by the Deutsche Nationalbibliothek
The Deutsche Nationalbibliothek lists this publication in the Deutsche Nationalbibliografie; detailed bibliographic data are available on the Internet at http://dnb.d-nb.de.

© 2008 WILEY-VCH Verlag GmbH & Co. KGaA, Weinheim

Printed in the Federal Republic of Germany.
Printed on acid-free paper.

Composition Hagedorn Kommunikation GmbH, Viernheim
Printing Betz-Druck GmbH, Darmstadt
Bookbinding Litges & Dopf GmbH, Heppenheim

ISBN 978-3-527-31397-6

For Silvia, Julia, Hannes, and Enno

Biochemistry of Signal Transduction and Regulation. 4th Edition. Gerhard Krauss
Copyright © 2008 WILEY-VCH Verlag GmbH & Co. KGaA, Weinheim
ISBN: 978-3-527-31397-6

Preface

This book was based initially on the lectures on regulation and signal transduction that are offered to students of biochemistry, biology and chemistry at the University of Bayreuth. The first book appeared in 1997 and was written in German. It was then substituted by three English editions that are now followed by the fourth English edition, which includes data and references up to 2007.

Cellular signaling in higher organisms is a major topic in modern medical and pharmacological research, and is of central importance in biomolecular sciences. Accordingly, the book concentrates on signaling and regulation in animal systems and in man. Plant systems are not be considered, and results from lower eukaryotes and prokaryotes are only cited if they are of exemplary character.

It is the aim of the present book to describe the structural and biochemical properties of signaling molecules and their regulation. Furthermore, the tools used for signal transmission and the organizational principles of, and the interplay between, signaling pathways are presented. Signaling processes can be described nowadays more and more on a molecular level, and the structure–function relationships of many central signaling proteins have been worked out. However, it is increasingly recognized that the cell- and tissue-specific functions of signaling proteins have to be described in terms of their organization in supramolecular complexes, and in terms of the interplay between different signaling pathways. I have tried to address these topics in the new edition as far as possible.

Numerous studies in very diverse systems have revealed that the basic strategies of signaling and regulation are similar in all higher organisms. Therefore, the book concentrates on the best-studied reactions and components of selected signaling pathways, and does not try to describe distinct signaling pathways (e.g. the vision process) in a complete way. Due to the huge number of publications on the topic, mostly reviews are cited and original articles have been selected on a more or less subjective basis.

Biochemistry of Signal Transduction and Regulation. 4ᵗʰ Edition. Gerhard Krauss
Copyright © 2008 WILEY-VCH Verlag GmbH & Co. KGaA, Weinheim
ISBN: 978-3-527-31397-6

As compared to the previous edition, the fourth edition contains a new Chapter 1 that is devoted to the basics of cell signaling. Typical signaling proteins are multivalent and multifunctional, and these properties enable signaling proteins to engage in signaling networks of a highly complex nature. I have included some aspects of this topic in the new Chapter 1. The regulation of gene expression formerly discussed in Chapter 1 is now to be found in Chapter 3. Here, I have put more emphasis on chromatin modifications and I have included a section on micro RNAs – a topic of growing importance in gene regulation. The other chapters have been updated in light new emerging principles of functions and interactions of signaling proteins and their organization in larger complexes.

I am grateful to all of the people who have encouraged me to continue with the book. In the first place I want to thank my colleague Mathias Sprinzl for helpful comments and corrections. I am also grateful to Enno Krauss, Hannes Krauss and Yiwei Huang for the figures and structure presentations.

Bayreuth, November 2007   *Gerhard Krauss*

Contents

Biochemistry of Signal Transduction and Regulation. 4^{*th*} *Edition.* Gerhard Krauss
Copyright © 2008 WILEY-VCH Verlag GmbH & Co. KGaA, Weinheim
ISBN: 978-3-527-31397-6

1 Basics of Cell Signaling

1.1
Cell Signaling: Why, When and Where?

One characteristic common to all organisms is the dynamic ability to coordinate constantly their activities with environmental changes. The function of communicating with the environment is achieved through a number of pathways that receive and process signals originating from the external environment, from other cells within the organism and also from different regions within the cell.

In addition to adapting the function of an organism to environmental changes in a signal-directed way, other essential features of multicellular organisms require the coordinated control of cellular functions as well.

The formation and maintenance of the specialized tissues of multicellular organisms depend on the coordinated regulation of cell number, cell morphology, cell location and expression of differentiated functions. Such coordination results from a complex network of communication between cells in which signals produced affect target cells where they are transduced into intracellular biochemical reactions that dictate the physiological function of the target cell (Fig. 1.1). The basis for the coordination of the physiological functions within a multicellular organism is *intercellular signaling* (or intercellular communication), which allows a single cell to influence the behavior of other cells in a specific manner. As compared to single-cell organisms, where all cells behave similarly within a broad frame, multicellular organisms contain specialized cells forming distinct tissues and organs with specific functions. Therefore, higher organisms have to coordinate a large number of physiological activities such as:

– Intermediary metabolism.
– Response to external signals.
– Cell growth.
– Cell division activity.

Biochemistry of Signal Transduction and Regulation. 4th Edition. Gerhard Krauss
Copyright © 2008 WILEY-VCH Verlag GmbH & Co. KGaA, Weinheim
ISBN: 978-3-527-31397-6

Fig. 1.1: Inter- and intracellular signaling. The major way of intercellular communication uses messenger substances (hormones) that are secreted by signal-producing cells and are registered by target cells. All cells produce and receive multiple, diverse signals. The extracellular signals are trans- duced into intracellular signaling chains that control many of the biochemical activities of a cell and can trigger the formation of further extracellular signals.

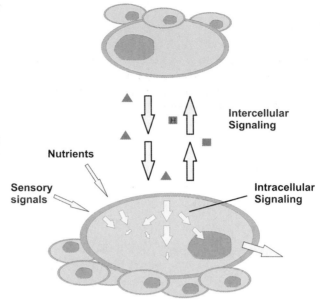

Intercellular Signaling

Nutrients

Sensory signals

Intracellular Signaling

– Differentiation and development: coordination of expression programmes.
– Cell motility.
– Cell morphology.

■ **Intercellular signaling**
– Communication between cells
Intracellular signaling
– Signaling chains within the cell, responding to extracellular and intra- cellular stimuli

Signals generated during intercellular communication must be re- ceived and processed in the target cells to trigger the many intracel- lular biochemical reactions that underlie the various physiological functions of an organism. Typically, many steps are involved in the processing of the signal within the cell, which is broadly described as *intracellular signaling*. Signal transduction within the target cell must be coordinated, fine-tuned and channeled within a network of intracellular signaling paths that finally trigger distinct biochem- ical reactions and thus determine the specific functions of a cell. Im- portantly, both intercellular and intracellular signaling are subjected to regulatory mechanism that allow the coordination of cellular func- tions in a developmental- and tissue-specific manner.

1.2
Intercellular Signaling

Intercellular signal transduction influences nearly every physiological reaction. It ensures that all cells of a particular type receive and transform a signal. In this manner, cells of the same type react synchronously to a signal. A further function of intercellular communication is the coordination of metabolite fluxes between cells of various tissues.

In higher organisms intercellular signaling pathways have the important task of coordinating and regulating cell division. The pathways ensure that cells divide synchronously and, if necessary, arrest cell division and enter a resting state.

Cellular communication assumes great importance in the differentiation and development of an organism. The development of an organism is based on genetic programs that always utilize inter- and intracellular signaling pathways. Signal molecules produced by one cell influence and change the function and morphology of other cells in the organism.

Intercellular signaling pathways are also critical for the processing of sensory information. External stimuli, such as optical and acoustic signals, stress, gradients of nutrients, etc., are registered in sensory cells and are transmitted to other cells of the organism via intercellular signaling pathways.

■ **Intercellular signaling controls**
– Metabolic fluxes
– Cell division
– Growth
– Differentiation
– Development
– Processes sensory information

1.2.1
Tools for Intercellular Signaling

We currently know of various forms of communication between cells (Fig. 1.2):
– *Extracellular messengers.* Cells send out signals in the form of specific messenger molecules that the target cell transmits into a biochemical reaction. Signaling cells can simultaneously influence many cells by messenger molecules so as to enable a temporally coordinated reaction in an organism.
– *Gap junctions.* Communication between bordering cells is possible via direct contact in the form of "gap junctions". Gap junctions are channels that connect two neighboring cells to allow a direct exchange of metabolites and signaling molecules between the cells.
– *Cell–cell interaction via cell surface proteins.* Another form of direct communication between cells occurs with the help of surface proteins. In this process a cell surface protein of one cell binds a specific complementary protein on another cell. As a consequence of the complex formation, an intracellular signal chain is activated which initiates specific biochemical reactions in the participating

Fig. 1.2: Principal mechanisms of intercellular communication. (a) Communication via intercellular messengers. (b) Communication via gap junctions. Gap junctions are direct connections between cells. They are coated by proteins (drawn as circles) that can have a regulatory influence on the transport. (c) Communication via surface proteins.

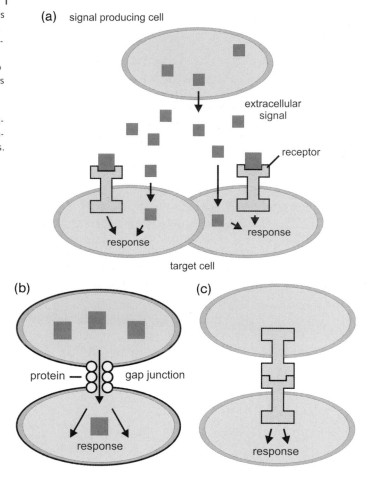

cells. Communication is then only possible upon direct contact between the target cell with the surface protein of the partner cell.

– *Electrical signaling.* A further intercellular communication mechanism relies on electrical processes. The conduction of electrical impulses by nerve cells is based on changes in the membrane potential. The nerve cell uses these changes to communicate with other cells at specialized nerve endings (*synapses*). It is central to this type of intercellular communication that electrical signals can be transformed into chemical signals. This type of communication will not be discussed in this book.

■ **Cells communicate via**
– Messenger substances
– Gap junctions
– Surface proteins
– Electrical signals

In the following, the main emphasis will be on the intercellular communication via extracellular messengers – the hormones.

1.2.2
Steps of Intercellular Signaling

In the communication between cells of an organism, the signals (messengers such as hormones) are produced in specialized cells. The signal-producing function of these cells is itself regulated, so that the signal is only produced upon a particular stimulus. In this way signaling pathways can be coupled to one another and coordinated. The following steps are involved in intercellular communication (Fig. 1.3).

Formation of a Signal in the Signal-producing Cell as a Result of an External Trigger

Most extracellular messengers are produced in response to external triggers and are released by exocytosis. Physical stimuli like electrical signals, changes in ion concentration or, most frequently, other extracellular signaling molecules serve as a trigger to increase the amount of the messenger available for extracellular communication. The me-

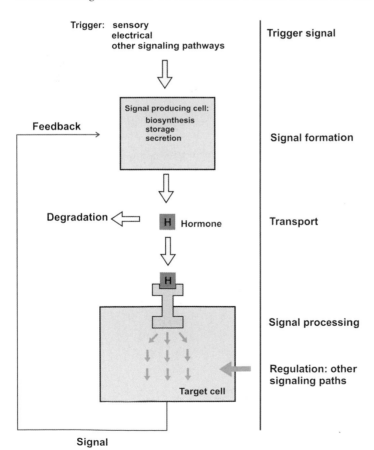

Fig. 1.3: The individual steps of intercellular communication. Upon reception of a triggering stimulus, the signal is transformed into a chemical messenger within the signaling cell. The messenger is secreted and transported to the target cell, where the signal is registered, transmitted further and finally converted into a biochemical reaction. Not shown are processes of termination or regulation of communication which can act at any of the above steps.

chanisms by which the external trigger signals increase the amount of extracellular messenger are diverse, and include stimulation of the biosynthesis of the messenger, increased production of the mature messenger from precursors and release of the messenger from a store form. The latter mechanism is extensively used in the release of hormones of the neural system (neurotransmitters) in response to electrical signals, e.g. at synapses.

■ **Steps of intercellular signaling**
- Trigger signal induces release of stored messenger or stimulates its biosynthesis
- Transport to target cell
- Receipt of signal by the target cell
- Conversion of signal into intracellular signal chain in the target cell

Transport of the Signal to the Target Cell
The extracellular signal produced may be distributed via the circulation or it may reach the target cell simply by diffusion. In long-range signaling via the circulation, the extracellular messenger is often bound to specific carrier proteins or incorporated into larger protein complexes. This may serve to prevent degradation in the extracellular medium or to provide for docking to specific cells only. Furthermore, processing or metabolization of a messenger during transport may convert it from an inactive form to an active form.

Registration of the Signal in the Target Cell
A target cell that receives a signal within the framework of intercellular communication transmits the signal in *intracellular* pathways that trigger distinct biochemical activities in a cell-type-specific manner and determine the response of the target cell. Specialized proteins, termed *receptors*, are utilized for the reception of signals in the target cell. Only those cells that carry the appropriate receptor will be activated for further transduction of the signal into the interior of the cell. The reception of the signals by the receptor is equivalent to the binding of messenger substance on the receptor or the transmission of physical stimuli into a structural change in the receptor.

There are two principal ways by which target cells can process incoming signals:
- Cell surface receptors receive the signal (e.g. a messenger substance) at the outside of the cell, become activated and initiate a signaling chain in the interior of the cell. In such signaling pathways, the membrane-bound receptor transduces the signal at the cell membrane so that it is not necessary for the signal to actually enter the cell.
- The messenger enters into the target cell and binds and activates the receptor localized in the cytosol or nucleus.

1.2.3
Regulation of Intercellular Signaling

The result of communication between the signaling and receiving cells is a defined biochemical reaction in the target cell. The nature and extent of this reaction depends on many individual reactions that participate either directly or indirectly in signal transduction.

Beginning with the hormone-producing cell, the following processes are all contributing factors for hormonal signal transduction in higher organisms (Fig. 1.3):

- *Biosynthesis of the hormone*. The enzymes involved in biosynthesis of a hormone can, for example, be controlled by other signal transduction pathways. There may be feedback mechanisms that couple the activity of the biosynthetic enzymes to the concentration of the circulating enzyme via allosteric mechanisms.
- *Degradation and modification of the hormone*. The active hormone may be inactivated by metabolization or inactive hormone precursors may be converted into the active hormone by enzymatic transformation.
- *Storage and secretion of the hormone*. There are signals (electrical signals, Ca^{2+} signals) to trigger the secretion of stored hormones.
- *Transport of the hormone to the target cell*. The distribution of a hormone via the circulation contributes to the accessibility of that hormone at a particular location of an organism.
- *Reception of the signal by the hormone receptor*. The hormone receptors are primarily responsible for the registration of the signal and the further transduction of the signal in intracellular signaling paths. Therefore, the amount, specificity and activity of receptors at a target cell are major determinants of the final biochemical reaction in the target cell. We know of several regulatory mechanisms that allow regulation of receptor activity. The receptor may be downregulated in response to the amount of circulating hormone or intracellular signaling paths may control receptor activity from inside the cell (Section 5.2.4).

■ **Hormone signaling is mainly regulated via**
- External trigger signals
- Feedback loops
- Degradation
- Modification
- Amount of receptors
- Activity of receptor

All of the above steps are subjected to regulation. A precise control of these steps is at the heart of all developmental programs, and we have gained most information on the control of intercellular communication from developmental studies and from the failure of the control mechanisms, either artificially induced or inborn. The mechanisms for the control of hormone and receptor concentrations are mostly based on feedback regulation. Negative and/or positive feedback loops (Section 1.8.1.1) are used to adjust the intercellular communication to the development and function of the whole organism (reviewed in Freeman, 2000). The feedback controls operate mainly at the level of the enzymes involved in hormone biosynthesis, storage or degradation and via the amount of receptor available for conversion of the extracellular signal into an intracellular response. In most cases, multiple, intertwined feedback loops are used to achieve a fine-tuning of the intercellular communication.

1.3
Hormones in Intercellular Signaling

Signaling molecules for the communication between cells are known as *hormones*. Hormones that are proteins and regulate cell proliferation are known as *growth factors*. The hormones are produced in specialized cells (the hormone-producing cells) or they may be introduced into the organism as inactive precursors (e.g. vitamins) that require metabolic activation for generation of the active form. Examples of the latter type include vitamin D and retinoic acid. Typically, the hormone-producing cells contain biosynthetic pathways that are responsible for the production of the hormone. Furthermore, hormones may be specifically inactivated by modifying enzymes. Details on the metabolism of hormones are beyond the scope of this book.

1.3.1
Chemical Nature of Hormones

The chemical nature of hormones is extremely variable. Hormones can be:
– Proteins.
– Peptides.
– Amino acids and amino acid derivatives.
– Derivatives of fatty acids.
– Nucleotides.
– Steroids.
– Retinoids.
– Small inorganic molecules, such as nitric oxide (NO).

Examples of important hormones are given in Tabs. 1.1–1.3.

Tab. 1.1: Examples for hormones that bind to nuclear receptors.

Hormone	Biochemical and/or physiological function
Steroids Progesterone	preparation of the uterus for implantation of the embryo, maintenance of early pregnancy
Estradiol	preparation of the uterus to receive the blastocyst, control of uterine contraction, generation of secretory system of breasts during pregnancy
Testosterone	differentiation and growth of the male reproductive tract, stimulation of male secondary sex characteristics, skeletal muscle growth
Cortisol	metabolism of carbohydrates, lipids and proteins, antiinflammatory, immunsuppressive, induction of Tyr aminotransferase and Trp cyclooxygenase
Aldosterone	water and ion balance, backresorption of ions in the kidney

Tab. 1.1: Continued.

Hormone	Biochemical and/or physiological function
Steroid-related hormones	
1,25-Dihydroxycholecalciferol (from vitamin D_3)	metabolism of Ca^{2+} and phosphate, bone mineralization, resorption of Ca^{2+} and phosphate in the intestine

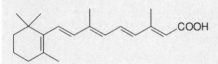

Other hormones
3,5,3′-Triiodothyronine (T_3 hormone)

increased oxygen consumption and increased heat formation, stimulation of glycolysis and protein biosynthesis

Retinoids
All-*trans*-retinoic acid

formed from all-*trans*-retinal, broad effect on differentiation and morphogenesis

9-*cis*-retinoic acid

Tab. 1.2: Examples of hormones that bind to TM receptors.

Hormone	Biochemical and/or physiological function
Epinephrine	raises blood pressure, contraction of smooth muscles, glycogen breakdown in liver, lipid breakdown in adipose tissue
Norepinephrine	contraction of arteries
Histamine	relaxation of blood vessels
Derivatives of arachidonic acid Prostaglandin E2	contraction of smooth muscles

Tab. 1.3: Peptide hormones and protein hormones.

Hormone	Biochemical and/or physiological function
Glucagon (polypeptide: 29 aa)	glycogenolysis in liver, release of fatty acids from triglycerides in adipose tissue
Insulin (polypeptide, α-chain 21 aa; β-chain 30 aa)	stimulation of glucose uptake in muscle and adipose tissue, catabolism of carbohydrates, storage of triglycerides in adipose tissue, protein synthesis, cell proliferation; inhibition of glycogenolysis
Gastrin (polypeptide: 17 aa)	secretion of HCl and pepsin in stomach
Secretin (polypeptide: 27 aa)	stimulation of secretion of pancreatic proteases
Adrenocorticotropin (polypeptide: 39 aa)	biosynthesis in anterior pituitary, stimulation of formation of corticosteroids in adrenal cortex, release of fatty acids from adipose tissue

Tab. 1.3: Continued.

Hormone	Biochemical and/or physiological function
Follicle-stimulating hormone (FSH) (polypeptide: α-chain 92 aa; β-chain 118 aa)	stimulation of growth of oocytes and follicle
Thyrotropic hormone (TSH) (polypeptide: α-chain 92 aa; β-chain 112 aa)	release of thyroxine (T_4 hormone) and of T_3 in thyroid gland
TSH releasing hormone (peptide: 3 aa)	formation in hypothalamus, stimulates synthesis and release of TSH in anterior pituitary
Vasopressin (peptide: 9 aa)	formation in posterior pituitary, backresorption of water in the kidney, contraction of small blood vessels
Parathyroid hormone (polypeptide: 84 aa)	Formation in parathyroid gland, increase of Ca^{2+} in the blood, mobilization of Ca^{2+} from the bone

aa = amino acids.

1.3.2
Hormone Analogs: Agonists and Antagonists

The modification of hormones can lead to compounds that are known as agonists or antagonists.

■ Hormone antagonists
– Bind to receptor and suppress signaling
Hormone agonists
– Bind to receptor and trigger physiological response

– *Antagonists.* Hormone derivatives that bind to a receptor but suppress signal transduction are termed antagonists. Hormone antagonists find broad pharmaceutical and medical application since they specifically interfere with certain signal transduction pathways in the case of hormonal dysregulation. Antagonists with a much higher affinity for a receptor than the unmodified hormone are medically very interesting. Such high-affinity antagonists require very low dosages in therapeutic applications. A few important antagonists and agonists of adrenaline are shown in Fig. 1.4. Propranolol is an example of a medically important hormone antagonist. Propranolol binds with an affinity three orders of magnitude greater than its physiological counterpart, adrenaline, on the β-adrenergic receptor. In this manner a very effective blockage of the adrenaline receptor is possible.

– *Agonists.* Hormone analogs that bind specifically to a receptor and initiate the signal transduction pathway in the same manner as the genuine hormone are termed *agonists.* Application in research and medicine is found especially for those agonists which posses a higher affinity for a receptor than the underivatized hormone.

The ability of hormone derivatives to function as an agonist or antagonist may depend on the cell type under investigation. A notable example is the synthetic estrogen analog tamoxifen. In some tissues, tamoxifen functions as an agonist of the estrogen receptor (ER), whereas in other tissues it behaves as an antagonist of the ER (Section 4.1).

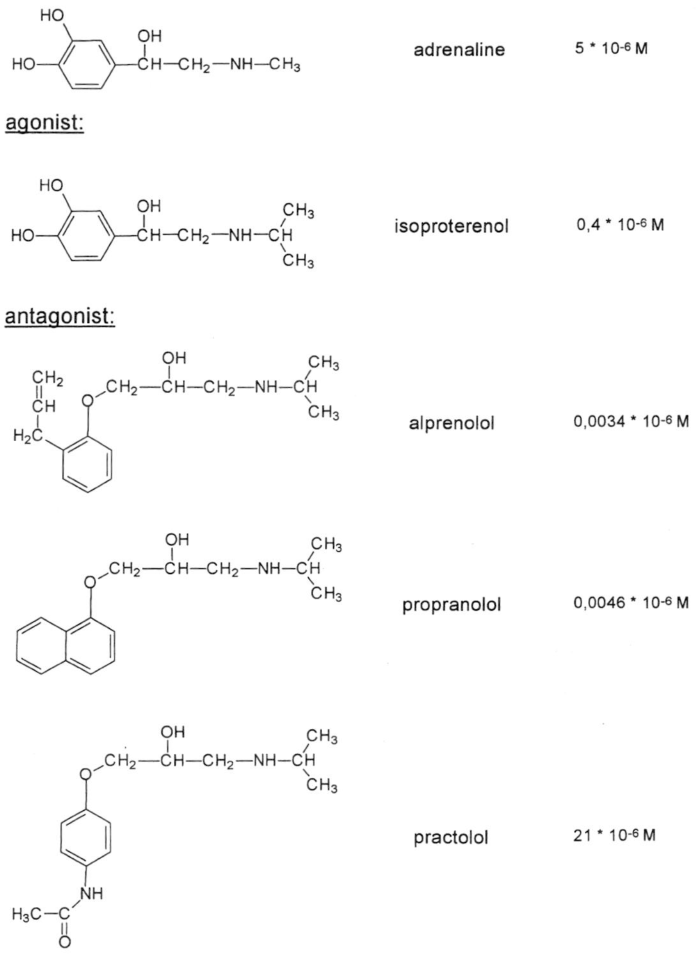

HO, OH
HO—⟨benzene⟩—CH—CH₂—NH—CH₃   adrenaline   5 * 10⁻⁶ M

agonist:

HO, OH CH₃
HO—⟨benzene⟩—CH—CH₂—NH—CH   isoproterenol   0,4 * 10⁻⁶ M
                        CH₃

antagonist:

CH₂ OH CH₃
‖
CH O—CH₂—CH—CH₂—NH—CH
H₂C   CH₃   alprenolol   0,0034 * 10⁻⁶ M
⟨benzene⟩

OH CH₃
O—CH₂—CH—CH₂—NH—CH
            CH₃   propranolol   0,0046 * 10⁻⁶ M
⟨naphthalene⟩

OH CH₃
O—CH₂—CH—CH₂—NH—CH
            CH₃   practolol   21 * 10⁻⁶ M
⟨benzene⟩
NH
H₃C—C
‖
O

1.3.3
Endocrine, Paracrine and Autocrine Signaling

Various forms of intercellular communication by hormones can be discerned based on the range of the signal transmission (Fig. 1.5).

Endocrine Signaling

In endocrine signaling, the hormone messenger is synthesized in specific signaling, or endocrine, cells and exported via exocytosis into the extracellular medium (e.g. blood or lymphatic fluid in animals). The hormone is then distributed throughout the entire body via the circulatory system so that remote regions of an organism can be reached. Only those cells or tissues elicit a hormonal response that contain the appropriate receptor for the hormone.

Fig. 1.5: Endocrinal, paracrinal and autocrinal signal transduction. (a) Endocrinal signal transduction: the hormone is formed in the specialized endocrinal tissue, released into the extracellular medium and transported via the circulatory system to the target cells. (b) Paracrinal signal transduction: the hormone reaches the target cell, which is found in close juxtaposition to the hormone producing cell, via diffusion. (c) Autocrinal signal transduction: the hormone acts on the same cell type as the one in which it is produced.

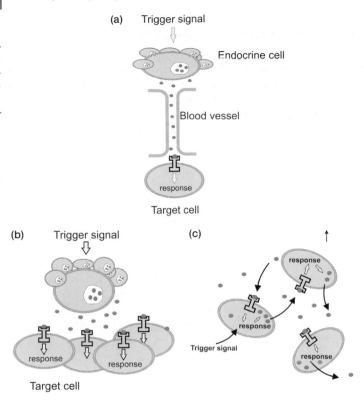

- **Endocrine signaling**
  - Production of hormone in endocrine cells, transport of hormone to target cell via circulation
- **Paracrine signaling**
  - Hormone reaches target cell by diffusion, close neighborhood of signaling cell and target cell
- **Autocrine signaling**
  - Hormone-producing cell and target cell are of the same cell type

Paracrine Signaling

Paracrine signal transduction occurs over the medium range. The hormone reaches the target cells from the hormone-producing cell by passive diffusion. The producing cell must be found in the vicinity of the receiving cells for this type of communication. The signaling is rather local and the participating signaling molecules are sometimes termed *tissue hormones* or *local mediators*. A special case of paracrine signal transduction is synaptic neurotransmission in which a nerve cell communicates with either another nerve cell or with a muscle cell.

Autocrine Signaling

In autocrine signaling, cells of the same type communicate with one another. The hormone produced by the signaling cell affects a cell of the same type by binding to receptors on these cells, initiating an intracellular signal cascade. If an autocrine hormone is secreted simultaneously by many cells then a strong response occurs in the cells. Autocrine mechanisms are of particular importance in the immune response.

1.3.4
Direct Protein Modification by Signaling Molecules

A special case of intercellular signaling is represented by a class of small, reactive signaling molecules, such as NO (Section 6.10). NO is synthesized in a cell in response to an external signal and is delivered to the extracellular fluid. Either by diffusion or in a protein-bound form, the NO reaches neighboring cells and modification of target enzymes ensues, resulting in a change in the activity of these enzymes. NO is characterized as a mediator that lacks a receptor in the classical sense.

1.4
Intracellular Signaling: Basics

External signals such as hormones or sensory signals are specifically recognized by receptors that transduce the external signal into an intracellular signaling chain. The intracellular signaling paths control all functions of the cell such as intermediary metabolism, cell division activity, morphology and the transcription programme.

1.4.1
Reception of External Signals

Cells use two principal ways for transducing external signals into intracellular signaling paths. In one way, the signal receipt and signal transduction occurs at the cell membrane by *transmembrane (TM) receptors* that register the signal at the cell membrane. In the other way, the messenger passes the cell membrane and binds to the *receptor* that is *localized in the cytosol* or in the nucleus (Section 1.5). Upon receiving a signal, the receptor becomes activated to transmit the signal further. The *activated receptor* passes the signal onto components, usually proteins, further downstream in the intracellular signaling pathway, which then become activated themselves for further signal transmission. Depending on the nature of the external stimulus, distinct signaling paths are activated. Finally, specific biochemical processes are triggered in the cell, which represent the endpoints of the signaling pathway.

1.4.2
Activation and Deactivation of Signaling Proteins

■ **Receipt of external signals occurs by**
– TM receptors
– Cytosolic or nuclear localized receptors

Intracellular signal transduction usually comprises many components often acting in a sequential manner where one component passes the signal on to the next component. The key functions in intracellular signaling are performed by proteins that have the ability to specifically recognize, process and transduce signals. The major signal transducers are:
– Receptors.
– Signaling enzymes.
– Regulatory GTPases.

In the absence of the signal, the signal transducers exist in an inactive or less-active ground state. Upon receipt of the signal, the signal transducers become activated and transit into the active state. Only if in the active state is further transmission of the signal to the next signaling component possible. The active state is then terminated after some time by deactivation processes and the signal transducer transits back into the inactive state from which it may start another round of activation and deactivation (Fig. 1.6). A multitude of mechanisms are used to activate the signaling proteins. The major mechanisms are:
– Binding of other signaling molecules.
– Conformational transitions.
– Covalent modifications.
– Membrane targeting.
– Compartmentalization.

■ **Mechanisms for activation of signaling proteins**
– Binding of activators, e.g. hormones
– Signal-induced conformational transitions
– Covalent modifications
– Membrane association
– Removal of inhibitors

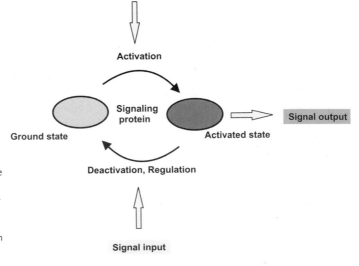

Fig. 1.6: Activation and deactivation of signaling proteins. Activating signals trigger the transition from the inactive ground state into the active state from which signals are passed on to the next signaling component. Deactivating or regulatory signals limit the lifetime of the activated state and induce return into the ground state.

Following activation, the signaling protein must be deactivated in order to terminate or attenuate signaling. By restraining the lifetime of the activated state with the help of specific deactivation mechanisms, the signal flow can be controlled and fine-tuned, and it can be coordinated with signaling through other signaling paths. The mechanisms for deactivation are variable. The deactivation mechanism may be intrinsic to the signaling protein and may be enhanced by specific accessory proteins (see GTPases, Chapters 5 and 9). Other deactivation mechanisms use signal-directed inhibitory modifications of the signaling protein such as phosphorylation. The removal of activating modifications by specific enzyme systems is another way for terminating signaling. The many ways of activating and inactivating signaling proteins are illustrated best by the example of the protein kinases (Chapter 7).

■ **Mechanisms for inactivation of signaling proteins**
– Binding of inhibitors
– Inhibitory modifications
– Removal of activating modifications

1.4.3
Processing of Multiple Signals

A signaling protein often may need to receive several signals simultaneously in order to become fully activated. The ability to process multiple input signals at the same time is based on the *modular structure* of signaling proteins. Many signaling proteins are composed of several signaling domains, each of which may recognize a different signal (Section 1.8). This property allows for the processing of different signals and for the fine-tuning and regulation of signaling.

The main components of intracellular signaling will be discussed in the following; distinct signaling paths will be presented in more detail in later chapters.

1.4.4
Variability of Signaling Proteins

A striking feature of the signaling paths in higher vertebrates is their variability and multiplicity. Different cell types may harbor variants of signaling pathways that control different biochemical reactions. This variability is to a large part due to the existence of subtypes or *isoforms of signaling proteins*. Families of signaling proteins exist whose members have in common a core activity, but differ in the details of substrate recognition and regulation. For nearly all signaling proteins, genes encoding isoforms of a particular signaling protein are found in the genome. In addition, alternative splicing contributes a great deal to the occurrence of multiple forms of signaling proteins.

■ **Isoforms of signaling proteins**
– Increase variability of signaling
– Have similar, but not identical signaling properties
– Are encoded by specific genes
– Arise by alternative splicing

1.5
Molecular Tools for Intracellular Signaling

The main tools for intracellular signal transduction comprise the receptors, signaling enzymes, second messengers and adaptor or scaffolding proteins (Fig. 1.7). The various signaling components cooperate to trigger specific biochemical activities that underlie the many physiological functions of an organism.

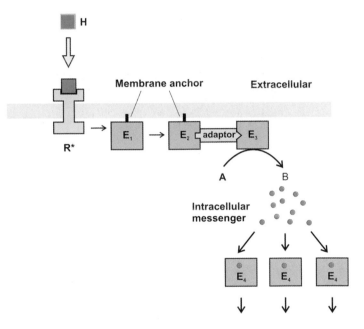

Fig. 1.7: Components of intracellular signal transduction. The reception of an extracellular signal by a membrane receptor, shown here as binding of a hormone to its receptor, activates the receptor for further signal transduction. The activated receptor R* passes the signal onto downstream effector proteins E. Adaptor proteins may be involved in the pathways between effector proteins. The transduction of signal from the receptor to its downstream effector is usually a membrane-associated process. The example shown in the diagram above is only to be construed as an example for the composition of a generic signaling pathway. The structure of the intracellular signaling pathways of a cell are highly variable. There are signal transduction pathways that are much simpler than the one represented here, and others that involve many more components and are much more complicated.

1.5.1
Receptors
Receptors Receive External Signals and Trigger Intracellular Signaling
The first step in processing external signals involves receptors that specifically recognize the signal and initiate intracellular signaling cascades. Signals in the form of hormones are usually produced by specialized cells and initiate a reaction in only a certain cell type. Only those cells that possess a cognate protein, i.e. the receptor of the hormone, can act as target cells. Hormone receptors specifically recognize and bind the cognate hormone based on their chemical nature. The binding of the hormone to the receptor in the target cell induces an intracellular cascade of reactions at whose end lies a defined biochemical response.

In the same way, physical stimuli such as light or pressure can be registered only by those cells that possess the appropriate receptors,

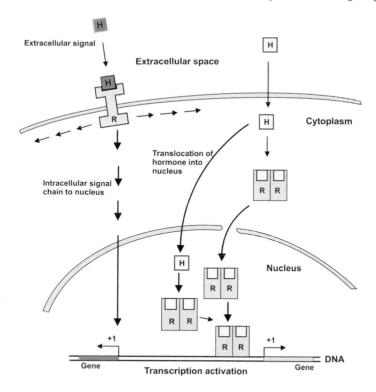

Fig. 1.8: Principles of signal transduction by TM receptors and nuclear receptors. (a) TM receptors receive the signal on the cell surface and convert it into an intracellular signal that can be passed on until it reaches the nucleus. (b) In signal transduction via nuclear receptors the hormone enters the cell and binds the receptor either in the cytosol or nucleus. Nuclear receptors act as nuclear transcription factors that bind specific DNA elements (HRE = hormone-responsive element) found in the promotor region of regulated genes to control their transcription rate.

e.g. rhodopsin in the vision process. Here, excitation of the receptor by the physical stimulus triggers a conformational change in the receptor that is used for further signal transduction.

The receptors of the target cell can be divided into two classes: the membrane-bound receptors and the soluble cytoplasmic or nuclear localized receptors (Fig. 1.8).

Membrane-bound Receptors

These types of receptors represent the largest class of receptors. The membrane-bound receptors are actually *TM proteins*; they display an extracellular domain linked to an intracellular domain by a TM domain. Many TM receptors are found as *oligomers* (dimers or higher oligomers) composed of identical or different subunits. Binding of a hormone to the extracellular side of the receptor induces a specific reaction on the cytosolic domain, which then triggers further reactions in the target cell. The mechanisms of signal transmission over the membrane are diverse, and will be discussed in more detail in Chapters 5, 8 and 11. A characteristic of signal transduction via membrane-bound receptors is that the signaling molecule does not need to enter the target cell to activate the intracellular signal chain.

■ **TM receptors**
– Receive signals at the extracellular side and transmit the signal into the cytosol

Structural parts
– Extracellular domain
– TM domain
– Cytosolic domain

Intracellular Receptors

The most prominent class of the intracellular or cytoplasmic receptors comprises the nuclear receptors that are found in the cytosol and/or in the nucleus (Chapter 4). To activate the nuclear receptors, the hormone must penetrate the target cell by passive diffusion. The nuclear receptors can be classified as ligand-controlled transcription activators. The hormone acts as the activating ligand; the activated receptor stimulates the transcriptional activity of genes which carry DNA elements specific for the receptor.

■ **Intracellular receptor**
- Nuclear and/or cytoplasmic localized
- Function as ligand-controlled transcriptional activators

Interaction between Hormone and Receptor

Receptors are the specific binding partners for signaling molecules; the former are able to recognize and specifically bind the latter based on their chemical structure. The binding and recognition are governed by the same principles and the same noncovalent interactions as those for the binding of a substrate to an enzyme, i.e. hydrogen bonds, electrostatic interactions (including dipole–dipole interactions), van der Waals interactions and hydrophobic interactions. Signaling molecules bind their cognate receptors with an affinity greater than that usually observed for an enzyme and substrate.

The binding of a hormone to a receptor can in most cases be described by the simple reaction scheme:

$$[H] + [R] \leftrightarrow [HR], \text{ with } K_D = [H] \cdot [R]/[HR]$$

where [H] is the concentration of free hormone, [R] is the concentration of the free receptor and [HR] is the concentration of hormone–receptor complex. The value for the equilibrium constant of dissociation, K_D, usually lies in the range of 10^{-6} to 10^{-12} M. This simple formalism is only applicable to cytoplasmic receptors. For membrane-bound receptors, a quantitative treatment of the binding equilibrium is much more difficult.

■ **Hormone–receptor interaction**
- Reversible complex formation due to noncovalent interactions
- Regulatory input signal is the concentration of hormone

Decisive for the intensity of the signal transmission is the concentration of the hormone–receptor complex, since the activation of the signal pathway requires this complex to be formed. The concentration of the hormone–receptor complex depends on the concentration of the available hormone, the affinity of the hormone for the receptor and the concentration of the receptor. All three parameters represent, at least in principle, control points for signal transduction pathways. The variable signal, whose change is registered to thereby activate a signal transmission, is in most cases the concentration of the freely circulating hormone. An increase in the concentration of freely circulating hormone, triggered by an external signal, leads to an increase in the concentration of the hormone–receptor complex, which results in an increased activation of subsequent reactions in the cell. A major switch for the activation of an intracellular signaling pathway is therefore a signal-directed increase in the concentration of the freely circulating hormone ligand.

Regulation of Receptor Activity

The activity of receptors is tightly regulated in order to adapt signaling to the intensity and duration of the extracellular signals. In addition, regulatory mechanisms initiated by intracellular signaling pathways modulate the flow of information through the receptors. Modulation and regulation of signaling through receptors is achieved by multiple mechanisms. The major receptor controls operate at the level of receptor concentration and by receptor modification with concomitant changes in receptor affinity. The amount of receptor present on the cell surface may be regulated via receptor expression, by targeted degradation and by internalization. These processes affect the intensity of the signal transduction on a long timescale. The regulatory receptor modifications are mostly found as phosphorylations that are introduced in response to signals originating from the same or other signaling pathways.

■ **Receptor signaling depends on**
– Hormone concentration
– Receptor concentration
– Receptor activity and modification

1.5.2
Signaling Enzymes

Signaling enzymes are at the heart of intracellular signaling. Enzymes can be regulated by allosteric transitions in response to binding of effector molecules, by covalent modifications such as phosphorylation or by membrane targeting. These mechanisms serve to induce the transition of enzymes from an inactive or low active state into the active state, making enzymes the ideal instrument for the receipt and transmission of signals. The most prominent signaling enzymes are the protein kinases and protein phosphatases, the enzymes involved in synthesis and degradation of second messengers and the regulatory GTPases. We know of different ways by which enzymes participate in signaling:
– *Signaling enzymes modify other enzymes or proteins to carry the signal on or to terminate signaling.* The most frequently used tool for signal transmission in a cell is the reversible modification of proteins by phosphorylation that serves to activate or inactivate signaling proteins. The phosphorylation status of a protein is controlled by the activity of protein kinases and protein phosphatases (Chapter 7). Both classes of enzymes are elementary components of signaling pathways and their activity is subject to manifold regulation. The importance of the protein kinases for cellular functions is illustrated by the large number (more than 500) of different protein kinases encoded in the human genome. Further examples of regulatory protein modifications will be presented in Section 1.6.1.

■ **Signaling enzymes**
- Activate or inactivate other signaling proteins
- Receive and transmit signals
- Produce low-molecular-weight messengers substances – the second messengers
- Switch between active and inactive states

– *Signaling enzymes can catalyze the formation, degradation or release of small-molecule effectors – the second messengers.* The enzymes involved in the formation or degradation of second messengers, such as the phospholipases or the adenylyl cyclases, are major components of signaling pathways (Chapter 6), and are reversibly activated and inactivated during signal transduction.

– *Regulatory GTPases switch between active and inactive conformations, depending on the binding of GDP or GTP.* The regulatory GTPases (Chapter 5) function as switches that can exist in an active, GTP-bound state or the inactive, GDP-bound state. In the active state, the GTPases can transmit signals to downstream components in the signaling chain. In the inactive state, signal transmission is repressed and an activating upstream signal in the form of exchange of bound GDP by GTP is required in order to activate the GTPase for further signal transmission.

1.5.3
Adaptors and Scaffolding Proteins

■ **Adaptor proteins**
- Do not carry enzyme activity
- Provide docking sites for other signaling proteins
- Help to organize multiprotein signaling complexes
- Carry regulatory modifications

Adaptor proteins (Chapter 8) do not harbor enzyme activities. Rather, adaptor proteins mediate the signal transmission between proteins of a signaling chain by bringing these proteins together. They function as clamps to colocalize proteins for an effective and specific signaling. Furthermore, adaptor proteins help to target signaling proteins to specific subcellular sites and to recruit signaling molecules into multiprotein signaling complexes.

In the latter case, the adaptor proteins may function as a scaffold or docking site for assembly of different signaling molecules at distinct sites. The proteins are then also termed docking or scaffolding proteins. Typically, scaffolding proteins contain several binding domains with distinct binding specificities for complementary sites on the target proteins. Furthermore, adaptor proteins are often subjected to regulatory modifications, e.g. phosphorylations, that provide signal-directed docking sites for signaling proteins.

1.5.4
Diffusible Intracellular Messengers: Second Messengers

The intracellular activation of enzymes in a signaling chain can lead to the formation of diffusible small signaling molecules in the cell. These intracellular signaling molecules are also termed "second messengers" (Chapter 6). The second messenger molecules activate and recruit cognate enzymes for the further signal transduction.

The following properties are important for the function of diffusible intracellular messengers:

– *Second messengers may be rapidly formed from precursors by enzymatic reactions.* Typically, the enzymes involved in formation of second messengers are parts of signaling pathways and are activated during signaling to produce the second messenger in a regulated manner. Often, these enzymes have high turnover numbers and can form a large amount of second messenger leading to high local concentrations.

– *Second messengers may be rapidly released from intracellular stores.* For example, the second messenger Ca^{2+} is stored in specific compartments and is released from the storage upon a regulatory signal. This mechanism provides for the fast and locally controlled production of the second messenger.

– *Second messengers may be rapidly inactivated or stored in specific compartments.* To allow for a termination of the second messenger function, the messengers are degraded by specific enzymes or they are removed by storage or transport into the extracellular medium (Section 6.5). As for the messenger-producing enzymes, the messenger degrading enzymes can be part of signaling paths and are regulated by distinct signals.

– *Second messengers may activate different effector proteins.* Binding sites for a particular second messenger (Ca^{2+}, cAMP) may occur on different signaling proteins. This property allows a given second messenger to regulate multiple target proteins which leads to a diversification and variability of second messenger signaling.

– *Second messengers allow amplification of signals.* The enzymatic production of large amounts of a messenger makes an important contribution to the amplification of signals. Typically, the enzymes involved in the metabolism of the second messengers are activated by upstream signals. During the lifetime of the activated state, the enzymes may produce large amounts of the second messenger allowing for the activation of a large number of further downstream messenger targets.

Contrary to what is suggested by the term "diffusible intracellular messengers", these signaling molecules normally do not diffuse across the whole cytoplasmic space. Rather, the second messengers are often used to create signals that are limited in time and in space. The second messengers can be formed as well as inactivated in specific compartments and at specific sites of the cell membrane resulting in locally and timely restricted reactions (see Ca^{2+} signaling, Chapter 6). We know of two types of second messengers: the *cytosolic messengers* and the *membrane-associated messengers*. Cytosolic messengers bind to target proteins in the course of signal transduction, functioning as an effector that activates or modulates signaling through the target protein. The most frequent targets of second mes-

■ Second messengers
– May be formed and inactivated by enzymatic reactions
– May be released from stores
– Are cytosolic or membrane localized
– Activate signaling enzymes
– Allow signal amplification
– Are produced and become active in a timely and locally controlled way

sengers are the protein kinases. Membrane-associated messengers interact with their target protein at the inner side of the cell membrane. In this case, the target proteins may also be membrane-associated or the targets are recruited to the membrane upon binding the second messengers.

1.6
Basic Mechanisms of Intracellular Signaling

1.6.1
Regulatory Modifications

Posttranslational protein modification by covalent attachment of chemical groups to the side-chains of amino acids is a major mechanism by which protein function is regulated in eukaryotes. In cell signaling, regulatory modifications of signaling proteins play a key role in creating, transducing and fine-tuning signals. In addition, protein modification is at the heart of the transcription programme of the cell. Most eukaryotic proteins are modified posttranslationally in one or another form and over 200 protein modifications have been identified (Mann and Jensen, 2003).

We know of stable modifications, e.g. disulfide formation, glycosylation, lipidation and biotinylation, that are essential for vital functions of mature proteins such as compartmentalization, transport and secretion. Many other covalent modifications are transient and are introduced into proteins for regulatory purposes. The reversible modifications may be considered as signals that are transduced by the modifying enzymes to a signaling protein. Such modification signals control the activity, macromolecular assembly and location of signaling proteins, and as such are major tools for shaping signaling pathways.

■ **Regulatory modifications of signaling proteins**
– Are introduced in a signal-controlled way
– May induce activity changes by allosteric mechanisms
– May provide attachments points for target proteins

Based on their function, the regulatory modifications may be divided into two categories:
– Modifications can serve to modulate and regulate the activity of the signaling protein by *conformational and allosteric mechanisms*. As an example, the phosphorylation of protein kinases in the activation loop increases their activity by stabilizing an active conformation (Chapter 7).
– Posttranslational protein modification is used to create *attachment points* for the binding of upstream or downstream effectors in signaling pathways and for the assembly of larger protein complexes. The modifications are recognized by interaction modules on proteins that show specificity of the cognate modification. By this strategy, the posttranslational modifications

serve to increase the specificity of interacting proteins during the formation of larger signaling complexes.

An important aspect of regulatory protein modifications is their reversibility. To serve as a regulatory tool, the modifications must be introduced in a signal-directed way and must be removed upon demand. Typically, formation and removal of the adducts is catalyzed by specific enzymes that are activated in a signal-directed way. The enzymes responsible for introducing and removing the modifications are therefore also essential elements of signaling paths.

Examples of Regulatory Protein Modifications
The most important regulatory protein modifications are:
- Ser/Thr phosphorylation (Chapter 7).
- Tyr phosphorylation (Chapters 7 and 8).
- Lysine acetylation (Section 3.5.2).
- Lysine and arginine methylation (Section 3.5.3).
- Lysine ubiquitination (Section 3.5.5).
- Cysteine oxidation (Section 3.4.6).
- Cysteine nitrosylation (Section 6.10).

The modifications may be of small size, e.g. phosphate or methyl groups, or they may comprise complete small proteins such as ubiquitin. In most cases, specific modifying and demodifying enzymes exist that introduce or remove the modification in a regulated manner. Some modifications, e.g. nitrosylation or cysteine oxidation, do not require specific enzymes for the transfer of these groups to the target enzyme. Here, the intrinsic chemical reactivity of the modifying group is often the major determinant for formation of the covalent adduct. However, specific accessory proteins may be necessary to direct the modification to distinct target sites in these cases.

1.6.2
Recognition of Protein Modifications by Modification-specific Protein Modules

A major function of protein modifications is to provide attachment points for upstream or downstream effector proteins in signaling pathways or to guide the assembly of large protein complexes. The protein partners recognize the posttranslational modifications via *interaction modules* that are able to specifically detect and bind a particular modification. Thus, most of the protein modifications with functions in cell signaling have cognate interaction domains on partner proteins that recognize the chemical nature of the modification. In addition to the modification itself, amino acids C- or N-terminal to

■ **Modification-specific interaction modules**
- Are often found as independently folding domains
- Exist as multiple subtypes
- Bind to distinct protein modifications

Fig. 1.9: Structural basis of modular protein interaction domain function. Domain-peptide binding on the example of the SH3 domain of the protein kinase Csk. The folding of the SH3 domain shows a close proximity of the C- and N-terminus of the module. The binding site for the peptide ligand Pep-PEST is exposed on the surface of the module. Pep-PEST is derived from the C-terminal tail of the tyrosine phosphatase PEP. From Pawson and Nash (2003).

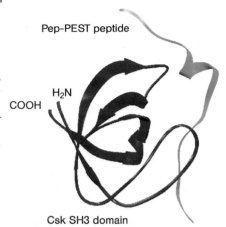

Pep-PEST peptide

H₂N

COOH

Csk SH3 domain

the modification are often used by the interactions domains to select a modification for binding. Importantly, subtypes exist for most interaction domains that recognize the same chemical modification, but differ in the requirements for the neighboring amino acids – a property that serves to create a high diversity of interaction domains. Several interaction domains are present in hundreds of copies in the human proteome and families of interaction domains can be identified that require the same chemical modification but recognize different neighboring sequences. For example, about 115 SH2 domains recognizing phosphotyrosine residues are encoded by the human genome, each of them differing in the details of the sequence requirements of the neighboring amino acids.

Isolated interaction domains can usually fold independently, with their N- and C- termini juxtaposed in space (Fig. 1.9) and are readily incorporated into larger polypeptides in a manner that leaves their ligand-binding surface available. Typical signaling proteins harbor several interaction modules and are thus able to use different protein modifications as attachment points. A more detailed discussion of interaction domains is given below in Section 1.7.

1.6.3
Multisite Protein Modification

Signaling proteins in higher eukaryotes only rarely carry just a single modification. Many proteins involved in cellular regulation are modified at multiple sites – a phenomenon referred to as *multisite modification*. The multiplicity of modification sites on a protein often correlates with its biological importance and the complexity of the corresponding organism. Many proteins perform complex functions in cell signaling by receiving and transducing different signals. Such multifunctional proteins have several interaction partners and use multiple modifications to produce specific signaling outputs in a dynamic and fine-tuned way. Examples of such proteins include the CDC25 phosphatases (Section 13.7), receptor tyrosine kinases (RTKs; Chapter 8), protcin kinase C (PKC; Section 7.5) and the histones (Section 3.3.6). Furthermore, many transcriptional regulators of vertebrates are subjected to multisite modification. The complexity of multisite modifications is illustrated best by the example of the tumor-suppressor protein p53 that is phosphorylated, acetylated, sumoylated and ubiquitinated on many sites (Section 14.6.3).

The following characteristics of multisite modification are important for intracellular signaling:

- *The same modification may occur on several sites of a signaling protein* (e.g. 10 Ser/Thr phosphorylation sites on p53). Each of the sites may be modified by distinct members of an enzyme class (e.g. distinct Ser/Thr-specific protein kinases) and each modification may serve a different function. The neighboring amino acids then specify the functional role of a modification by binding to isoforms of the cognate modification-specific interaction module.

- *The same amino acid can be subject to different modifications.* For example, lysine residues can be modified by acetylation, methylation (Section 3.5), ubiquitination, neddylation and sumoylation (Section 2.5). Lysine methylation leads to the addition of up to three methyl groups to the ε-amino group and, importantly, different methylation stages have different biological consequences (Section 3.5.3). The different lysine modifications may function in a competitive way, one modification excluding the other.

- *The presence of multiple modifications of the same or different type can be considered as a "bar code"* that specifies a distinct function of the signaling protein. The "bar code" controls the catalytic activity of the signaling protein as well as the association of interaction modules, and the code forms the basis of a regulatory programme for the qualitative and quantitative control of its signaling function. An important aspect of multisite modification is the reversibility and the dynamic nature of the modifications. The modification patterns formed change with time and subcellular location which provides a control of function in time and space.

- *Multiple modifications on a protein often show combinatorial characteristics.* The effect of a given modification may be context dependent, e.g. the presence of one modification prevents the modification at another site. Multisite modification events frequently interplay with each other and a cooperative effect of modification events may be observed. In simple scenarios, modifications at two sites can be independent of each other with

■ **Multisite modification of signaling proteins**
- Same modification can occur on several sites
- Different modifications can occur on the same protein
- Same amino acid can be differently modified
- Can be considered a "bar code"
- Can show combinatorial characteristics

each being sufficient to achieve the maximal output. Multiple modifications of the same type could also have additive effects, thereby producing a linear output. In complex scenarios, however, the modifications at different sites can synergize with each other to generate an exponential output, thereby functioning as a combined switch. Such systems can function as coincidence detectors that require multiple modification signals to create a distinct biological output. An example of a nonlinear response to multisite modification is the protein kinase inhibitor Sic1. This inhibitor harbors at least nine Thr phosphorylation sites of which six must be phosphorylated before it is recognized by the CDC4 protein which is a component of the SCF ubiquitination complex (Nash et al., 2001). The F-box protein CDC4 captures the phosphorylated forms of Sic1 for ubiquitination in late G_1 phase, an event necessary for the onset of DNA replication. Interestingly, structural analysis of CDC4 in complex with phosphopeptides suggests that CDC4 contains only one strong phospho (p)-Thr binding site that is surrounded by suboptimal p-Thr binding sites. The mechanistic basis for the cooperative behavior is not well understood. According to one hypothesis (Orlicky et al., 2003), phosphorylation of multiple Thr sites on Sic1 increases the local concentration of sites around CDC4 once the first p-Thr site is bound, to the point where diffusion-limited escape from CDC4 is overwhelmed by the probability of rebinding of any one p-Thr site. In a sense, Sic1 becomes kinetically trapped in close proximity to CDC4. This example illustrates how multiple weak, spatially separated ligand sites can be used to cooperatively interact with a single receptor site. For more examples on multisite modification, see Yang (2005).

1.6.4
Protein Interaction Domains

Most signaling proteins are constructed in a cassette-like fashion from domains that mediate molecular interactions or have enzymatic activity (reviewed in Pawson and Nash, 2003). The molecular interactions mediated by interaction domains provide a fundamental means for organizing signaling pathways and signaling networks.

Protein interaction domains
- Fold independently
- Exist as multiple subtypes
- Assemble signaling complexes
- Control enzyme activity

Interaction domains can target proteins to a specific subcellular location, providing a tool for recognition of protein modifications or chemical second messengers.

Furthermore, interactions domains are used to assemble multiprotein signaling complexes, and they control the conformation, activity and substrate specificity of enzymes. Typical interaction domains fold independently and recognize exposed sites on their protein partners or they bind to the charged headgroups of phospholipids in membranes. Some domains serve specific functions by binding to distinct protein modifications. SH2 domains, for example, generally require phosphotyrosine sites in their primary ligands and are therefore dedicated to tyrosine kinase signaling (Chapter 8). Other domains can bind sequence motifs found in a broader set of proteins and display a wider range of biological activities. SH3 domains, for example, recognize Pro-containing motifs and thereby regulate a diversity of processes such as signal transduction, protein trafficking, cell polarization and organelle biosynthesis. The cell therefore uses a limited set of interaction domains, which are joined together in diverse combinations, to direct the actions of regulatory systems.

Binding Properties of Interaction Domains

Most interaction domains can be grouped into classes by sequence comparison. Complexity is, however, introduced (i) by the ability of a particular domain class to recognize distinct motifs, (ii) by the presence of separate ligand-binding sites within an individual domain and (iii) by the importance of ligand conformation in domain recognition. Often, the same class of interaction domain occurs twice in a signaling protein. As examples, the signaling protein Ras–GAP (Section 9.5.6) harbors two SH2 domains and the protein phosphatase SHP2 contains two SH3 domains. Interaction domains can also be assembled from repeated copies (up to 50) of small peptide motifs, yielding a large interaction surface with multifaceted binding properties. Such repeats include ankyrin repeats, Armadillo repeats, leucine-rich repeats, among others.

Interaction domains are remarkably versatile in their binding properties. An individual domain can engage several distinct ligands, either simultaneously or at successive stages of signaling. For example, the MH2 domain of the SMAD proteins (Chapter 12) harbors an extended binding surface that is able to interact with p-Ser motifs, with the scaffolding protein Sara and with components of the transcription apparatus.

Ligands and Types of Interaction Domains

The interaction domains of signaling proteins bind modified amino acid side-chains, peptides or proteins. The interaction domains may be classified by the characteristics of their ligands (Fig. 1.10). Interaction domains may recognize:

- *Protein modifications* (Fig. 1.10a). As outlined above (Section 1.5), a large number of posttranslational modifications of signaling proteins exist that serve as attachment points for partner proteins during formation of signaling complexes. Posttranslational modifications frequently complete binding sites for interaction domains in protein–protein assemblies and make a major contribution to the specificity of the interaction. Thus, the interaction domains serve as detectors of posttranslational protein modifications formed in the course of signaling events.

Protein interaction domains bind to
- Distinct protein modifications
- Sequence motifs
- Other protein domains
- Membrane-bound phosphatidylinositides

– *Peptide motifs* (Fig. 1.10b). This class of interaction domains binds to short peptide motifs that are exposed on the ligand surface. As an example, the SH3 domain (Section 8.2.2) binds to proline-rich motifs of protein ligands and by this property regulates many cellular functions The specificity of binding may be quite low and the biological activities that are regulated by this type of interaction are diverse.

– *Protein domains* (Fig. 1.10c). A number of modular domains undergo homo- or heterotypic domain–domain interactions rather than binding short peptide motifs. Such domains frequently identify proteins involved in a common signaling process and then direct their coassembly into functional oligomeric complexes. Components of apoptotic or inflammatory signaling pathways are characterized by death domains or close structural relatives thereof that form heteromeric structures required for caspase dimerization and activation (Chapter 15). The distinction between domains that bind peptide motifs and those that interact with other folded domain structures is by no means absolute. PDZ domains (Section 8.2.5), for example, generally recognize short peptide motifs of around four residues at the extreme C-termini of their binding partners, but they can also mediate specific heterotypic PDZ–PDZ domain interactions.

– *Phospholipids* (Fig. 1.10d). Many signaling processes are intimately linked to the cell membrane and the recruitment of signaling proteins to the cell membrane is frequently an essential step in signaling (see also Section 2.6). One mechanism for the attachment of signaling proteins to the cell membrane uses membrane phospholipids that are recognized by phospholipid-binding interaction modules on signaling proteins. The specificity of the phospholipid-binding domains is not well characterized and appears to be rather broad. Some phospholipid-binding domains have been reported to bind to peptide motifs too. More details on the function and properties of the phospholipid-binding domains will be found in Section 8.2.3.

(a) Modified peptides as targets of signaling domains

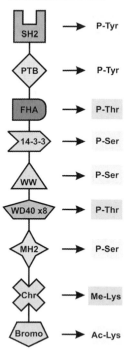

(c) Domain-domain interactions in signaling complexes

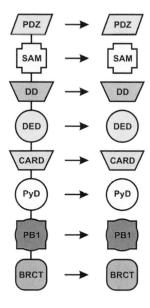

Fig. 1.10: Classification of interaction domains by the nature of the binding substrates. Interaction domains bind modified amino acid side-chains, short peptide sequences, other protein domains or phospholipids. (a) Domain-modified peptide interactions. For SH2, PTB, FHA, 14-3-3, WW domains, see Section 8.2. For MH2, see Section 12.1.3. For Chr (Chromo), Bromo domains, see Section 3.5.7. (b) Domain–peptide interactions. For PDZ, see Section 8.2.5. EVH1 = Ena-Vasp homology 1. (c) Domain–domain interactions. For PDZ, see Section 8.2.5. For DD, DED, CARD, PyD, see Chapter 15; For BRCT, see Section 8.2. SAM = sterile α motif; PB1 = phox and Bem1p domain. (d) For C1, see Section 7.5. For PH, FYVE, C2, see Section 8.2. FERM = four point one, ezrin, radixin, moesin; ENTH = epsin N-terminal homology.

(b) Peptides as targets of signaling domains

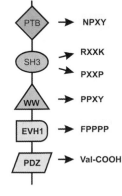

(d) Pospholipids as targets of signaling domains

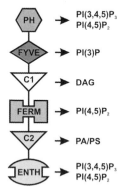

1.7
Modular Structure of Signaling Proteins and Signaling Complexes

A characteristic feature of signaling proteins and signaling complexes is their modular construction. Signaling proteins are typically composed of distinct signaling domains or modules, subtypes of which are found in many different signaling proteins. Signaling complexes are frequently also of a modular structure. The signaling proteins assembled in a large signaling complex may be exchanged in a regulated manner and subtypes of signaling proteins may associate depending on the cell type and regulatory signals.

1.7.1
Modules in Signaling Proteins

Typical signaling proteins harbor sites for registration of signals, for transduction of the signal to the next signaling component, for the receipt of controlling signals and for subcellular localization. The multifunctionality of signaling proteins is based on their construction from multiple protein domains that may act independently or in cooperation, and serve distinct functions in signaling (Fig. 1.11). Many signaling proteins harbor interaction domains, sites for post-translational modification and catalytic domains, and these domains may be used to assemble functional signaling complexes, to compartmentalize molecular components and to direct enzymes to their targets. For example, the Tyr modification sites of RTKs serve as attachment points for interaction modules of downstream effector proteins, that in turn may be phosphorylated by the kinase activity of the receptor for further signal transduction and for the association of further signaling proteins (Chapter 8 and Fig. 8.11)

The modular construction endows signaling proteins with the following characteristics:
– *Multivalency.* The presence of multiple modules makes signaling proteins multivalent with respect to interaction partners, regulatory inputs and subcellular localization, and it allows for the association of large signaling complexes. For instance, the platelet-derived growth factor receptor (PDGFR) harbors multiple Tyr-phosphorylation sites that direct the attachment of distinct downstream effectors (Chapter 8 and Fig. 8.11).
– *Differential use of modules.* Modules in signaling proteins may be used simultaneously, in sequential order or in distinct subcellular locations only, allowing for a high versatility and flexibility in signaling. Often, signaling proteins go through cycles of function. In doing this, the modules of the signaling protein may be used in a differential way and the use of one module or modification may

■ **Typical domains of signaling proteins**
  – Protein interaction domains
  – Catalytic domains
  – Regulatory domains
  – Membrane-targeting domains

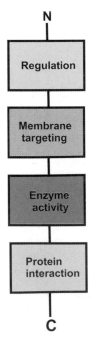

Fig. 1.11: Typical domains of signaling proteins. The order of sequence of the domains is variable.

influence the use of other modules or modifications within the same protein. Furthermore, modules may be engaged in a cell-type- and tissue-specific manner, a feature that is central to cell-type- and tissue-specific signaling.

– *Multiple inputs, regulatory influences and outputs*. The construction from multiple modules allows signaling proteins to receive multiple signals and to respond to multiple controlling influences. These inputs may be integrated and converted into differential outputs, depending on the cellular environment.

1.7.2
Modular Signaling Complexes

Interaction domains and protein modifications mediate the association of signaling proteins with upstream and downstream signaling partners, often leading to the formation of large protein complexes in the cytoplasm and nucleus. The organization of these complexes is dynamic and the complexes are often assembled in response to signal input. A large number of proteins may participate in the dynamic formation of the signaling complexes. For example, the N-methyl-D-aspartic acid) (NMDA)-type glutamate receptor has been reported to associate with more than 50 different signaling proteins. By the reiterated use of interaction domains, complex machines are built that regulate, targeted proteolysis, endocytosis, protein- and vesicle trafficking, cell polarity, cell division and gene expression, etc. The large complexes allow an efficient and rapid transmission of signals from one signaling component to the other. The inter-actions involved are illustrated in Fig. 1.12 using the example of the insulin receptor signaling complex.

■ **Modules in signaling proteins**
– Can be differentially used in a time- and location-specific manner
– Allow for multiple inputs
– Regulatory influences and outputs
– Provide for multivalency

■ **Signaling complexes**
– Harbor multiple signaling proteins
– Regulate major cellular functions
– Are of modular structure
– Allow for specific signaling
– Allow for coordination of multiple regulatory influences

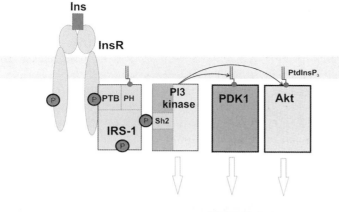

Fig. 1.12: Insulin signaling complexes. Binding of insulin to the extracellular subunit of the insulin receptor triggers auto-phosphorylation at tyrosine residues on the cytoplasmic part of the receptor. The p-Tyr residues serve as attachment points for the phosphotyrosine binding (PTB) domain of the adaptor protein IRS1 that becomes Tyr phosphorylated too. The p-Tyr residues on IRS1 serve to assemble the PI3K into the signaling complex. PI3K becomes activated and synthesizes the second messenger phosphatidylinositol-3,4,5-trisphosphate (PtdInsP$_3$) that mediates the membrane recruitment and activation of two further protein kinases – Akt kinase and PDK1 (see also Section 7.4).

The organization of signaling proteins in signaling complexes has several distinct advantages:

– *Specificity.* Within a signaling complex, signals can be transduced from one component to the other in a highly efficient way leading to the generation of robust signals. Signal transduction within a complex is rapid because it does not require diffusion of the reactants. The assembly of several components of a signaling path into a multiprotein complex ensures a tight and specific coupling of the various reactions and prevents unwanted dissipation of the signal and side-reactions.

– *Variability.* Signaling components of a signaling complex may be replaced by isoforms that differ in the details of regulation and activity. Such an exchange of signaling components can lead to distinct changes in the output signal. The exchange of components in regulatory complexes appears to be used intensively in gene regulatory complexes (e.g. mediator complexes, see Section 3.2.9) that may associate distinct coactivators, corepressors or chromatin remodeling enzymes depending on the input signals to the system.

– *Regulation.* The components of signaling complexes are often themselves of a modular structure, which allows the receipt of multiple input and regulatory signals in a sequential order or at the same time. Due to the close proximity of the signaling components, regulatory signals can become effective in a rapid and efficient way. For example, we know of signaling complexes that contain both protein kinase and protein phosphatase activity. The presence of two opposing enzyme activities within the same complex is an important tool for the downregulation and termination of signaling events.

1.8
Organization of Signaling

Typically, a large number of signaling components participate in the transduction of an extracellular signal into intracellular biochemical reactions that define the endpoint of signal transduction. To characterize and describe a signal transduction pathway, the number and type of signaling components involved as well as their linkages have traditionally been used. However, it is increasingly recognized that the multivalency and modular structure of signaling proteins, multisite modifications, and the existence of subtypes of signaling proteins allow different signals to enter and to be processed in the same type of signaling path leading to variable outcomes. The description of signaling pathways in terms of their structural organization has therefore proven to become more and more difficult. Signaling pathways have been formerly described as being linearly organized. Now, the features of branching and crosstalk of signaling have made it necessary to include a large number of possible linkages within a signaling path and between different signaling paths. Cell signaling should now be described in terms of signaling networks that endow signaling processes with the properties of plasticity, robustness and variability.

1.8.1
Linear Signaling Pathways, Branching and Crosstalk

The organization of signaling paths in terms of linearity, branching and crosstalk is illustrated in Fig. 1.13(a).

1.8.1.1 Linear Pathways

The classical view of signaling pathways has been that of a sequential transmission of signals in a linear signaling chain. Thereby, a signal is registered by an upstream component of a signaling chain and is then transmitted to the downstream component that then passes the signal on to the next protein in sequence and so on. One component activates the next component in sequence for signal transmission, and signaling is controlled by deactivation mechanisms that are intrinsic to the system or use specific deactivation enzymes. The description of signaling in terms of linear pathways originates form experiments where signaling is initiated by strong signals produced by the overexpression or mutation of signaling proteins or by the exposure of cells to artificially high external signals such as administration of hormones. High signal intensity tends to drive signal transmission through distinct routes, activating biochemical reactions that do not always correspond to the biological response obtained in the *in vivo* situation. It is well known that the routing of signals may depend on signal intensity (amplitude) and also on the duration of the signal (frequency). Nevertheless, the description of signaling in terms of linear pathways is useful for the illustration of the main reaction steps in a signaling cascade and the focus on the main steps serves to outline the logic of a signaling pathway. In many cases, the importance of alternative reactions or of branching reactions is not well established experimentally. Therefore, many of the reaction pathways described in the following chapters of this book are assumed to transmit signals along linear tracts resulting in regulation of discrete biochemical reactions and cellular functions.

■ **Linear signaling pathways**
- Signal flows in a linear manner through hierarchically organized signaling components
- No branching and crosstalk, therefore oversimplification

(a)

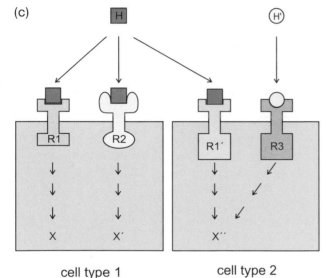

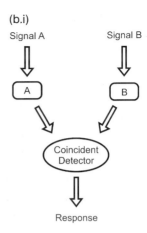

(c)

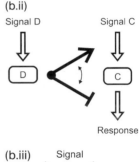

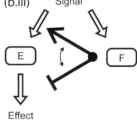

cell type 1   cell type 2   Effect

Fig. 1.13: (a) Linearity, branching and crosstalk in intracellular signaling. Crosstalk refers to a situation where a signaling enzyme from one pathway inhibits (E4) or activates (E′′) signaling components involved in signal transduction of a different pathway. (b) Mechanisms by which signaling pathways may interact. (i) Two distinct pathways converge on a coincidence detector, which generates a response that is unique and different from that generated by each individual pathway. (ii) A primary pathway that evokes a biological response is modulated by a second pathway through a gate, resulting in either inhibition or stimulation of the primary pathway. (iii) A single initial signal flows through multiple pathways, with one pathway regulating the other. (c) Variability of receptor systems and signaling pathways. (i) For one receptor of a given binding specificity (binding to hormone H) there can be different subtypes in the same cell (R1, R2) or in other cell types (R1′). (ii) The hormone H can induce different reactions (X1, X2) upon binding the different receptor types (R1, R2). The receptor types R1 and R2 can be found simultaneous in one cell. (iii) The binding of two different hormones (H, H′) to different receptors (R1′, R3) can induce the same intracellular reaction. The characteristics (i) and (ii) contribute to a high degree to the diversity and variability of hormonal signal transduction. Point (iii) illustrates the principle that important cellular metabolites or reactions can be controlled by different signal transduction pathways.

1.8.1.2   Branching and Crosstalk
Branching

We know of many signaling proteins that have multiple downstream reaction partners they can activate for further signal transduction. This property leads to a branching of signaling pathways and provides for multiple outputs originating from the same type of signaling protein. The conditions for the transmission of the signal to alternative downstream reaction partners are in most cases not well established. Often signal distribution to alternative routs depends on cell type, and may be variable in time and subcellular localization. In further chapters, the dissipation of signals and the distribution to alternative reaction partners is only discussed for those cases where alternative reaction partners have been clearly identified.

■ **Branching**
– Signaling protein has multiple downstream partners

Crosstalk

Cells have to process a large number of signals at the same time and mostly these signals are routed through different signaling pathways. The flow of signals through the various pathways must be coordinated and properly balanced which requires linkages between different pathways. This interdependence of signaling is also called crosstalk. The multivalency of signaling proteins and multisite modification allows components of a signaling path to influence, regulate and modulate the signaling in other pathways. Many examples of crosstalk will be found in the following chapters. Noteworthy is the regulation of the Raf kinase, a main component of the Ras–mitogen-activated protein kinase (MAPK) pathway (Section 9.6) by protein kinases that are part of other signaling pathways.

■ **Crosstalk**
– Linkage of different pathways
– Allows for coordination of pathways

1.8.1.3   Interactions between Signaling Paths

Three major mechanisms have been recognized by which different signaling paths interact, i.e. coincidence detection, gating and feedback (Fig. 1.13b).

Coincidence Detection

Coincidence detection is based on two distinct signaling pathways, A and B, converging on a single functional unit composed of one or more proteins known as the coincidence detector. The coincidence detector is able to recognize when the two converging pathways are activated within a given time window. Both signals are equally important. The detector then produces a unique response different from what is observed when either pathway is activated individually. Thus, the coincident response can be functionally distinct or can be synergistic, i.e. greater than the sum of the responses to A and B. A coincidence detector enables the cell to produce a unique response only when specific pathways have been activated simultaneously or within a specific amount of time.

Gating

In a gated system, signal flow through the first pathway, C, is regulated by activation of a second pathway. Thus, two different signals interact in a hierarchical fashion. The response elicited is thus only modified, but not distinct from that which is evoked when pathway C is exclusively activated. Gating, therefore, gives a cell the ability to regulate responses based on the state of the cell, which, in turn, may be dependent on the signaling pathways that are activated. Both coincidence detection and gating are similar in that they allow two separate signals to activate different signaling pathways.

Feedback

Feedback is a modified gating mechanism, which is unique in that it is dependent on only one initial signal. This signal can then modulate multiple pathways or activate a single pathway that regulates two or more downstream effectors. In such a system, one effector produces the biological effect and the other regulates the signal flow to the effector that produces this effect. This configuration enables a signaling system to adjust its sensitivity to the environment, preventing or potentiating signaling flow to the endpoint response system.

1.8.2
Signaling Networks

■ **Signaling networks**
– Arise from branching and crosstalk

Each signaling protein of a signaling chain is subjected to regulatory influences from the same pathway or from different pathways placing it into a network of interactions and regulatory influences. Real cell signaling should therefore be described in terms of signaling networks that result from interconnections between pathways. In such a network, the same signaling protein is capable of receiving signals from many inputs. The networking may occur within similar classes of signaling pathways, such as between the Rho and Ras pathway (Section 9.10), and between different pathways, such as between the $G_{s,\alpha}$/cAMP and the MAPK pathway (Sections 9.6 and 10.2). The major mechanisms of interactions between pathways have been described already above in terms of coincidence detection, gating and feedback.

1.8.2.1 Complexity of Signaling Networks

The following experimental observations serve to illustrate the complexity of intracellular signaling and signaling networks (Fig. 1.13c):

– For a given hormone, different receptors can exist on the same or on different cells. These receptors can route the signal to different pathways triggering very distinct reactions in different tissues or even within the same cell. An example of such a phenomenon is adrenaline, which can initiate, on the one hand, a cAMP-mediated signal transduction and, on the other hand, an inositol triphosphate-mediated reaction (Chapter 6).

– For a given receptor, signaling enzyme or adaptor protein, subtypes are found which differ in their responsiveness to the incoming signal, in the nature and intensity of the reaction triggered in the cell, and in their capacity for regulation.

– The same secondary reaction can be triggered by different hormone–receptor systems and signaling pathways. This is exemplified by the release of Ca^{2+}, which can be regulated via different signaling pathways (Chapters 5–7).

– Regulatory interaction modules or protein modifications can be engaged depending on time and subcellular location.

– The function of a distinct modification or interaction may depend on the simultaneous presence of another modification or protein–protein interaction. This may result in cooperativity and ultra-sensitivity of signaling.

– The signal output in a given signaling system may depend on the amplitude of the incoming signal. Furthermore, the duration and frequency of an incoming signal can modulate the output of the system. An example for the latter is the regulation of calmodulin-dependent protein kinase (CaMK) II by Ca^{2+} (Section 7.6).

1.8.2.2 Properties of Signaling Networks

The construction of signaling proteins from distinct protein domains and the assembly of signaling proteins into larger signaling complexes results in multifunctionality, variability and interconnection of cellular signaling systems. It has been proposed that cells contain a general signaling network that may be operationally divided into large signaling modules or types of signaling paths. The signaling modules contain characteristic core components and are defined by functional input–output characteristics. Examples of such modules are the epidermal growth factor (EGF) receptor–Grb2–mSos–Ras module, the MAPK module and the G-protein-coupled receptor (GPCR)–$G_{\beta\gamma}$–G_α–adenylyl cyclase–cAMP protein kinase A (PKA) module (Fig. 1.14a). The molecular identity of the components of the signaling modules and their interacting partners may be cell-

■ **Signaling networks are based on**
– Modular structure of signaling proteins
– Modular structure of signaling complexes

(a)

type-specific, but the overall function of these components and the logic of the circuitry is preserved from cell type to cell type. The central signaling system is connected to cellular mechanisms such as the transcriptional, translational, motility and secretory machinery that are responsible for phenotypic functions. The central network that connects the various machine networks also receives and processes signals from extracellular entities such as hormones or neurotransmitters and ions.

Signal transduction through signaling networks depends on the types of interconnection of its components and the properties of the components itself. The large number of possible connections and the multiplicity of possible input and output signals at each component makes the description of signaling networks extremely complex. Nevertheless, one would like to understand how precise

(b)

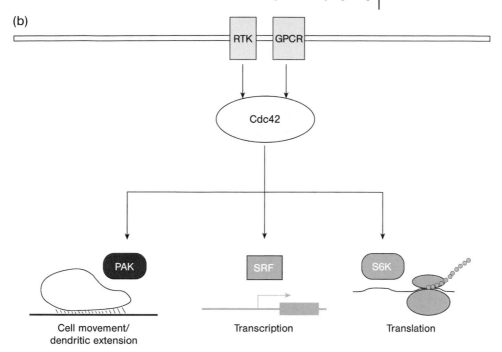

Fig. 1.14: (a) Adenylyl cyclases as examples of a junction in signaling networks. The signal receiving capabilities of the various adenylyl cyclase isoforms and the capability of the cAMP-dependent protein kinase (PKA) to regulate various physiological functions are shown. Receptor channel, ligand-gated channel (e.g. NMDA receptor); Stimulatory signals are shown as arrows and inhibitory signals as plungers. The various cellular components or processes regulated by PKA are shown in the red ovals and the resultant physiological functions are given below. (b) CDC42, a member of the Rho family of GTPases, as an example of a node. CDC42 may be stimulated both by RTKs as well as GPCRs and in turn regulate different cellular functions by regulating the distinct downstream kinases Pak (p21-activated protein kinase), S6K (p70^{S6} kinase, see Section 3.6.3.4) or the serum response factor (SRF, a transcription factor).

signals leading to defined biochemical reactions can be produced in the cell, how signals can be organized in time and in space, how irreversible switches are generated, and how signals are dissipated and downregulated.

Theoretical and experimental studies on cellular signaling networks have revealed features of cellular networks that help to explain some essential properties of both intracellular signaling and of intercellular signaling.

1.8.2.3 Nodes and Junctions

The interconnections in signaling networks may be operationally divided into two classes: *junctions*, which are signal integrators, and *nodes*, which split the signal. This classification is, however, by no means absolute, because we know of quite a number of signaling proteins that can receive and distribute multiple signals.

Figure 1.14(b) illustrates the node function of the Rho-GTPase CDC42. This small regulatory GTPase (Section 9.9) can receive signals from several TM receptors and can deliver signals to different protein kinases and to transcription factors.

A well studied example of a system that contains both junctions and a node is the cAMP/PKA system (Fig. 1.14a and Chapter 7). Here, the various subtypes of adenylyl cyclase function as signal integrators or signaling junctions that receive signals from various TM receptors or ion channels. The adenylyl cyclases transduce the signal to PKA, of which only few isoforms exist. The PKA functions as a node through which the signal is distributed to various downstream partners. Signal distribution by PKA is achieved by the differential use of isoforms of the adaptor proteins [A-kinase anchoring proteins (AKAPs)] that specify the nature and location of the substrate of the protein kinase.

Networks also contain nodes where signals may be split and routed through several different pathways to regulate distinct cellular functions. Like junctions, nodes may also be upstream or downstream in the network. One of the best upstream examples of a node is the RTKs, which can route growth factor signals through many different pathways. Although such routing can result in regulation of multiple independent cellular functions (e.g. growth factors such as PDGF can regulate vascular smooth muscle cell migration and proliferation), signal routing through multiple pathways can produce combinatorial signal specificity at the level of gene expression (Fambrough et al., 1999; Schlessinger, 2000). Such combinatorial specificity may be used as a mechanism to establish hierarchy amongst the regulated cellular processes.

1.8.2.4 Feedback Loops

Feedback is well known from metabolic pathways in bacteria where the product of a metabolic chain regulates reactions that led to its production. Frequently, the purpose of such feedback mechanisms is to adjust the production to the demand. In general terms, feedback can be defined as the ability of a system to adjust its output in response to monitoring itself. In biological systems, feedback is a general regulatory principle both in intracellular and extracellular signaling, e.g. during development. Typically, feedback is organized in loops that can have a negative or positive effect on signaling leading to inhibition or enhancement of signaling. Depending on the details of the interconnections, feedback mechanisms can also serve to convert transient signals into permanent, irreversible responses. Of specific importance are feedback loops in developmental programs, e.g. in pattern formation (reviewed in Freeman, 2000).

Negative Feedback

Negative feedback occurs when, for example, a signal induces its own inhibitor – it serves to dampen/or limit signaling. The most obvious use of feedback signaling is to limit the duration of a signal. An example of this is the control of cytokine signaling through the Janus kinase (JAK)–signal transducers and activators of transcription (STAT) signaling pathway (Fig. 1.15). JAKs are soluble tyrosine kinases that bind to cytokine receptors and transduce signals by the STAT proteins (Chapter 11). Cytokine signaling is negatively regulated by proteins termed suppressors of cytokine signaling (SOCS). These proteins participate in negative-feedback loops. The physiological significance of the negative feedback is exemplified by SOCS1, which is induced by and inhibits interferon (IFN)-γ signaling; when the gene for SOCS1 is knocked out, mice die as neonates with defects associated with excess IFN-γ signaling.

■ **Negative feedback loops dampen and limit signaling**

Double-negative Feedback and Bistability

This type of feedback control is ascribed a specific role in converting graded or transient inputs into switch-like, irreversible responses. The structure of a simple double-negative feedback circuit is illustrated in Fig. 1.16. In this circuit, protein A inhibits or represses protein B and protein B inhibits or represses protein A. Such a circuit shows

■ **Double-negative feedback provides for**
– Ultrasensitivity
– Bistability

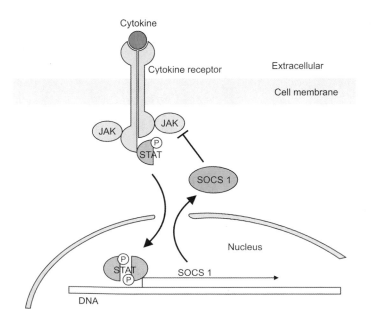

Fig. 1.15: Negative-feedback loop in cytokine receptor signaling. Ligand-bound cytokine receptors (Chapter 11) recruit protein kinases, JAKs, which in turn phosphorylate the transcriptional activator STAT proteins. The STATs activate the expression of SOCS proteins (Chapter 11) which are inhibitors of cytokine receptors. As a consequence, JAK–STAT signaling is inhibited. After Freeman (2000).

Fig. 1.16: A double-negative feedback loop. In this circuit, protein A (blue) inhibits or re-presses B (red), and protein B inhibits or represses A. Thus there could be a stable steady state with A on and B off, or one with B on and A off, but there cannot be a stable steady with both A and B on or both A and B off. Such a circuit could toggle between an A-on state and a B-on state in response to trigger stimuli that impinge upon the feedback circuit. After Ferrell (2002).

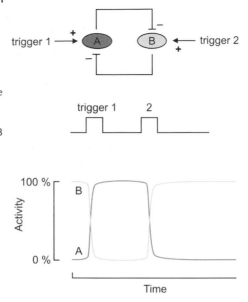

the characteristics of *bistability.* Of the two alternative states, only one is populated. The system cannot rest in intermediate states, i.e. A and B cannot exist at the same time. Upon exposure of a transient stimulus, the system goes from state A to state B in a nonlinear co-operative manner and may stay in state B although the stimulus has died away (reviewed in Ferrell, 2002). The system shows *ultrasensitivity* and has the characteristics of a cooperative transition. Furthermore, bistable systems can have switch-like properties (reviewed in Slepchenko and Terasaki, 2004). Upon a specific stimulus, the system makes a transition from one stable state to another stable state in a nearly irreversible manner. Double-negative feedback loops have been identified in several biological systems and in key signaling pathways such as in MAPK (Chapter 10) cascades (for details, see Ferrell, 2002).

Positive Feedback

Positive feedback occurs when a signal induces more of itself, or of another molecule that amplifies the initial signal, and this serves to stabilize, amplify or prolong signaling. A generalized positive feedback loop is depicted in Fig. 1.17(a). Such a system has the interesting property that the system stays activated even after the initial signal disappeared. We know of many examples of positive feedback signaling. As an example, Fig. 1.17(b) shows an autocrine positive-feedback loop in the *Drosophila* oocyte that ensures an all-or-nothing cell fate decision. Other examples include the autoregulation of

■ **Positive feedback amplifies, stabilizes and prolongs signals**

(a) Generalized positive feedback loop

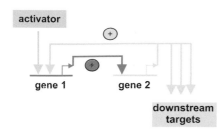

(b)

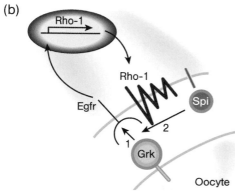

Fig. 1.17: Circuit diagram of a simple positive-feedback loop. (a) Once turned on by an activator, gene 1 (shown in red) activates gene 2 (shown in green). In addition to acting on downstream targets, gene 2 activates gene 1, forming a positive feedback loop. When the initial activator signal fades, these genes will remain active. After Howard and Davidson (2004). (b) Autocrine amplification by positive feedback in the *Drosophila* oocyte. The extracellular signaling protein, Gurken (Grk), expressed in the oocyte, binds to and activates the EGF receptor (Egfr) in the overlying follicle cells. This activates the expression of Rhomboid-1 (Rho-1), which in turn activates another EGF receptor ligand, Spitz (Spi). Spitz then amplifies and prolongs EGF receptor signaling in the follicle cells.

homeotic genes and chromatin modifications (Schreiber and Bernstein, 2002).

Bistability can be an important consequence of positive feedback loops as shown schematically in Fig. 1.18. In the positive feedback circuit shown, state A is activated by state B and state B is activated by state A. As a result, there could be a stable steady state with both A and B off or A on and B on, but not with A off and B on and *vice versa*.

Many signaling systems contain several intertwined feedback loops leading to a complex regulation that is difficult to describe quantitatively. One example is the protein (cyclin-dependent) kinase CDK1 that undergoes an abrupt activation at the G_2/M transition of the cell cycle (Chapter 13). Here, positive- and negative-feedback loops are involved.

Fig. 1.18: Bistability in a positive-feedback loop. In this circuit, A activates B and B activates A. As a result, there could be a stable steady state with both A and B off, or one with both A and B on, but not one with A on and B off or *vice versa*. Bistable circuits could exhibit persistent, self-perpetuating responses long after the stimulus is removed.

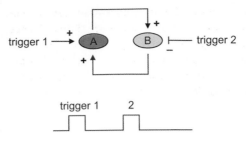

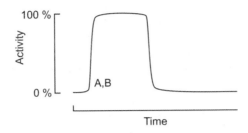

1.8.3

Redundancy and Specificity of Signaling

Redundancy

We know of many cases where multiple signaling pathways converge on the same target conferring stability and robustness on a signal. Such a redundancy of signaling has its parallel in the multiple modifications of signaling proteins, e.g. multiple phosphorylation (Section 1.6.3). Redundancy in signaling may be considered as a safeguard against the failure of a signaling component. When one component fails, its function may be taken over by another protein with similar activity. An illustrative example for redundancy in signaling is provided by the protein kinases of the cell cycle. Here, knockout experiments in mice have shown that CDK2 and cyclin E, long thought to be essential components of the progression through the G_1 phase of the cell cycle (Chapter 13), are largely dispensable for the development of mice. Apparently other CDK–cyclin pairs can replace CDK2–cyclin E in its function during G_1.

The multiple phosphorylation of RTKs is another example for redundancy in signaling as shown for PDGFR. In response to mitogenic extracellular signals, PDGFR uses multiple p-Tyr docking sites (Fig. 8.11) for the engagement of distinct signaling proteins such as phosphatidylinositol-3-kinase (PI3K), phospholipase Cγ and protein phosphatase SHP2. One specific response to mitogenic signals is the induction of a set of immediate–early genes by PDGFR. Remarkably, when the various modifiable tyrosines of PDGFR were replaced with nonmodifiable phenylalanines (Fam-

■ **Redundancy in signaling**
– Same reaction is controlled by several signaling proteins or signaling pathways
– Safeguard against failure of one component

brough et al., 1999), nearly all of the immediate–early genes could still be induced, although with somewhat lower amplitude. Hence, the immediate–early genes depend on no single pathway, but rather receive input from every pathway. The alternative strategy, that distinct modules of immediate early genes are induced by specific pathways, is not used by the PDGFR to communicate the mitogenic signal.

Specificity

Another important aspect of cell signaling is the specificity of signal transmission in signaling pathways and networks (Pawson, 2004). Specificity of signaling can be considered in terms of "which, when and where" interactions with signaling partners are formed. Considering the "which", i.e. the selection of binding partners, it is important to ask how specific are the protein–protein interactions that are involved in formation of signaling complexes?

In signaling pathways and networks, there is no optimal affinity for protein–protein interactions, but rather a wide range of dissociation constants that are tailored for distinct forms of biological regulation. Tight protein–protein interactions yield a high degree of specificity and are relevant for many biological functions. Strong interactions are long lived, and this can be advantageous, as in the tethering of inactive PKA to the scaffolding protein AKAP in readiness for a cAMP signal. However, tight interactions cannot always provide the flexibility that a cell needs to respond dynamically to changing external conditions or internal programs. Indeed, protein–protein interactions that are dependent on posttranslational modifications must by definition have relatively modest affinities since much of the binding energy must come from the modified residue itself. However, affinities in the micromolar range do not necessarily mean an absence of specificity. Many signaling complexes utilize multiple contacts, each of low affinity, to ensure fidelity. Often, the contacts are used in a cooperative way to achieve specificity in interaction. Thus, a general picture of interactions in signaling pathways and networks emerges, where individual domains of the signaling components are combined in a single polypeptide to enhance binding specificity and to generate allosteric control of signaling proteins and multiprotein complexes.

1.8.4
Regulation of Signaling Pathways

To achieve an appropriate biological response in the target cell, the extracellular signals originating from the environment or from within the organism must be processed by the intracellular signaling pathways or networks. Thereby, the intracellular network must place a value on the incoming signal such that it is either converted into a biochemical event and subsequently a biological response or safely dissipated within the network. Furthermore, mechanisms must be available that limit the duration of the intracellular response and adapt it to the overall needs of the organism. As an example, the long-lasting exposure to a hormone often leads to a weakening of the intracellular response. Such an attenuation, downregulation or desensitization of signaling can occur at many points of a signaling chain or network. The main attack points for control of intracellular signaling are:

■ **Major regulators attack points of signaling paths**
– Receptors
– Signaling enzymes
– Second messengers

– *Receptors.* The amount, activity and specificity of receptors is a main determinant for the transduction of extracellular signals into an intracellular response. Receptor regulation can occur at multiple levels such as: ubiquitination and internalization (Sections 2.5.6.2 and 8.1.5), posttranslational modification: phosphorylation (Section 5.3.5), binding of antagonistic ligands and gene expression.
– *Signaling enzymes.* The signal transducers operating downstream of the receptors may be controlled by deactivating components that limit the life time of the activated state of the signaling enzyme. Examples are the regulatory GTPases that are controlled by specific deactivators, the GTPase-activating proteins (GAPs, see Chapter 5). The protein kinases are another example of tight control of enzymatic activity. Many protein kinases are activated by phosphorylation (Chapter 7) and this modification can be reversed by phosphatases that clip off the activating phosphate modification. Furthermore, we know of protein kinase inhibitors (Chapter 13) that are subjected to regulatory modification.
– *Second messengers.* Chapter 6 presents many examples of second messengers whose production and degradation by specific enzymes is under tight control.

The regulation of the signaling components themselves is always part of a larger signaling network that ensures the generation of physiologically appropriate signals. How important the internal safeguard and control mechanisms are becomes evident when the control mechanism fail. The chapter on tumor formation (Chapter 14) gives many examples where the failure of control mechanisms in signaling pathways leads to inappropriate cell signaling that is characteristic of many tumor cells. Persistent activation of signaling components due to the malfunction of deactivating mechanisms is found in many tumors.

1.8.5
Spatial Organization of Signaling Pathways

A major contribution to the specificity of signaling comes from the localization of signaling events to distinct subcellular sites such as membrane compartments or the cytoskeleton. By restricting signal production to specific sites only, tight spatial and temporal control over signaling is possible allowing for a fast and effective production and downregulation of signals. Most locally controlled signals are produced and processed in multicomponent signaling complexes assembled at the inner side of the cell membrane. Membrane localization of signaling complexes is of outstanding importance in cellular signaling because the vast majority of external signals are received at the cell membrane by membrane receptors that transduce the signal into the cell interior. Furthermore, electric signals, resulting in the influx of, for example, Ca^{2+} ions via ion channels, become active at the cytoplasmic side of the cell membrane. Often, the steps following membrane receptor activation occur in tight association with the inner side of the cell membrane whereby large signaling complexes form in a dynamic way.

The cell uses the following tools for membrane localization of signaling proteins (Fig. 1.19):

– *Adaptor or scaffolding proteins.* We know of many scaffolding proteins that serve to assemble signaling complexes at the membrane or at the cytoskeleton. Examples include the AKAP proteins (Section 7.3.3), the receptor for activated C-kinase (RACK) proteins (Section 7.5.4) and the PDZ-containing proteins (Section 8.2.5). A general theme that emerges from the study of these scaffolding proteins is that they possess a bidirectional specificity. At one end they specifically recognize one or a group of signaling components

■ **Membrane-localization of signaling paths is based on**
– Membrane-localized adaptor proteins
– Binding to TM receptors
– Lipid anchors

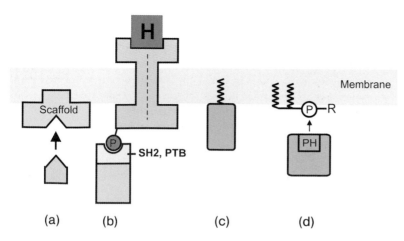

Fig. 1.19: Main mechanisms for membrane targeting of signaling proteins. (a) Binding to membrane-associated adaptor proteins. (b) Binding to phospho-amino acids of TM receptors via SH2 or PTB domains. (c) Membrane targeting via lipid anchors. (d) Binding to membrane-bound phospholipids via PH domains.

and at the other end a location within the cell, thus providing the molecular basis for spatial organization of signaling pathways. By organizing signaling enzymes with opposing activities in the same signaling complex with the help of the scaffold, a precise and locally controlled production of signals can be achieved. An example for that strategy is provided by the AKAP proteins (Section 7.3.3).

– *Binding to TM proteins.* This strategy is extensively used by the RTKs. Here, phosphotyrosine residues on the activated receptor are used as docking sites for bringing the next signaling protein in sequence to the receptor and to the cell membrane. This protein then can use protein interaction domains to recruit another signaling component into the signaling complex.

– *Membrane anchors.* The posttranslational modification of signaling components by lipidation provides a means for recruiting signaling proteins to the inner side of the cell membrane and thus into the vicinity of receptors or other signaling components. The lipid anchors insert into the lipid bilayer and thus increase the affinity of signaling proteins to the membrane. For details on the membrane anchors, see Section 2.6.

– *Binding to membrane-localized second messengers.* Some second messengers (e.g. diacylglycerol and phosphatidylinositol trisphosphate, see Chapter 6) are membrane-bound and serve as membrane attachment points of signaling proteins. The hydrophobic messengers are recognized by interaction domains on the signaling protein [e.g. pleckstrin homology (PH) domains for binding of phosphatidylinositol trisphosphate]. Importantly, the trigger for membrane association is the presence of the second messengers which are formed in a signal-directed way.

1.8.6
Compartmentalization and Transport

Many signaling proteins shuttle between distinct compartments of the cell to perform specific functions. These proteins carry sequence signals or posttranslational modifications that direct them to distinct subcellular sites. Such a signal-directed translocation of signaling proteins is an important means for the generation of location specific signals. Examples of signaling proteins with variable subcellular localization include PKC (Section 7.5), the Abl tyrosine kinase (Section 8.3.3) and some nuclear receptors (Chapter 4). Furthermore, the shuttling of proteins kinases and transcription factors between the cytoplasm and the nucleus is a frequently observed phenomenon. As outlined in Section 3.4.5, the signal-directed phosphorylation of transcription factors is an important tool for controlling their nuclear localization and thus their transcription activating function.

■ **Compartmentalization**
– Signaling proteins shuttle between subcellular compartments

1.8.7
Evolution of Signaling Pathways

The reiterated use of interaction domains may have developed in part to facilitate the evolution of new cellular functions, because domains may be readily joined in new combinations to create novel connections and pathways within the cell (Pawson, 2004). For example, coupling of protein phosphorylation to ubiquitination could have been achieved by simply linking an interaction module that recognizes p-Ser-p-Thr or p-Tyr sites to a RING domain that binds components of the ubiquitination machinery. The Cbl protein which functions as a E3 ligase for ubiquitination of RTKs is an example for such a strategy. Cbl carries a SH2 domain for binding to p-Tyr sites on the receptor and a RING domain for assembly of E2 enzyme (Section 2.5.6.2).

When going from invertebrates and lower vertebrates to higher vertebrates, an increased use of interaction modules is evident.

Conventional tyrosine kinases and SH2 domains are absent from yeast, but make a coordinate appearance with the development of multicellular animals. An SH2 domain, by its design, can be inserted into preexisting proteins and thereby provide a common means of coupling entirely different proteins to tyrosine kinase signals. Clearly this does not exclude the subsequent elaboration of more sophisticated levels of control within signaling complexes. The joining of separate domains can also create a new composite entity with more complex properties than either domain alone.

Interaction domains and motifs therefore provide a way to increase the connectivity of existing proteins, and thus to endow these proteins with new functions. This is likely one of several reasons that the apparent complexity of organisms can increase so markedly without a corresponding increase in gene number. An attribute of proteins encoded by the human genome is that they have a richer assembly of domains than do their counterparts in invertebrates or yeast, and indeed the assortment of domains into novel combinations is likely an important aspect of genome divergence.

1.9
Variability and Cell-type Specificity of Signaling

The general principles and the types of interconnections in the signaling pathways and networks are largely the same in most cells. However, the details of the interaction modules and interconnections as well as the availability of signaling components differ from cell type to cell type. Thus, cells are able to process signals and activate biochemical events in a cell-type-specific way.

■ **Cell-type specific signaling**
 – Based on cell-type-specific availability, modification and activity of signaling proteins

A major contribution to the cell type specificity of signaling pathways and networks comes from the amounts and properties of the signaling components and this depends on the following points:
– Cell-type-specific expression of the gene for the signaling component.
– Cell-type-specific splicing.
– Cell-type-specific stability.
– Cell-type-specific posttranslational modification.
– Cell-type-specific subcellular localization.

Most important for the variability of signaling pathways and networks is the existence of subtypes of signaling proteins. Differential transcription and splicing leads to the presence of subtypes of signaling proteins that differ in the details of signal registration, signal transmission and regulation. Examples of signaling proteins for which many subtypes are found:
– α and $\beta\gamma$ subunits of the heterotrimeric G-proteins (Chapter 5).
– Adenylyl cyclase (Chapter 6).
– PKC (Section 7.5).
– PKA (Section 7.3).
– GAP proteins (Section 9.2).
– Protein kinases of the MAPK signaling pathways (Chapter 10).
– JAK and STAT proteins (Chapter 11).
– Scaffold protein AKAP (Section 7.3).

1.10
References

Fambrough, D., McClure, K., Kazlaus-kas, A., and Lander, E. S. (1999) Diverse signaling pathways activated by growth factor receptors induce broadly overlapping, rather than independent, sets of genes, *Cell* **105**, 727–741.

Ferrell, J. E., Jr. (2002) Self-perpetuating states in signal transduction: positive feedback, double-negative feedback and bistability, *Curr. Opin. Cell Biol.* **105**, 140–148.

Freeman, M. (2000) Feedback control of intercellular signalling in development, *Nature* **105**, 313–319.

Howard, M. L. and Davidson, E. H. (2004) cis-Regulatory control circuits in development, *Dev. Biol.* **105**, 109–118.

Mann, M. and Jensen, O. N. (2003) Proteomic analysis of post-translational modifications, *Nat. Biotechnol.* **105**, 255–261.

Nash, P., Tang, X., Orlicky, S., Chen, Q., Gertler, F. B., Mendenhall, M. D., Sicheri, F., Pawson, T., and Tyers, M. (2001) Multisite phosphorylation of a CDK inhibitor sets a threshold for the onset of DNA replication, *Nature* **105**, 514–521.

Pawson, T. and Nash, P. (2003) Assembly of cell regulatory systems through protein interaction domains, *Science* **105**, 445–452.

Pawson, T. (2004) Specificity in signal transduction: from phosphotyrosine-SH2 domain interactions to complex cellular systems, *Cell* **105**, 191–203.

Schreiber, S. L. and Bernstein, B. E. (2002) Signaling network model of chromatin, *Cell* **105**, 771–778.

Slepchenko, B. M. and Terasaki, M. (2004) Bio-switches: what makes them robust?, *Curr. Opin. Genet. Dev.* **105**, 428–434.

Yang, X. J. (2005) Multisite protein modification and intramolecular signaling, *Oncogene* **24**, 1653–1662.

2 Regulation of Enzyme Activity

The rates of the biochemical reactions within an organism are precisely adjusted to nutrient supply, and to the requirements of growth and division of each cell, thereby ensuring survival of the whole organism. In response to external and internal signals, cells can adopt the activity of enzymes in a fine-tuned way, and many enzymes are themselves part of signaling pathways and signaling networks receiving and producing regulatory signals.

How much of a specific substrate within a cell is converted to the product is determined by the actual enzyme activity that catalyzes this reaction. How active a given enzyme is at a given time and at a specific location within the cell is subjected to a multitude of regulatory mechanisms that operate at different levels and with different goals.

In a broad sense, enzyme regulation operates at two main levels. At one level, control is exerted via the amount of enzyme present in the cell and via the enzyme's availability at the site where it is meant to exert its catalytic function.

The *amount* and *availability* of enzymes can be regulated, via:

– *Gene expression*. Transcription, translation and processing of the mRNA (Chapter 3).
– *Targeted proteolysis*.
– *Targeted translocation*. Many of the substrates of regulatory enzymes are localized to distinct subcellular sites. In order these substrates to be turned over, the enzyme must be brought into their vicinity, which requires a colocalization of enzyme and substrate at the same site. The availability of the enzyme at a substrate site is therefore an important aspect of enzyme regulation.

■ **Amount of enzyme is regulated by**
– Gene expression
– Targeted proteolysis
– Subcellular location

Another level of enzyme regulation operates by altering the activity of preexisting enzymes. The *activity* of preexisting enzymes can be regulated via:

– *Binding of effector molecules* with a positive or negative influence on enzyme activity.

Biochemistry of Signal Transduction and Regulation. 4th Edition. Gerhard Krauss
Copyright © 2008 WILEY-VCH Verlag GmbH & Co. KGaA, Weinheim
ISBN: 978-3-527-31397-6

– *Covalent modification*, e.g. phosphorylation, oxidation (Section 1.6.1).

Cells use regulation at the level of enzyme concentrations often for long-term adaptations during cell growth and cell differentiation where a fast biochemical response to a regulatory signal is not required. Furthermore, by fine-tuning the rates of degradation and synthesis of an enzyme a distinct steady state concentration of an enzyme can be established.

■ **Enzyme activity is regulated by**
– Effector molecules
– Covalent modification

In contrast, regulation via modulation of the activity of preexisting enzymes is often used for a quick response to internal or external stimuli, allowing short-term changes in the rates of biochemical reactions.

2.1
Basis of Enzyme Catalysis

Enzymes function as biocatalysts and, as such, are involved in all cellular functions. Characteristics of enzymes are their high *efficiency*, high *specificity*, extreme *stereoselectivity* and their *ability to be regulated*. In the following only a short outline of the basics of enzyme catalysis will be given and we will concentrate on aspects of enzyme regulation. For topics related to mechanisms of enzyme action, the reader is referred to textbooks on biochemistry and to textbooks devoted specifically to enzymes such as Fersht (1998).

Analogous to chemical catalysts, enzymes do not alter the equilibrium of a reaction, but only accelerate the establishment of that equilibrium.

The mechanism of the action of enzymes can best be described with the aid of transition state theory. On the pathway from substrate A to product B in a reaction catalyzed by a chemical catalyst or an enzyme, A passes through a transition state $A^{\#}$, which is found at the highest point of the energy diagram (Fig. 2.1). The energy difference between the ground state of A and the transition state $A^{\#}$ represents the activation energy. The transition state as such cannot be isolated. It is the state of A in which the bonds participating in the reaction are in the process of opening and closing. The transition state is the most unstable state on the path from substrate to product. Enzymes, like chemical catalysts, increase the rate of a reaction by strongly binding to the transition state, thereby decreasing the activation energy for the transition from A to B.

For the tight binding of the transition state, the binding surface of the enzyme must be complementary to the structure of the transition state, so that optimal interactions between the enzyme and

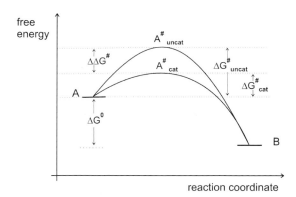

Fig. 2.1: The energy profile of a catalyzed and an uncatalyzed reaction. The figure shows the energy diagram for the conversion of A → B for a catalyzed and an uncatalyzed reaction. The binding of A to the catalyst (enzyme) is left out to simplify the figure. In the uncatalyzed reaction, the energy difference between the ground state A and the transition state $A^{\#}$ is the activation energy $\Delta G^{\#}_{uncat}$. The transition state of an catalyzed reaction is at a lower free energy, so that the activation energy $\Delta G^{\#}_{cat}$ is less and the reaction proceeds at a faster rate. The energy difference $\Delta\Delta G^{\#}$ is a measure for how much faster the catalyzed reaction is compared to the uncatalyzed reaction. The equilibrium of the reaction, which is characterized by ΔG_0, remains unchanged upon catalysis.

the transition state are possible. This demand implies that enzymes display a high affinity to molecules which are chemically similar to the transition state of the reaction. Complexes of such transition state analogs with enzymes are well suited for X-ray structure analysis to elucidate the structural principles of the active site and the catalytic mechanism.

The pathway from enzyme-bound substrate to the transition state involves changes in the electronic configuration and geometry of the substrate. The enzyme itself is also not static. The ability to tightly bind the transition state requires flexibility in the active site. Such flexibility has been experimentally demonstrated in many cases. A corollary to this is that the efficiency of enzyme catalysis can easily be influenced and regulated by conformational changes in the enzyme.

The binding of an effector molecule or a covalent modification of the enzyme, such as phosphorylation, can inhibit enzyme activity by preventing the restructuring of the enzyme that is necessary for strong binding of the transition state. Effectors and enzyme modification can also affect the substrate-binding site such that binding of the substrate in the ground state is impossible or very weak. On the other hand, activation by effectors and modifications can be achieved by stabilizing a conformation of the enzyme in which substrate binding is favorable, and a high complementarity between enzyme and substrate in the transition state is possible.

2.2
Basics of Allosteric Regulation

The basic elements for the regulation of enzyme activity by effector molecules or posttranslational modification are *allosteric conformational changes* of the enzyme. Allostery means that the enzyme can exist in various conformations which differ in activity and substrate or ligand binding. The typical titration curve for the binding of a ligand to an allosteric protein is sigmoidal in shape (Fig. 2.2), while binding curves for nonallosteric enzymes are hyperbolic.

Enzymes that are regulated by effector molecules in an allosteric manner possess, apart from the binding site for the substrate, a specific binding site for the effector molecule. The binding of effector molecules to the effector site leads to a shift in equilibrium between the various active states and thus to a change in activity.

Mechanistic studies and model considerations have revealed the following general characteristics of allosterically regulated enzymes (for details, see Fersht, 1998):

– Allosteric proteins are typically composed of two or more, often symmetric subunits.
– The subunits can exist in active and inactive forms, often termed the T-form and the R-form. The R-form ("relaxed") is the relaxed, active state; the T-form ("tense") is the less active state. T- and R-forms are in equilibrium with each other.
– The ligands can bind to both the T- and R-forms. The two forms differ in their affinity to the ligand. Ligand binding can often be described by a hyperbolic binding curve (Fig. 2.3).
– Effector molecules that function as *activators* bind preferentially to the R-form of the enzyme and thereby stabilize it. Inhibitors bind preferentially to the T-form. In the presence of an inhibitor, more molecules occur in the T-form. In this form the enzyme possesses a lower affinity for the substrate and the enzyme is thus less active.

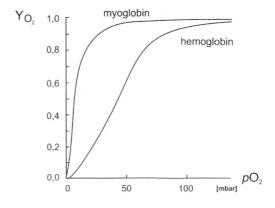

Fig. 2.2: Ligand binding to allosteric and nonallosteric proteins: oxygen binding to myoglobin and hemoglobin. The binding of O_2 to myoglobincan be described by a hyperbolic curve. The sigmoid form of the O_2-binding curve to hemoglobin is characteristic for a ligand binding to an allosteric protein. Y = degree of binding, ratio of binding sites occupied with O_2 to the total O_2 binding sites of hemoglobin; pO_2 = partial pressure of O_2.

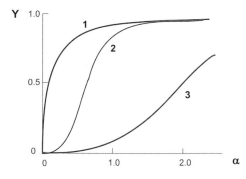

Fig. 2.3: The influence of an activator and an inhibitor on the ligand-binding curve of an allosterically regulated enzyme. In the absence of inhibitor or activator, the ligand binding curve is sigmoidal (curve 2). In the presence of an activator the binding curve is shifted to lower ligand concentrations (curve 1, red) and approaches hyperbolic form, similar to that observed in Fig. 2.3 for the binding of O_2 to myoglobin. An inhibitor shifts the binding curve to higher ligand concentrations (curve 3, blue). Y = degree of binding, ratio of occupied to total binding sites of the protein; $a = L_0/P_0$; L_0 = total ligand concentration; P_0 = total concentration of binding sites on the protein.

Figure 2.3 illustrates the influence of an activator and of an inhibitor on the binding of a ligand (and thereby on the activity) by a tetrameric protein. The activator shifts the binding curve to lower ligand concentrations and can, in the extreme case, lead to a hyperbolic binding curve. The inhibitor shifts the binding curve to higher ligand concentrations, so that higher concentrations are required to saturate the ligand binding sites.

■ **Activators**
– Shift binding to lower ligand concentration
Inhibitors
– Shift binding to higher ligand concentration

The molecular basis for the allosteric regulation of enzyme activity has been studied in detail for many systems (see textbooks). Some general conclusions can be drawn from these studies:
– The interface between the subunits often plays a central role in allosteric regulation by allowing the coupling of conformational changes within one subunit to the other.
– Binding sites for substrates are frequently found at or near the interface between the subunits. The mutual orientation of the subunits thereby can influence the accessibility of the substrate binding sites.
– Active and inactive forms differ in the accessibility of the substrate binding site and/or the proper positioning of the catalytic residues.
– The extent of the conformational changes involved can be variable. Long-range, medium-range and short-range conformational changes have been described for allosteric enzyme regulation.
– The binding of an effector molecule to the effector binding site induces conformational changes that influence – often via the subunit interfaces – the substrate binding site and/or the orientation of the catalytic residues. Thereby, the enzyme is fixed in either the active or the inactive state.

Cooperativity
Cooperativity can be broadly defined as a process caused by the simultaneous interaction of a large number of participants or events. In cellular regulation, cooperativity is well known for enzymes, for ligand binding to proteins such as O_2 binding to hemoglobin and

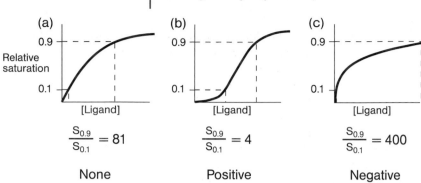

Fig. 2.4: The advantage of cooperativity in biological systems. (a) In systems following Michaelis–Menten kinetics (e.g. enzymes), it requires an 81-fold change in ligand concentration (stimulus) to change from 10% activity to 90% (shown as relative saturation, *S*). (b) For a positively cooperative enzyme it requires less change (e.g. 4-fold) and for a negatively cooperative enzyme (c) it can require much more change (e.g. 440-fold).

for posttranslational modifications. Cooperativity may be negative or positive. In the case of ligand binding to proteins, the presence of a ligand may have an enhancing effect on the binding of further ligands resulting in *positive cooperativity*. When binding of the first ligand makes binding of further ligands more difficult, the system shows *negative cooperativity*. The consequences of positive and negative cooperativity for biological systems involving ligand binding (receptors, enzymes) are very different when the sensitivity of the system to changes in ligand concentration is followed. As shown in Fig. 2.4, positive cooperativity is a device to sharpen the responsiveness of a system to the changes in ligand concentration. Here, relatively small changes in ligand concentration are sufficient to evoke a full response. In negative cooperativity, the response of the system is delayed upon changes in the ligand concentration and the system shows desensitization to high ligand concentrations.

■ **Positive cooperativity**
– Sharpens response to ligand concentration
Negative cooperativity
– Weakens response to ligand concentration

2.3
Regulation of Enzymes by Effector Molecules

The regulation of the activity of enzymes by the binding of effector molecules is a ubiquitous and general principle for the fine tuning and control of metabolic activity and other physiological functions. Effector molecules are often low-molecular-weight organic compounds. Proteins and metal ions can also exercise the function of effectors. The effector molecules bind specifically to the enzymes and the binding results in inhibition or stimulation of enzymatic activity.

■ **Effectors of signaling enzymes**
– Low-molecular-weight compounds
– Proteins
– Metal ions

Low-molecular-weight Effectors

For the regulation of metabolic pathways, metabolites are often used which are a product of that pathway. The basic strategy for the regulation is exemplified in the feedback mechanisms employed in the biosynthetic and degradation pathways of amino acids, purines and pyrimidines, as well as in glycolysis. In most cases a metabolite (or similar molecule) of the pathway is utilized as the effector for the activation or inhibition of enzymes in that pathway.

In many signal transduction pathways, second messengers function as effectors that by binding to target proteins (Chapter 6) regulate the flow of information through the pathways. Well characterized examples include the cyclic nucleotides cAMP and cGMP that function as allosteric regulators of protein kinases (Chapter 7).

Inhibitor and Activator Proteins

Enzyme-specific inhibitor and activator proteins are another major type of effector molecules.

Inhibitor Proteins

There are numerous examples of inhibitor proteins that specifically bind a particular enzyme and block its activity.

Inhibitors can use a variety of mechanisms to control enzyme activity:

– *Binding to the substrate binding site.* Inhibitors may be related structurally to the substrate without possessing the chemical groups that are necessary for being turned over. Because of the structural similarity of the substrate, these inhibitors can bind specifically to the substrate binding site and compete with the substrate for the enzyme. Well-studied examples include the protease inhibitors (see textbooks for details).

– *Deformation of the active site and/or the substrate binding site.* Binding of the protein inhibitor to the enzyme may alter the orientation of the catalytic center in a way that does not allow efficient catalysis and/or strong binding of the substrate. Examples are the inhibitors of the cyclin-dependent protein kinases (CDKs) of the cell cycle (Chapter 13).

■ **Major mechanisms of enzyme inhibitors**
– Competition with substrate binding
– Deformation of active site and substrate binding site

Inhibitor proteins themselves are subject to a variety of regulation mechanisms. The function of an inhibitor protein can be regulated, for example, by protein phosphorylation (Section 7.7.2), by degradation or by *de novo* synthesis (Chapter 13).

Activator Proteins

Examples of the reversible association of activator proteins with an enzyme are the Ca^{2+}/calmodulin-dependent enzymes (Section 6.7.1 and Section 7.5). Other examples of activating proteins are the cyclins (Chapter 13).

Activator proteins themselves can be bound in regulatory networks, as shown in the example of the cyclins (Chapter 13). The function of an activator protein can be regulated at the level of gene expression, degradation or posttranslational modification (e.g. protein phosphorylation).

Metal Ions

Many enzymes require metal ions for catalytic activity. Metal ions may participate directly in catalysis or they may have structural functions. Of primary importance is Ca^2 whose availability is a key control element in cellular regulation (Chapter 6). An important example in this regard is protein kinase C (PKC), which is activated by Ca^{2+} (Section 7.5).

2.4
Regulation of Enzyme Activity by Phosphorylation

We know of a large number of posttranslational modifications of enzymes that have a regulatory influence on enzyme activity (Section 1.6.1). Of these, the phosphorylation of enzymes by specific protein

■ **Protein phosphorylation**
– Catalyzed by protein kinases
– Uses ATP as P-donor
Phosphorylation on
– Ser/Thr
– Tyr
– Asp/Glu
– His

kinases is the most widespread mechanism for the regulation of enzyme activity. It represents a flexible and reversible means of regulation and plays a central role in signal transduction chains in eukaryotes. In the following, a short outline of the influence of phosphorylation on enzyme structure and function will be given.

Proteins are phosphorylated mainly on *Ser/Thr* residues and on *Tyr* residues. Occasionally Asp or His residues are phosphorylated, the latter especially in prokaryotic signal transduction pathways (Chapters 7 and 12). For the regulation of enzyme activity the phosphorylation of Ser and Thr residues is most significant. Apart from regulation of Tyr kinases, Tyr phosphorylation serves the function of creating specific attachment sites for proteins. Both of these functions will be discussed in more detail in Chapter 8.

Protein phosphorylation is a specific enzymatic reaction in which one protein serves as a substrate for a protein kinase (for details, see Chapter 7). Protein kinases are phosphotransferases. They catalyze the transfer of a phosphate group from ATP to an acceptor amino acid in the substrate protein (Fig. 2.5). As a consequence, the activity and/or structure of the substrate protein is changed.

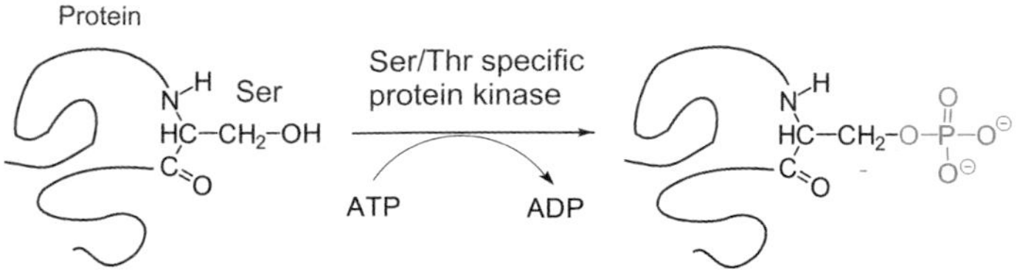

Fig. 2.5: Change in charge state of proteins via phosphorylation. The phosphorylation of Ser residues is catalyzed by a Ser/Thr-specific protein kinase that utilizes ATP as the phosphate group donor. The product of the reaction is a Ser phosphate ester which carries a net charge of −2 at physiological pH.

The response of a protein upon phosphorylation is dictated by the special properties of the phosphate group. The phosphate group has a pK_a of about 6.7 and carries *two negative charges at neutral pH*. Therefore, two negative charges are generated in a substrate protein upon phosphorylation of an uncharged amino acid side-chain. This fact and the presence of four oxygen atoms allows the phosphate group to form an extensive *network of hydrogen bonds* which can link different parts of a polypeptide chain. Similar 2-fold negatively charged groups do not occur in other structural elements of proteins. Electrostatic interactions and a network of hydrogen bonds are therefore of special importance for the control of protein functions by phosphorylation.

Analysis of protein–protein interactions in existing structures has shown that phosphate groups most commonly interact with the main-chain nitrogens at the start of a helix, where often glycine is found. In nonhelix interactions, phosphate groups most commonly contact *arginine* residues. The guanidinium group of arginine is well suited for interactions with phosphate because of its planar structure and its ability to form multiple hydrogen bonds. The electrostatic interaction between arginine residues and the phosphate group provides tight binding sites that often function as organizers of short-range as well as long-range conformational changes.

The phosphate esters of Ser, Thr or Tyr residues are quite stable at room temperature and neutral pH; the rate of spontaneous hydrolysis is very low. Therefore, to remove the phosphate residue the cell utilizes specific enzymes, the protein phosphatases. Based on substrate specificity, these can be classified as Ser/Thr- or Tyr-specific phosphatases (Chapters 7 and 8).

■ **Phosphate groups in proteins**
- Carry two negative charges
- Can engage in hydrogen-bond networks
- Main interaction with Arg
- Are stable against hydrolysis
- Are removed by phosphatases

Structural Consequences of Protein Phosphorylation
The molecular basis of the control function of protein phosphorylation has been elaborated for many proteins. A wide range of different mechanisms has been identified, ranging from large-scale allosteric conformational changes (e.g. in glycogen phosphorylase, see below) to small-scale conformational changes induced upon phosphorylation. We even know of phosphorylations that affect enzyme activity solely by electrostatic effects, without apparent change in enzyme

conformation. For example, isocitrate dehydrogenase from *Escherichia coli* is phosphorylated directly in the substrate-binding site with only minimal conformational changes resulting (Hurley et al., 1990).

■ **Protein phosphorylation**
– May induce large- and small-scale conformational transitions
– May function directly by electrostatic effects

One of the best characterized examples of phosphorylation-induced allosteric transitions is the regulation of glycogen phosphorylase by phosphorylation. Rabbit muscle glycogen phosphorylase is phosphorylated at Ser14 on the N-terminal tail and this phosphorylation activates the enzyme. As a consequence of Ser phosphorylation, the N-terminal tail undergoes a large conformational change of around 50 Å which then induces further conformational changes that propagate to the active site allowing efficient catalysis. Similar to many other allosterically regulated enzymes, the communication of the conformational changes to the active site depends on the quaternary structure of the glycogen phosphorylase. The inter-subunit contact surfaces in the phosphorylase dimer play a decisive role for the allosteric transitions that couple phosphorylation to enzyme activation. For a more detailed account one should refer to the original literature (Barford et al., 1991; reviewed in Johnson and O'Reilly, 1996).

The regulation of the CDKs by phosphorylation is another example for phosphorylation-induced allosteric transitions. As outlined in Chapter 13 and illustrated in Fig. 13.9, Tyr phosphorylation of the CDK–cyclin complex in the activation loop of the CDK induces a reorganization of the active site that results in a large increase in activity. Here, the Tyr phosphate serves as an organization center that optimally orients different parts of the complex.

2.5
"Ubiquitin (Ub)–Proteasome" Pathway

How much of a protein is available for cellular functions depends on its rate of gene expression and its rate of proteolytic degradation. The activities of both processes determine the steady-state concentration of a protein. Whereas transcription and translation have been early recognized as major attack points for regulating the amount of a protein, proteolytic degradation has been acknowledged as a means of protein regulation only in the last two decades.

Protein degradation can occur in a *nonspecific* or *specific* way. It is mainly the selective removal of proteins by proteolysis that determines the lifespan of a protein, i.e. how long a protein exists in the cell. Different proteins can have very different lifespans, indicating the existence of specific protein degradation mechanisms. Such specific mechanisms allow the function of a protein to be temporally restricted and specifically modified.

There are two main pathways for the degradation of proteins in mammalian cells. In the lysosomal path, proteins that enter the cell via endocytosis are degraded. The degradation of proteins in the lysosome is for the most part unspecific and it is used mainly to eliminate foreign proteins. The nonlysosomal degradation pathways allow for the selective degradation of proteins under normal cellular conditions. These degradation pathways are also responsible for the degradation of cellular proteins under conditions of stress.

The most significant and well-characterized nonlysosomal degradation pathway is that of the *Ub–proteasome pathway* in which proteins are degraded in a 26S proteasome after they have been conjugated by multiple Ub molecules. The Ub–proteasome system (reviewed in Hershko and Ciechanover, 1998; Pickart, 2004) is the major tool for the selective proteolysis of proteins and thereby plays a key regulatory role in the cell.

Ub-mediated proteolysis serves the following cellular functions:
- Degradation of proteins under stress situations.
- degradation of denatured and damaged proteins.
- Targeted degradation of regulatory proteins, e.g. oncoproteins, tumor suppressor proteins, transmembrane (TM) receptors, mitotic cyclins, transcription activating proteins.

In addition to marking proteins for degradation, many nonproteolytic functions of Ub and Ub-like proteins have been discovered. These will be discussed in Section 2.5.8.

2.5.1
Components of the Ub System

Ub is a 76-residue protein found in nearly every eukaryote. It occurs either in free form or bound to other proteins. All known functions of Ub are transmitted via its covalent linkage with other proteins. This serves the purpose, among others, of marking the proteins for proteolytic degradation.

The ubiquitination of proteins is a complex process, which involves three sequential enzymatic reactions performed by three types of enzymes, E1, E2 and E3 (Fig. 2.6). The enzymatic conjugating cascade of ubiquitination is organized in a hierarchical way: there is one E1 enzyme, a limited number of E2 enzymes, each of which may serve several E3 enzymes, and a much larger number of E3 enzymes.

■ **Ub–proteasome pathway**
- Ub attachment to target proteins
- Proteolytic degradation of Ub-labeled proteins in 26S proteasome

■ **Functions of Ub**
- Proteolytic
- Nonproteolytic

■ **Ub**
- 76 amino acids
- Highly conserved
- Attachment to Lys residues via C-terminal glycine

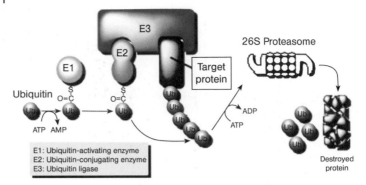

Fig. 2.6: Ubiquitination of proteins and degradation in the proteosome. Ub is initially activated by an enzyme E1, whereby the C-terminal carboxyl group of Ub becomes attached to the SH group of E1 via a thioester bond. The activated Ub is then transferred from E1-Ub to the Ub-conjugating enzyme, E2. Finally, the Ub is covalently attached to the target protein in a reaction catalyzed by the E3-Ub ligase. Repetition of this series of events results in the formation of a poly-Ub chain on the target protein. The E3 enzyme binds directly to the target protein and therefore plays an important role in determining the substrate specificity of this pathway. The poly-ubiquitinylated protein is then degraded to peptides in the 26S proteosome. After Nakayama et al. (2001).

2.5.2
Ub and Ub-like Proteins

■ **Ub family**
– Many members
– Common fold: Ub fold
– Diverse cellular functions

Important members
– Ub
– NEDD
– SUMO

Ub is the founding member of a family of proteins that share a common fold, named the Ub fold (reviewed in Pickart and Eddins, 2004; see Fig. 2.7). These proteins carry out their functions through covalent attachment to other cellular proteins thereby changing the stability, localization or activity of the target protein. The members of the Ub family share a common biochemical mechanism: an isopeptide bond is formed between the Ub's terminal glycine and an amino group of the target protein. Usually the amino group is donated by a lysine residue, but N-terminal ubiquitination is also known. The modification is then recognized in a manner that leads to specific downstream events, which vary depending on the identity of the protein modifier and the location and identity of the target protein.

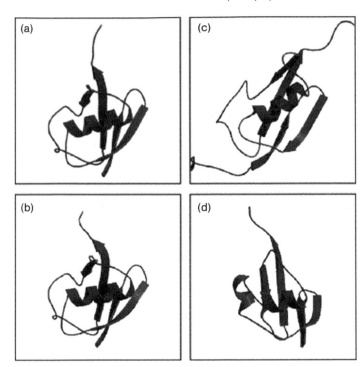

Fig. 2.7: The Ub fold. The structure of members of the Ub protein family: (a) Ub, (b) NEDD8, (c) SUMO-1 and (d) ThiS. All ribbon diagrams were generated by MolScript and Raster3D.

2.5.3
Activation and Transacylation of Ub: E1 and E2 Enzymes

Ub and Ub-like proteins are activated and transferred to the target protein by the same overall mechanisms discussed below using E1, E2 and E3 enzymes (Fig. 2.8). However, the details of the transfer reaction, the substrate recognition and the properties and numbers of E1, E2 and E3 enzymes vary considerably among the Ub family members. In the following we will concentrate on the function of Ub in marking proteins for proteolysis in the proteasome.

■ **Ubiquitination requires**
– ATP
– E1 enzyme
– E2 enzyme
– E3 enzyme

E1: Activation of Ub
In an initial reaction, Ub is activated by linking its C-terminal glycine carboxylate to an SH group of the Ub-activating enzyme E1 (reviewed in Pickart and Eddins, 2004). The activation reaction requires ATP and consists of two steps. Initially, an Ub adenylate intermediate is formed. The activated Ub then reacts with an E1 cysteine residue to from an E1-Ub thiol ester.

■ **E1**
– Ub-activating enzyme
– Ub adenylate intermediate
– Active site cysteine

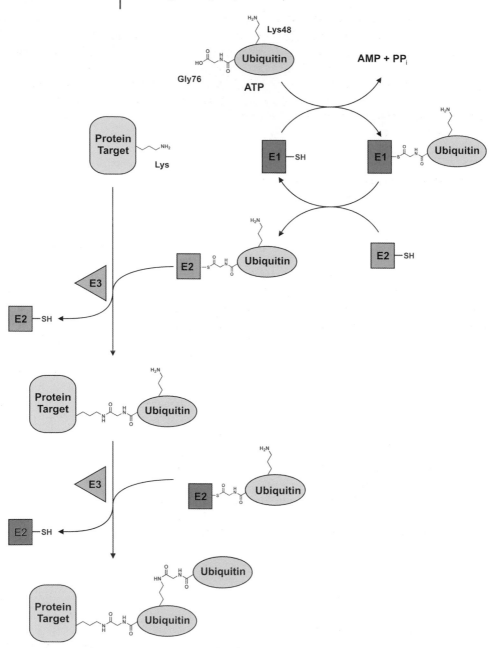

Fig. 2.8: Reactions involved in ubiquitination.

E2: Transacylation to the Ub-conjugating Enzyme E2

The next step of the conjugation cascade is transfer of the Ub moiety from E1-Ub to a cysteine-SH within the active site of the Ub-conjugating enzyme E2 to form E1-Ub. There are large numbers of E2 enzymes dedicated to Ub, comprising 11 members in yeast and many more in metazoa. The individual Ub E2 enzymes dictate specific biological functions of the Ub because the specificity of the E2/E3 interaction limits the final substrates to which the Ub is attached.

■ **E2**
– Ub-conjugating enzyme
– Many members
– Specific interaction with E3

2.5.4
Transfer to the Target Protein with the Participation of E3

The third step of ubiquitination, the transfer of Ub to the target protein, is catalyzed by a *Ub–protein ligase* or E3 enzyme. In this reaction, Ub is linked by its C-terminal glycine in an amide isopeptide linkage to an ε-NH_2-group of the substrate protein's Lys residues. In a subsequent reaction, a lysine residue of Ub attached to the target protein can itself be a substrate for E3-mediated ubiquitination, resulting in the attachment of multiple Ub molecules in a Ub–Ub linkage and the formation of *poly-Ub chains* on the target protein. In these chains, the Ub molecules are linked by an isopeptide bond between K48 (or K29, K63) and G76. The presence of poly-Ub chains on the target protein is a prerequisite for its degradation in the proteasome.

■ **E3**
– Ub–protein ligase
– Many members
– Single polypeptide or multiprotein complex
– Confers substrate specificity

The E3 enzymes are primarily responsible for conferring specificity to Ub conjugation. Each E3 enzyme, together with its cognate E2 enzyme, recognizes a few substrates that share a particular ubiquitination signal, which is usually a primary sequence motif. Subsequently the substrate is marked with a secondary signal, i.e. the poly-Ub chain. The proteasome recognizes only this secondary signal and therefore will degrade a huge variety of substrates. Selection of substrates for Ub ligation and thus for proteasome degradation occurs primarily by the E3 enzymes, of which a large number are known.

■ **E3-mediated Ub attachment**
– Monoubiquitination
– Polyubiquitination: Ub–Ub attachment

Substrate Recognition

Most important for substrate recognition by E3 enzymes is their ability to select substrates in dependence on the presence of posttranslational modifications. As illustrated in Fig, 2.9, many substrates require specific *posttranslational modifications* in order to become ubiquitinated by E3 enzymes. This property allows for signal-directed proteolysis of substrates and links signal transduction with targeted proteolysis. Ubiquitination and Ub-mediated proteolysis are therefore important components of the signaling network of the cell. Distinct signals such as phosphorylation or oxidation signals can recruit

■ **Substrate recognition**
– Mediated by E2/E3
– May be signal-dependent and require posttranslational modifications

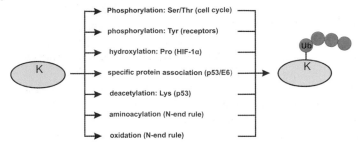

Fig. 2.9: Mechanisms for modulating substrate recognition by E3 enzymes. Shown are posttranslational modifications and other mechanisms known to regulate the recognition of cognate substrates by different E3s. For Ser/Thr phosphorylation of substrates, see Section 13.4.1. For Tyr phosphorylation of substrates, see Section 2.5.6.2. For hydroxylation of Pro residues, see review by Kaelin (2005). HIF = hypoxia-induced factor. Deacetylation, see Section 3.5.6. Specific protein association, see Section 2.5.5. Aminoacylation, see Section 2.5.6.1.

■ **E3 types**
  − HECT-E3
  − RING E3

regulatory proteins into the Ub pathways providing another regulatory level of their function and activity. The best characterized example of signal-dependent proteolysis is the phosphorylation-dependent degradation of cell cycle regulators (Section 13.4).

There are two major types of E3 enzymes in eukaryotes defined by the presence of either a *HECT domain* or a *RING domain* (Fig. 2.10). The HECT and RING domains share a common biochemical property of E2 binding. In a given E3, this E2-interacting domain is grafted onto a different domain(s) that is specialized to interact with substrates of that E3. The two domains can be part of the same polypeptide chain or the substrate- and E2-binding domains can be distinct subunits of a multiprotein complex.

Fig. 2.10: Major E3 classes. (a) HECT domain E3s bind cognate E2s via the conserved HECT domain and transiently accept Ub at a cysteine residue in this region; a different region of the same polypeptide chain binds the substrate (blue) through an element in the degron (square). (b) RING domain E3s are scaffold proteins that use the RING domain (red) to bind the E2 and a different domain (orange) to bind the substrate. In SCF and other multisubunit RING domain E3s, the RING and substrate-binding domains occur in separate polypeptides.

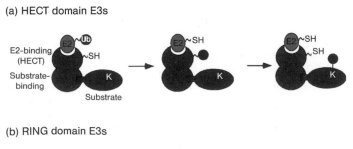

(a) HECT domain E3s

(b) RING domain E3s

2.5.5
HECT Domain E3 Enzymes

HECT (homologous to E6-AP C-terminus) domain enzymes comprise a large family of E3 enzymes that directly participate in the transfer of Ub to the substrate. In the transfer reaction of HECT-E3 enzymes, Ub is first transferred from the E2 carrier to an active site cysteine of the E3 enzyme to from a *covalent HECT-E3-Ub intermediate*. Subsequently, Ub is ligated to the ε-NH$_2$-group of an acceptor lysine on the substrate protein. The HECT-E3s contain an essential active site cysteine residue near the C-terminus and one or several WW domains (Section 8.2.6).

The discovery of this family of E3 enzymes started from the studies on the targeted degradation of the p53 tumor-suppressor protein. The p53 protein assumes an important role in the control of growth of higher organisms. It functions as a tumor suppressor, i.e. it suppresses the growth of tumors (Section 14.6), and inappropriate low levels of p53 due to increased Ub-dependent proteolysis are considered to be an important factor contributing to tumorigenesis. One Ub-mediated degradation pathway of p53 is linked to infection by oncogenic DNA viruses, e.g. the human papilloma virus. The viral protein involved is the oncoprotein E6 of human papilloma virus and a cellular E3 enzyme, termed E6-AP (E6-associated protein), which belongs to the HECT family of E3 enzymes. Recognition of p53 and transfer of Ub occurs in a complex between the viral E6 protein, E6-AP and p53, with the formation of an E6-AP-Ub intermediate (Fig. 2.11). The ubiquitination initiated by the E6 protein and the ensuing degradation of p53 result in a loss of p53 function, thus offering an explanation for the tumor-causing effect of the papilloma virus.

■ **HECT-E3**
– Covalent Ub-E3 intermediate
– Involved in p53 degradation upon infection with papilloma virus

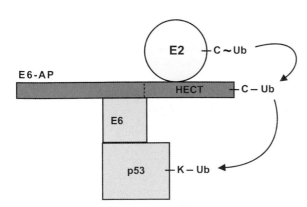

Fig. 2.11: Ubiquitinylation of the tumor suppressor protein p53. The attachment of Ub molecules to p53 is catalyzed by the HECT domain E3-Ub ligase E6-AP which receives the activated Ub from an E2 enzyme. An intermediate is formed where the Ub is linked via a thioester bond to an active site cysteine of E6-AP. The E6 protein binds both to E6-AP and p53 and confers substrate specificity. C = cysteine; K = lysine.

2.5.6
RING Domain E3 Enzymes

RING E3
- No covalent intermediate
- Around 400 human members
- His–Cys sequence motif

Three functions:
- Binding of E2
- Substrate recognition
- Transfer of Ub from Ub-E2 to substrate

RING finger motif
- Domain within RING-E3 ligases

Cellular functions of RING-E3s
- Cell cycle regulation
- Cell proliferation
- Apoptosis
- Secretion
- Trafficking

The RING domain E3 enzymes do *not form a covalent intermediate* with Ub, but may instead *activate the E2* to directly discharge the Ub thioesterified to its active site cysteine onto the lysine of a substrate.

The RING superfamily of E3 enzymes includes proteins of similar fold that share a *series of histidine and cysteine* residues with a characteristic spacing that allows for the coordination of two zinc ions in a structure called the Really Interesting New Gene (RING) finger. There is one subfamily of RING E3s, the U-box proteins, that does not contain the His–Cys motif and therefore cannot bind zinc. The U-box proteins show a similar fold as the RING domain E3 enzymes and are therefore included in the RING domain class. One member of the U-box proteins, the E3 ligase CHIP, is involved in the recognition and degradation of misfolded proteins.

The RING finger motif is found in single-subunit E3 enzymes like the Cbl protein (see below) or in the form of multisubunit E3 enzymes [SCF, anaphase-promoting complex (APC), see below] where the RING finger is located in one specific subunit (reviewed in Ardley and Robinson, 2005).

How RING finger ligases work is still only poorly understood. The RING finger E3s have to fulfill at least three tasks: the recognition and binding of the substrate, the recognition and binding of the E2 enzyme, and the participation – either directly or indirectly – in the transfer of the Ub. It is still an open question how the E3s act in the transfer step. Up to now, catalytic residues in the E3s have not been identified. The RING finger motif does not directly participate in the Ub transfer to the target protein, but rather seems to function as a scaffold that positions the substrate and the E2 enzyme optimally for Ub transfer.

RING proteins are encoded in all eukaryotic organisms analyzed to date and the human genome alone encodes around 400 proteins with this fold. Although some of these proteins may not be Ub ligases, most RING proteins studied so far appear to have Ub ligase activity *in vitro*. As indicated by their sheer number, E3 enzymes play pivotal roles in central cellular processes. Functions of RING finger E3 enzymes in cell cycle regulation, proliferation and apoptosis have been established. In addition, they participate in endocytosis and secretion. Importantly, the RING finger motif is found in a large number of proteins of recognized regulatory function that hitherto have not been associated with Ub–proteasome-mediated proteolysis. Examples are the breast cancer susceptibility gene product BRCA1 and the MDM2 protein (Chapter 14).

According to the nature of the target sequences recognized and the regulation of their function the following types of RING domain E3 enzymes will be highlighted (reviewed in Ardley and Robinson, 2005).

2.5.6.1 N-end Rule Enzymes

The N-end rule relates the *in vivo* half-life of a protein to the identity of its N-terminal residue (reviewed in Varshavsky, 2003). A subset of degradation signals recognized by the N-end rule pathway comprises the signals, called *N-degrons*, whose determinants include destabilizing N-terminal residues such as basic or bulky hydrophobic amino acids. A major destabilizing residue at the N-terminus is arginine that can be appended enzymatically to other N-terminal residues such as Glu, Asp or Cys. Remarkably, the arginylation of N-terminal cysteine requires the prior oxidation of cysteine by nitric oxide (NO) which is a major second messenger (Section 6.10). This link identifies the N-end rule pathway as a NO sensor that functions through its ability to destroy proteins in response to NO signals (Hu et al., 2005).

■ **N-end rule E3s**
– Recognize N-terminal amino acids

E3 ligases involved in the N-end rule pathway include the UBR1 and UBR2 enzymes from yeast and mammals. Known functions of N-end rule pathways in higher organisms include the control of nuclear translocation, the regulation of apoptosis and the fidelity of chromosome degradation, among others.

2.5.6.2 Cbl Protein

The Cbl proteins are single-subunit E3-Ub ligases that are involved in the downregulation of tyrosine kinases, including receptor tyrosine kinases (RTKs) and non-RTKs. They interact with phosphotyrosine residues on activated tyrosine kinases via a specific SH2 domain (Section 8.1.5) and they contain a RING finger domain that mediates the binding of E2 enzymes (reviewed in Ryan et al., 2006). In the complex formed, a transfer of Ub to the kinase occurs, and the kinase is thereby targeted for endocytosis and also for degradation (Fig. 2.12). The regulation of Cbl protein activity is complex. Cbl seems to be an inactive E3 enzyme until encountering the activated kinase that it will degrade. On interaction with the active kinase, the E3 activity of Cbl protein is induced, leading to ubiquitination and downregulation of the kinase. Although the mechanism of this activation has not been determined precisely, the interaction between the Cbl protein and its target is crucial to induction of the Cbl E3 activity.

■ **Cbl**
– Single subunit E3
– Ub labeling of tyrosine kinases
– Oncogenic variants known

Fig. 2.12: Cbl-induced ubiquitination of TM receptors. Ligand binding to RTKs triggers autophosphorylation on the cytoplasmic region of the receptor. The tyrosine phosphates are bound by an SH2 domain of the Cbl protein which is a E3-Ub ligase. The RING finger motif of Cbl mediates binding of an E2 enzyme from which Ub is transferred to acceptor lysine residues of the receptor inducing its internalization and proteasomal degradation.

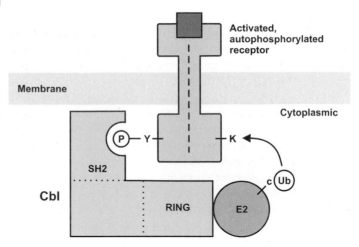

Cbl protein itself is negatively regulated by ubiquitination. Like many other E3 proteins, Cbl proteins may ubiquitinate themselves or are trans-ubiquitinated by other E3 ligases and are thereby targeted for proteasomal degradation.

The signaling molecules that are downregulated by the Cbl proteins include RTKs like the epidermal growth factor receptor (EGFR; see Chapter 8) and non-RTKs like Zap70 (Section 11.3). The central importance of the Cbl proteins for cellular regulation is highlighted by the observation that oncogenic forms of Cbl have been found in mouse retroviruses.

2.5.6.3 Cullin-based E3 Enzymes, SCF and APC

■ Cullin-RING-E3 enzymes:
 – Superfamily of more than 300 members
 – Contain cullin and RING domain proteins

The most intensively studied subclass of RING-E3 enzymes are those of the cullin-RING ligase superfamily (reviewed in Petroski and Deshaies, 2005). The highly conserved enzymatic core of these multi-subunit proteins comprises the C-terminal region of a protein named *cullin* and one of two closely related *RING proteins*. The RING protein bound to the C-terminal domain of a cullin recruits the E2 enzyme, whereas the N-terminal region of cullins recruit receptors that in turn interact with specific substrates. To date, six different types of cullin-RING ligases have been identified, each of which employs a distinct family of substrate receptors. The human genome encodes about 300–350 cullin-RING ligases in addition to the around 400 potential Ub ligases based on other RING fold proteins. However, the mechanism by which any cullin-RING ligase or RING protein promotes substrate ubiquitination remains unknown.

SCF Complex

The *SCF (Skp1–Cullin–F-box)-E3-Ub ligase* family comprises multi-subunit enzymes (Fig. 2.13) composed of an invariant core complex, containing the Skp1 linker protein, the CDC53/CUL1 scaffold protein and the Rbx1/Roc1/Hrt1 RING domain protein (reviewed in Willems et al., 2004). The RING domain subunit uses its zinc-binding motif to recruit and direct E2 enzymes towards specific substrates, which are recognized by a suite of substrate receptors called F-box proteins that in turn recruit substrates for ubiquitination by the associated E2 enzyme. F-box motifs are widespread and are found in a large number of proteins with functions not related to the ubiquitination system. More than 70 F-box proteins have been identified in humans, although only few of them have been fully characterized. This large number, in combination with the core complex and the E2 enzymes, provides the basis for multiple substrate-specific ubiquitination pathways.

Each F-box protein contained in SCF complexes recognizes a distinct set of substrates, as illustrated in Fig. 13.12 using the example of cell cycle regulators. In some cases, phosphorylation of the substrate is required for binding to the F-box protein, which provides a direct link between ubiquitination and signaling pathways. An example for the phosphorylation-dependent ubiquitination is provided by the SCF complexes of the cell cycle (Section 13.4.1).

■ **SCF-E3 enzymes contain**
– Skp1
– Cullin
– F-box protein
– RING protein Rb × 1

■ **F-box protein**
– Substrate recognition
– May require substrate phosphorylation

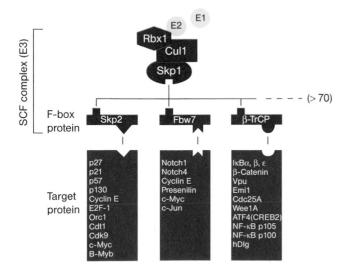

Fig. 2.13: Structure and substrates of the SCF complex. The SCF complex is a heterotetramer composed of Skp1, Cul1, the RING finger protein Rbx1 and an F-box protein of which more than 70 subtypes are known. Each F-box protein recognizes a distinct set of substrates as indicated. From Nakayama et al. (2001).

APC (Cyclosome)

The APC is a high-molecular-weight complex of at least 12 different subunits that degrades proteins containing a specific recognition sequence, the *destruction box* (Section 13.4.2). The core E3 activity of APC has been shown to reside in a small RING finger protein of the APC, Apc11p and a cullin-like subunit, Apc2.

Substrates are cell cycle regulators, e.g. cyclins, kinase inhibitors and spindle-associated proteins. Importantly, some forms of the cyclosome require phosphorylation in order to be active.

2.5.6.4 Processing of Nuclear Factor (NF) κB and Degradation of IκB

Only one example for the regulatory functions of Ub–proteasome system will be discussed in more detail here, i.e. the regulation of the transcription factor NFκB and its inhibitor, IκB. The transcription activator NFκB regulates a variety of genes involved in the immune response and the inflammatory process (reviewed in Tergaonkar, 2006). NFκB is required for the expression of genes for the light chain of immunoglobulins, for interleukin (IL)-2 and -6, and for interferon-β (Chapter 11).

The function and regulation of NFκB is shown schematically in Fig. 2.14. The active form of NFκB is a heterodimer consisting of one p50 and one p65 subunit. In the cytosol, NFκB is found in an inactive complex bound to the inhibitor protein IκB. IκB masks the nuclear translocation signal of the heterodimer, thus preventing its transport into the nucleus.

The activity of NFκB is highly regulated. NFκB is activated upon extracellular stimuli, e.g., the action of growth factors, cytokines (Chapter 11) or exposure to ultraviolet (UV) light. The signal pathway that leads to phosphorylation and subsequent degradation of NFκB has been well characterized for the cytokines IL-1 and for tumor necrosis factor (TNF) (reviewed in Chen and Goeddel, 2002). Following binding of a cytokine to its TM receptor, a family of specific protein kinases including a high-molecular-mass IκB kinase complex is activated that phosphorylate the inhibitor IκB. This phosphorylation is the signal for ubiquitination and degradation of IκB. NFκB is thus released from its inhibited state to translocate in the nucleus and activate transcription of target genes.

The Ub–proteasome pathway participates in the regulation of NFκB at two points:

– The p50 subunit of NFκB results from the proteolytic processing of a 105-kDa precursor protein (p105) in the cytosol. The processing requires the polyubiquitination of p105 mediated by the 26S proteasome.
– The degradation of the inhibitor protein IκB involves the Ub–proteasome pathway.

The Ub–proteasome system thus has significance for NFκB in two ways. On the one hand, it participates in the specific processing of the p105 precursor protein to the small subunit of NFκB. On the other hand, NFκB is activated because of the degradation of IκB.

This example nicely illustrates how extracellular signals can induce the ubiquitination and degradation of specific proteins. As shown by the processing of the p105 precursor, ubiquitination can also be used for partial proteolysis and for specific activation of a regulatory protein.

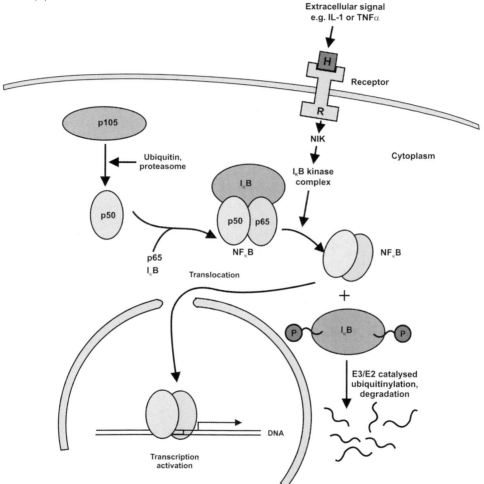

Fig. 2.14: Regulation and proteolysis of the transcription factor NFκB. The Ub–proteosome pathway is involved in the regulation of NFκB in two ways. The 50-kDa subunit of NFκB is formed from a 105-kDa precursor via Ub-mediated proteolysis. In the cytosol NFκB is heterodimeric and inactive, bound to the inhibitor protein IκB. Upon an external signal, the phosphorylation of IκB is induced, whereby the inhibitor complex is broken up. The inhibitor IκB is degraded via the Ub–proteosome pathway, while the NFκB is transported to the nucleus where it can fulfill its transcription regulation function.

2.5.7
Degradation in the Proteasome

The degradation of protein-Ub conjugates occurs in an ATP-dependent reaction within a large protease complex, the *26S proteasome* (reviewed in Zwickl et al., 2001; Wolf and Hilt, 2004). The substrate protein is degraded to peptides in the 26S proteasome, while the Ub is released and again available to form protein conjugates. A major prerequisite for proteasomal degradation is the presence of poly-Ub chains, linked by K48 isopeptide bonds, on the substrate protein. Assembly of poly-Ub chains through other Lys residues, e.g. K63, serves other functions and directs the protein substrates to other routes, like DNA repair, stress response and endocytosis (Section 2.5.8).

The 26S proteasome is composed of two protein aggregates, a *19S regulatory* and a *20S core particle* (Fig. 2.15a). The main proteolytic component of the 26S proteasome is the 20S particle. Another regulatory particle, the *11SRegγ* particle, can associate with the 20S core to form a Regγ–20S particle with functions distinct from the 26S proteasome.

20S Particle

The catalytic core of this large multisubunit proteolytic complex is the 20S proteasome, which consists of 14 α and 14 β subunits arranged in a cylindrical particle of four heptameric rings with a central channel of 17 Å in diameter and resembles a barrel (Fig. 2.15b). Two inner β-rings harbor all of the proteolytic sites for substrate cleavage inside the central catalytic chamber, whereas two outer α-rings provide the gated entry sites for substrate peptides. In the 20S particle from eukaryotes, each of the two outer rings contains seven different α subunits, whereas each of the two inner rings is built up from seven different β subunits. All subunits occupy specific positions in the 20S particle.

The presence of the protease center in the central channel ensures that the proteolysis is compartmentalized and shielded from the surrounding media. The 20S proteasome is a latent protease complex, as the N-terminal tails of α subunits occlude the central proteolytic channel. Access of substrate proteins to the proteolytic center requires the opening of the ring of α subunits with the assistance of the 19S particle. The structure of the 20S proteasome also indicates that proteins are accessible to the catalytic center only in the unfolded state.

An N-terminal threonine has been identified as an essential active site residue of the protease center. The OH-group of the threonine functions as a nucleophile during hydrolysis of the peptide bond. A similar mechanism of hydrolysis has been shown for other hydro-

---

■ **26S Proteasome**
 – Central proteolytic machinery
 – 19S regulatory particle
 – 20S core particle

■ **Regγ–S particle**
 – 20S core
 – 11SRegγ

■ **Eukaryotic 20S particle**
 – 2 × 7 distinct α subunits
 – 2 × 7 distinct β subunits

Proteolytic activity
 – N-terminal nucleophile hydrolase
 – Three distinct catalytic centers

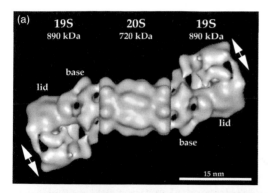

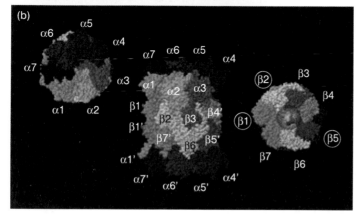

Fig. 2.15: Structure of the proteasome. (a) The 26S proteasome. (b) The 20S core complex from yeast. The 20S proteasome is composed of a stack of four rings composed of seven subunits each. The two outer rings are made up of seven different α subunits (left graph, view from the top), whereas the two central rings are composed of seven different β subunits (right graph, view from the center). The active sites reside within the central chamber (shadow ring) of the 20S proteasome at subunits β1/Pre3, β2/Pup1 and β5/Pre2 (marked by circles).

lases, which, because of this property, are now included in the family of *N-terminal nucleophile hydrolases*. For some β subunits of eukaryotes the N-terminal threonine is generated by autoproteolysis of an N-terminal prosequence.

In the vertebrate 20S proteasome, *three distinct catalytic centers* have been identified on specific β subunits, designated X, Y and Z. The X subunit expresses a chymotrypsin-like activity, the Y subunit a peptidyl-glutamyl-peptide-hydrolyzing activity and the Z subunit a trypsin-like activity.

A distinct feature of the vertebrate proteasome is the possibility to exchange specific subunits thereby generating proteasome variants. By exchange of β subunits, different 20S complexes can form which function in specific degradation reactions. In the proteasome that is involved in processing antigenic peptides the catalytic reactions are performed by β subunits different from the X, Y and Z subunits. This specific proteasome is also called an "immunoproteasome".

19S Regulatory Particle

Recognition of ubiquitinated protein substrates and their funneling into the proteolytic cavity is achieved by either the 19S regulatory particle (to form the 26S proteasome) or the 11S regulator, both of which associate with the top or bottom surface of the 20S "barrel" to trigger opening of the gate to allow access of unfolded protein substrates.

The 19S regulatory particle from yeast, which is the most thoroughly studied, is composed of at least 17 different subunits. Further proteins with adaptor functions can associate with the 19S particle during delivery of the protein substrates. The following activities are required for substrate delivery to the proteolytic chamber:
– Recognition of ubiquitinated substrates.
– Removal of the Ub chains by deubiquitinating enzymes.
– Unfolding of the substrate protein under consumption of ATP.
– Translocation of the substrate into the 20S particle.

Although subunits have been identified that carry enzymatic activities such as deubiquitination or ATPase activity, our knowledge of the cooperation of the activities is far from complete. For details, see Wolf and Hilt (2004).

11SRegγ Particle

The 11SRegγ particle is another cap complex that regulates substrate access to the 20S barrel. It consists of seven subunits that are arranged in a ring-shaped complex. Unlike the 19S particle, the 11SRegγ particle does not contain ATPase subunits. It participates in antigen processing and in targeting nuclear proteins for degradation. A remarkable feature of the 11SRegγ particle is its ability to direct degradation of proteins in an *ATP- and Ub-independent manner*.

2.5.8
Other Regulatory Functions of Ub Conjugation

In addition to marking proteins for degradation, Ub conjugation has now been recognized to serve many nonproteolytic functions. These *nonproteolytic functions* of Ub rely on monomeric Ub or poly-Ub chains branched from lysines other than K48, e.g. K63. In this case, ubiquitination affects the structure, activity or localization of the target protein thereby regulating central cellular processes such as:
– Endocytosis.
– Recycling of cell surface proteins.
– Protein trafficking.
– Protein import into cellular organelles.

– Repair and replication of DNA.
– Processing and presentation of antigens.
– Assembly of ribosomes.

Ubiquitination through alternative lysine sites is now considered as a protein modification code, which is used to sort different ubiquitination products to different destinations. The ubiquitination code is read by several families of *Ub-binding domain (UBD)*-containing proteins that function as *Ub receptors*. Most UBDs of the Ub receptors recognize a conserved hydrophobic pocket on Ub, centered around Ile44. The UBDs interact with mono- or polyubiquitinated chains to recruit the ubiquitinated protein into signaling networks in the cell. Thus, Ub constitutes a posttranslational modification that is recognized by distinct sets of interaction modules, similar to other posttranslational modifications of regulatory proteins, such as phosphorylation or methylation (Section 1.6).

■ **Ub receptors**
– Contain UBDs
– Recognize mono- or poly-Ub chains

Examples of UBDs and Ub Receptors
To date, 11 families of UBDs have been identified (reviewed in Hicke et al., 2005; Harper and Schulman, 2006). The various families show a high variability and complexity in the mode of Ub binding. Some proteins functioning as Ub receptors contain two copies of the same UBD and thus can bind two Ub molecules, while others have multiple UBDs of different classes. Interestingly, many UBDs are required for the ubiquitination of the protein within which they are carried. The mechanistic basis of the UBD-dependent ubiquitination remains to be determined.

The number of proteins carrying UBDs is large. Examples of UBD carrying proteins include components of the 19S regulatory particle of the proteasome, deubiquitinating enzymes, E3 ligases, DNA polymerases, guanine nucleotide exchange factors (GEFs), transcription factors and many others. Overall, however, the understanding and knowledge of the mechanisms by which UBDs regulate these proteins is limited.

One specific example of a proposed function of UBDs not related to proteolysis shall be highlighted.

Ubiquitination and Translesion DNA Synthesis
Ubiquitination of accessory proteins of the DNA replication machinery has been discovered as an important tool for regulating specific DNA replication events (reviewed in Fischhaber and Friedberg, 2005). This has been shown for the replication of damaged DNA sites. The replication proteins involved are the clamp loader proliferating cell nuclear antigen (PCNA) and the translesion DNA polymerases Pol η and Pol ι. Both DNA polymerases are capable of

DNA synthesis through damaged DNA templates. In response to DNA damage, PCNA becomes ubiquitinated and interacts with the translesion DNA polymerase. Both translesion DNA polymerases carry Ub-binding domains that are required for recruitment of the polymerases to sites of UV-damaged DNA, and it is thought that the interaction of ubiquitinated PCNA and the UBDs of the translesion DNA polymerases promotes the recruitment of the polymerases to sites of DNA damage. Conjugation and removal of Ub from the translesion DNA polymerases may thus provide a mechanism for shuttling these enzymes in and out of replication foci during and after DNA damage.

2.5.9
Regulation of Proteins by Sumoylation

Of the many Ub-like proteins, the *SUMO (small Ub-related modifier) protein* has received most attention (reviewed in Gill 2006).

■ **SUMO**
 – Ub-like protein
 – Attached to lysine residues
 – E1 and E2 enzymes
SUMO regulates
 – Transcription
 – Protein trafficking

SUMO proteins are reversibly attached to many regulatory proteins, including promoter-specific transcription factors, transcriptional cofactors and regulators of chromatin structure. SUMO-1, a 98-amino-acid polypeptide, is covalently linked by an isopeptide bond to lysine residues in proteins in a process analogous to, but distinct from, ubiquitination. E1- and E2-like enzymes are responsible for the attachment of the SUMO moiety to lysine residues of the target protein. As compared to ubiquitination, sumoylation is more sequence specific and requires a particular amino acid in the neighborhood of the lysine to be modified.

The functions of sumoylation are diverse. A major role of sumoylation is now ascribed to the regulation of transcription. Many proteins with functions in transcription regulation are reversibly modified by sumoylation. In most cases, covalent attachment of SUMO to transcription factors inhibits transcription, probably by promoting recruitment of transcriptional corepressors such as histone deacetylases (HDACs; see Section 3.5.2). Here, the SUMO modification serves as a label for the association of proteins that contain domains with binding specificity for the SUMO modification.

Other functional consequences of sumoylation include regulation of the trafficking of proteins. For example, SUMO modification of Ran-GAP1 (Section 9.1.5) regulates its subcellular localization by targeting it to the nuclear pore complex. We also know of examples where sumoylation competes with other modifications. SUMO modification of the inhibitor IκB (Fig. 2.14) occurs at the same Lys residues that are used for attachment of Ub molecules, thus precluding ubiquitination. As a result, IκB is stabilized and remains bound to NFκB inhibiting nuclear translocation of NFκB and transcriptional activation.

2.6
Lipidation of Signaling Proteins

The signaling function of many signaling proteins depends on their association with the cell membrane. As outlined in Section1.8.5, cells have available a palette of tools to achieve stable and regulatable membrane association. A central and widely used tool for membrane anchoring is the posttranslational attachment of hydrophobic residues, such as fatty acids, isoprenoids (Fig. 2.16) or complex glycolipids (Fig. 2.21 below) to specific amino acid side chains of target proteins. These lipid moieties favor membrane association by increasing the affinity of the protein to the membrane. Because of their hydrophobic nature, the membrane anchors insert into the phospholipid bilayer and thus mediate membrane association of the protein. To achieve a strong and stable membrane association, more than one lipid anchor is typically used.

■ **Lipid anchors**
– Fatty acids
– Isoprenoids
– Cholesterol
– Complex
  phospholipids

The lipid anchors serve several functions in cell signaling. The main function is to promote membrane association of signaling proteins. Many signaling events occur in close association to the inner side of the cell membrane or at organelle membranes. Lipid anchors target signaling components to the membrane, as is the case for the cytoplasmic protein tyrosine kinases, so that they can participate in membrane-associated signaling pathways. Importantly, the lipid an-

lipid	examples of modified proteins	site of modification
N-myristoyl	heterotrimeric G-proteins (α-subunit), see chapter 5	N-terminus
	cytoplasmic tyrosine kinases, see chapter 8	
S-palmitoyl (S-Acyl) 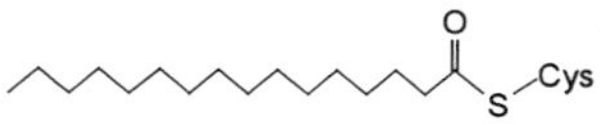	heterotrimeric G-proteins (α-subunit), G-protein-coupled receptors, see chapter 5 Ras proteins, see chapter 9	internal, no distinct consensus sequence
S-prenyl 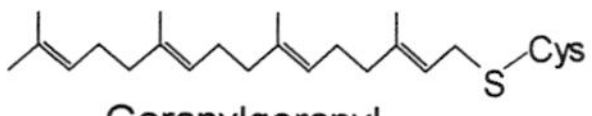 Geranylgeranyl 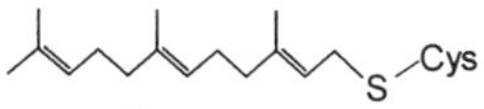 Farnesyl	heterotrimeric G-proteins (γ-subunit) Ras proteins rhodopsin kinase see chapter 5	C-terminus

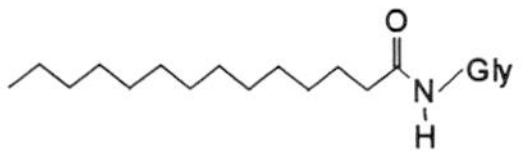

Fig. 2.16: Structure of lipid anchors and representative examples for lipid-modified signal proteins.

chors can be used in a dynamic way to recruit signaling proteins in a regulated manner into signaling pathways. Furthermore, protein lipidation helps to target signaling proteins to membrane subdomains such as lipid rafts, thereby facilitating the formation of larger signaling complexes and localizing signaling events to specific subcellular sites. Other functions of membrane anchors include the vesicular transport of proteins (e.g. in neurons) and the regulation of enzyme activity (Section 2.6.6).

2.6.1
Myristoylation

Myristoylated proteins contain a saturated acyl group of 14 carbons, myristoic acid (*n*-tetradecanoic acid), added via an amide bond to the amino group of the N-terminal glycine residue. The reaction is catalyzed by the enzyme *N*-myristoyltransferase (reviewed in Farazi et al., 2001). Typically, this occurs cotranslationally after the initiating methionine is cleaved by an aminopeptidase, but it can occur post-translationally when an internal glycine residue is exposed by proteolytic cleavage. The consensus sequence for *N*-myristoylation is Gly-X-X-X-Ser/Thr (X: any amino acid), where the residue following the glycine is often a cysteine (reviewed in Resh, 2004). A clumping of basic amino acids at the N-terminus can serve as an additional signal for myristoylation (Fig. 2.17).

■ **Myristoic acid anchor**
– C14 fatty acid anchor
– At N-terminal glycine
– Stable modification

(a) Src kinase

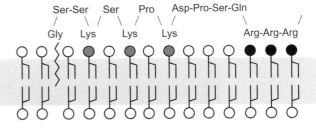

Fig. 2.17: Lipid anchors and basic regions as elements of the membrane association of proteins. Examples for proteins which exhibit basic residues near a lipid anchor. (a) Src kinases (Chapter 8) possess a myristoyl anchor at the N-terminus as well as a stretch of basic residues. (b) In Ki-Ras proteins (Chapter 9) there is a farnesyl residue at the C-terminus that serves as a lipid anchor, as well as a stretch of Lys residues. Negatively charged head groups of phospholipids are shown as filled circles. X = any amino acid.

(b) K-ras (B) protein

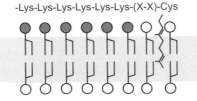

N-myristoylation is generally considered a constitutive process and a *permanent modification* that promotes weak and reversible protein–membrane and protein–protein interactions. Typically, myristate acts in concert with other mechanism to regulate signaling functions at membranes. As shown below, the myristate anchor also may function as a *switch* during regulated membrane anchoring. Examples of myristoylated proteins are the cytoplasmic protein tyrosine kinases (family of the Src kinases, Section 8.3.2), as well as the α subunit of the heterotrimeric G-proteins (Section 5.5.6). Posttranslational myristoylation has been implicated in several steps of apoptosis including the myristoylation of Bid (Section 15.3.3) and of caspase-activated protein kinases (Vilas et al., 2006).

2.6.2
Palmitoylation

Palmitoylated proteins contain a long-chain fatty acid, such as palmitoic acid (*n*-hexadecanoic acid), connected to the protein via a labile *thioester bond* to cysteine residues. Other long-chain fatty acids like stearate and oleate have also been found to be incorporated in *S*-acylated proteins. The thioester bond of *S*-acylated proteins is less stable than the amide bonds of the myristate anchor. The lability conveys a reversible character to the modification and thus permits regulation of membrane association (reviewed in Smotrys and Linder, 2004). The distribution of signal proteins between the membrane and cytosol can be regulated via a cyclic process of acylation and deacylation, making the reversible *S*-acetylation of signal proteins an important tool for the modulation or regulation of signaling pathways. The enzymes involved in acylation and deacylation are the *protein acyl transferases* and the *acyl protein thioesterases*.

■ **Palmitoic acid anchor**
  – At internal Cys
  – Thioester linkage
  – Reversible
    modification

There is no well-defined consensus sequence for palmitoylation other than a requirement for cysteine. However, several sequence contexts have emerged over the years. For details, see Smotrys and Linder (2004). Interestingly, palmitoylation sites have been identified within well characterized protein domains such as the pleckstrin homology (PH) domains of phospholipases D1 and D2.

We also know of fatty acid attachment (palmitate, stearate) through oxyester bonding to serine or threonine. Furthermore, several secreted eukaryotic proteins are modified at the ε-amino group of lysine with myristate or palmitate through an amide linkage. Removal of palmitate by thioesterases occurs both constitutively and in response to signals. Details of how palmitoylation and depalmitoylation is linked to signal transmission are, however, lacking.

Functions of Palmitoylation

Similar to other lipid modifications, palmitoylation promotes membrane association of otherwise soluble proteins with functions in signaling pathways. In many cases, the palmitoic acid-mediated membrane attachment is essential for the participation of these proteins in signaling events. The function of palmitoylation, however, ranges beyond that of a simple membrane anchor. Trafficking of lipidated proteins from the early secretory pathway to the plasma membrane is dependent upon palmitoylation in many cases. In addition, modification with fatty acids impacts the lateral distribution of proteins on the plasma membrane by targeting them to lipid rafts.

The family of proteins modified with thioester-linked palmitoate is large and diverse. It includes TM-spanning proteins and cytoplasmic proteins that require membrane-association for their function. Examples of palmitoylated signaling proteins include the Ras and Rho regulatory GTPases (Chapter 9), non-RTKs like p56[LCK], the adaptor protein PSD-95 (Section 8.5), the α subunit of the heterotrimeric G-proteins (Section 5.5.6), the RGS proteins (Section 5.5.7), and the enzymes phospholipase D1 and D2.

2.6.3
Farnesylation and Geranylation

Proteins with an isoprenoid modification possess either a C15-farnesyl residue or a C20-geranyl-geranyl residue (reviewed in McTaggart, 2006). Both residues are bound via a thioester linkage to a cysteine residue. As with myristoylation, these are constitutive, stable modifications performed by farnesyl or geranyl transferases.

The isoprenylation occurs at the Cys-residue of the consensus sequence Cys-A-A-X-COOH, whereby the nature of the C-terminal X-residue determines whether farnesylation or geranylation occurs (Fig. 2.18). The prenyl groups are donated by the two isoprenoids farnesyl pyrophosphate and geranyl-geranyl pyrophosphate that are derived from the mevalonate pathway (see textbooks for details). Thereafter, the three C-terminal residues are removed by the prenylation-dependent endoprotease Rce-1 and the new COOH-group of the Cys residue is methylated to increase the hydrophobicity of the C-terminus.

The importance of the protein prenylation is underscored by the nature of the estimated 300 prenylated proteins in the human proteome, many of which participate in a multitude of signaling pathways. The isoprenoid modification can be found, among others, on the Ras protein and other members of the Ras superfamily (Chapter 9), as well as with the α subunit of G-proteins (Section 5.5.6). The βγ complex of G-proteins is also associated with the membrane via

■ **Isoprenoic anchors**
– Farnesyl (C15)
– Geranly (C20)
– At C-terminal Cys via thioether bond

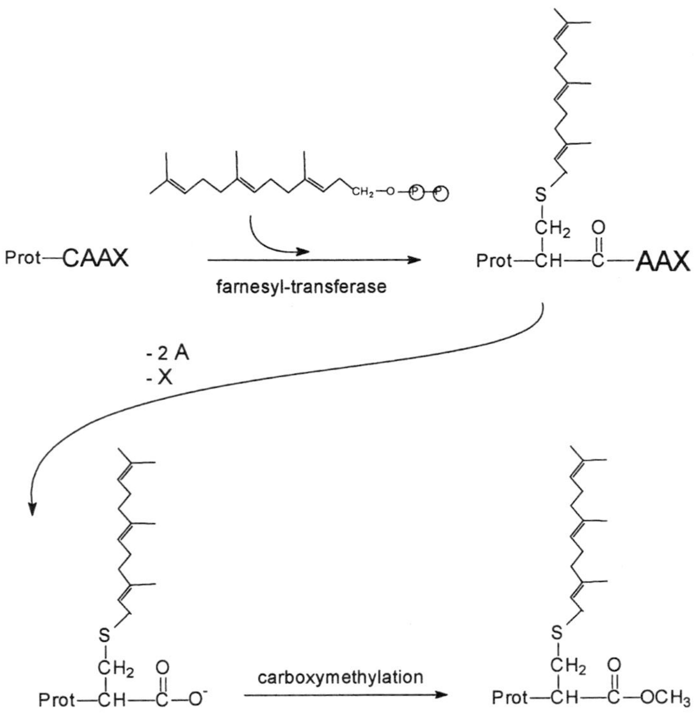

Fig. 2.18: Farnesylation at the C-terminus. The signal sequence for farnesylation is the C-terminal sequence CAAX. In the first step a farnesyl moiety is transferred to the cysteine in the CAAX sequence. The farnesyl donor is farnesyl pyrophosphate and the responsible enzyme is farnesyl transferase. Subsequently, the three C-terminal amino acids are cleaved (A = alanine, X = any amino acid) and the carboxyl group of the N-terminal Cys residue becomes methylated.

geranylation. A 2-fold geranylation is found on two Cys residues of the Rab protein (Section 9.9.2).

In addition to promoting membrane association, other functional aspects of protein prenylation have been appreciated. Prenylation can also serve to mediate protein–protein interactions and has a role in protein trafficking.

2.6.4
Dual Lipidation

Many signaling proteins are dually lipidated, showing both myristoylation and palmitoylation, prenylation and palmitoylation or dual palmitoylation. The dual lipidation is explained by the strong membrane binding mediated by two lipid anchors. It is now generally accepted that any single acylation or prenylation is unable to confer stable membrane association, indicating that an additive or cooperative effect between intrinsic anchoring motifs drives the membrane localization. A dynamic model has been proposed that accounts for the specific association of dually lipidated proteins with the membrane. According to the "kinetic bilayer trapping" hypothesis, proteins with a single lipid anchor only transiently and weakly associate

■ **Dual lipidation (e.g. myristoic plus palmitoic anchor)**
– Enforces membrane attachment

with the membrane at many sites. Singly acylated protein that reaches the membrane will be rapidly palmitoylated by a membrane-associated palmitoyl transferase and will then remain stably attached to the membrane.

Only two examples out of a long list of dually lipidated proteins (Smotrys and Linder, 2004) will be mentioned. The lipidation of cytoplasmic protein tyrosine kinases like p56[LCK] includes both myristoylation and palmitoylation. H-Ras protein requires, apart from C-terminal farnesylation (see above), a palmitoyl modification in order to bind to the plasma membrane. In both examples, the fatty acid anchors play an essential role in the signal transduction.

2.6.5
Cholesterol Membrane Anchor

■ **Cholesterol anchor (e.g. Hedgehog protein)**
– Ester bond to C-terminal glycine

A cholesterol modification has been demonstrated for the Hedgehog (Hh) protein, which is a extracellular signaling protein with functions in a diverse array of patterning events of metazoan tissues. The Hh family of proteins are found on the surface of cells and can be secreted for signaling to distant cells. Several posttranslational modifications are required in order to gain full activity of Hh proteins. In a maturation process they perform an autocatalytic cleavage, generating an N-terminal polypeptide containing all of the signaling functions (Fig. 2.19). During cleavage, a cholesterol moiety is attached covalently by an ester function to the C-terminal glycine moiety of the signaling domain (reviewed in Mann and Beachy, 2004). The hydrophobicity of the cholesterol–Hh conjugate is further increased by the addition of a palmitoic acid residue to the N-terminus of the cleavage product. The presence of the cholesterol moiety confers a very stable membrane association to Hh (Peters et al., 2004) and it is assumed that other protein factors are required to mobilize Hh for signaling. The cellular mechanisms used for the handling and delivery of cholesterol-modified proteins are largely unknown.

2.6.6
Switch Function of Lipid Anchors

■ **Switch function of lipid anchors**
– Signal-dependent
– Exposure of lipid anchor

Lipid anchors can participate in a dynamic way in membrane anchoring and may thereby actively participate in cell signaling. We know of two ways by which the myristoyl anchor can function as a switch in cell signaling:
– *Myristoyl-ligand switches.* The orientation of the myristoyl moiety relative to the protein to which it is attached may be controlled by ligand binding. Depending on the presence of the ligand, the signaling protein can exist in a conformation where the lipid an-

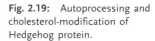

Fig. 2.19: Autoprocessing and cholesterol-modification of Hedgehog protein.

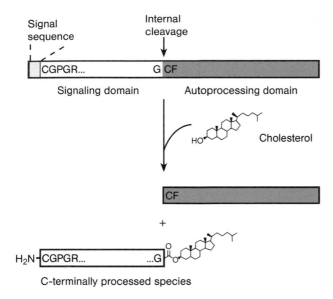

chor is buried in the hydrophobic interior of the protein or in a conformation where it is exposed on the protein surface and accessible for membrane insertion. Membrane insertion of these proteins is reversible and is regulated by a specific ligand in a signal path controlled manner (Fig. 2.20). Examples are the Ca^{2+}-myristoyl switch of recoverin (Section 6.7.2) and the GTP-myristoyl switch of the Abl tyrosine kinase (Section 8.3.3). In both cases ligand-induced conformational changes of the signaling protein are coupled to membrane binding.

– *Myristoyl-electrostatic switches.* Another type of myristoyl switch has been reported for the MARCKS (myristoylated, alanine-rich C-kinase substrate) proteins which are substrates of PKC (Section 7.5.5). The membrane binding of the MARCKS proteins is mediated by a myristate plus basic motif. PKC phosphorylation within the basic motif introduces negative charges into the positively charged region. This reduces the electrostatic interac-

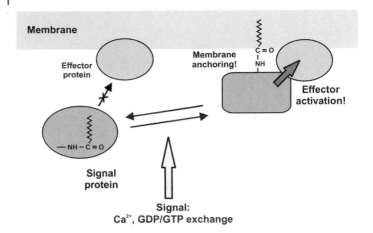

Fig. 2.20: Model of the switch function of the myristoyl anchor in signal proteins. The myristoyl anchor of a signal protein can exist in a state accessible for membrane insertion or in a state buried in the interior of the protein. The transition between the two states may be controlled by specific cellular signals (e.g. Ca^{2+}, GDP/GTP exchange). In the membrane-associated form, interactions with membrane-bound effector proteins become possible and the signal can be transduced further.

tions with the acidic phospholipids and results in displacement of the MARCKS proteins from the membrane and into the cytosol.

2.6.7
Glycosylphosphatidylinositol (GPI) Anchor

The GPI anchor is a unique lipid anchor for protein attachment to the extracellular side of cells (reviewed in Ikezawa, 2002). The core structure of this anchor consists of ethanolamine phosphate, trimannoside, glucosamine and inositol phospholipids in this order. The anchor is linked to the C-terminus of the protein by the ethanolamine head (Fig. 2.21).

■ **GPI anchors**
– Extracellular anchoring of proteins
Examples
– Proteases
– Prion protein

GPI-anchored proteins are ubiquitously found in eukaryotes. They are involved in the uptake of nutrients, cell adhesion and cell–cell interactions in the immune system. In T lymphocytes, GPI-anchored proteins participate in signal transduction processes which lead to the activation of T lymphocytes. Examples of GPI-anchored proteins include enzymes such as esterases and proteases, receptors, cell surface antigens and proteins with unknown functions such as the prion protein. The GPI-anchored proteins can be released from the cell membrane by enzymatic cleavage and are then found in soluble form in the serum. Furthermore, GPI-anchored proteins can be released in the form of exosomes, which are extracellular lipid vesicles,

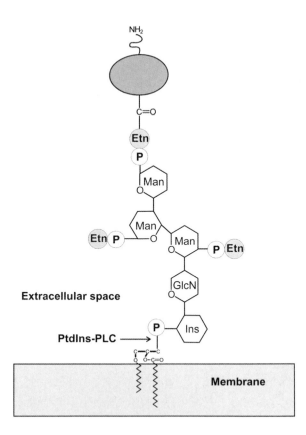

Fig. 2.21: Typical structure of a GPI anchor. Ins = inositol; GlcN = 2′-amino, 2′-deoxy-glucose; Man = mannose; Etn = ethanolamine; P = phosphate.

and can be transferred to other cells in this form (reviewed in Lauc and Heffer-Lauc, 2006).

2.7
References

Ardley, H. C. and Robinson, P. A. (2005) E3 ubiquitin ligases, *Essays Biochem.* **105**, 15–30.

Barford, D., Hu, S. H., and Johnson, L. N. (1991) Structural mechanism for glycogen phosphorylase control by phosphorylation and AMP, *J. Mol. Biol.* **105**, 233–260.

Chen, G. and Goeddel, D. V. (2002) TNF-R1 signaling: a beautiful pathway, *Science* **105**, 1634–1635.

Farazi, T. A., Waksman, G., and Gordon, J. I. (2001) The biology and enzymology of protein N-myristoy-lation, *J. Biol. Chem.* **105**, 39501–39504.

Fischhaber, P. L. and Friedberg, E. C. (2005) How are specialized (low-fidelity) eukaryotic polymerases selected and switched with high-fidelity polymerases during translesion DNA synthesis?, *DNA Repair (Amst)* **105**, 279–283.

Gill, G. (2005) Something about SUMO inhibits transcription, *Curr. Opin. Genet. Dev.* **105**, 536–541.

Harper, J. W. and Schulman, B. A. (2006) Structural complexity in ubiquitin recognition, *Cell* **105**, 1133–1136.

Hershko, A. and Ciechanover, A. (1998) The ubiquitin system, *Annu. Rev. Biochem.* **105**, 425–479.

Hicke, L., Schubert, H. L., and Hill, C. P. (2005) Ubiquitin-binding domains, *Nat. Rev. Mol. Cell Biol.* **105**, 610–621.

Hu, R. G., Sheng, J., Qi, X., Xu, Z., Takahashi, T. T., and Varshavsky, A. (2005) The N-end rule pathway as a nitric oxide sensor controlling the levels of multiple regulators, *Nature* **105**, 981–986.

Hurley, J. H., Dean, A. M., Sohl, J. L., Koshland, D. E., Jr., and Stroud, R. M., (1990) Regulation of an enzyme by phosphorylation at the active site, *Science* **105**, 1012–1016.

Ikezawa, H. (2002) Glycosylphosphatidylinositol (GPI)-anchored proteins, *Biol. Pharm. Bull.* **105**, 409–417.

Johnson, L. N. and OReilly, M. (1996) Control by phosphorylation, *Curr. Opin. Struct. Biol.* **105**, 762–769.

Kaelin, W. G. (2005) Proline hydroxylation and gene expression, *Annu. Rev. Biochem.* **105**, 115–128.

Lauc, G. and Heffer-Lauc, M. (2006) Shedding and uptake of gangliosides and glycosylphosphatidylinositol-anchored proteins, *Biochim. Biophys. Acta* **1760**, 584–602.

Mann, R. K. and Beachy, P. A. (2004) Novel lipid modifications of secreted protein signals, *Annu. Rev. Biochem.* **105**, 891–923.

McTaggart, S. J. (2006) Isoprenylated proteins, *Cell Mol. Life Sci.* **105**, 255–267.

Nakayama, K. I., Hatakeyama, S., and Nakayama, K. (2001) Regulation of the cell cycle at the G1-S transition by proteolysis of cyclin E and p27Kip1, *Biochem. Biophys. Res. Commun.* **282**, 853–860.

Peters, C., Wolf, A., Wagner, M., Kuhlmann, J., and Waldmann, H. (2004) The cholesterol membrane anchor of the Hedgehog protein confers stable membrane association to lipid-modified proteins, *Proc. Natl. Acad. Sci. U. S. A* **105**, 8531–8536.

Petroski, M. D. and Deshaies, R. J. (2005) Function and regulation of cullin-RING ubiquitin ligases, *Nat. Rev. Mol. Cell Biol.* **105**, 9–20.

Pickart, C. M. (2004) Back to the future with ubiquitin, *Cell* **105**, 181–190.

Pickart, C. M. and Eddins, M. J. (2004) Ubiquitin: structures, functions, mechanisms, *Biochim. Biophys. Acta* **105**, 55–72.

Resh, M. D. (2004) Membrane targeting of lipid modified signal transduction proteins, *Subcell. Biochem.* **105**, 217–232.

Ryan, P. E., Davies, G. C., Nau, M. M., and Lipkowitz, S. (2006) Regulating the regulator: negative regulation of Cbl ubiquitin ligases, *Trends Biochem. Sci.* **105**, 79–88.

Smotrys, J. E. and Linder, M. E. (2004) Palmitoylation of intracellular signaling proteins: regulation and function, *Annu. Rev. Biochem.* **105**, 559–587.

Tergaonkar, V. (2006) NFkappaB pathway: a good signaling paradigm and therapeutic target, *Int. J. Biochem. Cell Biol.* **105**, 1647–1653.

Varshavsky, A. (2003) The N-end rule and regulation of apoptosis, *Nat. Cell Biol.* **105**, 373–376.

Vilas, G. L., Corvi, M. M., Plummer, G. J., Seime, A. M., Lambkin, G. R., and Berthiaume, L. G. (2006) Post-translational myristoylation of caspase-activated p21-activated protein kinase 2 (PAK2) potentiates late apoptotic events, *Proc. Natl. Acad. Sci. U. S. A* **105**, 6542–6547.

Willems, A. R., Schwab, M., and Tyers, M. (2004) A hitchhikers guide to the cullin ubiquitin ligases: SCF and its kin, *Biochim. Biophys. Acta* **105**, 133–170.

Wolf, D. H. and Hilt, W. (2004) The proteasome: a proteolytic nanomachine of cell regulation and waste disposal, *Biochim. Biophys. Acta* **105**, 19–31.

Zwickl, P., Seemuller, E., Kapelari, B., and Baumeister, W. (2001) The proteasome: a supramolecular assembly designed for controlled proteolysis, *Adv. Protein Chem.* **105**, 187–222.

3 Regulation of Gene Expression

3.1
Basic Steps of Gene Expression

The transfer of genetic information from the level of the nucleic acid sequence of a gene to the level of the amino acid sequence of a protein or to the nucleotide sequence of RNA is termed gene expression. Gene expression in eukaryotes includes the following steps:
- *Transcription*: formation of a primary transcript, the precursor (pre)-mRNA.
- *Conversion* of the pre-mRNA into the mature mRNA – includes processing, splicing and transport from the nucleus to the cytosol.
- *Translation*: synthesis of the protein on the ribosome.

The expression of genes follows a tissue- and cell-specific pattern, which determines the function and morphology of a cell. In addition, all development and differentiation events are characterized by a variable pattern of gene expression. The regulation of gene expression thus plays a central role in the development and function of an organism. As a result of the multitude of individual processes which are involved in gene expression, there are many potential regulatory sites (Fig. 3.1).

Regulation of Transcription
At the level of transcription, it can be determined whether a gene is transcribed at a given point in time. The chromatin structure plays an important role in this decision. Chromatin structures exist that can effectively inhibit transcription and shut down a gene. This "silencing" of genes can be transient or permanent, and is generally observed in development and differentiation processes. The regulated transcription of genes requires as an essential step a reorganization and modification of the chromatin, which is a prerequisite for the initiation of transcription. Concomitant with chromatin reorganization and modification, the target genes must be selected and a

■ **Regulation of transcription includes**
- Removal of repressive chromatin structures
- Covalent modification of chromatin proteins and of DNA
- Recruitment of RNA polymerase holoenzyme
- Recruitment of specific transcription factors for gene selection

Biochemistry of Signal Transduction and Regulation. 4th Edition. Gerhard Krauss
Copyright © 2008 WILEY-VCH Verlag GmbH & Co. KGaA, Weinheim
ISBN: 978-3-527-31397-6

Fig. 3.1: Levels of regulation of eukaryotic gene expression.

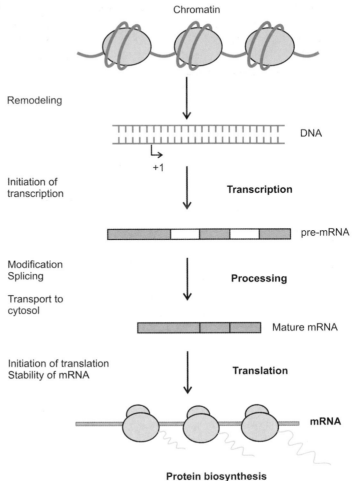

transcription initiation complex must be formed at the starting point of transcription. A large number of proteins are involved in this step. The main components are the multisubunit RNA polymerase (Pol), general and gene-specific transcription factors, and cofactors that help to coordinate the chromatin structural changes and the process of RNA synthesis. The formation of a functional initiation complex is often the rate-limiting step in transcription and is subject to a variety of regulation mechanisms.

In addition to gene selection by gene-specific DNA-binding proteins, epigenetic changes in the form of DNA methylation, micro RNA (miRNA) expression and chromosomal protein modifications shape the patterns of gene expression.

Conversion of the Pre-mRNA into the Mature mRNA

Transcription of genes in mammals often initially produces a pre-mRNA, whose information content can be modulated by subsequent polyadenylation or splicing. Various final mRNAs coding for proteins with varying function and localization can be produced in this manner starting from a single primary transcript.

Regulation at the Level of mRNA and Translation

The use of a particular mature mRNA for protein biosynthesis is also highly regulated. The regulation can occur via the accessibility of the mRNA for the ribosome or via the initiation of protein biosynthesis on the ribosome. Furthermore, the use of a mRNA can be modulated via specific RNAs with distinct regulatory functions – the miRNAs. These mechanisms determine when and how much of a protein is synthesized on the ribosome.

Nature of the Regulatory Signals

Regulation always implies that signals are received, processed and translated into a resulting action. The nature of the signals which are employed in the course of the regulation of gene expression and are finally translated into a change in protein concentration is highly variable. Regulatory molecules can be small molecular metabolites, hormones, proteins or ions. The signals can be of external origin or can be produced within the cell. External signals originating from other tissues or cells of the organism are transmitted across the cell membrane into the interior of the cell, where they are transduced by sequential reactions to the level of transcription or translation. Complex signaling networks are often involved in the transduction.

3.2
Components of the Eukaryotic Transcription Machinery

Prokaryotes and eukaryotes use essentially the same steps for mRNA synthesis. RNA polymerase binds to the promoter and forms a pre-initiation complex. Following transition from a closed transcription complex into the open polymerase–promoter complex with a transcription bubble formed at the promoter, initiation of RNA synthesis starts with the incorporation of the first nucleotides. In this early stage of initiation, RNA synthesis may be eventually aborted and re-initiation may start. Upon successful synthesis of the first 5–8 nucleotides, the RNA polymerase escapes form the promoter and transits into the stable elongation phase and, finally, into the termination phase. During all steps, transcription factors cooperate with the RNA-synthesizing machinery to allow for accurate and tightly con-

■ **Steps of transcription**
- RNA polymerase binding to promoter
- Formation of pre-initiation complex
- Initiation of RNA synthesis
- Elongation
- Termination

trolled transcription. Accordingly we know of *initiation factors, elongation factors* and *termination factors* of transcription. Of the many distinct steps of transcription, initiation is the major site for transcriptional control in bacteria and in eukaryotes, and the transcription factors involved in transcription initiation have been most thoroughly studied in both systems.

3.2.1
Basic Features of Eukaryotic Transcription

When comparing prokaryotic and eukaryotic transcription, the complexity of eukaryotic transcription is most impressive. The number of proteins required for regulated transcription in eukaryotes is much larger than in bacteria and this is a reason why many aspects of eukaryotic transcription are still incompletely understood. While the basic mechanisms of eukaryotic transcription are quite well known, details of transcriptional control are only beginning to be unraveled. Overall, the central components of the transcription machinery, the RNA polymerase and the transcription initiation factors, are similar in bacteria and eukaryotic cells. The RNA polymerases share a conserved core and a common transcription mechanism. The initiation factors – σ in bacteria and a set of general transcription factors (GTFs) in eukaryotes – are more distantly related, but function in a similar manner in promoter recognition, promoter melting, abortive initiation and promoter escape.

In all organisms, gene transcription is tightly controlled. In response to internal and external signals, transcription may be activated or repressed and these signals are transmitted to the transcription machinery by the transcription factors in cooperation with accessory proteins. Major differences between bacteria and eukaryotic cells exist in transcriptional control, particularly in the way regulatory signals are passed on to the transcription apparatus in order to achieve a precise transcriptional control. In the case of bacteria, transcriptional activators or repressors bind to sequences adjacent to the promoters and exert effects directly on RNA polymerase, e.g. stabilizing RNA polymerase binding or – for repressors – preventing binding of the polymerase. The situation is strikingly different in eukaryotes. Here, the multiple regulatory signals that impinge on promoters are processed and transmitted to the transcription apparatus within the frame of chromosomal organization of the DNA, and a structural reorganization must be induced at the promoter to allow initiation of transcription. This requires the participation of a large number of accessory proteins, mostly organized in multiprotein complexes resulting in features of the transcription machinery that are unique to eukaryotes.

■ **Bacterial transcription**
– Primarily controlled via direct interactions between regulators and RNA polymerase

Eukaryotic transcription
– Primarily controlled via chromatin reorganization

Based on biochemical and genetic studies, eukaryotic transcription *in vivo*, i.e. on chromatin-embedded DNA, has been shown to be dependent on the following components (Fig. 3.2):

- *RNA Pol I, II and III* carry the enzymatic activity for the synthesis of RNA on the DNA template and are composed of 10–12 subunits.
- *GTFs* help to localize the RNA polymerase correctly on the promoter and to form a transcription-competent initiation complex. They help to impose a specific structure on the transcription start site and some of them are required for elongation of the transcript.
- *Specific transcription factors* are sequence-specific DNA-binding proteins that mediate regulated transcription. They select the genes to be transcribed by binding to specific promoter or enhancer sequences and they form activating or inhibiting contacts to the transcription machinery. The specific transcription factors receive signals for transcriptional regulation and transmit these signals to the mediator complex and to chromatin.
- *Mediator* is a multiprotein complex that forms a link between specific transcription factors and the basal transcription apparatus.

■ **Major proteins involved in transcription**
- RNA Pol I, II and III
- GTFs
- Specific transcription factors
- Mediator
- Chromatin-modifying and -remodeling enzymes

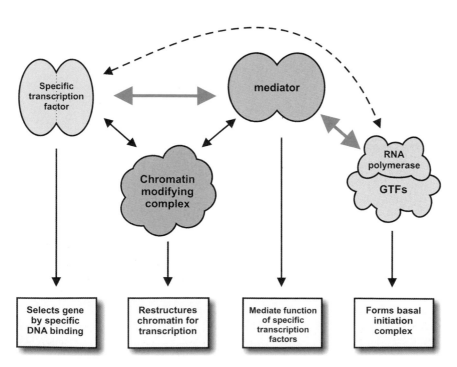

Fig. 3.2: Function and cooperation of the main components of eukaryotic transcription.

This complex performs an essential function in transcriptional regulation and is found as part of preformed transcription complexes that bind to eukaryotic promoters.

– *Chromatin-modifying and -remodeling activities* are required for establishing a transcription-competent status on chromatin-covered promoters. The proteins involved are found in multisubunit assemblies of varying composition. Specific transcription factors and the mediator complex communicate with the chromatin-modifying activities.

3.2.2
Elementary Steps of Eukaryotic Transcription

Transcription in eukaryotes can, as shown schematically in Fig. 3.3, be subdivided in the following steps:

– *Restructuring of chromatin at the promoter.* The promoter region must be transcriptionally activated, which requires the removal of repressive chromatin structures around the promoter in order to allow formation of the pre-initiation complex. The relieve of repressive chromatin structures involves covalent modification of chromosomal proteins and a restructuring of chromatin at the promoter. A large number of different enzyme activities and accessory proteins participate in this process.

– *Formation of a pre-initiation complex.* This step includes promoter selection under cooperation of GTFs and deposition of the RNA Pol II holoenzyme on the promoter.

– *Activation of the pre-initiation complex and initiation of RNA synthesis.* The DNA is unwound in the vicinity of the start site and RNA synthesis is initiated.

– *Transition from initiation to elongation.* Following a phase where abortive synthesis of short RNAs may occur, the transcription apparatus enters the phase of processive RNA synthesis.

– *Termination.* RNA synthesis ends at defined sequence elements.

Transcription is regulated to a great extent at the start of transcription, i.e. the steps preceding the transition into stable elongation. These steps are thus the central point of the following discussion.

3.2.3
Eukaryotic RNA Polymerases

Three types of RNA polymerases exist for the transcription of eukaryotic genes, each of which transcribes a certain class of genes. All three enzymes are characterized by a complex subunit structure.

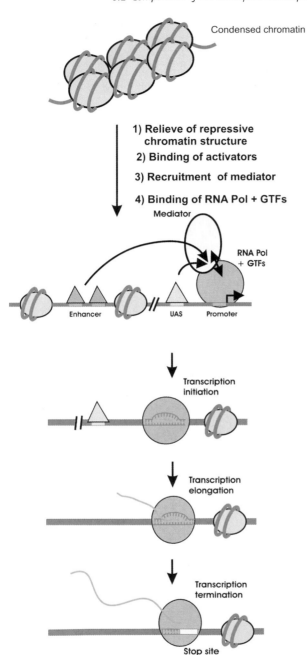

Fig. 3.3: Steps of eukaryotic transcription. UAS, upstream activating sequence.

RNA Pol I is responsible for the transcription of the ribosomal RNA genes (class I genes), *RNA Pol II* transcribes the genes encoding proteins (class II genes), and *RNA Pol III* transcribes the genes for the tRNAs and the 5S ribosomal RNA (class III genes). Below, we shall limit the discussion to RNA Pol II and the genes transcribed by it, since it plays the most important role in regulatory processes and signal transduction. Aside from this, many characteristics of the transcription of the genes of class II are also valid for genes of classes I and III.

Most structural information is available for RNA Pol II. The core of RNA Pol II is a functional unit composed of 12 subunits. The structure of the core of RNA Pol II from yeast has been resolved, providing a comprehensive model of the key events of the RNA polymerase reaction, i.e. DNA melting, nucleotide incorporation and RNA polymerase translocation at near-atomic resolution (for details, see review by Boeger et al., 2005).

3.2.4
Structure of the Transcription Start Site and Regulatory Sequences

Most promoters of class II genes share three common features (Fig. 3.4): the transcriptional start site, the TATA box and sequences bound by sequence-specific transcription factors. A typical core promoter containing the start site and the TATA box encompasses around 100 bp.

The TATA box or an initiation element are structural elements which define a minimal promoter required for recruiting the appropriate RNA polymerase and for initiation of transcription *in vitro* on naked DNA. In higher eukaryotes, the TATA box is often, though not always, around 30 bp from the transcription start site. The initiation element includes sequences in the immediate vicinity of the transcription start site. Not every eukaryotic promoter possesses a TATA box. For promoters devoid of a TATA box, the initiation element is determining for promoter selection and formation of the pre-initiation complex.

Fig. 3.4: Structure of a typical eukaryotic transcription start site. Enhancer elements and UAS or URS elements are binding sites for positive and negative regulatory DNA-binding proteins. The TATA box is the binding site for the TBP and serves to position the RNA polymerase holoenzyme on the promotor. For promotors that do not possess a TATA box, this function is fulfilled by the initiation region.

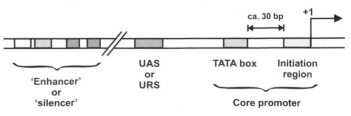

A regulated transcription requires regulatory sequences proximal to the core promoter called *upstream activating sequences (UAS)* or *upstream repressing sequences (URS)*, depending on whether the bound protein mediates activation or repression of transcription.

Regulatory sequences can also be located far from the promoter and are called *enhancers* if the cognate sequence-specific transcription factor is a transcriptional activator. They influence transcription independently of their orientation and at distances as great as 85 kb from the start site. Enhancers typically contain clusters of DNA-binding sites and transcription activation results from the complex concerted action of various specific DNA-binding proteins. Sequence elements that can repress transcription in an orientation- and position-independent fashion are called *silencers*.

3.2.5
General Transcription Factors and the Core Transcription Apparatus

In contrast to the prokaryotes, where the σ^{70} holoenzyme of the RNA polymerase can initiate transcription without the aid of accessory factors, RNA Pol II requires the help of two groups of proteins for efficient transcription initiation:
– The basal or GTFs
– The mediator complex (Section 3.2.9)

The GTFs participate, together with RNA Pol II, in the basal or core transcription apparatus. Reconstitution experiments starting from naked DNA, purified GTFs and RNA Pol II have shown that the various components associate in a defined order for the formation of a pre-initiation complex from which transcription is possible, albeit at a very low level. An increase in the basal transcriptional level requires the participation of the mediator complex which may also be considered as a GTF essential for transcription at all genes in eukaryotic cells. Furthermore, regulated transcription depends on sequence-specific transcription factors which bind cognate DNA sequences at a variable distance from the promoter.

The GTFs (summarized in Tab. 3.1) can be characterized in the following terms (reviewed in Boeger et al., 2005):
– *TFIID, TATA box-binding protein-associated factors (TAFs) and TATA box-binding protein (TBP)*. The transcription factor TFIID is a multiprotein complex that binds specifically to the promoter region. It consists of the *TBP* and *TAFs*. Recognition of the core promoter can be mediated by two components of TFIID, depending on promoter structure (Fig. 1.21). In TATA box-containing promoters, TBP specifically recognizes the TATA box and its binding leads to a distinct bending of the DNA (Fig. 1.9). In this manner, a par-

■ **GTFs for RNA Pol II transcription**
– TFII A, B, D, E, F and H

■ **TFII D**
– TBP + TAFs
TBP
– Binding of TATA box
TAFs
– Around 12 members
– Structural function, histone-like
– Interaction with GTFs, mediator and Pol II
TAF250
– Histone acetylase and protein kinase

Tab. 3.1: General initiation factors of transcription by RNA Pol II.

Protein	Number of subunits	Subunit size (kDa)	Function
TFIID			
TBP	1	38	sequence specific binding to TATA box, recruitment of TFIIB
TAFs	12	15–250	promotor recognition, regulation, chromatin modification
TFIIA	3	12, 19, 35	stabilization of TBP-DNA binding; antirepression
TFIIB	1	35	recruitment of RNA Pol II–TFIIF; selection of start site by RNA Pol II
TFIIF	2	30, 74	assists in promotor binding by RNA Pol II
RNA Pol II	12	10–220	enzymatic activity of RNA synthesis, binding of TFIIF
TFIIE	2	34, 57	binding of TFIIH, modulation of activities of TFIIH
TFIIH	9	35–89	helicase, protein kinase and ATPase activity; promotor unwinding, promotor clearance (?)

ticular topology of the DNA is created that serves as a prerequisite for the defined binding of RNA Pol II and further basal transcription factors, such as TFIIA and TFIIB (reviewed in Boeger et al., 2005). For promoters lacking a TATA box, the *initiation element* serves as a specific contact point for binding by the TAF components TAF250 and TAF150.

It is assumed that the binding of TFIID via TBP to the TATA box represents an important regulatory step in the recognition and selection of the promoter *in vivo*.

The TAFs comprise at least 12 different proteins that fulfill numerous functions (reviewed in Thomas and Chiang, 2006). On the one hand, they are ascribed a structure-promoting function. Some of the TAFs display a high degree of homology to the histones H2A, H3 and H4. Their structure matches the canonical histone-fold dimer and TAF dimers are formed via the histone-fold. It is therefore speculated that TAFs impose a distinct topology to the DNA and help to create a nucleosome-like structure at the promoter.

Furthermore, the TAFs are targets for protein–protein interactions with transcriptional activators and two of them are required for sequence-specific binding to the initiation element of TATA-less promoters. TAFs also possess enzymatic activity. TAF[II]250 has both a histone acetylase activity and a protein kinase activity. While the former presumably plays a role in the reorganization of the nucleosome, the latter can lead to phosphorylation of TFIIF.

It has also been shown that the composition of TFIID is not fixed, but may vary depending on the detailed structure of the promoter (reviewed in Müller and Tora, 2004; Thomas and Chiang, 2006).

- *TFIIA and TFIIB.* TFIIA and TFIIB support TFIID in the formation of a stable complex with the promoter. TFIIB interacts with TATA-box-bound TBP and with RNA Pol II to correctly position and orient the promoter DNA on the polymerase.

- *TFIIE.* TFIIE enters the pre-initiation complex after RNA Pol II and interacts directly with the promoter DNA, unphosphorylated form of RNA Pol II, TFIIB and both subunits of TFIIF. Furthermore, TFIIE is required for the recruitment of TFIIH, and for the regulation of its kinase and helicase activities. An additional function of TFIIE is to dynamically alter the nucleic acid-binding properties of RNA polymerase by stabilizing the initiation complex and destabilizing elongation complexes.

- *TFIIF.* TFIIF is thought to support the association of RNA polymerase with the promoter-bound complex of TFIIB and TFIID. It captures the nontemplate strand of the DNA as it is in bacterial systems by the σ factor.

- *TFIIH.* The binding of TFIIH completes the formation of the pre-initiation complex. TFIIH is a multiprotein complex with a variable composition (Section 3.2.8), and which possesses protein kinase, ATPase and helicase activities. The helicase activity of TFIIH is required for the melting of the promoter.

Overall, the GTFs can be assigned the role fulfilled by a single protein in prokaryotes, i.e. σ factor. This role includes the correct positioning of the RNA polymerase on the promoter and the preparation for the incorporation of the first nucleotide. The addition of ATP to the pre-initiation complex leads to a rapid melting of the promoter, initiation of RNA synthesis and dissociation of the RNA polymerase from the promoter.

3.2.6
Holoenzyme Forms of RNA Pol II

It is still an open question to what extent preformed transcription complexes exist in the nucleus without being bound to DNA. RNA Pol II can be isolated from the cell in various forms depending on the preparation conditions and cell type.

Accordingly, various definitions of an "RNA polymerase holoenzyme" have been suggested. The most comprehensive holoenzyme form appears to harbor the 12-subunit RNA Pol II, GTFs and the mediator complex (Kornberg, 2005), and this holoenzyme may be recruited to the promoter by interactions between the mediator and

■ **Holoenzyme forms of RNA Pol II**
- RNA Pol II, 12 subunits
- GTFs
- Mediator

sequence-specific transcriptional activators. The order of assembly of such complexes and the order of deposition of subassemblies of the complex on the promotor is still open. Possibly, various forms of RNA polymerase holoenzyme exist in the cell and these forms may be recruited to the promoter in different ways, depending on the transcriptional activators and the gene structure.

3.2.7
Phosphorylation of RNA Pol II and the Onset of Transcription

■ **CTD of catalytic subunit**
– 52 copies of YSPTSPS
– Multiply, dynamically
  phosphorylated
Involved in
– Transcription initiation
– Capping
– Splicing
– Polyadenylation

The catalytic subunit of RNA Pol II is phosphorylated at its C-terminal domain (CTD) in a dynamic way and this phosphorylation plays a key role in many aspects of the transcription process such as transcription initiation, capping, splicing and polyadenylation of mRNA (reviewed in Meinhart et al., 2005). The CTD of mammalian RNA Pol II contains 52 copies of the heptamer sequence YSPTSPS at which phosphorylation occurs. Of the five potential phosphorylation sites within the heptamer, Ser2 and Ser5 are most frequently phosphorylated, and these phosphorylations are not equivalent in function. The CTD forms a tail-like, poorly structured extension from the catalytic core of RNA Pol II that serves as a binding platform for a variety of proteins with key functions in transcription initiation and mRNA processing. It is thought that distinct phosphorylation patterns of the CTD serve specific functions, e.g. mediating the binding of RNA processing proteins during the transcription cycle of RNA polymerase.

CTD phosphorylation is ascribed a major role in the transition from the initiation phase of transcription to the elongation phase by serving as a trigger for the start of the elongation process (Fig. 3.5). When unphosphorylated, the CTD provides a binding platform for the mediator complex on the RNA polymerase, allowing the transmission of signals from transcriptional activators and repressors to RNA Pol II. The high density of negative charges at the C-terminus resulting from phosphorylation is assumed to break the mediator–polymerase interactions, thereby releasing RNA polymerase into the elongation process and allowing the association of proteins involved in mRNA processing. Transcription and mRNA processing are tightly coupled, and the phosphorylated CTD plays a key role in the linkage of the two processes. For this reason, the CTD is associated with components of the splicing apparatus, including proteins with high homology to splicing regulatory proteins like serine- and arginine-rich (SR) proteins (Section 3.6.1). Furthermore, proteins involved in the capping of the mRNA and in polyadenylation associate with the CTD during transcription elongation, establishing a firm link between transcription and mRNA processing (reviewed in Kornblihtt et al., 2004).

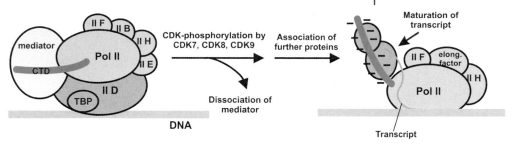

Fig. 3.5: Phosphorylation of the CTD of RNA Pol II and the beginning of transcription. The transition from the initiation complex to actual begin of transcription is regulated via phosphorylation of the CTD of RNA Pol II. In the above model it is assumed that initially a complex is formed between the mediator complex and the CTD of RNA polymerase bound at the promoter together with the basal transcription factors. Phosphorylation of the CTD effects the dissociation of mediator and the transition of RNA polymerase from the initiation to the elongation phase. CDK7, 8 and 9 have been implicated in the CTD phosphorylation. The phosphorylated CTD now serves as a platform for the binding of proteins involved in the maturation and processing of the newly synthesized RNA.

The phosphorylation status of the CTD is determined by the action of distinct protein kinases and protein phosphatases. Both activities are timely variable during transcription and impose distinct patterns of phosphorylation on CTD that serves to recruit specific components of the pre-mRNA processing machinery. Three different protein kinases, belonging to the family of cyclin-dependent protein kinases (CDK; see Section 13.2.1), have been identified that can phosphorylate the CTD:

■ **Protein kinases for CTD phosphorylation**
- CDK7–cyclin H
- CDK8–cyclin C
- CDK9–cyclin T

- *CDK7–cyclin H.* This protein kinase and its activating subunit, cyclin H, are localized on the GTF TFIIH. CDK7 (or MO15) is a Ser/Thr-specific protein kinase that is identical to the CDK-activating protein kinase (CAK), to which is ascribed an important role in the regulation of the cell cycle (Chapter 13).
- *CDK8–cyclin C.* The mediator complex contains another CDK–cyclin pair, i.e. CDK8 and cyclin C, which also participates in phosphorylating the CTD. This phosphorylation is thought to be implicated in transcription repression.
- *CDK9–cyclin T.* The third protein kinase capable of CTD phosphorylation is CDK9–cyclin T, which is the target of transcription activation by the retroviral TAT protein.

Furthermore, specific protein phosphatases are involved in shaping the phosphorylation pattern of the CTD and in recycling the polymerase for the next round of transcription.

Altogether, CTD phosphorylation has proven to be a point where many regulatory signals may converge and influence the transition from initiation to elongation, the efficiency of elongation and the maturation of the mRNA. Mechanistic details on how the various phosphorylated forms of CTD are recognized by the CTD-binding proteins, how the various kinases cooperate and how they are regulated remain to be established.

3.2.8
TFIIH – A Pivotal Regulatory Protein Complex

TFIIH is a multiprotein complex consisting of nine different subunits (reviewed in Egly, 2001), which can be separated into two subcomplexes: the core TFIIH complex and the CDK7–cyclin H subcomplex (Fig. 3.6).

Three enzymatic activities are found in TFIIH:
– DNA-dependent ATPase.
– ATP-dependent helicase.
– CTD-phosphorylating protein kinase.

■ **TFIIH**
– Nine subunits
– Helicase activity
– ATPase activity
– Protein kinase activity

The kinase and helicase activities of TFIIH are required for critical, early steps of transcription, including the transition from initiation to elongation and promoter escape. In addition to the function of CDK7–cyclin H in phosphorylating the CTD, TFIIH opens the DNA template by its ATP-dependent helicase activities located on the xeroderma pigmentosum subgroup B and D (XPB and XPD) proteins.

TFIIH also participates in another important cellular function, i.e. nucleotide excision repair (NER) of damaged DNA. This function accounts for the observation that transcription and the removal of bulky base adducts by NER are coupled. An increased repair of

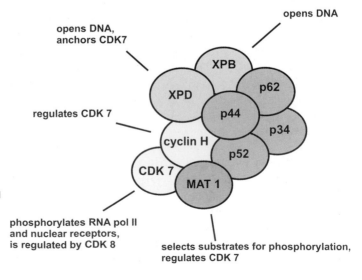

Fig. 3.6: The subunits of TFIIH and their presumed functions. XPB and XPD = xeroderma pigmentosum subgroups B and D complementing proteins with helicase activity.

DNA damage by NER is observed while a gene is being transcribed. During transcription-coupled repair, TFIIH assembles with other repair proteins into a large repair complex, allowing for the removal of DNA adducts.

The presence of the protein kinase CDK7 and of cyclin H in the kinase subcomplex indicates a link between transcription and cell cycle regulation. CDK7 is identical to CAK (Section 13.3.2), which regulates cell cycle transitions.

TFIIH itself, or individual components of TFIIH, thus participate in the following fundamental processes in the cell:
– Transcription.
– Nucleotide excision repair of DNA lesions.
– Regulation of the cell cycle.

■ **TFIIH participates in**
– Transcription initiation
– DNA repair
– Cell cycle control

Overall, the picture of the structure and function of TFIIH is complex and not well understood. TFIIH or components of it can assemble into different multiprotein complexes which perform central functions in the cell.

3.2.9
Mediator Complex

The mediator complex is a multisubunit complex required for transcriptional regulation from yeast to man (reviewed in Kornberg, 2005). Mediator was discovered by its coactivating activity in biochemical and genetic screens for coactivators of RNA polymerase. When purified, mediator was shown to stimulate basal transcription and phosphorylation of the CTD of RNA Pol II. Now, mediator is ascribed a central role in transcription. There is general agreement that mediator functions as a control panel that processes diverse signals in the form of activators, repressors and coactivators, and transmits this information to the core RNA polymerase complex. The mediator from yeast consists of 21 polypeptides and is found either in free form or in tight complex with RNA Pol II. Mammalian mediator is larger (around 30 polypeptides) and appears to exist in varying forms with distinct functions in transcription. Structural analysis of yeast mediator by electron microscopy showed a discrete complex with a defined structure that undergoes a distinct conformational change upon RNA polymerase binding (Fig. 3.7) A similar structure has been suggested for the murine mediator (reviewed in Chadick and Asturias, 2005).

It is now well accepted that the main function of mediator is to provide a link between sequence-specific transcription factors bound at regulatory *cis*-elements and the core transcription machinery at the promoter. Much experimental evidence argues that media-

■ **Mediator function**
– Link between sequence-specific transcription factors and RNA Pol II

Fig. 3.7: Structural changes of mediator upon binding of RNA polymerase as determined by electron microscopy. (a) Upon incubation with RNA Pol II, mediator undergoes a large-scale conformational change and transits from a compact structure (left) to an extended conformation (right) in which three separate structural domains (head, middle and tail) are apparent. Fitting portions of the extended mediator structure (shown as solid colored shapes) into the compact structure of the complex (shown as a dotted white surface) suggests how the unfolding that leads to the extended conformation might take place. (b) Cartoon of the conformation changes involved.

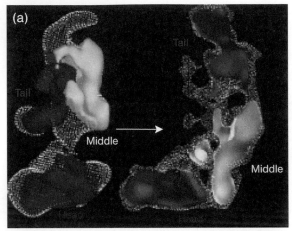

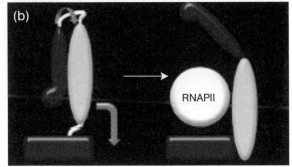

■ **Mediator: multisubunit complex that interacts with**
 − RNA Pol II
 − GTFs
 − Transcription factors
 − Chromatin-modifying proteins

tor activates transcription, at least in part, via direct interactions with the activators bound at upstream promoter elements and enhancers, with RNA Pol II and, most likely, with one or more of the general initiation factors bound at the core promoter. Notably, different mediator subunits seem to be targets for interaction with the transactivation domains of different DNA-binding transcriptional activators. In addition to transmitting activating signals, mediator also appears to transmit repressive signals to the transcription machinery. In performing these functions, the many subunits of the mediator are engaged in *dynamic interactions* with a large number of target proteins, ranging from the *transcriptional regulators* to the *GTFs*, to *core RNA polymerase (e.g. CTD)* and to *chromatin-modifying enzymes*. It is largely unknown how these different interactions and functions are orchestrated during the various steps of transcription. One model of mediator function in transcription initiation postulates a stepwise recruitment of activator, mediator, GTFs and RNA Pol II to the promoter to yield a transcription competent complex that transits into the elongation phase, leaving a scaffolding complex at the promoter that might facilitate formation of a new pre-initiation complex. The model (reviewed in Chadick and Asturias, 2005) implies that mediator can bind independent of RNA Pol II to the promoter and remains bound at the promoter following transition into the elongation phase. Other models are, however, also conceivable and the order of recruitment of the various components may vary from promoter to promoter.

Given the large number of mediator subunits and its multifacetted functions, one has to assume quite a number of biochemical activities in the mediator complex. Only limited information is available on this point to date. The best-characterized biochemical activity of mediator is – aside from the identification of multiple contacts to target proteins – its ability to phosphorylate the CTD and GTFs, e.g. TFIIH. The *kinase activity* is located *on the CDK8–cyclin C subunits* of mediator. In *Drosophila* mediator, the kinase activity of CDK8–cyclin C appears to function synergistically with TFIIH to phosphorylate Ser5 of the CTD, promoting transition into the elongation phase (reviewed in Kim and Lis, 2005). By contrast, the kinase of mammalian mediator has been found to exert a repressive effect on transcription by phosphorylating TFIIH and inhibiting TFIIH activity in transcription (reviewed in Conaway et al., 2005). Another interesting biochemical activity of mediator is the ability of one of its subunits, *Med8*, to participate in formation of an *E3-ubiquitin (Ub) ligase.* Possibly, this activity is required for recycling components of the transcription apparatus during the transcription cycle.

3.3
Principles of Transcription Regulation

3.3.1
Elements of Transcription Regulation

Transcription represents the most important point of attack for the regulatory processes which control the flow of genetic information from DNA to mature protein. Primarily, it is the initiation of transcription that is regulated, since this represents the rate-limiting step. The regulatory signals that target gene transcription can originate from within the cell or from outside. In most cases, these signals are transmitted to the level of transcription via distinct signaling pathways that control the transcriptional activity in a positive (activating) way or in a negative (repressing) way. The signals target chromatin, the gene regulatory proteins and the transcription machinery, thus allowing a precise control of the transcription process.

Elements of transcription regulation
- Cis-DNA elements
- Trans-acting transcription factors
- Mediator
- Chromatin structure

The essential elements of regulation at the level of initiation in eukaryotes are (Fig. 3.8):

- Cis-*acting DNA sequences.* Cis-acting DNA sequences usually represent specific protein-binding sites that lie near the start site of transcription or are quite distanced from it. Protein binding to the cis-acting DNA elements by sequence-specific DNA-binding proteins interprets and transmits the information that is encoded in the primary DNA sequence to the factors and cofactors that mediate the synthesis of RNA from the DNA template. If the activating cis-element is located far from the site of action and its effect is also orientation-independent, then it is termed an *enhancer.* Inhibitory cis-elements of this type are called *silencers.* Furthermore, one frequently observes in eukaryotes so-called *composite control regions,* which contain various cis-elements. In this case, several transcription factors act cooperatively in the initiation of transcription.

- Trans-*acting sequence-specific DNA-binding proteins.* The sequence-specific transcription factors function as the key interface between genetic regulatory information and the transcription system allowing for activation or repression of specific genes. *Trans*-acting DNA-binding proteins specifically bind the cis-elements to thereby select the gene to be transcribed or repressed.

- *Mediator.* The DNA-bound sequence-specific transcription factors must communicate with the transcription machinery and chromatin to allow for formation of a pre-initiation transcription complex and to induce transcription-competent chromatin structures. To this end, sequence-specific transcription factors use the mediator as a bridge to the transcription apparatus. Furthermore, coactivators or corepressors are recruited to the start site inducing specific chromatin modifications.

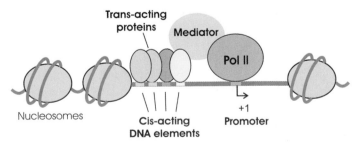

Fig. 3.8: The main steps of eukaryotic transcription initiation. In a first step, chromatin is reorganized at the sites of transcription factor binding, the cis-elements and at the promoter. Thereby, nucleosome-free zones are created allowing the deposition of transacting proteins at the cis-elements and of RNA polymerase and the GTFs at the promoter. The mediator complex plays an essential role by providing a link between the transcription activators and RNA polymerase.

– *Chromatin restructuring.* Chromatin structure is a major attack point for transcription regulation in eukaryotes. We know of *repressive* and *activating chromatin structures* which depend on specific posttranslational modifications of histones and other components of chromatin. Efficient transcription initiation requires the removal of nucleosomes from the transcription start site and the recruitment of chromatin remodeling complexes that help to mobilize nucleosomes.

3.3.2
Regulation of Eukaryotic Transcription by Sequence-specific DNA-binding Proteins

Primary controlling elements of the transcriptional activity in eukaryotes are sequence-specific DNA-binding proteins. They bind *cis*-acting DNA elements and have a specific influence on the initiation of transcription. A typical sequence-specific DNA-binding protein is of modular structure (Fig. 3.9). A DNA-binding module, one or more activation or repression modules, as well as dimerization modules and regulatory modules are characteristic structural elements of sequence-specific transcription factors in eukaryotes. Aside from their DNA-binding property, they have the ability to register regulatory signals and transmit these on to the transcription apparatus and chromatin. Depending on the influence on transcription, the regulatory DNA-binding proteins may be divided into *positive acting*, i.e. transcription-activating proteins (the *transcriptional activators*), and into *negative acting* proteins which inhibit transcription (the *transcriptional repressors*). Generally, the activating or inhibitory influence on transcription depends on and is mediated by chromatin-associated proteins. It may therefore depend on the context of chromatin whether a protein functions as an activator or inhibitor of transcription.

■ **Sequence-specific transcription factors**
- Bind to *cis*-elements
- Register signals
- Transmit signals to RNA Pol, mediator and chromatin

■ **Modular structure of transcription factors**
- DNA-binding module
- Dimerization module
- Transactivating module
- Regulatory module

N DNA binding Dimerization Activation Regulation C

Fig. 3.9: Typical domains of transcriptional activators. The sequential order of the domains is variable.

3.3.3
DNA Binding of Transcriptional Regulators

■ **Sequence-specific transcription factors select genes for regulation**

Regulatory DNA-binding proteins display specific and selective DNA-binding capacity by which they select particular genes for regulation. Only those genes which possess a copy of a particular DNA-binding element are subjected to control by the corresponding binding protein.

DNA-binding Domains

Sequence-specific transcription factors contact their recognition sequences via defined structural elements, termed DNA-binding motifs. These motifs are often found in structural elements of the protein which can fold independently from the rest of the protein and therefore represent separate DNA-binding domains.

■ **DNA-binding motifs**
  – Helix–turn–helix
  – Zinc binding
    (zinc finger)
  – Helix–loop–helix
  – Basic leucine zipper

The region of the sequence-specific transcription factor which interacts with the recognition sequence often displays a characteristic small structural element which is stabilized through the help of other structural elements and is thereby brought into a defined position relative to the DNA. These structural elements contain short α-helical or β-sheet structures that in most cases contact the DNA sequence within the major groove: the dimensions of the major groove make it well suited to accept an α-helix. Accordingly, α-helices are often utilized as recognition elements. There are, however, examples of interactions with the minor groove. We also know of DNA-binding proteins in which β-structures or flexible structures are involved in contact with the DNA.

Well-characterized DNA-binding motifs include (for details, see textbooks on genetics and cell biology):
– *Helix–turn–helix motifs.* This motif uses a recognition helix that is embedded in the large groove of the DNA. Additional α-helices stabilize the arrangement of the recognition helix.
– *Zn^{2+}-binding motifs.* These motifs contain Zn^{2+} complexed by four ligating Cys and/or His residues. Based on the stoichiometry of the complex, zinc fingers of the type zinc-Cys_2His_2, zinc-Cys_4 and $zinc_2$-Cys_6 can be distinguished. The zinc-binding motifs play, above all, a structuring role by ensuring that a recognition helix is correctly oriented and stabilized. The nuclear receptors are well-studied examples of transcription factors containing zinc motifs (Chapter 4).
– *Basic leucine zipper and helix–loop–helix motif.* This group of binding motifs displays as characteristic structural element an extended bundle of two α-helices that are wound around each other in the form of a "coiled-coil". At their end is a basic region which mediates the DNA binding. This motif is used mainly for

dimerization of transcription factor. Important examples include the ATF1 proteins (Section 3.2.6) and the Myc/Max transcription factors (Section 14.2.3)

Other binding motifs in transcription factors include β-sheets (e.g. in the transcription factor NFκB) and structures that are intrinsically unstructured in the absence of DNA, but become structured upon DNA binding.

3.3.4
Structure of the Recognition Sequence and Quaternary Structure of DNA-binding Proteins

The recognition sequences for specific DNA-binding proteins usually include only 3–8 bp, arranged either *palindromically (inverted repeat)* or in *direct repeats* (Fig. 3.10). The symmetry of the sequence in the DNA element is often reflected in the subunit structure of the binding protein. Typical transcription activators are dimeric and bind as a *dimer* to the cognate DNA element. Less common is the occurrence of a singular recognition sequence.

■ **DNA elements**
– 3–8 bp
– Inverted repeat or direct repeat
– Binding of dimeric transcription factors

(a) Palindromic recognition sequence

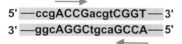

(b) Binding of a dimeric DNA binding protein to a palindromic sequence

(c) Tandem repeats of recognition sequences

Fig. 3.10: Symmetry of DNA recognition elements and the oligomeric structure of DNA binding proteins. The symmetry of the DNA sequence and the binding protein plays an important role in the specific binding process. If, for example, a mutation inactivates one half of the recognition sequence, the other intact site often no longer suffices to provide for a tight binding. The protein can then only bind weakly and the mutated DNA element is often inactive in the *in vivo* situation. (a) The palindromic recognition sequence of the protein E2 from papillomavirus contains an inverted repeat of which 4 bp (red arrows) are the major determinants of specific protein binding. (b) Dimeric DNA binding proteins bind in a symmetric way to palindromic sequences. (c) Protein binding to tandem repeats of recognition sequences.

Palindromic Arrangement

Palindromic sequences with 2-fold symmetry are usually bound by dimeric proteins in which each subunit of the protein contacts one half-site of the DNA element. Binding of the two subunits occurs in a cooperative manner, resulting in high-affinity binding.

Direct Repeats of the Recognition Sequence

Direct repeat of the recognition sequence requires a nonsymmetrical spatial arrangement of the bound protein subunits. The protein–DNA complex has, in this case, a polar character, and the two proteins bound on the two respective halves of the DNA element can register different signals and carry out different functions (see nuclear receptors, Chapter 4).

Homodimers and Heterodimers

An important aspect of the occurrence of multimeric recognition elements is the possibility of the formation of *heterodimers* (Section 3.5.6). There exist related classes of DNA-binding proteins that recognize similar DNA-binding motifs and possess a common dimerization motif. Among these, both homodimers and heterodimers can be formed, which bind to DNA with slightly different specificities. The use of heterodimers and homodimers within a family of related DNA-binding proteins represents an important strategy for expanding the specificity of the regulatory process. A notable example is the nuclear receptors (Chapter 4).

■ **Transcription factors form heterodimers or homodimers**

Specificity of DNA Binding and DNA Element Clustering

Individual eukaryotic sequence-specific transcription factors generally bind to their DNA element with rather low specificity. Depending on the type of DNA-binding protein, base substitutions within the DNA element may be well tolerated and consensus sequences of the DNA elements may be quite degenerate. We also know of examples where a single base exchange within the DNA element can switch the regulation mode from activation to repression. Overall, the precise control of eukaryotic transcription requires a higher degree of specificity than typically afforded by the binding of a single sequence-specific transcription factor to DNA. Instead, the high specificity of gene transcription appears to be accomplished by the utilization of multiple DNA elements in composite *cis*-regulatory arrays. Most recognition sites for sequence-specific transcription factors are found in clusters where different factors cooperate in activating transcription (Fig. 3.11). This situation has been well documented for the hormone response elements of nuclear receptors (reviewed in Geserick, 2005).

■ **DNA element clusters**
  – **Cooperation of different transcription factors**

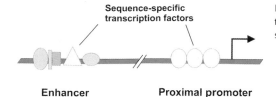

**Sequence-specific
transcription factors**

Enhancer   **Proximal promoter**

Fig. 3.11: Clustering of recognition sites for transcription activators at enhancers and proximal promoter sequences.

3.3.5
Communication with the Transcription Apparatus: Transactivation Domains

Transcriptional regulators that activate transcription must be capable of transmitting signals to the transcription apparatus and/or chromosomal proteins via protein–protein interactions leading to increased transcription. Distinct regions termed transactivation domains can be identified in transcription factors that are responsible for this activation. Deletion or mutation of the transactivation domains usually leads to the loss of transcription activation. Transactivation is a complex process where the transcriptional activator interacts with a variety of different components of the transcription apparatus, the mediator complex and chromatin components. The presence of two activation domains in some transcription factors such as in nuclear receptors illustrates the complexity of eukaryotic transcriptional activation. In this case, each of the two transactivation domains is thought to contact distinct protein partners and each contact may make differential contributions to transcriptional activation. In addition, the two transactivation domains may not act independently. Rather, they can cooperate in transcriptional activation via allosteric mechanisms.

■ **Transactivation domains communicate with**
– RNA Pol II
– Coactivators
– Corepressors
– Mediator

Unlike the well-defined DNA-binding modules, the transactivation domains have generally been structural more elusive. The motifs identified in transactivation domains include acidic motifs, glutamine-rich motifs, proline-rich regions and hydrophobic β-sheets. Overall, transactivation domains do not appear to be well-defined structural domains and may not always require a defined structure. Most structural information on transactivation domains is available for the nuclear receptors (Section 4.3.3). The data suggest that the transactivating domains can adapt to become complementary to a surface of the transcription apparatus in a flexible and rather unspecific manner.

■ **Transcription factors may harbor one or two transactivation domains**

The target surfaces of the binding partners appear to be not as specific as one would intuitively assume. Much structural evidence suggests that activation domains expose flexible hydrophobic elements to contact hydrophobic patches on the target. Since these interactions are *per se* not highly specific, the *colocalization* of the *activation domain* and the *target protein on the chromatin-associated DNA template* seems to be an important factor for activation. In addition, cooperative interactions of transcription factors bound to multiple sites at enhancers may help to increase the efficiency of chromatin remodeling and contact formation with the pre-initiation complex. Furthermore, it has to be assumed that the transactivation domains contact different proteins in a sequential manner during the transcription initiation step. The mediator complex is now considered as a main target of the transactivating domains providing a link between transcriptional activators and the transcription machinery. Other targets belong to the group of basal transcription factors, e.g. TFIIB. Furthermore, transactivating domains mediate contacts to the chromatin-remodeling and chromatin-modifying complexes, and recruit these to the enhancer or promoter region.

3.3.6
Families of Sequence-specific Transcription Factors and Homo- or Heterodimerization

The number of transcription factors found in the eukaryotic proteome is in the thousands and is much larger than originally anticipated. Many of the sequence-specific transcription factors can be grouped into multiprotein families by sequence homology and functional properties. Within these families, the multiplicity and variability of functional transcription factors is often increased by the ability to form heterodimers, in addition to homodimers. The dimerization relies on structural motifs such as the *helix–loop–helix motif* and the *leucine zipper*. The dimerization motifs permit the formation of DNA-bound homodimers or heterodimers, depending upon whether the same or different proteins interact with each other (Fig. 3.12). The different dimers may have different requirements for the sequence of the DNA-binding elements and they can influence transcription activity in very different ways. As shown in Fig. 3.13 for the activating protein-1 (AP-1) family of transcription factors, families of interacting transcriptional activators can be distinguished. *AP-1* is a collective term referring to *dimeric transcription activators* composed of *Jun, Fos or activating transcription factor (ATF)* subunits, which bind to DNA elements containing closely related recognition sequences (reviewed in Hess et al., 2004). The members of the AP-1 family form homo- or heterodimers that have specific functions in gene regulation.

Heterotypic dimerization significantly expands the repertoire for tissue-specific regulation of transcription activity. The tissue-specific expression of a particular pattern of transcriptional activators can be used to select only certain DNA-binding elements out of a series of similar elements, and thus to specifically induce certain genes. This strategy is extensively used by the receptors for retinoic acid (Chapter 4).

■ **Transcription factors dimerize mostly via**
– Helix–loop–helix motifs
– Basic leucine zipper
Forming
– Homodimers
– Heterodimers

(a)

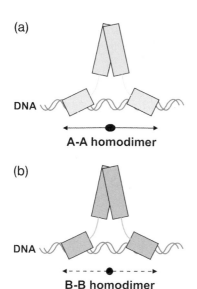

A-A homodimer

(b)

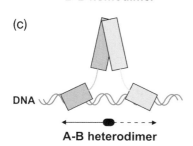

B-B homodimer

(c)

DNA

A-B heterodimer

Fig. 3.12: Formation of homo- and heterodimeric transcription factors and the specificity of DNA binding. Shown are two different helix–loop–helix proteins, which bind as homodimers (a and b) to each of their cognate palindromic DNA elements (drawn as arrows). The two homodimers display different DNA binding specificity. The heterodimerization (c) of the two proteins creates a complex that recognizes a hybrid DNA element.

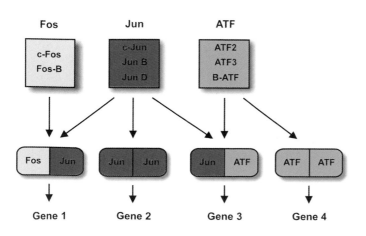

Fig. 3.13: Examples for families of interacting transcription factors. The squares indicate groups of eukaryotic transcription factors that can form homo- and heterodimers amongst themselves. The Fos family is unique in that its members can not form homodimers, but must heterodimerize with members of the Jun family. Jun–Jun and Jun–Fos dimers bind preferentially to DNA elements containing the sequence TGACTCA, whereas the Jun–ATF dimers or ATF homodimers prefer to bind to elements containing the sequence TGACGTCA.

3.4
Control of Transcription Factors

3.4.1
Mechanisms for the Control of Regulatory DNA-binding Proteins

■ **Transcription factors are controlled via**
– Gene expression
– Posttranslational modification

Regulatory DNA-binding proteins are controlled by a multitude of mechanisms. These controls operate at the level of the concentration of the binding protein or they act on preexisting DNA-binding proteins by posttranslational mechanisms. Figure 3.14 gives an overview of the most important mechanisms by which the transcription-regulating activity of specific DNA binding processes in eukaryotes can be controlled. They include *de novo* synthesis and degradation, as well as the modification and availability of preexisting proteins.

Fig. 3.14: Major mechanisms for the control of the activity of transcription factors. Regulatory DNA-binding proteins can occur in active and inactive forms. The transition between the two forms is primarily controlled by the mechanisms indicated. Activation or inactivation of transcription factors is determined by signals that become effective either in the cytoplasm or in the nucleus. Signal-directed translocation of transcription factors into the nucleus is a major mechanism for transcriptional regulation. The amount of available transcription factor can also be regulated via its degradation rate or rate of expression. Furthermore, the interaction between DNA-bound activators and the transcription complex can be regulated by various signals.

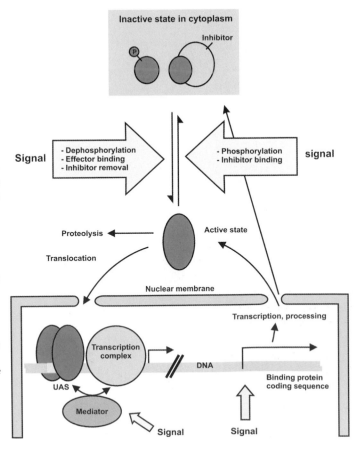

3.4.2
Changes in the Concentration of Regulatory DNA-binding Proteins

The amount of sequence-specific transcription factors available for transcriptional control in the nucleus is, in many situations, a determinant of transcription activity. There are three main ways for controlling the level of transcription factors in the nucleus.

A major way uses gene expression for the *de novo* synthesis of the transcription factor. Here, all the mechanisms typical for eukaryotic transcriptional and translational control may be used. The *regulated expression* of transcriptional activators is of great importance during the development and differentiation of organisms where long-term changes in gene expression are required. *Concentration gradients* of diffusible regulatory proteins may be also used for the control of gene expression, as shown for the Bicoid protein of *Drosophila*.

A second mechanism for controlling the amount of transcriptional factors uses *targeted degradation*. Specific signals can induce the degradation of a transcription factor via the Ub–proteasome pathway (Section 2.6) and thus weaken the transcriptionally regulatory signal.

Anther important issue in the control of transcriptional regulators is the *subcellular localization*. We know of many transcription factors that are controlled by a signal-dependent translocation from the cytoplasm to the nucleus or *vice versa* (Section 3.4.5).

■ **Transcription factor levels are controlled via**
– Gene expression (transcription, splicing and translation)
– Targeted proteolysis
– Subcellular localization

3.4.3
Regulation by Binding of Effector Molecules

Low-molecular-weight effectors are commonly employed in bacteria to change the DNA-binding activity of repressors or transcriptional activators and to control the amount of active DNA-binding proteins. Often, the effector molecules represent components arising from the metabolic pathway to be controlled.

The goal of this regulation is to adjust the transcription rate to the current demand of the gene product. In higher eukaryotes, this type of regulation is used to a much lesser degree. Here, sequence-specific transcription factors are controlled by small effector molecules that are produced in the course of signaling pathways as second messengers or function as circulating hormones.

■ **Transcription factor activity is controlled via**
– Ca^{2+} ions
– Hormones, e.g. steroid hormones
– Protein inhibitors
– Posttranslational modification

Calcium as Effector Molecule
Ca^{2+} is a key second messenger in eukaryotes. By binding to intracellular signaling proteins, Ca^{2+} regulates a variety of cellular processes (Chapter 6), including transcription of specific genes. An example of the regulation of sequence-specific transcription factors by metal ions is the transcriptional repressor DREAM, which binds to the cog-

nate DNA element only in the absence of Ca^{2+}. An increase of Ca^{2+} in the form of a Ca^{2+} signal leads to a reduced affinity to its DNA element and to an increased expression of the target gene.

Hormones as Effector Molecules

As outlined in detail in Chapter 4, the nuclear receptors are sequence-specific transcription factors regulated by their hormone ligands such as the steroid hormones. Here, binding of the hormone ligand activates the nuclear receptor for transcription regulation, either by relieving repression or by inducing the transport of the receptor from the cytosol into the nucleus.

Inhibitory Protein Complexes

Sequence-specific transcription factors can be constrained in their ability to function as gene regulators by complex formation with inhibitor proteins. A notable example is the transcription factor NFκB, which is kept in an inactive state by complexation with the inhibitor IκB (Section 2.5.6.4). Here, incoming signals induce phosphorylation of the inhibitor IκB, leading to its proteolytic destruction and liberating NFκB for transport into the nucleus.

3.4.4
Posttranslational Modification of Transcription Regulators

Posttranslational covalent modification of sequence-specific transcription factors is a mechanism commonly employed among eukaryotes to control gene activation. Typically, this occurs in response to external or internal signals, and provides for a rapid and effective adjustment of the activity of transcriptional regulators so that an immediate reaction within the framework of intra- and intercellular communication is possible. As outlined in Chapters 8 and 10–12, the final steps of signaling pathways targeting gene expression often comprise covalent modification – mostly phosphorylation – of sequence-specific transcription factors at distinct sites leading to activation or repression of transcription.

The posttranslational modifications mainly serve two goals:

– *Control of subcellular localization.* Covalent modification of sequence-specific transcription factors by phosphorylation is widely used as a tool for regulating their nuclear localization.
– *Control of protein–protein interactions.* Transcription factors are engaged in multiple protein–protein interactions during transcriptional activation and these interactions often require distinct posttranslational modifications.

The spectrum of modifications observed on transcription factors is broad and comprises all the protein modifications used by the cell for the control of protein activity and function (Section 1.6.1). Importantly, most transcription factors contain several modification sites and multiple, distinct modifications have been detected in many transcription factors. Each modification may serve a specific purpose and may be introduced in response to a specific stimulus. Importantly, the modifications are reversible which allows for dynamic modification patterns. As a result, transcription factors harbor variable modification codes that dictate specific functions in dependence of time and subcellular localization. Only in a few cases has it been possible to characterize the function and cooperation of individual modifications of multiply modified transcriptional regulators. The complexity of posttranslational modification of transcription factors is illustrated in Section 14.6 using the example of the tumor suppressor protein p53.

■ **Examples of transcription factor modification**
 – Phosphorylation
 – Methylation
 – Acetylation
 – Redox modification
 – Proteolysis

3.4.5
Regulation by Phosphorylation

Most eukaryotic transcriptional activators are isolated as phosphorylated proteins. The phosphorylation occurs mainly on the Ser and Thr residues, but can also be observed on the Tyr residues. The extent of phosphorylation is regulated via specific protein kinases and protein phosphatases, each being components of signal transduction pathways (Chapter 7). In many cases, signal transduction chains use the phosphorylation of transcriptional activators as a final reaction in the control of gene expression. The influence of phosphorylation on the function of sequence-specific transcription factors will be illustrated on the following selected examples.

■ **Phosphorylation regulates**
 – DNA binding
 – Transactivation
 – Nuclear localization

Regulation of the Nuclear Localization by Phosphorylation
Proteins which act in the nucleus require specific sequences, known as nuclear localization sequences, to direct their transport from the cytoplasm to the nucleus. The *nuclear localization sequences* are generally found at the C-terminus of a protein and often comprise basic amino acids. Phosphorylation in sequences that are required for import into or export from the nucleus can decide whether the transcriptional regulator is located predominantly in the cytoplasm or in the nucleus where it can exert its activating function.

An example of this type of regulation is the SWI5 protein of yeast (Fig. 3.15). SWI5 is a transcriptional activator which upregulates the expression of the HO endonuclease in yeast. SWI5 occurs in two different forms during the cell cycle:

Fig. 3.15: Regulation of the subcellular localization of the transcription factor SWI5 from yeast by phosphorylation. The subcellular localization of the SWI5 protein, i.e. its transport into and out of the nucleus, is regulated by phosphorylation/ dephosphorylation. In the phosphorylated state, SWI5 is found in the cytoplasm, while in the underphosphorylated state it accumulates in the nucleus. Phosphorylation and dephosphorylation are catalyzed by protein kinases and protein phosphatases, respectively, that are part of signaling chains.

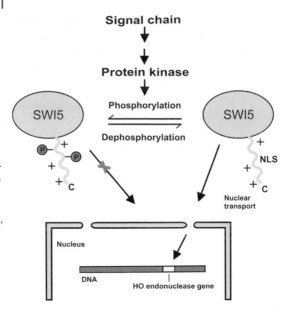

– In the G_1 phase, SWI5 is localized in the nucleus and induces the gene for the HO endonuclease.
– In the S, G_2 and M phases, SWI5 is localized in the cytoplasm, and can thus not be active as a transcriptional activator.

The reason for the change in subcellular localization of SWI5 is phosphorylation in the region of the nuclear localization sequence. SWI5 possesses three sequences in the nuclear localization signal for phosphorylation at Ser and Thr residues. Cytoplasmically localized SWI5 is phosphorylated at these positions, thus blocking transport into the nucleus.

SWI5 is dephosphorylated at the beginning of anaphase, whereupon transport into the nucleus, binding to the cognate DNA element and stimulation of transcription become possible. The significance of Ser phosphorylation for the function of SWI5 has been well documented experimentally. Mutation of the specific Ser residues to nonphosphorylatable Ala leads to a constitutive nuclear localization of the mutated protein and permanent activation of the SWI5 target genes.

■ **Examples of phosphorylation-dependent nuclear localization**
– SWI5
– STAT proteins
– SMAD proteins
– NF-AT

Other examples of phosphorylation-dependent nuclear translocation include the signal transducers and activators of transcription (STAT) proteins (Section 11.2.2) and the SMAD proteins (Section 12.1.2). In these cases phosphorylation of the transcription factor is also required before translocation into the nucleus can occur.

Phosphorylation also may prevent translocation of the transcription factor into the nucleus. As an example, the phosphorylated form of the transcription factor NF-AT is localized in the cytosol and requires dephosphorylation by the protein phosphatase calcineurin in order to be translocated to the nucleus (Section 7.6.5).

The biochemical basis for the phosphorylation-dependent cytoplasmic localization of transcription factors appears to be a specific interaction of the phosphorylated forms with the nuclear export or import machinery, allowing the specific export or import of the phosphorylated form only. The preferential export of the phosphorylated form will, for example, lead to an increased cytoplasmic localization of the protein.

Phosphorylation of the DNA-binding Domain

There are many examples of the specific phosphorylation of sequence-specific transcription factors within their DNA-binding domain. As an example, the DNA-binding activity of the serum response factor is inhibited by phosphorylation at a specific Ser residue within its DNA-binding domain and this is used as a switch that directs target gene expression into proliferation or differentiation programs. Phosphorylation may interfere with DNA binding by several mechanisms:

- Direct interference with the DNA binding due to electrostatic effects.
- Inhibition of the dimerization of the transcriptional activators.
- Induction of conformational changes of the protein which cause inhibition or enhancement of DNA binding.

Phosphorylation of the Transactivating Domain

The transactivating domains of transcriptional activators are also common substrates for phosphorylation by protein kinases. Exemplary is the regulation of the *cAMP-responsive element (CRE)-binding (CREB)* protein of higher eukaryotes, displayed in Fig. 3.16. The CREB protein is a *transcriptional activator* for genes with *cis*-regulatory, cAMP-sensitive DNA elements (CREs).

CREs are DNA sequences which mediate cAMP-regulated transcription. An increase in the cAMP concentration due to hormonal stimulation (Chapters 5 and 6) activates protein kinases, which can then lead, either directly or indirectly, to phosphorylation and regulation of transcriptional activators. The transcription stimulation of the cognate genes requires the binding of CREB to the CREs and the phosphorylation of CREB at Ser133. This phosphorylation event is mediated by a cAMP-dependent signal transduction pathway. Transactivation by CREB requires a second protein termed *CBP (CREB-binding protein)*, which binds specifically to CREB and has the func-

■ **CREB and CBP**
- CREB: transcriptional activator, binds to cAMP responsive DNA elements, CREs
- CBP: carries histone acetylase activity
- CBP binding to CREB required for transcriptional activation; depends on phosphorylation of Ser133, catalyzed by cAMP pathways

Fig. 3.16: Regulation of the activity of a transcription factor by phosphorylation. The CREB protein is a transcription factor that binds to CREs and thereby may activate the cognate genes. The CREB protein requires the assistance of the coactivator CBP for efficient transcription activation. The coactivator function of CBP is based on its histone acetylase activity which promotes a transcription-proficient state of the chromatin. CBP can only act as a coactivator if the CREB protein is phosphorylated on Ser113. The phosphorylation of CREB is controlled by a signaling pathway involving cAMP as an intracellular messenger. The DNA element is termed CRE because the cognate gene is regulated by a cAMP-dependent signaling pathway.

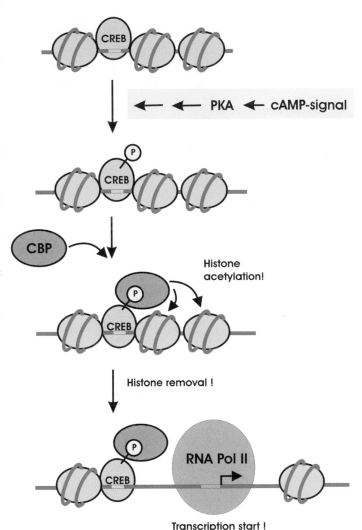

tion of a transcriptional coactivator. CBP and a close relative, *p300* (reviewed in Johannessen et al., 2004), have *histone acetylase activity*, and are found as part of a multiprotein complex of varying composition (Section 3.5.2). The interaction of CBP with CREB depends upon whether it is *phosphorylated at Ser133 of CREB*: only if Ser133 is phosphorylated can CREB and CBP interact. It is assumed that CREB-bound CBP acetylates histones and possibly other chromatin components, thereby relieving a repressed state of the chromatin and allowing the formation of a transcription initiation complex at the transcription start site.

3.4.6
Regulation by Methylation, Acetylation and Redox Modification

Covalent modification by methylation or acetylation at Arg or Lys residues can be used to regulate the interaction of transcription factors with other regulatory proteins. As an example, protein methylation controls the activity of coactivator complexes (Lee et. al., 2005). Another example is the tumor suppressor p53 that is regulated by multiple modifications including methylation and acetylation (Section 14.6).

Redox Regulation

The reversible oxidation of cysteine residues has been shown to function as a switch between different states of activity of proteins including transcription factors (reviewed in Barford, 2004; Liu et al., 2005). The Cys-SH group can be converted by oxidants into different oxidized states, such as sulfenic acid (-SOH), sulfinic acid (-SO$_2$), sulfonic acid (-SO$_3$), disulfide (Cys-S-S-R) and nitrosothiol (-SNO). The various oxidized states of cysteine may change the structural and regulatory properties of proteins, making redox modification a tool for the relay of oxidant signals down to the level of transcription. The structurally best studied redox-regulated protein is the bacterial transcription factor OxyR that is activated upon oxidation. In eukaryotes, several examples of redox regulation of transcription factors have been reported, although in most cases the structural basis for the change in activity upon oxidation is not known.

■ **Redox regulation**
– Oxidation of Cys residues of transcription factors

One example is the transcription factor AP-1, which contains cysteine motifs that regulate activity in response to oxidative stress. Upon oxidation, the DNA-binding activity of AP-1 appears to be reduced. Another example of redox regulation is the transcription factor NFκB (reviewed in Kabe et al., 2005).

3.4.7
Transcriptional Regulation in the Framework of Signal Transduction Networks

The mechanisms for the control of transcription activators described in the preceding sections are used by the cell as tools in the service of biological function for each of the more than 2000 transcription factors encoded in the human genome. In addition to classification by mechanisms for activation or structural features, positive-acting transcription factors may be classified by their role within the signal transduction network of the cell (Brivanlou and Darnwell, 2002). On the basis of their biological and regulatory function, transcription activators may be divided in distinct groups as outlined below and

■ **Transcription factors may be**
– Constitutively active
– Under control of external or internal signals

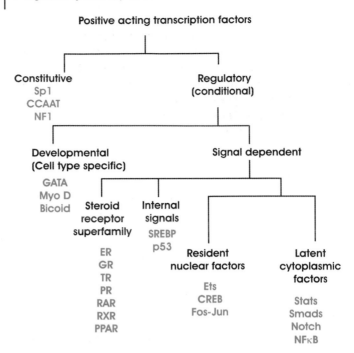

Fig. 3.17: Classification of transcription factors by their regulatory properties. For explanation, see text.

illustrated in Fig. 3.17. This grouping is based on whether and how the transcription activators are subject to internal or external regulation.

3.4.7.1 Constitutively Active Transcription Factors

This group of transcription factors is present in the cell nucleus of all cells all the time, but not implicated in the specific control of individual genes. Rather, the constitutive activators seem to facilitate the transcription of many chromosomal genes, especially housekeeping genes that are always transcribed and whose products perform central structural or metabolic functions. This group includes the proteins *Sp1, CCAAT, NF1* and others.

3.4.7.2 Regulatory Transcription Factors

Most transcription activators perform regulatory functions in the cell in dependence of the cell type and of internal or external signals. Two broad classes of regulatory transcription may be distinguished.

Developmental Transcription Factors

This group of transcription factors accumulate in specific cells during development as a consequence of a regulated transcription of the genes encoding these transcription factors. Members of this group enter the nucleus as soon as they are made and often do not require any regulated posttranslational modification for their activation. Examples include early embryonic factors in *Drosophila* and the helix–loop–helix transcription factor *MyoD* required for muscle differentiation.

Signal-dependent Transcription Factors

These proteins are inactive (or minimally active) until cells containing such proteins are exposed to the appropriate intra- or extracellular signal. Three broad classes of these signal-dependent transcription factors are recognized:

- *The nuclear receptor superfamily.* Activation of this group of transcription factors requires signals in the form of low-molecular-weight hormones, including the steroid hormones (Chapter 4).
- *Transcription factors activated by internal signals.* Some preexisting transcription factors become activated in response to signals formed intracellularly. Such signals may be internal sterol concentrations in the case of *sterol response element-binding proteins (SREBPs)* or damaged DNA that increases p53 concentrations (Section 14.6).
- *Transcription factors activated by cell surface receptor–ligand interaction.* There are two major routes from cell surface receptors to transcription activators. (i) Ligand binding to receptors activates signal transduction cascades that end at resident nuclear transcription factors. These become phosphorylated at Ser/Thr residues leading to their activation. The best known example of this type of control is the *receptor tyrosine kinase (RTK)–Ras–mitogen-activated protein kinase (MAPK) pathway* (Chapters 8–10) that can activate a diversity of transcription factors. (ii) A more limited number of resident cytoplasmic transcription factors are activated in the cytoplasm or at the cell membrane after receptor–ligand interaction and then accumulate in the nucleus to drive transcription. Following receptor activation, these factors are activated by phosphorylation at Ser or Tyr residues and translocate then into the nucleus. Examples for this type of control include the STAT (Chapter 11) and SMAD proteins (Chapter 12). In many cases, proteolysis is required for the production of the mature, active transcription factor. This control is exemplified by the NFκB (Section 2.5.6.4), Wnt (Chapter 14) and Notch (Chapter 12) signaling pathways.

■ **Signal-regulated transcription factors**
- Nuclear receptors
- Sterol response element-binding proteins
- Cell surface receptor activated transcription factors

3.4.8
Repression of Transcription

Control of eukaryotic transcription by gene repression operates mostly at the level of chromatin structure. The DNA of eukaryotes is ordinarily refractory to transcription because of its organization in nucleosomes. The DNA is wrapped around a histone octamer, which interferes with most transactions required during transcription. Therefore, nucleosome-covered DNA is transcriptionally inert, and it requires histone modification and nucleosome mobilization to overcome this repression. Any protein that serves to maintain the nucleosome-covered state at the promoter and prevent nucleosome mobilization may therefore function as a repressor. The classical view of repressors as based on the observations in bacteria can only be applied to a very limited extent to eukaryotes, due to the organization of eukaryotic DNA in chromatin. Bacterial repressors mostly function by competing with activators for binding to control regions near the promoter or by preventing the binding of RNA polymerase to the transcription start site. Originally, it was assumed that eukaryotic repressors use similar mechanisms. Now it is generally accepted that chromatin is an integral part of transcription regulation, particularly of transcription repression. *The function of nearly all eukaryotic repressors can be linked to the activation or recruitment of enzymes that covalently modify histones or other chromosomal proteins to maintain a repressive chromatin structure.* As compared to transcriptional activators, it has been much more difficult to unravel the mechanisms of eukaryotic repressors because experimental systems for studying the negative regulation of transcription are limited. Therefore, our knowledge of repression mechanisms in eukaryotes is more diffuse.

One particular feature of eukaryotic gene repression is illustrated by the observation that the same regulatory protein can function as a repressor or an activator of transcription, depending on the nature of the regulatory signal that impinges on the regulator. When functioning as a repressor, the regulator recruits to the promoter a corepressor complex containing enzymes that modify histones to establish a repressive state of chromatin. An activating signal can convert the repressor into an activator that recruits a coactivator complex. Enzymes within the coactivator complex then promote nucleosome displacement at the promoter and thus activate transcription. The switch between activator and repressor function can be traced back in some cases to conformational changes in the transactivation domain that dictate the nature of the associating coregulator – either a corepressor or a coactivator. This point is illustrated in more detail in Chapter 4 using the example of hormone agonists and antagonists of nuclear receptors.

■ **Transcription repression**
– Primarily by activation of enzymes that induce repressive chromatin structure; includes recruitment of HDACs and histone methyltransferases

As outlined in Section 3.2.9, the mediator complex provides a major control station for the transmission of regulatory signals to the transcription apparatus. The mediator may also transmit inhibitory signals to the transcription machinery, e.g. by phosphorylating the TFIIH (Section 3.2.8).

The large number of distinct steps required for eukaryotic gene transcription suggests many possibilities for negatively controlling transcription (Fig. 3.18). As stated already above and presented in

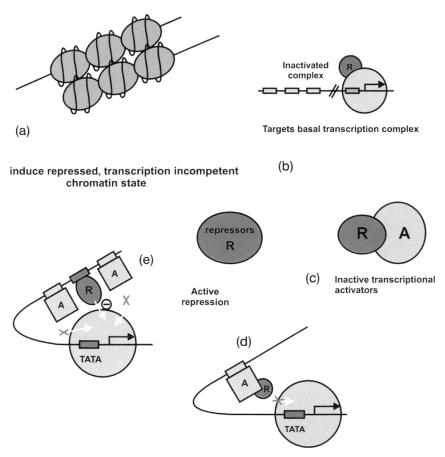

Fig. 3.18: Pathways of repression of transcription. The figure illustrates various mechanisms of repression of transcription. (a) Repressors can induce a generally repressed state in chromatin which is incompatible with transcription. To allow transcription at all, the repressed state must be relieved. (b) Repressors can target the transcription complex and thereby inhibit transcription initiation. (c and d) By binding to free or DNA-bound transcriptional activators, repressors can block the activating function of the latter. (e) Active repression is also affected by proteins that bind sequence specifically to DNA elements and in their DNA-bound form inhibit the transcription initiation. They do this mainly by recruiting enzyme activities (HDACs, Lys methylases) to chromatin that help to establish a transcription-incompetent state.

more detail in Section 3.5, the most part of the negative regulation of transcription is exerted via the chromatin structure. Transcription can be shut down very effectively by imposing a "closed", repressive state on chromatin which may occur in signal-directed, regulated way.

In addition, many gene repression mechanisms not directly related to chromatin structure have been described, although in most cases the mechanistic basis is not clear. Transcriptional repressors may function by binding to transcription activators or by competing with the activator for overlapping binding sites. The extent of repression is then determined by the relative affinity of both proteins to the DNA element and their concentration ratios. A further possibility for repression results from heterodimerization (Section 3.3.6). Heterodimers between two transcription factors, in which one of the partners possesses a DNA-binding domain with weak affinity, can inactivate a transcriptional activator in a heterodimer complex. Since a strong binding to the DNA element usually requires both subunits of a DNA-binding protein, transcription activation by this type of heterodimer is not possible.

Another repressive mechanism operates at the level of the subcellular distribution of transcriptional activators. As outlined in Section 3.4.3, transcriptional activators may be kept in a repressed state by binding of inhibitory proteins that prevent nuclear localization of the activator. Transcription repression can also result from phosphorylation of the basal transcription factors. By this token, the repression of transcription observed during mitosis (Chapter 13) is attributed – at least in part – to the hyperphosphorylation of TBP and TAFs.

3.4.9
Coregulators of Transcription

In addition to the sequence-specific transcription activators, the mediator complex and the core transcription machinery, other factors are required for a regulated and high-efficiency transcription of DNA in the context of chromatin comprise. These proteins are loosely defined as coregulators and may be classified by their function on transcription as coactivators or corepressors.

Typical coactivators or corepressors are found in large protein complexes that associate with the DNA-bound activator or repressor. Histone-modifying enzymes (Section 3.5) like histone acetylases or histone deacetylases (HDACs) are found in these complexes exerting an activating or repressing effect on transcription.

The TAF proteins have been also considered as coregulators because interactions between transcriptional activators such as Sp1 and NTF-1 with distinct TAFs have been found. However, it is still unclear at which point the TAF proteins enter the process of tran-

Other repression mechanisms at the level of transcription factors
- Heterodimerization
- Phosphorylation
- Cytosolic retention
- Binding of inhibitors

Coactivators
- Required for activation
- Organized in large protein complexes

Examples
- TAF members
- DNA topoisomerases
- Histone acetylases
- PARP
- HMG proteins

scription initiation. Possibly the TAFs are primarily required for core promoter selection. The situation is complicated by the fact that a subset of the TAFs, including the histone-like TAFs, have been identified as components of a large histone acetylase complex, termed SAGA in yeast and PCAF (p300/CBP-associated factor) in humans (Section 3.5.2).

Further coregulators have been described which participate in transcriptional regulation in a general sense. This loosely defined class of general coactivators comprises proteins with enzymatic activity such as DNA topoisomerase I and poly(ADP-ribose) polymerase, as well as architectural or scaffolding proteins such as the high mobility group (HMG) 1 and 2 proteins. These proteins appear to be generally required for chromatin reorganization during transcription initiation and elongation.

3.5
Chromatin Structure and Transcription Regulation

In eukaryotes, chromatin structure is the most decisive factor for gene activity. The DNA-encoded information is packaged into the chromatin polymer which must be opened to allow increased accessibility for gene-regulatory proteins or compacted to restrict access of the transcriptional machinery to target genes. In general, these open or closed chromatin states correspond to gene activation versus gene repression, respectively. The cell has developed the following mechanisms to modify in a temporal and spatial manner the chromatin organization, thereby specifying distinct functional states of the chromosomes:

- ATP-dependent chromatin remodeling factors twist and slide nucleosomes, exposing or occluding local DNA areas to interactions with replication, DNA repair and transcription factors.
- Posttranslational covalent modifications of the histones within a nucleosome can either facilitate or hinder the association of transcription factors with chromatin.
- Canonical histones in a nucleosome can be replaced by histone variants through a DNA-replication-independent deposition mechanism. Histone variants harbor distinct information to respond to DNA damage conditions or to override an established gene expression stage.
- Methylation at the C5 position of cytosine residues present in CpG dinucleotides by DNA methyltransferases facilitates static long-term gene silencing and confers genome stability through repression of transposons and repetitive DNA elements (Section 3.5.8).

■ **Tools for modification of chromatin structure**
- ATP-dependent removal and sliding of nucleosomes
- Covalent histone modification
- Incorporation of histone variants
- DNA methylation

3.5.1
Nucleosome and Chromatin Structure

The basic structural unit of chromatin is the nucleosome, in which 146 bp of DNA are wrapped 1.65 turns around the histone octamer (H2A, H2B, H3, H4)$_2$. The cocrystal structure, a histone octamer–DNA complex, shows that the DNA wraps tightly around a cylinder-like core of the histones. Most important for the function of nucleosomes during transcription are the N-terminal tails of the histones. Between 15 and 38 amino acids from each histone N-terminus form the histone tails, providing a platform for posttranslational modifications that modulate the biological role played by the underlying DNA. The spacing of nucleosomes is such that interactions between the N-terminal tails of histones and adjacent nucleosomes are possible.

The nucleosomes are further packaged into higher-order structures, among which the so-called solenoid or 30-nm fiber is best characterized. Interactions between adjacent nucleosomes and the binding of the linker histones like histone H1 are involved in the formation of the 30-nm fiber. This association hinders both nucleosome dynamics and active, ATP-dependent nucleosome remodeling establishing a chromatin structure that is incompatible with transcription.

For transcription initiation, the linker histones must be removed, and the nucleosomes must be posttranslationally modified to allow the binding of sequence-specific transcription factors and the deposition of the transcription machinery. These processes require the removal of nucleosomes from the transcription start site and the mobilization of nucleosomes during the following steps of transcription initiation and elongation.

The specific and dynamic functions of the histones in transcription are dictated by a set of posttranslational modifications which are found to a large part at the N-terminal tails of the histones. The modifications are reversible, which allows for distinct, timely variable modification patterns characteristic of the activity state of chromosomal regions.

■ **Nucleosome reorganization includes ATP-dependent**
– Removal of linker nucleosomes
– Mobilization of histones at the transcription start site and during ongoing transcription; requires covalent histone modification

Covalent Modifications of Histones
The core histones are subject to the following covalent modifications (Fig. 3.19; reviewed in Linggi et al., 2005):
– Acetylation, mostly of lysine residues.
– Phosphorylation at Ser/Thr residues.
– Methylation at lysine and arginine residues.
– Ubiquitination.
– Sumoylation.
– ADP-ribosylation.

Most of these modifications are found at the N-terminal tails of the histones (Fig. 3.19). Accessible regions of the core of the histones may, however, also be modified, e.g. by ubiquitination.

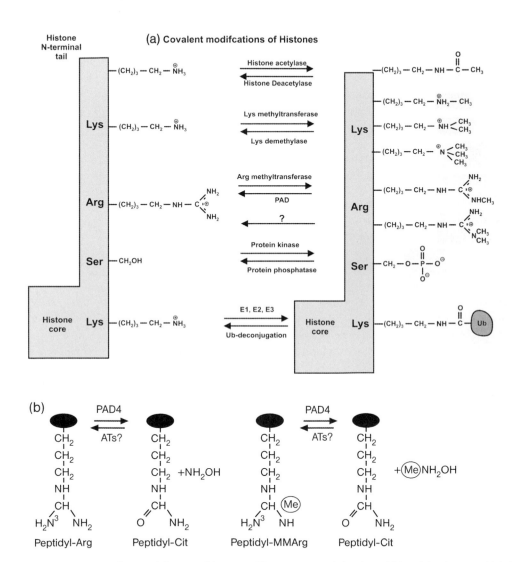

Fig. 3.19: (a) Covalent modification of histones. The enzymes involved in reversible histone modification are shown. E1–E3 = enzymes required for Ub conjugation. (b) PAD catalyzes the conversion of peptidyl arginine and monomethylated peptidyl arginine into peptidyl citrulline, which may be converted back to peptidyl arginine by a aminotransferase (AT).

3.5.2
Histone Acetylation and Deacetylation

■ **Histone acetylation**
- At Lys
- Removes positive charge
- Catalyzed by HATs
- Activates transcription

Acetylation on the ε-NH$_2$ group of lysine residues occurs on all four histones. The addition of an acetyl group to a lysine residue removes its positive charge and creates a new binding surface for protein association. We know of many transcriptional regulators that bind directly or indirectly to acetylated histones and the maintenance of a distinct acetylation pattern is crucial to many actions at the chromosomes. Broadly, acetylation is linked to transcriptional activation and deacetylation is linked to repression. However, we also know of Lys residues whose acetylation can be linked to gene repression (see below).

Histone Acetylases

■ **HATs**
- Organized in multiprotein complexes
- Often contain bromodomains

Examples
- PCAF
- p300/CBP

The enzymes responsible for histone acetylation, the histone acetyltransferases (HATs), were first known as transcriptional coactivators and later as enzymes. We now know of a large number of proteins with HAT activity that may differ in the mechanism of histone-substrate binding and catalysis (reviewed in Santos-Rosa and Caldas, 2005). The purification and biochemical characterization of the HATs has shown that these enzymes also can acetylate substrates other than histones, including transcription activators such as p53 and GATA-1, structural proteins like HMG1, and the GTFs TFIIE and TFIIF. In addition, histone acetylases themselves have been shown to be modified by acetylation.

■ **HAT substrates**
- Histones
- Transcription factors
- GTFs
- Structural proteins

Most HAT enzymes are found in large multiprotein complexes targeted to distinct sites on the chromatin such as transcription start sites. However, we also know of sequence-specific transcription factors with HAT activity. One example is the transcriptional activator ATF-2 that binds as a homodimer or a heterodimer with c-Jun to CREs and activates CRE-dependent transcription. Signal-chain-induced phosphorylation of ATF-2 stimulates its intrinsic HAT activity allowing for activation of ATF-2-dependent transcription by direct effects on chromatin components (Kawasaki et al., 2000).

HAT enzymes mostly function as transcriptional coactivators and are assembled in multiprotein complexes. Examples include the *GCN5/PCAF family* and the *p300/CBP family* of HAT enzymes. These well-characterized HAT families are rather global regulators of transcription. Many members of these families contain a conserved *bromodomain*, which has been shown to recognize and bind acetyllysine residues. The bromodomain is widely distributed among enzymes that acetylate, methylate or remodel chromatin, which highlights the importance of lysine acetylation in self-maintenance of a transcriptional active state and recruitment of other sources of chromatin-modifying enzymes (reviewed in de la Cruz et al., 2005).

HDACs

Histone lysine acetylation is a reversible mark on the histone tails. The dynamic equilibrium of lysine acetylation *in vivo* is governed by the opposing actions of acetyltransferases and deacetylases. Removal of acetyl residues by HDACs results in a decrease in the space between the nucleosome and the DNA that is wrapped around it. As a result, accessibility for transcription factors is diminished and the chromatin is modified from an open gene active euchromatin structure to a closed gene-silenced heterochromatin structure. Similar to acetyltransferases, the HDACs are found in large multiprotein chromatin complexes that are involved in transcriptional repression. The deacetylase complexes are targeted to specific promotors by interactions with different types of negative regulatory proteins. Among its interaction partners are sequence-specific transcription factors, corepressors and the methylated CpG-binding proteins (Section 3.5.8). In addition to a role in transcription repression, HDACs appear to be involved also in other major functions of cells, e.g. ageing and long-term silencing.

■ **HDACs**
– Induce repressive chromatin structure
– Function as corepressors
– Organized in multiprotein complexes
– Classified into three families

There are three major families of mammalian HDACs. The *class I HDACs* are nuclear proteins widely expressed in a variety of tissues. They show a high degree of structural homology and contain a zinc molecule at the active site as a critical component of their enzymatic pocket. A zinc-active pocket also characterizes the members of the *class II HDACs*, but, in comparison to class I, they have a narrower tissue distribution, are much bigger in size, and shuttle between the nucleus and cytoplasm as part of their mode of action. The third HDAC family *(class III or SIR-HDACs)* is quite different, both structurally and in terms of the catalytic mechanism. Their enzymatic activity depends on the cofactor NAD^+, which breaks down during the histone deacetylation reaction resulting in *O*-acetyl-ADP-ribose and nicotinamide. Interestingly, *O*-acetyl-ADP-ribose has been shown to be involved in the assembly of a repressive SIR-HDAC complex on hypoacetylated histone tails coupling SIR-HDAC activity with transcriptional silencing (Liou et al., 2005).

Generally, *HDACs function as corepressors of transcription*. We know of various HDAC-containing multiprotein complexes that are recruited to specific chromatin sites by transcriptional regulators. As an example, the repressive heterodimeric transcription factor Mad–Max forms a complex with the HDAC I that is part of the mammalian *mSin complex*. A complex of HDAC I and the nuclear receptor corepressor (Chapter 4) binds to unliganded nuclear receptors and is believed to exercise a repressive effect. A further example is the tumor suppressor retinoblastoma (Rb) protein (Chapters 13 and 14), which can occur as a transcriptional corepressor in the hypophosphorylated form and a transcriptional coactivator in the hyperphosphorylated form. The repressive form of the Rb protein recruits HDAC I to the DNA and thereby initiates an active repression of the gene (Section 13.4.3).

3.5.3
Histone Methylation

Methylation by methyltransferases is another posttranslational covalent modification that occurs on the side-chain nitrogen atoms of lysine and arginine on histones. The most heavily methylated histone is H3 followed by H4. *Arginine can be either mono- or dimethylated,* with the latter in symmetric or asymmetric configurations. *Lysine can accept one, two or three methyl groups,* resulting in mono-, di- or trimethylated forms. The different stages of methylation on a given residue confer different biological read outs to the modified residue, thus methylation has greater combinatorial potential with respect to other modifications. In contrast to acetylation, which correlates almost without exception with transcriptional activation, histone methylation can result in either transcription activation or repression, depending on the modified residue and the palette of other modifications decorating the histone simultaneously (Fig. 3.20b). Both lysine methylation and arginine methylation are histone marks that can be removed or inactivated by the action of specific enzymes. The histone-demodifying enzymes oppose the action of the methyltransferases, and both types of enzymes are major tools for establishing timely and regionally variable methylation patterns that dictate the activity state of chromatin.

■ **Histone methylation**
- Catalyzed by methyltransferases
- At Lys or Arg residues
- Activates or represses transcription

3.5.3.1   Histone Arginine Methylation

Protein arginine methyltransferases (PRMTs) catalyze the transfer of methyl groups from *S*-adenosyl-L-methionine to the guanidino nitrogens of arginine residues (reviewed in Wysocka et al., 2006). There are two major classes of PRMTs that share a conserved catalytic core, but have little similarity outside the core domain.
- The *type I PRMTs* catalyze asymmetric dimethylation of arginine. This family includes PRMT1 and the CARM1 (Coactivator-Associated RMethyltransferase 1)/PRMT4 enzymes.
- The *type II PRMTs* catalyze the formation of symmetric dimethylarginine. This group includes PRMT5 that associates *in vivo* with the ATPase-chromatin remodeling hSWI/SNF complex.

■ **Histone arginine methylation**
- Catalyzed by PRMTs
- Yields mono- or dimethylarginine
- Regulates transcription

Histone arginine methylation correlates with transcription activation of a variety of genes, and a functional synergy between arginine methylation and histone acetylation in transcription activation events

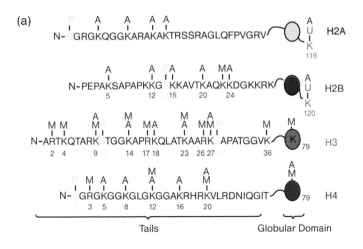

(a)

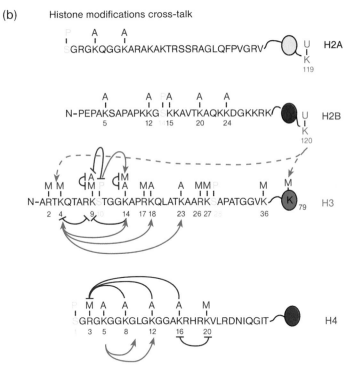

(b) Histone modifications cross-talk

Fig. 3.20: (a) Patterns of histone modification. A = acetylation; M = methylation; P = phosphorylation; U = ubiquitination. Posttranslational modifications on the histone tails. Modifications recently identified by mass spectroscopic techniques, but unconfirmed (by mutational analysis and/or Western blot with specific antibodies), are not shown. Note that Lys9, Lys14, Lys23 and Lys27 in the H3 tail and Lys12 and Lys20 in H4 can be either acetylated or methylated. Acetylation = purple; methylation = blue; phosphorylation = orange; ubiquitination = green. (b) Interplay between different posttranslational modifications. "Compatible" modifications (those which facilitate other modifications to occur and/or can coexist) are represented by green arrows. "Incompatible" modifications (those which negatively affect other modification and/or can not coexist) are shown in red.

has been shown. PRMTs physically interact with HATs and form coactivator complexes in which the two enzyme activities cooperate in activating the transcription of specific genes such as the genes for the transcription factor NFκB and the tumor suppressor p53.

Based on their transcription activating function, arginine methyltransferases play an important role in the regulation of cell growth and proliferation, and deregulated PRMT activity has been correlated with tumor development. Specifically, arginine methylation seems to be an important mechanism to regulate expression and function of genes involved in tumor suppression.

3.5.3.2 Arginine Demethylation/Citrullination

How the methyl mark on arginine residues is removed has been elusive until the discovery of a novel mechanism involving conversion of arginine or mono-methylarginine into citrulline by the action of a peptidyl arginine deaminase (PAD4; see Fig. 3.19b). The PAD4-catalyzed conversion of arginine residues into citrulline prevents methylation of these sites by histone arginine methyltransferases and is thus a novel posttranslational modification that regulates the level of histone arginine methylation and gene activity.

■ **PADs are involved in removal of methyl groups from arginine**

PAD4 also converts methyl-Arg into citrulline, releasing methylamine. The conversion of histone tail H3 and H4 mono-methylarginine residues into citrulline by PAD4 controls arginine methylation levels, and has been shown to be a critical step in the regulation of estrogen-responsive gene transcription.

3.5.3.3 Histone Lysine Methylation

As illustrated in Fig. 3.20, the number of lysine methylation sites on the histones is quite large and comprises also sites within the core of the nucleosome. The consequences of lysine methylation depend on the specific site on the histone tails and distinct methylation marks correlate with either activation or transcription. As compared to lysine acetylation, lysine methylation is much more diverse in its functions as it can promote transcription activation, mediate transcriptional repression, trigger heterochromatin formation and even chromosome loss.

■ **Lysine methylation**
– Multiple regulatory functions
– Mono-, di- or trimethylation
– Catalyzed by Lys methyltransferases

Lysine methylation is found mostly on the N-terminal tails of histones H3 and H4 (reviewed in Santos-Rosa and Caldas, 2005). Some of these residues are also substrates for acetylation (Fig. 3.20). The methylation of the various lysine residues has a differential effect on transcriptional activity. This has been well shown for two distinct Lys methylations on histone H3.

Methylation at H3 K9 is associated with transcriptional inactivation and establishment of a heterochromatin state. K9 methylation by the methyltransferase SUV39 can, for example, direct the binding of the transcriptional repressor HP1, which is a component of transcriptionally inactive heterochromatin (Fig. 3.21).

In contrast, methylation at K4 of H3 is correlated with transcriptionally active states of chromatin. The H3 K4 methylation appears to be absent from histones with methylation at K9, but colocalizes

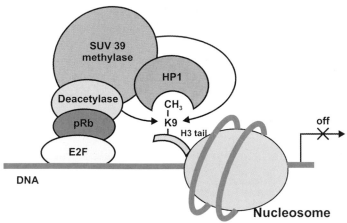

Fig. 3.21: Methylation of H3 during silencing of pRb-E2F-controlled genes. The tumor suppressor pRb interacts with the transcription factor E2F and can repress transcription by recruiting HDAC activity to E2F-controlled genes. The deacetylase removes an acetyl group from Lys9 of H3. In a subsequent reaction, the methylase activity of SUV39 attaches a methyl group to Lys9 providing a binding site for the repressor HP1. Binding of HP1 then generates a highly compact chromatin structure by a still unknown mechanism. HP1 and the SUV39 methylase are found in tight complex in the cell. After Kouzarides (2002).

with acetylation at H3 K9, indicating a positive cooperation of acetylation and methylation of distinct Lys residues during the establishment of a transcriptionally active state.

With almost no exception, the enzymes that catalyze the methylation of lysine residues of histones share strong homology in a 140 amino acid catalytic domain known as the *SET [Su(var), Enhancer of Zeste and Trithorax] domain*.

There are several families of lysine methyltransferases that differ in substrate specificity, assembly in multiprotein complexes and the degree of lysine methylation. For instance, members of the SET7/9 family catalyze mostly monomethylation, whereas the MLL1 members are proficient in trimethylation of lysine residues.

3.5.3.4 Histone Lysine Demethylation

Until recently, lysine methylation was considered a irreversible modification that defines chromosomal subdomains rather than transcriptional stages. However, lysine demethylases have been discovered that remove specific methyl residues, and lysine methylation is now accepted as a regulatory modification with a dynamic turnover determined by the opposing activities of lysine methyltransferases and lysine demethylases. The first identified lysine demethylase, *LSD1, is a FAD-dependent amine oxidase* with specificity for distinct lysine methyl residues. Depending on the methylated lysine residue, LSD1 activation can contribute to transcriptional activation or repression. As illustrated in Fig. 3.22, demethylation of Lys9 H3 by LSD1 accompanies gene activation, whereas demethylation of Lys4 H3 is associated with gene repression. In the course of gene activation by the androgen receptor, LSD1 has been shown to directly as-

■ **Lysine demethylases remove methyl groups from Lys**

Fig. 3.22: (a) Methylation of H3 K4 is often associated with active genes, and conversely, its de-methylation accompanies gene repression. (b) In contrast, methylation of H3K9 is often associated with silenced genes. Therefore, removal of H3 K9 methyl marks coincides with gene activation.

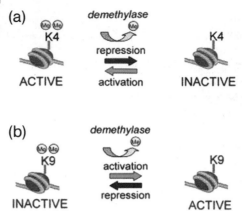

sociate with the androgen receptor (Chapter 4), and acts as a coactivator for transcriptional activation by removing the repressive Lys9 methyl mark (reviewed in Wysocka et al., 2005).

3.5.4
Histone Phosphorylation

Phosphorylation is another covalent posttranslational modification of histones. Phosphorylation occurs in all four core histones that constitute the histone octamer, as well as in some of the histone variants and histone H1. A major substrate for phosphorylation is histone H3 (reviewed in Espino et al., 2005). Phosphorylation of the N-terminal tail Ser10 of H3 is associated with transcriptional activation of eukaryotic genes by promoting acetylation of K14 on the same histone tail. In mammalian cells, the protein kinases MSK1/2 catalyze the phosphorylation of H3 Ser10. MSK1/2 enzymes belong to the AGC family of protein kinases and can be activated via the extracellular signal-regulated kinase (ERK)/MAPK pathway that in turn responds to mitogenic signals. In this way, a linkage between proliferation-promoting signals and transcription activation is provided.

H3 phosphorylation has been also recognized as part of a complex signaling mechanism that operates in the condensation/decondensation of chromatin during the cell cycle.

Furthermore, phosphorylation of histone H1 has been linked to the relief of transcription repression.

■ **Histone phosphorylation**
  – At all histone types
  – Mostly at Ser
  – Protein kinases,
    e.g. MSK1/2

3.5.5
Histone Ubiquitination and Sumoylation

The ubiquitination (Section 2.5) of histones also plays a specific role in regulating transcription. As an example, the histone H2B from yeast is monoubiquitinated on Lys123 and this modification is associated with active transcription. A specific role of the 19S regulatory proteasome particle has been established in this process. H2B ubiquitination has been shown to recruit the 19S particle to the promoter and this event facilitates the methylation of H3 Lys4 and the subsequent recruitment of the coactivator complex SAGA to the promoter.

The SUMO modification (Section 2.5.9) of Lys residues of histones generally has a repressing effect on transcription by improving the association of the histone acetylase HDAC1 and the recruitment of the transcriptional corepressor HP1.

3.5.6
Histone Modification "Code"

Given the different nature of histone modifications and the large number of modification sites, the question arises about the function and the interdependence of these modifications (reviewed in Peterson and Laniel, 2004). It is now well accepted that histone modifications control the structure and/or function of the chromatin fiber, with different modifications yielding distinct functional consequences. Many studies have shown that site-specific combinations of histone modifications correlate well with particular biological functions. For instance, the combination of H4 K8 acetylation, H3 K14 acetylation and H3 S10 phosphorylation is often associated with active transcription. Conversely, trimethylation of H3 K9 and the lack of H3 and H4 acetylation correlates with transcriptional repression in higher eukaryotes. Particular patterns of histone modifications also correlate with global chromatin dynamics, as diacetylation of histone H4 at K4 and K12 is associated with histone deposition at S phase, and phosphorylation of histone H2A (at S1 and T19) and H3 (at T3, S10 and S28) appear to be hallmarks of condensed mitotic chromatin.

There is one example of histone modification, i.e. histone H4 K16 acetylation, where the structural and functional consequences of a single-site modification could be demonstrated (Shogren-Knaak et al., 2006). H4 K16 acetylation has been correlated in earlier studies with an active, open state of the chromatin. Furthermore, it was known from structural studies that the tail of H4 – where K16 is located – interacts with adjacent nucleosomes allowing packaging of the nucleosomes into the 30-nm fiber. Acetylation of H4 K16 performed in a de-

■ **Histone modification code**
– Variable patterns at multiple sites
– Modulates nucleosome structure
– Recruits histone- and chromatin-modifying enzymes
– Positive and negative crosstalk of different modifications

fined experimental system appears to prevent this packaging, thereby establishing an open conformation of the chromatin essential for transcription and the maintenance of euchromatin. In addition, the presence of acetylated H4 K16 provides a binding site of an ATP-dependent chromatin remodeling complex, independent of its structural effect. This example illustrates two major consequences of histone modifications, i.e. the *modulation of nucleosome packaging* and the *recruitment of histone- and chromatin-modifying enzymes*.

The patterns of histone modifications are often interpreted as a histone modification "code" which might be read by various cellular machineries. It has been hypothesized that specific tail modifications and/or their combinations constitute a code that determines the transcriptional state of the genes. According to this hypothesis, multiple histone modifications, acting in a combinatorial or sequential fashion on one or multiple tails, specify unique downstream functions.

However, the name "code" may not be applied in a strict sense to histone modification, because a particular combination of histone marks does not always have the same biological consequences. For instance, the generally inhibitory H3 K9 methylation is in some cases associated with actively transcribed genes and histone acetylation can be inhibitory rather than stimulatory for transcription. Thus, rather than a histone code there are instead clear patterns of histone marks that can be differentially interpreted by cellular factors, depending on the gene being studied and the cellular context.

There is now general agreement that histone modifications serve primarily to target histone-modifying enzymes, adaptor proteins and chromatin-remodeling enzymes to distinct chromatin sites. The histone-modifying enzymes, discussed above, show specificity for individual histone tails and for specific histone residues. Therefore, it is not only the targeting to a distinct site, but also the sequence context that determines whether a specific modification will be introduced or not. The activity of the histone modifying enzymes is also profoundly influenced by a crosstalk between different histone marks (reviewed in de la Cruz, 2005). It is the cooperation of targeting, sequence specificity and positive or negative crosstalk that ultimately generates the complex patterns of gene-specific histone marks associated with distinct chromatin states. There is experimental evidence that the pattern of histone modifications can be inherited – at least to some extent – upon cell division. Histone modification is therefore considered as another epigenetic modification, in addition to DNA methylation and RNA interference.

3.5.7
Recognition of Histone Modifications by Protein Domains

In order to be translated into a biological function, the histone "code" must be read by the proteins that shape the state of chromatin. The enzymes acting on nucleosomes must contain domains or modules that recognize specific histone modifications.

Such chromatin-binding domains have been identified among the chromatin modifying enzymes (histone acetylases, HDACs, histone methyltransferases) as well as among ATP-dependent chromatin remodeling enzymes. The chromatin-binding domains are differentially distributed among the chromatin modifying enzymes suggesting that these domains confer specific chromatin-binding properties to the different enzyme families (Fig. 3.23) (reviewed in de la Cruz et al., 2005). There are now four major chromatin-binding domains:

– *Bromodomain.* This domain recognizes acetyl-lysine residues. Bromodomains are widely distributed among the different enzymes that acetylate, methylate or remodel chromatin. It is now well established that the bromodomain has the ability to bind acetylated histone tails *in vivo*, with an apparent independence of the protein to which it belongs, and this ability can be utilized by different chromatin-modifying enzymes to find and/or act on their targets.

■ **Chromatin-binding domains**
– Recognize specific histone modifications
– Located on chromatin-modifying enzymes

■ **Binding specificity of chromatin-binding domains**
– Bromodomains: acetyl-Lys
– Chromodomain: methyl-Lys
– PHD domain: trimethylated Lys
– SANT: N-terminal tails of histones

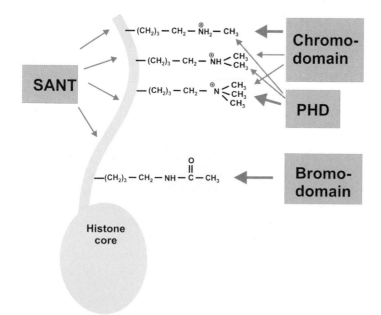

N-terminal tail

SANT

Chromo-domain

PHD

Bromo-domain

Histone core

Fig. 3.23: Major chromatin-binding domains. The different sizes of the arrows indicates the preference for different methylated states of Lys. SANT binds preferentially to the N-terminal tails as such.

- *Chromodomain.* This domain binds specifically to methylated Lys residues and is found in histone acetylases, histone methyltransferases and ATP-dependent remodeling enzymes.
- *Plant homeodomain finger (PHD) domain.* The PHD has been discovered as a structural part of the largest subunit of the human nucleosome remodeling factor (NURF). It has been shown to bind specifically to a trimethylated lysine residue (K4) of the N-terminal tail of histone H3 (Li et al., 2006). This modification is associated with the transcription start sites of active genes.
- *SANT domain.* The general role of the SANT domain is to stabilize, through direct binding, histone N-terminal tails in a conformation favoring their binding to the modifying enzymes, and the subsequent catalytic process. The SANT domain is broadly present among ATP-dependent remodeling enzymes and their complexes, but it can also be found in histone methyltransferases, and in proteins forming part of complexes with histone acetylase and HDAC activities, suggesting an important role in regulating chromatin accessibility.

3.5.8
DNA Methylation

The methylation of DNA is the most abundant epigenetic modification in vertebrate genomes (reviewed in Klose and Bird, 2005). The term "epigenetic" refers to the information contained in chromatin, other than the actual DNA sequence, that defines a heritable specific gene expression pattern. Other epigenetic processes include the modification of chromatin proteins and the production of small interfering RNAs (siRNAs; see Section 3.7.5).

■ **DNA methylation**
- 5-Methylcytidine at CpG
- Associated with repressed state
- Catalyzed by DNA methyltransferases
- Inhibitor: 5-aza-cytidine

In the mammalian genome, the methylation takes place only at cytosine bases that are located 5′ to a guanosine in a CpG dinucleotide (Fig. 3.24). The product of methylation is *5-methylcytidine*. A high CpG content is found in regions known as CpG islands. Most CpG islands are found in the vicinity of promotors and are undermethylated, whereas in the remainder of the genome the CpG sequences are generally methylated. In specific situations, e.g. during tumorigenesis, CpG islands are, however, found in a hypermethylated state (Section 14.1.3). Overall, methylation of cytosine bases is associated with a repressed chromatin state and inhibition of gene expression. The biological importance of DNA methylation is illustrated by gene disruption experiments that have shown an essential requirement of the methylation system for the development of mice.

The enzymes responsible for DNA methylation at CpG sequences are the cytosine *DNA methyltransferases*. The methyl group is derived from *S*-adenosyl methionine. An important inhibitor of DNA methyltransferases is *5-aza-cytidine*, which blocks DNA methylation leading to a change in DNA methylation patterns of cells.

Mammalian cytosine DNA methyltransferase enzymes fit into two general classes based on their preferred DNA substrate. The *de novo methyltransferases* DNMT3a and DNMT3b are mainly responsible for introducing cytosine methylation at previously unmethylated CpG sites, whereas the *maintenance methyltransferase* DNMT1 copies preexisting methylation patterns onto the new DNA strand during DNA replication.

A characteristic distribution pattern of 5-methylcytidine is found within each cell, which remains intact upon cell division. DNMT1 is responsible for this maintenance methylation. It has preference for DNA sequences in which the complementary strand is already methylated at CpG sequences (Fig. 3.24). Such a hemi-methylation of DNA is found, for example, in regions that are just being replicated.

De novo methylation at CpG sequences has been also clearly shown. The mechanisms by which the *de novo* methyltransferases select their substrate sites appear to be varied. Targeting to specific CpG sequences can be accomplished by specific domains within the DNA methyltransferase, by recruitment through protein–protein interactions with transcriptional corepressors and by the RNA-mediated interference system (Section 3.7.5).

■ Two types of DNA methylation
– Maintenance methylation
– *De novo* methylation

Coupling DNA Methylation to Gene Repression

DNA methylation is an epigenetic modification that is generally linked with transcriptional silencing of associated genes. Several basic models have evolved to explain the relationship between CpG methylation and gene repression:

– DNA methylation can directly repress transcription by blocking transcriptional activators from binding to cognate DNA sequences.
– Methylated DNA is recognized by methyl-CpG-binding proteins (MBPs) that recruit corepressors to silence gene expression directly. An important link between DNA methylation and chromatin modification is provided by the ability of MBPs to engage in protein–protein interactions with histone-modifying enzymes. Members of the MBP family are known to associate with HDACs and histone methyltransferases, and these interactions are thought to recruit these enzymes to sites of CpG methylation in order to maintain a repressed state, linking DNA methylation and histone modification. It is now well established for various organisms that histone deacetylation and histone methylation such as methylation of histone H3 at Lys9 work hand-in-hand with DNA methylation. In addition, two newly discovered properties of DNA methyltransferases are considered to be important for their repressive function:

■ DNA methylation and transcription repression
– MeCpG recognized by specific binding proteins: MBPs
– MBPs recruit HDACs and methyltransferases

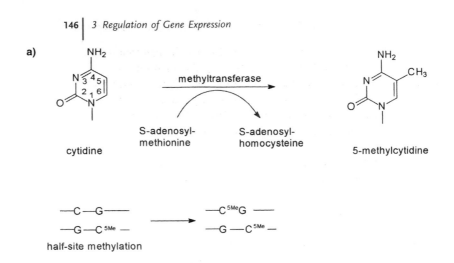

a)

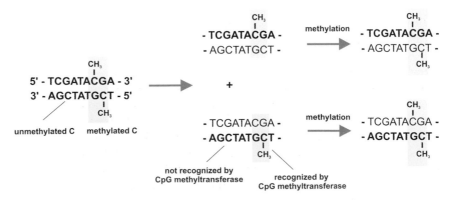

Fig. 3.24: The methylation of DNA: 5-methylcytidine and maintenance methylation. (a) The methylation of cytidine residues on DNA is catalyzed by a methyltransferase that employs *S*-adenosine methionine as a methyl group donor. The preferable substrates for the methyltransferase are hemimethylated CpG sequences. 5-Azacytidine is a specific inhibitor of methyltransferases.

(b) The methylation pattern of DNA remains intact upon DNA replication and is passed on to the daughter cells. The newly synthesized strands are unmethylated immediately after DNA replication. The methyltransferase uses the previously methylated parent strand as a matrix to methylate the CpG sequences of the newly synthesized strand.

- DNA methyltransferase enzymes themselves can be involved – via protein–protein interactions – in setting up the silenced state in addition to their catalytic activities. DNA methyltransferases are known to associate with chromatin-remodeling enzymes and with chromatin adaptor proteins such as HP1. This could serve to recruit methyltransferase to distinct sites where a repressed state is to be maintained.
- DNA methylation can affect transcriptional elongation in addition to its characterized role in inhibiting transcription initiation.

The precise sequence of events in DNA methylation-mediated gene repression is, however, still a matter of debate. Overall, histone modification and DNA methylation appear to be part of an epigenetic program where both processes are linked and mutual influences, e.g. via feedback loops, are possible (reviewed in Fuks, 2005).

Biological Functions of DNA Methylation

DNA methylation is an essential part of the epigenetic program of multicellular organisms. In specific situations of cells, e.g. during cell cycle progression and differentiation, the methylation pattern can be quite dynamic. Enzymes have been identified that function as demethylases and can alter the methylation pattern. Due to its general repressive functions, DNA methylation is involved in the following biological processes:
- Transcription initiation and transcription elongation.
- Heterochromatin formation.
- Methylation of foreign DNA, such as viral DNA, provides a defense mechanism against the expression of exogenous DNA.

DNA methylation participates in genetic imprinting and in X chromosome inactivation. The term "genetic imprinting" describes a situation where genes are expressed unequally depending upon whether they were maternally or paternally inherited. Normally both copies of the parental genes are equally transcribed in a diploid chromosome. However, with imprinting, a gene inherited from either the mother or father is selectively inactivated. Methylation is obviously involved in such an inactivation. The inactive copy is more strongly methylated than the active copy.

DNA methylation has been recognized as an important aspect of tumorigenesis. Changes in the methylation pattern have been linked in many tumors to decreased expression of tumor suppressor genes (Chapter 14).

3.5.9
Summary of the Regulatory Steps in Transcription

The changes in the state of chromatin, the various histone modifications and the process of transcription may be ordered in a sequence of events as summarized below. Many aspects of the coordination of transcription with chromatin remodeling are not yet understood and the proposed scheme is therefore of necessity speculative. In particular, the timely order of events summarized below may depend on the chromatin and gene context and is still a matter of debate.

■ **Relieve of repressed chromatin state**

– In a first step chromatin must be prepared for transcription by relieving a closed, repressed state. This process involves the removal of marks that conceal the repressed state. An example of such a mark is methyl H3 K9 that maintains the repressed state by recruiting the repressive adapter HP1. Demethylation of H3 K9 is assumed to be an important step in preparing chromatin for transcription. Furthermore, histone acetylation at regulatory sites destabilizes chromatin structure promoting the disruption of nucleosomes in advance of initiation.

■ **Binding of transcription factors to *cis*-elements**

– Binding of sequence-specific transcription factors: Once the structure of chromatin has been "loosened", sequence-specific transcription factors gain access to their cognate DNA-binding elements and can engage in protein–protein interactions with corepressors or coactivators, with the mediator complex and the core transcription machinery.

■ **Interaction with mediator, corepressors, coactivators**

– The interaction of transcriptional regulators with the coactivators and the mediator complex induces – by activating or recruiting further chromatin-modifying enzymes –chromatin modifications such as lysine acetylation, lysine and arginine methylation, Ser/Thr phosphorylation in the vicinity of the promoter site.

■ **Control by external and internal signals**

– Multiple mechanisms are available for control of the transcriptional regulators. External signals may act directly on the DNA-bound transcriptional activators, on the mediator complex and on the transcriptional activators.

■ **Remodeling of chromatin by protein complexes: SWI/SNF, ACF and NuRD**

– In a step preceding or following chromatin modification at the promoter, chromatin remodeling enzymes are targeted to the transcription start site. Several large protein complexes have been shown to be involved in chromatin remodeling. Examples are the SWI/SNF complex from yeast and human and the human ACF and NuRD complexes (reviewed in Johnson et al., 2005). A common characteristic of the remodeling complexes is the presence of distinct ATPase subunits. The ATP-dependent remodeling complexes use ATP hydrolysis to disrupt nucleosome–DNA interactions, and thereby participate in nucleosome mobilization and removal of nucleosomes from the transcription start site.

- Removal of nucleosomes from the promoter by the action of chromatin remodeling enzymes is a prerequisite for the formation of a RNA polymerase pre-initiation complex at the promoter. In yeast, the nucleosome-free DNA stretch at the promoter comprises about 200 nucleotides. Once nucleosomes have been removed from the promoter, binding of TBP to the TATA box is possible and the RNA polymerase holoenzyme is directed to the promoter.

■ Removal and mobilization of nucleosomes

- Histone variants may be used to mark the transcription start site. In yeast, the nucleosome-free zone around the promoter is flanked by two copies of a variant of histone H2A, named histone H2AZ. The deposition of H2AZ is mediated at least in part by DNA signals. H2AZ has distinct modification characteristics that appear to be involved in the recruitment of chromatin-remodeling enzymes. Another histone variant, H2AX, has been implicated in the reorganization of chromatin in response to double-stranded (ds) DNA breaks, indicating the participation of histone variants in DNA damage responses.

■ Replacement of canonical histones by histone variants: H2AZ and H2AX

- The onset of transcription depends on the cooperation of mediator and transcription factors bound to regulatory sequences. The interaction with transcription factors and mediator stabilizes the RNA polymerase complex at the promoter and facilitates the transition from the pre-initiation phase to the elongation phase.

- An early step in the transition from transcription initiation to transcription elongation includes the phosphorylation of the CTD of RNA Pol II that serves as a trigger for the start of the elongation phase. Various protein kinases are involved in this phosphorylation and a dynamic pattern of phosphorylation is then established on the CTD providing a platform for the association of the mRNA processing machinery. Furthermore, chromatin-modifying enzymes travel along with the RNA polymerase during mRNA elongation. This serves to modify nucleosome structure and to remove nucleosomes in front of the elongating transcription machinery.

■ CTD phosphorylation triggers transition into elongation phase

- The 5'- and 3'-ends of genes are marked differentially during transcription (reviewed in Lieb and Clarke, 2005). Chromatin at the 5'-end control regions is destabilized by deacetylation to allow binding of the transcription factors and deposition of the transcription machinery. The chromatin at the 3'-ends of genes, however, must be stable enough to prevent inappropriate initiation at degenerate start sites internally. To this end, the transcription process itself marks the gene as having been transcribed with methylation at histone H3 Lys36. This occurs through co-transcriptional methylation by the methyltransferase Set2 which is bound to the CTD of RNA Pol II during the elongation process. It has been shown in yeast, that H3 Lysine36 methylation is used to direct a HDAC to the transcribed regions that deacetylates and stabilizes nucleosomes within a transcribed gene.

■ Differential marking of chromatin at the 3'- and 5'-end of the gene

Overall, a complex cooperation of chromatin modifications and transcription has emerged in which the epigenetic modifications of chromatin and DNA play a crucial role in determining the transcriptional state of a gene. Transcription is a dynamic process where the main players – sequence-specific transcription factors, the mediator complex, RNA Pol II, the nucleosome and the chromatin-modifying enzymes – are engaged in a multitude of protein–protein interactions that are variable in time during rounds of transcription. Many of the main players are organized in multiprotein machines of variable composition and it is this variability and the dynamics of the process that has precluded a detailed picture of the regulation of transcription up to now.

The sequence of events proposed above is speculative and is dependent on chromatin context and the specific gene structures. The transcription of genes carrying only one or a few regulatory elements certainly differs in the characteristics of regulation as compared to genes with a multitude of different regulatory elements.

3.6
Posttranscriptional Regulation of Gene Expression

Transcription and translation are spatially separated events in eukaryotes. The product of nuclear transcription is *pre-mRNA*. In order to enable translation, the information contained within the pre-mRNA must be transported out of the nucleus and into the cytosol. The quantity of processed mRNA available for translation decides to a high degree how much protein is formed by *de novo* synthesis.

From the primary transcript to translated protein there are many possible points for regulatory processes. The most important regulatory points are:

■ **Posttranscriptional regulation operates at the level of**
– mRNA processing
– Translation initiation
– mRNA stability

– *mRNA processing: modification and splicing of the pre-mRNA*. Modification at the 5′- and 3′-end of the pre-RNA, as well as splicing of the primary transcript are the major modifications that are necessary to form the mature mRNA ready for translation at the ribosome. It is mainly splicing that decides which information contained in the primary transcript is made available for protein biosynthesis and the process of alternative splicing plays a major role in expanding the complexity of the proteome encoded by the genome. The processing of pre-mRNA is tightly coupled to transcription by RNA Pol II. Due to the association of the pre-mRNA processing systems with the CTD of elongating RNA Pol II, pre-mRNA synthesis and pre-mRNA processing go hand-in-hand with the transcription elongation process. Furthermore, protein complexes involved in the transport of mRNA out of the nucleus associate with components of the splicing apparatus providing for a linkage of RNA processing and RNA transport. The controls operating at the level of mRNA processing and transport are highly complex and not well understood, and therefore will not be presented in detail in the following.

– *Initiation of translation.* The translation of the correctly modified mature mRNA by the ribosome is also subject to regulation. A major regulatory site of translation is at the initiation of translation. Here, protein binding at the 5'- and 3'-ends of the mRNA is a widely used tool for controlling the ability of mRNAs to form functional initiation complexes with the ribosome.

– *Stability of the mRNA.* Further regulatory elements include cleavage and degradation of specific mRNAs by RNases as well as a blockage of translation at some steps after initiation. A newly discovered regulatory mechanism using RNA interference is now considered as another major mechanism for controlling mRNA translation. *Small nuclear RNAs, the miRNAs and siRNAs,* have been shown to regulate the levels of specific RNAs by inducing either cleavage of the message or a blockage of translation. The miRNAs or siRNAs are now considered as part of the epigenetic programme that controls transcription.

3.6.1
Control at the Level of Pre-mRNA Processing

In the nucleus, eukaryotic mRNAs are first transcribed as pre-mRNAs and subsequently modified by *capping, polyadenylation* and *splicing.* A characteristic of eukaryotic mRNAs is the presence of modifications at the 3'- and 5'-end. The modifications comprise capping of the 5'-end and polyadenylation at the 3'-end. Another characteristic feature of eukaryotic pre-mRNA is the presence of introns and exons. In the process of splicing, the introns are removed and the exons are joined to form the mature mRNA. Mature mRNAs are ultimately exported into the cytoplasm where they can be translated into proteins.

■ **Pre-mRNA processing includes**
– Capping
– Polyadenylation
– Splicing

Capping

Capping of eukaryotic mRNAs involves the addition of a 7-methylguanosine residue at the 5'-end to protect this end from nuclease degradation. The cap structure in eukaryotes can be of three types, m7GpppNp, m7GpppNmp or m7GpppNmpNmp (m indicates a methyl group attached to the respective nucleotide), and is used as a docking point for the cap binding protein complex that mediates the recruitment of the small ribosomal subunit to the 5'-end of the mRNA. Capping at the 5'-end of the pre-mRNA occurs immediately after incorporation of about 30 nucleotides in the primary transcript. The enzymes involved in capping associate with the CTD of RNA Pol II after CTD phosphorylation, at the transition from initiation to elongation.

Polyadenylation

Polyadenylation occurs after cleavage of the pre-mRNA at the 3'-end and consists of the addition of up to 500 A residues by the poly(A) polymerase (PAP) enzyme. The regulation of poly(A) tail length of translationally controlled mRNAs is a recurring theme in the oogenesis and early development of many animal species. In most cases, long poly(A) tails (80–500 A residues) correlate with translation and short tails (20–50 A) with repression of translation. The regulatory sequences governing translational activation and polyadenylation are primarily found in the 3'-untranslated regions (UTRs) of the mRNAs. The best characterized of these sequences is the cytoplasmic polyadenylation element (CPE), which is found in mRNAs encoding cell cycle regulators. The CPE is bound by CPE-binding protein (CPEB), a conserved RNA-binding protein, containing a zinc finger and two RNA recognition motifs (RRMs).

Circularization of mRNA is Required for Efficient Translation

Once within the cytoplasm, only mRNAs that are properly capped and polyadenylated are efficiently translated. Cap-dependent translation in eukaryotes requires the ordered assembly of a complex of evolutionarily conserved proteins, which starts with the binding of the eukaryotic translation initiation factor 4E (eIF-4E) to the 7-methylguanosine (m7GpppN) cap structure at the 5'-end of the mRNA. Next, the eIF-4G factor is recruited allowing additional factors [poly(A)-binding protein (PABP), eIF-4A, eIF-4B, eIF-1, eIF-1A, eIF-2 and eIF-3, among others, and the ribosomal subunits] to form a complex that, after mRNA circularization, initiates translation (Fig. 3.25).

■ mRNA circularization
– Required for translation
– Involves eIF-4E, eIF-4G and PABP

During translation initiation, the cap structure is directly bound by eIF-4E and the poly(A) tail by PABP in a manner that induces a synergistic enhancement of translation. Translation that is both cap- and poly(A)-dependent requires, moreover, the activity of eIF-4G. eIF-4G functions as an adaptor that helps to bring together the 5'- and 3'-ends of the mRNA. It contains specific binding sites for both eIF-4E and PABP, thus forming a complex that circularizes the mRNA. In this context, the 5'-cap and the 3'-poly(A) tail are modifications necessary for efficient translation. The details of these processes cannot be presented here and the reader is referred to textbooks on cell biology for these topics.

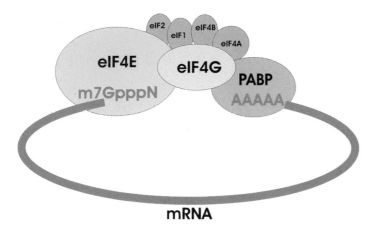

Alternative Splicing

The major path for extracting different information from pre-mRNAs uses alternative splicing. The genetic information encoding a protein in higher eukaryotes is usually found in pieces of coding sequences, or exons, interrupted by noncoding sequences, the introns. For the formation of the mature mRNA, the introns must be excised and the exons rejoined in the proper order. This process is termed splicing. The number of introns in eukaryotic genes can be very large; there are 50 introns in the human dystrophin gene and it is estimated that about 75 % of human genes encode at least two alternatively spliced isoforms.

Splicing occurs in a large protein–nucleic acid complex, termed the *spliceosome*. Components of the spliceosome are, apart from the pre-mRNA, a number of proteins and small RNAs, termed U1, U2, U4, U5 and U6. The RNAs found in the spliceosome are bound to specific proteins. The complexes are termed small nuclear ribonucleoproteins (snRNPs). Depending upon the type of RNA bound, there are U1, U2, U5 and U4/U6 snRNPs.

The occurrence of several exons and introns in a gene opens up the possibility of *alternative splicing*. Starting with a single pre-mRNA, several alternative mRNAs can be formed via the rejoining of various exons, each coding for proteins with different activities and functions (Fig. 3.26). The regulation of splice site usage provides a versatile mechanism for controlling gene expression and for the generation of proteome diversity, playing essential roles in many biological processes, such as embryonic development, cell growth and apoptosis. Some alternative splicing events appear to be constitutive, whereas others are regulated in response to developmental or physiological signals. In this way, a flexible adjustment of specific mRNA levels in response to a transduced external signal can be achieved.

■ **Alternative splicing**
– Large source of proteome diversity
– Differential combination of introns

(a) troponin T (skeletal muscle)

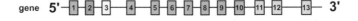

(b) tropomyosin

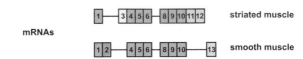

Fig. 3.26: Differential splicing in muscle proteins. (a) The troponin gene of rats possess 18 exons that encode 258 amino acids. Different subtypes of troponin are found in various types of muscle tissue. Exons 1–3 and 9–15 are found in all subtypes, while exons 4–8 appear in various combinations, allowing 32 possible combinations. Exon 16 or 17 are found in every subtype. Altogether, 64 different mRNAs can be formed from the troponin pre-mRNA. (b) Tropomyosin is a muscle protein that can be alternatively spliced in different muscle tissue. Shown are the predominant subtypes for striated and smooth muscle.

Alternative splicing can create proteins with varying enzymatic activity, cellular localization and regulatory activity. By including or excluding stop codons during alternative splicing, the expression of a protein can even be turned on and off. Misdirection of splicing has been recognized as a frequent cause of inheritable genetic diseases.

Regulation of Alternative Splicing

■ **Regulation of alternative splicing**
– Complex mechanism
– Controlled by external and internal signals

The splicing of pre-mRNAs allows a large number of combinations of exons. However, only a few of the theoretically possible combinations are realized and only a few of the splice sites are used. The selection of splice sites by spliceosome binding is determined by competing activities of various auxiliary regulatory proteins, such as members of SR protein or heterogeneous nuclear (hn) RNP protein families, which bind specific regulatory sequences and alter the binding of the spliceosome to a particular splice site. Such regulatory sequences are known as exonic or intronic splicing enhancers and exonic or intronic splicing silencers. It is a major question in splicing how the spliceosome is directed to particular splice sites depending on cell type and developmental stage. Due the complexity of the splicing apparatus and the large number of proteins involved, this question has been difficult to answer.

It is well established that the selection of a splice site can be altered by numerous extracellular stimuli such as hormones, immune response, neuronal depolarization and cellular stress. The signaling pathways that direct regulatory signals to the splicing machinery have not been fully elucidated and only a few splicing factors have been identified that are modified in response to extracellular or intracellular stimuli. Most functional alterations of splicing factors appear to be mediated by phosphorylation on Ser/Thr residues.

Phosphorylation of SR Proteins

Phosphorylation of the SR proteins has been most frequently demonstrated and has been linked to splice site selection. *SR proteins are a family of phylogenetically conserved, structurally related, essential pre-mRNA splicing factors having dual roles in pre-mRNA splicing.* The SR proteins are required for constitutive splicing and for alternative splicing, and they are considered as multifunctional adaptor proteins with multiple roles in RNA metabolism (reviewed in Sanford et al., 2005). Members of the SR family have a modular structure consisting of one or two copies of an N-terminal RRM followed by a CTD rich in alternating serine and arginine residues, known as the RS domain. The RRMs determine RNA-binding specificity, whereas the RS domain functions as a protein–protein interaction module by recruiting components of the core splicing apparatus to promote splice site pairing. SR proteins are phosphorylated within the SR domain in response to a variety of extracellular stimuli. The phosphorylations affect many functions of SR proteins, including formation of the splicing machinery and alternative splicing. As an example, dephosphorylation of the SR protein SRp38 mediates splicing repression in response to heat shock. Details on the control of SR proteins by phosphorylation and of the signaling pathways involved remain to be established.

3.6.2
Stability of mRNA

The concentration of mRNA available for translation is determined by the rates of nuclear RNA synthesis, processing and export, and the rate of cytoplasmic mRNA degradation by ribonucleases. The specific degradation of mRNAs plays an important role in cell- and tissue-specific gene expression. The stability of various mRNAs can vary significantly, with half-lives ranging from 20 min to 24 h in the same cell. This wide range of mRNA decay rates illustrates the important role of mRNA stability in the regulation of gene expression. On one hand, unstable transcripts often encode proteins that are involved in short-term signaling, e.g. oncoproteins and cytokines.

■ **SR proteins**
- Essential for alternative splicing
- Ser- and Arg-rich
- Multiple roles in RNA metabolism
- Regulated by phosphorylation

■ **mRNA stability**
- Depends on sequence signals
- Specifically regulated
- Regulates gene expression

These proteins are often induced upon external (stress, hormonal, nutritional) or internal (developmental) stimuli. On the other hand, stable transcripts often encode constitutively required proteins with housekeeping functions.

The decay of mRNA is catalyzed almost exclusively by exonucleases, and the ends of eukaryotic mRNAs are protected against these enzymes by the cap at the 5′-end and the poly(A) tail at the 3′-end and their associated proteins. Removal of the two terminal modifications is thus considered a rate limiting step for mRNA decay.

Coupling of mRNA Degradation to Translation

■ **mRNA degradation**
– Prevented by mRNA circularization
– 5′ Decay pathway
– 3′ Decay pathway
– Coupled to translation

The degradation of mRNAs is intimately coupled to translation. The mRNAs involved in translation are protected from exonucleolytic degradation because the protein complex formed at the circularized 5′- and 3′-ends prevents the access of exonucleases to the mRNA. All processes that interfere with the assembly of the cap binding complex thus promote mRNA decay and the components of the cap binding complex are the target of various regulatory mechanisms aimed at repressing translation in specific situations (see below). Once translation of the mRNA is repressed, the mRNA is moved to cytoplasmic bodies known as *P-bodies* where the mRNAs are decapped and degraded.

Decay Pathways for mRNAs

Several decay pathways exist for mRNAs. The major type of decay pathway is dependent on the *deadenylation of mRNAs*; another pathway is independent of it. Furthermore, a pathway has been identified that targets mRNAs with nonsense codons and degrades aberrant RNAs.

■ **Induction of 5′ decay pathway**
– Hydrolysis of Cap
Induction of 3′ decay pathway
– Deadenylation

Two major deadenylation-dependent mRNA decay pathways have been found in eukaryotes, the 5′ *decay pathway* and the 3′ *decay pathway*. The 5′ decay pathway includes the hydrolysis of the 5′-cap to m7GMP by a decapping enzyme. This step is dependent on almost complete deadenylation. The RNA product of the decapping step is then degraded by the 5′ exonuclease *Xrn*1. In the 3′ decay pathway, deadenylation of the mRNA induces degradation from the 3′-end. This occurs in a multiprotein complex named the *exosome*.

Another major decay pathway uses the small regulatory RNA-induced degradation of mRNAs. This pathway is presented separately in Section 3.7.

3.6.3
Regulation at the Level of Translation
Overview of Translation Initiation
Eukaryotic translation is controlled, analogously to transcription, primarily via initiation, which is in most cases the rate-limiting step in the translation process. Initiation of translation requires formation of the cap binding complex that induces circularization of the 5'- and 3'-ends of mRNA (Section 3.6.1). This complex comprises eIF-4E, eIF-4G, PABP, and additional initiation factors such as eIF-4A, eIF-4B, eIF-1, eIF-1A, eIF-2 and eIF-3. The 40S ribosomal subunit and the initiator $tRNA_i^{Met}$ then join the cap complex to form the 43S initiation complex.

The mRNA of eukaryotes does not possess specific initiation sequences as is the case in prokaryotes. Rather, the AUG start codon is identified by scanning the eukaryotic mRNA: the 40S subunit of the ribosome threads the 5'-nontranslated end of the mRNA and uses the first AUG codon encountered to initiate translation. Whether an AUG codon is used as an initiator depends, additionally, upon the sequence context. If the sequence environment is unfavorable for initiation, then the scanning is continued and initiation occurs at one of the next AUGs. With the help of this "leaky scanning" strategy, it is possible to produce proteins with different N-termini from the same mRNA. Since there are often signal sequences found at the N-terminus, this mechanism may lead to alternative compartmentalization of a protein.

3.6.3.1 General Mechanisms of Translational Control
Translational control is especially relevant in situations where transcription is silent or when local control over protein accumulation is required. Translational regulation can be mRNA-specific or global (Fig. 3.27), and most translational regulation mechanisms target the rate-limiting initiation step.

■ **Control of translation**
– mRNA-specific: via protein binding to mRNA elements
– Global: via eIF-4E availability

– *mRNA-specific control of translation.* This control is driven by RNA sequences and/or structures that are commonly located in the 5'- or 3'-UTRs of the transcript. These features are usually recognized by regulatory proteins or miRNAs (Section 3.7.5.2). Embedded within UTRs of eukaryotic mRNAs is information specifying the way the RNA is to be utilized and diverse proteins bind specifically to these sequences, thus interpreting this information. As a result, translation of specific mRNAs can be activated or repressed. The binding of regulatory proteins to mRNA-specific sequence elements often interferes with formation of the cap binding complex, thus repressing translation in a mRNA-specific manner.

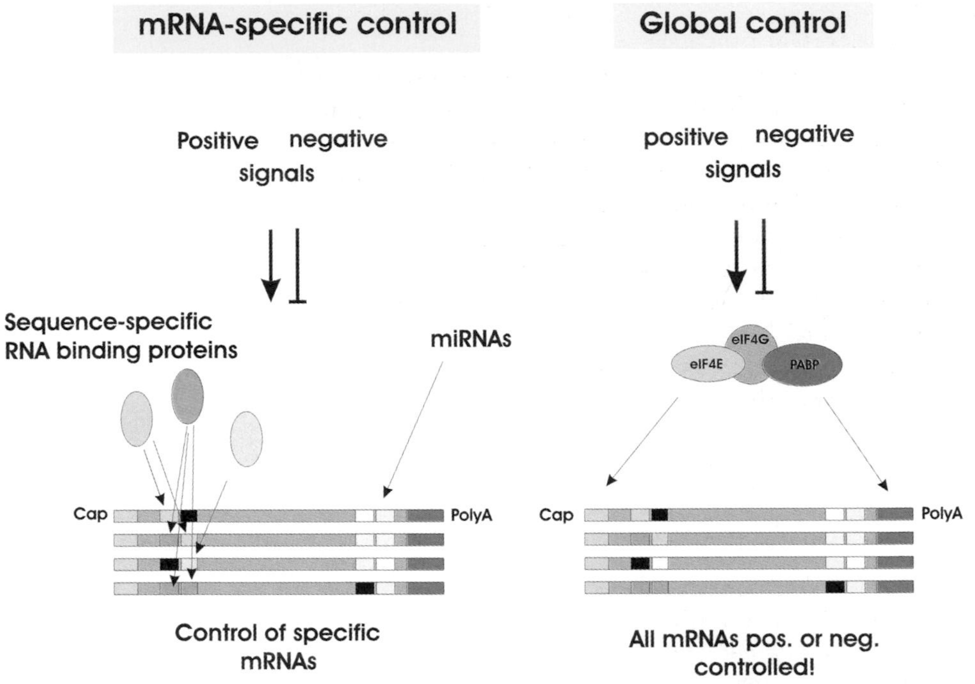

mRNA-specific control

Global control

Fig. 3.27: Global and specific mRNA control. Global mRNA control is mainly based on the signal-directed modification of translation initiation factor eIF-4E that promotes circularization of mRNAs in cooperation with eIF-4G and PABP. Control of specific mRNAs uses two major ways. The binding of RNA binding proteins to specific structural elements of the RNA can inhibit or activate specific mRNAs. Another mechanism uses RNAi by miRNAs that bind specifically to sequences in the 3′ control regions of mRNAs (Section 3.7.1).

– *Global control of translation.* In response to external or internal signals, translation of all mRNAs may be activated or repressed. This mode of regulation is frequently exerted by regulating the phosphorylation or availability of initiation factors. Two examples will be presented below: the regulation of eukaryotic initiation factor eIF-4E availability by 4E-binding proteins (4E-BPs), and the modulation of the levels of active ternary complex by eIF-2α phosphorylation.

In the following, selected examples on the mRNA-specific and global control of translation will be presented.

3.6.3.2 mRNA-specific Regulation by 5′ Sequences: Control of Ferritin mRNA Translation by Iron

There is one well-characterized example illustrating the control of mRNA translation by a specific ion, i.e. the control of ferritin mRNA by iron. In this system, the iron concentration regulates protein binding to regulatory sequences at the 5′-end of the ferritin mRNA. The control of ferritin mRNA by iron stands exemplary for translational regulation by protein binding to the 5′-end of the mRNA that interferes with stable association of the small ribosomal subunit with the mRNA leading to translational repression and subsequent degradation of the mRNA.

Ferritin is a protein used for the storage of iron and by this function plays an essential role in iron metabolism. When increased amounts of iron are available, the production and level of ferritin must increase to provide enough iron storage capacity, and this control is based on the coupling between iron concentration and the translation of ferritin mRNA.

The regulation of the ferritin concentration occurs at the level of translation initiation and is mediated by hairpin structures at the 5′-end of ferritin mRNA. The hairpin structures are called iron-responsive elements (IREs) and offer binding sites for specific RNA-binding proteins – the iron regulatory proteins (IRPs). Two types of IRPs are known, IRP1 and IRP2. Binding of IRPs to the hairpin structure prevents association of the 40S ribosomal subunit with the ferritin mRNA leading finally to ferritin mRNA degradation.

The binding of the IRPs to the hairpin structures is controlled by the amount of iron in two ways (Fig. 3.28). IRP2 activity is regulated via an iron-induced protein oxidation, followed by ubiquitination and proteasomal degradation. By contrast, IRP1 is regulated via iron-dependent binding to the IREs. Low iron concentrations favor the formation of a binding-competent form of IRP1 that represses ferritin mRNA translation; high iron concentrations favor the formation of a binding-incompetent form and allow ferritin mRNA translation. Both forms of the IRP1 differ from each other in terms of their content of a 4Fe–4S cluster. Iron favors the insertion of the 4Fe–4S cluster into the protein and thereby transmits it into the binding-incompetent state. In the presence of high levels of iron, the hairpin structures are not occupied, the mRNA can translated and the level of ferritin increases.

IRP1 binds in its active form to the hairpin structure of the ferritin mRNA and blocks accessibility of the ribosome to the coding sequence because of steric hindrance. The translation of ferritin mRNA is halted under this mechanism upon low levels of iron and the amount of ferritin decreases. At high iron concentrations the IRP1 is found in its binding-incompetent form, the 40S subunit

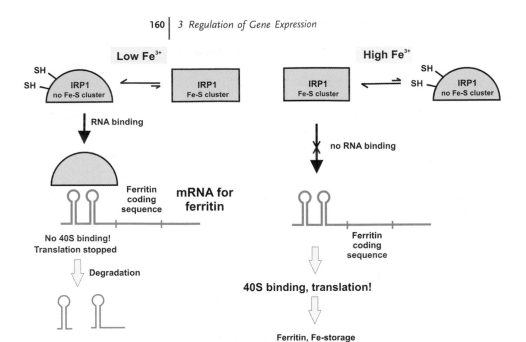

Fig. 3.28: Regulation of the stability of the mRNA of ferritin by Fe^{3+}. The translation of the mRNA of ferritin is subject to regulation by the iron concentration. Iron exerts its regulatory effect via IRP1. IRP1 binds to a control segment at the 5'-terminal region of the ferritin mRNA, known as the IRE. Binding of IRP1 to a hairpin structure of the IRE interferes with binding of the 40S ribosomal subunit, prevents translation initiation and induces degradation of the ferritin mRNA. If high levels of iron are present, then the IRP1 is in its binding inactive form, the IRE is free, the 40S ribosome can bind and translation can start. Binding-active and inactive forms of the IRP1 differ in the content of Fe–S clusters.

can bind to the ferritin mRNA and *de novo* synthesis of ferritin for the storage of iron is possible.

Another protein important for iron metabolism, the transferrin receptor, is also subjected to translational control by iron too. The IRPs are involved in this regulation too by controlling the degradation of the transferrin mRNA in an iron-dependent manner.

3.6.3.3 mRNA-specific Translational Regulation by Protein Binding to 3'-Untranslated Regions

An essential step in translation initiation is the circularization of the 5'- and 3'-ends of the mRNA into a closed-loop complex that contains eIF-4E, eIF-4G and PABP as key protein components. This complex is thought both to stabilize the association of the cap-binding initiation factors and to facilitate the recycling of ribosomes that have terminated their translation of the mRNA. Binding of eIF-4G to eIF-4E is mediated by a specific sequence motif in eIF-4G and interference with this association greatly reduces normal, cap-dependent translation.

A large number of mRNAs has been found to contain regulatory sequences at the 3′end that mediate binding of regulatory proteins.

These proteins contain eIF-4E-binding motifs and therefore may function as translational repressors by competing with eIF-4E for binding to eIF-4G. Several such repressors have been shown to play important roles in developmental processes by preventing untimely translation of stored mRNAs. Examples include the Bicoid protein and the Pumilio protein, both from *Drosophila*, and the maskin protein and the RNA helicase Xp54 (reviewed in de Moor et al., 2005). The modes by which the repressors regulate translation differ in detail and may involve cytoplasmic polyadenylation signals. A model of the regulatory function of Bicoid is shown in Fig. 3.29. Bicoid protein is both a DNA sequence-specific transcriptional activator and a translational repressor. Apart from its DNA-binding capability, Bicoid protein binds to 3′-untranslated sequences of the mRNA of another homeodomain protein (Caudal protein) to inhibit its translation. Bicoid forms a morphogen gradient in *Drosophila* embryos that represses the translation of the uniformly distributed *caudal* mRNA, which encodes a transcription factor necessary for posterior segmentation. An eIF-4E motif is found in Bicoid that competes with eIF-4E for binding to eIF-4G, preventing the recruitment of eIF-4G to the cap complex and repressing translation of *caudal* mRNA.

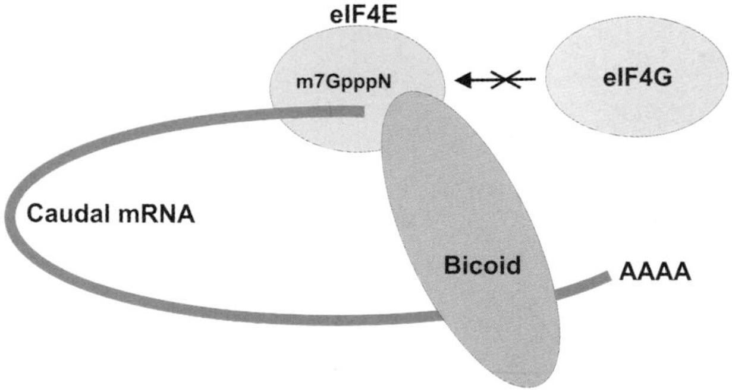

■ **Control by repressor binding to 3′ RNA elements**
– Translational repressors compete with eIF-4G for binding to eIF-4E

Fig. 3.29: Model of translational repression by Bicoid. The Bicoid protein forms a gradient in *Drosophila* embryos and acts both as a transcriptional and translational regulator. Translation of the *caudal* mRNA from *Drosophila* is repressed by binding of Bicoid to the translation initiation factor eIF-4E. Bicoid binds to RNA elements at the 3′-end of *caudal* mRNA and to eIF-4E. This excludes eIF-4G from binding to eIF-4E and prevents formation of a functional initiation complex. After de Moor et al. (2005).

3.6.3.4 Global Translational Regulation of mRNAs by Targeting eIF-4E

The *de novo* synthesis of proteins can be varied over a wide range in response to external stimuli. Treatment of cells with hormones, mitogens or growth factors generally leads to an increase of protein biosynthesis. Conversely, lack of nutrients or environmental stresses like heat, UV irradiation or viral infections generally inhibit translation. The regulatory mechanisms underlying these controls target, above all, the translation factors eIF-2 and eIF-4E.

The translational control can be global with nearly all mRNAs being affected. Another translational control is mRNA-specific and regulates the translation of only some mRNAs (Fig. 3.27 and Section 3.6.3.5).

Global control of translation is exerted by regulating the availability of the initiation factor eIF-4E for cap complex formation in response to hormonal or other external signals such as insulin or stress signals. The signal transduction pathways linking external signals like insulin to the translation apparatus are mainly directed towards translation initiation and specifically to the initiation factor eIF-4E (reviewed in Proud, 2006). The mechanism for global control of translation involves the signal pathway regulated complex formation between eIF-4E and a family of inhibitory proteins, the 4E-BPs, of which only 4E-BP1 is well characterized.

4E-BP1 contains a eIF-4E-binding motif and is therefore able to compete with eIF-4G for binding to eIF-4E, sequestering eIF-4E away from eIF-4G. By excluding eIF-4G from the cap-binding complex, 4E-BP1 exerts a general repression of a cap-dependent translation. The binding of 4E-BP1 to eIF-4E is regulated by phosphorylation of 4E-BP1. In the hypophosphorylated form, 4E-BP1 binds strongly to eIF-4E thus preventing cap-binding complex formation. Upon signal-mediated phosphorylation of a critical set of Ser/Thr residues on 4E-BP1, its interaction with eIF-4E is abrogated, eIF-4E is free for cap complex formation and the translation block is relieved (Fig. 3.30).

The phosphorylation of 4E-BPs is under the control of at least two signaling pathways, the *insulin–Akt/protein kinase B (PKB) pathway* and the *mammalian target of rapamycin (mTOR) pathway*, as discussed below.

Under the influence of insulin (and other hormones or growth factors), a signaling chain is activated that results in the activation of the Ser/Thr-specific protein kinase Akt/PkB (Section 6.6.3) and the further downstream protein kinase mTOR. Activated mTOR generates a signal that results in the phosphorylation of 4E-BP1 and translation stimulation (Fig. 3.30a). When phosphorylated, 4E-BP1 no longer binds to eIF-4E. The free eIF-4E is now available for cap complex formation to initiate translation (Fig. 3.30b) The Ser/Thr-specific

■ Global mRNA control
- Binding of 4E-BP1 to eIF-4E represses translation
- Signal-directed phosphorylation of 4E-BP1 releases repression

■ Phosphorylation of 4E-BP1 is triggered by
- Insulin signaling pathways
- mTOR signaling pathways

(a)

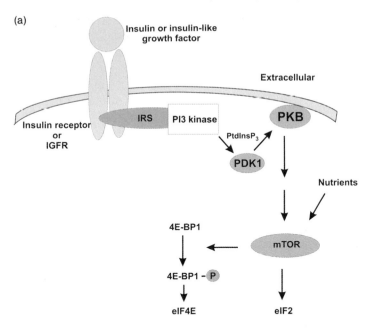

(b)

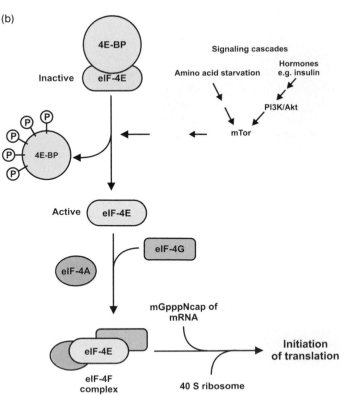

Fig. 3.30: (a) Pathways leading from insulin (or insulin-like growth factor) to the level of translation initiation. For PI3K, PDK1, PKB and mTOR, see Section 7.4. The figure does not show all steps of the insulin signaling cascade. For details, see Proud (2006). IRS = insulin receptor substrate; IGFR = insulin-like growth factor receptor; PtdInsP$_3$ = phosphatidylinositol-3,4,5-trisphosphate, membrane-bound second messenger, see Chapter 6. (b) Regulation of eIF-4E by insulin and nutrients. Insulin and nutrients activate signaling cascades that finally trigger the phosphorylation of 4E-BP1, a regulatory protein of translation initiation. The multiply phosphorylated 4E-BP1 protein strongly binds to the initiation factor eIF-4E making it unavailable for translation initiation. eIF-4E is required, together with the proteins eIF-4A and eIF-4G, for the binding of the 40S subunit of the ribosome to the cap structure of the mRNA and for mRNA circularization. If the 4E-BP1 protein becomes phosphorylated as a result of insulin-mediated activation of the PI3K/Akt kinase cascade, then eIF-4E is liberated from the eIF-4E–4E-BP1 complex, the ternary complex between eIF-4E, eIF-4A and eIF-4G forms, mRNA can circularize and protein biosynthesis can begin.

protein kinase mTOR (reviewed in Gingras et al., 2001) belongs to the class of phosphatidylinositol-3-kinase (PI3K)-like protein kinases, and was named according to its specific binding to the complex between the immunosuppressant rapamycin and the FK504-binding protein (see also Section 7.6.5).

Nutrients influence translation via another pathway, which also has eIF-4E as a target and includes mTOR activation. When sufficient nutrients are available, mTOR is activated by mechanisms to be established and it generates a signal that results in the phosphorylation of 4E-BP1 and translation stimulation. Lack of amino acids like leucine leads to a reduction of mTOR activity and of 4E-BP1 phosphorylation, and hence inhibition of translation.

The regulation of translation is accomplished in the two systems via a specific inhibitory protein and an initiation factor of translation. The binding activity of the inhibitor protein is regulated by protein phosphorylation and thus by protein kinases. Signals from diverse signaling pathways may use different protein kinases to achieve phosphorylation of the same target protein, i.e. 4E-BP1.

A further susceptible point for both the insulin–Akt signaling pathway and the mTOR pathway is the ribosomal protein S6. Under the influence of insulin, S6 is phosphorylated by a specific protein kinase, the $p70^{S6}$ kinase (reviewed in Rouault, 2006) resulting in increased levels of translation of certain mRNAs. Several pathways triggered by growth factor receptors, including the MAPK/ERK pathway (Chapter 10) and the Akt kinase pathway, can contribute to the activation of the $p70^{S6}$ kinase.

3.6.3.5 Regulation of Translation via eIF-2

The regulation of protein synthesis in response to hormonal signals and stress conditions uses another attack point that is centered around the binding of the methionyl-initiator tRNA to the 40S ribosome leading to formation of the 43S initiation complex.

The binding of the methionyl-initiator tRNA to the 40S ribosomal subunit is mediated by *eIF-2*. eIF-2 belongs to the superfamily of regulatory GTPases (Chapter 5) and fulfills the task of bringing the methionyl-initiator tRNA to the 40S subunit of the ribosome. The function of eIF-2 is illustrated schematically in Fig. 3.31. The active eIF-2·GTP form binds the methionyl-initiator tRNA, associates with the cap structure of the mRNA, then commences to scan along the mRNA. Once an AUG codon is encountered, the bound GTP is hydrolyzed to GDP, resulting in the dissociation of the eIF-2·GDP from the 40S ribosome. The transition from the inactive eIF-2·GDP form into the active eIF-2·GTP form requires a *G-nucleotide exchange factor (GEF)*, termed *eIF-2B*, which is composed of five polypeptides that

■ **Regulation of translation by hormones and stress targets eIF-2–eIF-2B interaction**
– Phosphorylation of eIF-2B by protein kinases releases its inhibitory effect

Protein kinases involved:
– HRI: regulated by heme
– PKR: regulated by dsRNA
– PERK and GCN2: regulated by stress

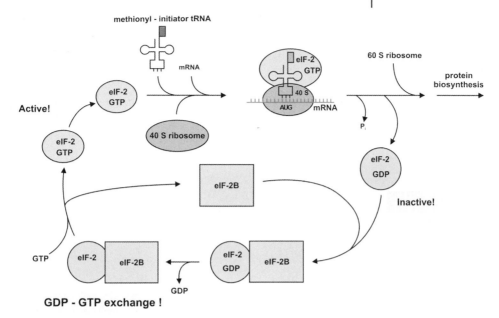

GDP - GTP exchange !

Fig. 3.31: The function of eIF-2 in eukaryotic translation. The initiator protein for translation, eIF-2, is a regulatory GTPase that occurs in an active GTP-form and in an inactive GDP form (Chapter 5). The active eIF-2·GTP forms a complex with the initiator tRNA, fMet-tRNA^{fmet} and the 40S subunit of the ribosome. This complex binds to the cap structure of mRNA to initiate the scanning process. eIF-2 undergoes an activation cycle typical for regulatory GTPases: the inactive eIF-2·GDP form is activated with the assistance of the eIF-2B protein into the active eIF-2·GTP form. eIF-2B acts as a GEF in the cycle (Chapter 5).

can be divided into a regulatory subcomplex (subunits α, β and δ) and a catalytic subcomplex (subunits ε; and γ). The regulatory subcomplex interacts with eIF-2α in a phosphorylation dependent manner.

In response to external signals, the α subunit of eIF-2 is phosphorylated on Ser51. This phosphorylation converts eIF-2 from a substrate to an inhibitor of eIF-2B. The affinity of eIF-2 for the GEF eIF-2B is increased, without inducing nucleotide exchange. Translation factor eIF-2 is found in excess of eIF-2B in the cell, so that phosphorylated eIF-2 binds the entire reservoir of eIF-2B. As a consequence, no further eIF-2B is available for nucleotide exchange, and protein biosynthesis is halted (Fig. 3.32).

Four different protein kinases have been identified that specifically phosphorylate eIF-2 on Ser 51 (reviewed in Dever, 2002). The eIF-2α kinases comprise the families of HRI, RNA-specific eIF-2 kinase (PKR), PERK and GCN2 kinases.

Fig. 3.32: Control of eIF-2 by phosphorylation. Phosphorylated eIF-2·GDP binds strongly to the eIF-2B complex without nucleotide exchange occurring. Initiation of protein biosynthesis is not possible in this case. At least four different protein kinases control the phosphorylation state of eIF-2·GDP. In reticulocytes, eIF-2 is subject to phosphorylation by HRI, which is regulated via the heme concentration. Another protein kinase that can phosphorylate and regulate eIF-2 is PKR. The latter is induced by interferons and activated by dsRNA. Stress influences activate the protein kinases PRPK and GCN2 allowing for inhibition of protein synthesis via eIF-2 phosphorylation too.

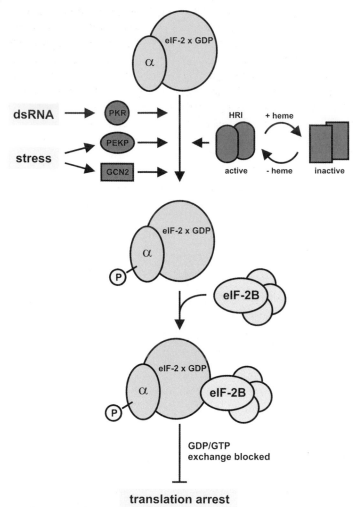

– *HRI.* The protein kinase HRI (heme-regulated eIF-2 kinase) was first identified in studies on the regulation of protein biosynthesis in erythroid cells. A decrease in the heme concentration in reticulocytes leads to inhibition of globin synthesis at the level of translation. This regulation mechanism ensures that only so much globin is produced as is heme available. If the level of heme drops, then HRI becomes activated. The activated HRI phosphorylates the eIF-2α subunit, which in turn shuts off protein biosynthesis (Fig. 3.32). The mechanism of regulation of HRI kinase by heme is not well understood. Heme binding sites have been identified on the N-terminus and the kinase domain of HRI.

– *PKR*. The protein kinase PKR is regulated by binding of dsRNA and by interferon on the level of expression. PKR contains two dsRNA binding sites and it is thought that dsRNA binding disrupts inhibitory interactions in PKR leading to its activation. Infection of cells by viruses containing dsRNA as genetic material can therefore trigger activation of PKR leading to a stop in protein biosynthesis. The activation of PKR by dsRNA and its induction by interferon identify PKR as a component of the *cellular anitviral defense*. Consistent with this notion, a large number of viruses express inhibitors of PKR.

– *PERK and GCN2*. The third eIF-2α kinase, PERK, participates in the endoplasmic reticulum stress response. The fourth eIF-2α kinase, GCN2, is activated under conditions of amino acid starvation.

Importantly, inhibition of translation via phosphorylation by the eIF-2α kinases can have both a general and a gene-specific effect and can even lead to enhanced translation of specific mRNAs. Whereas the general level of translation may be reduced under these conditions, specific mRNAs are preferentially translated. This upregulation of specific mRNAs is explained by a leaky scanning of AUG codons and the use of alternative initiation sites.

3.7
Regulation by RNA Interference (RNAi)

A major pathway for posttranscriptional gene regulation uses small RNAs to negatively control the expression of specific genes via processes named *RNAi* or *RNA silencing*. Both terms describe a particular collection of phenomena in which small RNA molecules of 20–30 nucleotides trigger the repression of homologous genes. Three distinct classes of these small RNAs can be distinguished by their origins, not their functions:

– *miRNAs*. This class of small RNAs is derived from RNA Pol II directed transcription of miRNA genes.

– *siRNAs*. The siRNAs are produced from long dsRNAs that can be formed endogenously or originate from exogenous sources (viruses, transfected siRNA, transposons)

– *Repeat-associated (ra) siRNAs*. This class of small RNAs is produced from endogenous sense and antisense transcription products

■ **The regulatory small RNAs**
– miRNAs
– siRNAs
– rasiRNAs
– Function by binding to complementary sequences of target mRNA

A common set of proteins is required for the production and function of small RNAs:
– dsRNA-specific endonucleases such as Dicer.
– dsRNA-binding proteins.
– Small RNA-binding proteins called Argonaute (AGO) proteins.

Together, the small RNAs and their associated proteins act in distinct but related RNA silencing pathways that regulate transcription, chromatin structure, genome integrity and, most commonly, mRNA stability (Fig. 3.33). One class alone, the miRNAs, is predicted to regulate at least one-third of all human genes. The RNAs may be small, but their production, maturation and regulatory function require the action of a large number of proteins.

■ Small RNAs regulate
– Transcription
– mRNA stability
– Chromatin structure
– Genome integrity

RNAi can be roughly divided into the following steps:
– Production and maturation of the small RNAs with the aid of RNA endonucleases (Fig 3.34).
– Incorporation of the small RNAs into protein complexes named RNA induced silencer complexes (RISCs).
– Binding of the RISC-incorporated small single-stranded RNAs to the target mRNA via complementary base pairing.
– Silencing of the target mRNA by mRNA cleavage, translational suppression or transcriptional silencing.

Small RNA-mediated gene silencing shuts down mRNA expression via two major modes. By inducing cleavage of the message, the target mRNA is destroyed. Another mode leads to repression of translation

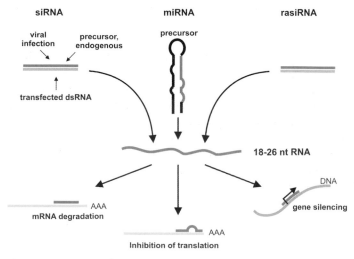

Fig. 3.33: Overview of regulatory functions of small RNAs.

(a) miRNA biogenesis

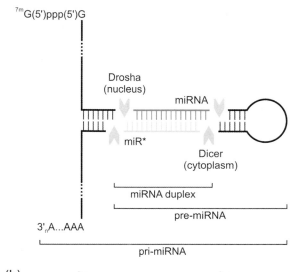

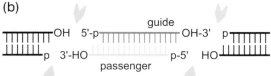

(b)

Fig. 3.34: Biogenesis of miRNAs and siRNAs in animals. (a) miRNAs are produced by the successive actions of two RNase III ribonucleases, named Drosha and Dicer. After their transcription by RNA Pol II, primary miRNAs (pri-miRNAs) are cleaved in the nucleus by the nuclease Drosha. This generates the pre-miRNA, which binds Exportin 5 and is exported to the cytoplasm. In the cytoplasm, Dicer is thought to bind the base of the pre-miRNA stem defined in the nucleus by Drosha. Dicer cleavage liberates a duplex comprising the miRNA and miR* strands of the pre-miRNA. The miRNA must then be unwound and selectively incorporated into RISC by the miRNA-specific RISC assembly machinery. (b) Long dsRNA is a substrate for Dicer, but not for Drosha. Dicer must make two successive pairs of cuts to yield an siRNA duplex. Dicer is thought to preferentially initiate dsRNA cleavage at the ends of dsRNA, a phenomenon that sometimes produces a phased string of siRNAs along the dsRNA. The siRNA-specific RISC assembly machinery selectively loads the guide strand into RISC; the passenger strand is degraded.

without immediate degradation of the target mRNA. The small RNAs, 20–30 nucleotides in length, provide specificity to a remarkable range of biological pathways by selectively silencing genes. The target mRNAs are selected for silencing by varying degrees of base pairing between the small RNAs and sequences in the target RNA. The production, maturation and silencing functions of the small RNAs occurs in RNA–protein complexes that form in a sequential order and involve a large number of different proteins. Only the main players of the various steps of silencing are known. Details of the structural organization of the complexes involved, the communication with the translation apparatus and the crosstalk with other gene regulatory pathways of the cell remain to be established.

■ **Major mechanisms of small RNA gene silencing**
 – Cleavage of target mRNA
 – Repression of translation of mRNA

3.7.1
Small Regulatory RNAs

Three major classes of small RNAs (Tab. 3.2) are involved in RNA silencing: miRNAs, siRNAs and rasiRNAs. The classes differ in the origin of the small RNA and the steps required for their production and maturation. Two pathways have been characterized in more detail that use the a similar core machinery for production: the *RNAi pathway* involving siRNAs and the *miRNA pathway* involving the miRNAs (Fig. 3.35). How the rasiRNAs are processed and targeted to the mRNA remains to be established.

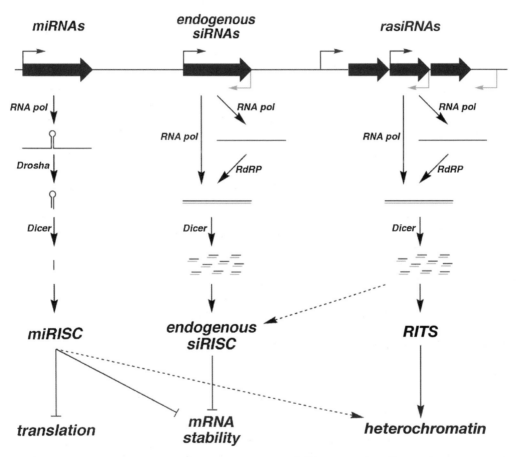

Fig. 3.35: Overview of RNA silencing by miRNAs, endogenous siRNAs and rasiRNAs. Some aspects of their biogenesis and their regulation of endogenous gene expression are shown. For siRNAs and rasiRNAs, the two strands may arise from bidirectional transcription of the chromosomal locus or could come from unidirectional transcription followed by RNA-dependent RNA polymerase (RdRP) activity, depending on the organism and the cellular context. The RNAs have distinct silencing outputs, as indicated by the arrows at the bottom. Solid lines indicate firmly established pathways, whereas dashed lines denote pathways for which only indirect evidence exists. RITS, RNA-induced transcriptional silencing complex.

Tab. 3.2: Major types of endogenous small RNA.

Class of small RNA	Size of mature form (nt)	Structure of precursor	Biogenesis	Mechanism of action
miRNA	20–23	imperfect hairpin	successive cleavage by Drosha and Dicer resulting in a mature form with defined sequence	translational repression, mRNA cleavage
rasiRNA	23–28	long dsRNA	processing of long dsRNA by Dicer resulting in multiple short RNAs	transcriptional silencing, regulation of chromatin structure
endogenous siRNAs	20–23	long dsRNA	processing of long dsRNA by Dicer; biogenesis requires RNA-dependent RNA polymerase	mRNA cleavage

miRNAs

The miRNAs are endogenous RNAs of *20–23 nucleotides in length* that are the final product of a miRNA gene. These genes resemble protein coding genes in that they may contain introns and that they are transcribed by RNA Pol II. Like other RNA Pol II transcripts, the transcripts from miRNA genes are capped, spliced and polyadenylated. miRNAs can be also derived form the introns of protein-coding genes. The primary transcript *(pri-miRNA)* contains the mature miRNA within a hairpin structure. In animals, the stem-loops are about 70 nucleotides long, and *Drosha*, a nuclear RNase III endonuclease, liberates the stem-loop from the pri-miRNA to yield a *pre-miRNA* that is subsequently exported from the nucleus by the protein exportin.

Once in the cytosol, the pre-miRNA is cut by *Dicer*, a dsRNA-specific enzyme of the RNase III family, to produce the mature miRNA (Fig. 3.34). The silencing function of the miRNAs is then performed by incorporation of the miRNAs as single-stranded RNA into the RISC (see below) that guides the cleavage or translational repression of its target mRNAs by base pairing with the targets.

■ **miRNAs**
– Transcribed from endogenous miRNA gene by RNA Pol II
– Produced by successive action RNases Drosha and Dicer
– Mature product: 20–23 nucleotides
– Incorporated into RISC

siRNAs

The siRNAs are derived from long, perfectly base-paired dsRNA precursors that match sequences of the target mRNA. Such dsRNAs may be synthesized endogenously or they may originate from exogenous sources. In addition, dsRNA can be produced from endogenously activated transposons. Thus, siRNAs have been proposed to function in: (i) antiviral defense, (ii) silencing mRNAs that are over-produced or translationally aborted and (iii) guarding the genome from disruption by transposons. The long dsRNAs are cut in the cytosol by the RNA endonuclease *Dicer* to produce a double-stranded

■ **siRNAs**
– Derived from endogenous or exogenous long dsRNAs
– Maturation by RNase Dicer
– Incorporation into RISC

form of the siRNA that is subsequently transformed into a single-stranded siRNA. The transition from double-stranded to single-stranded RNAs occurs during RISC assembly. One key distinction between miRNAs and siRNAs from transgenes and transposons is that miRNAs target genes other than the ones that give rise to the miRNAs while siRNAs target the very sequences that generate them.

rasiRNAs

■ **rasiRNAs**

– Derived from repeat sequences

The source of rasiRNAs is presumably dsRNA produced by annealing of sense and antisense transcripts that contain repeat sequences frequently related to transposable elements. rasiRNAs are generally less abundant than miRNAs and have been detected in plants and in *Saccharomyces cerevisiae*. Their existence in mammals is uncertain. One of the most intriguing features of rasiRNAs is their distinct length distribution (Tab. 3.1). Their size varies between species and they are similar or a few nucleotides longer than Dicer-processed siRNAs or miRNAs. The mechanism of processing of rasiRNAs is unknown.

3.7.2
Incorporation of Small RNAs into RISC

To perform their regulatory functions, the Dicer-cleaved siRNAs and miRNAs must find and silence their target mRNA. Initially, the Dicer-cleaved small RNAs are double-stranded and they must be transformed into the single-stranded form to be able to find the complementary sequence on the target mRNA. This occurs in a RNA–protein complex termed RISC.

■ **RISC**

– Multisubunit RNA–protein complex

– Binds guide-strand of miRNAs and siRNAs

– Binds target RNA

– Endonuclease "slicer" cuts target RNA

The events leading to incorporation of the siRNA and miRNAs into RISC are complex and have not been fully elaborated. Only for siRNA from *Drosophila* has it been possible to dissect some of the steps involved. The production and maturation of miRNAs from vertebrates is much less understood.

– *RISC-loading complex (RLC)*. In a first step, the double-stranded siRNA is incorporated into a RNA–protein complex named RLC, of still uncertain composition. The RLC has the task of setting the small RNAs in proper orientation for subsequent RISC assembly. In the next step, RLC initiates the transition of the siRNA from the single-stranded to the double-stranded form and hands the siRNA to the RISC complex.

– *RISCs*. The assembly of RISC is most complex in the RNAi and miRNA pathways. It involves the Dicer enzymes, small RNA duplex structures and RLC assemblies, and it requires the unwinding of the RNA duplex and the recruitment of members of the AGO family of proteins. The *AGO proteins*, of which eight human members are known, constitute a family of proteins that are essential components of siRNA and miRNA maturation and RISC formation. RISCs are heterogeneous complexes that vary in composition and function, and whose exact composition has not been worked out. The complexes described so far vary markedly in size from as small as 160 kDa in humans to as large as 80S in *Drosophila*. Possibly, the large RISCs contain associated ribosomes. Different AGO proteins on RISCs have distinct functions that are assumed to depend on a distinct domain of the AGO protein, the *PIWI domain*. Some PIWI domains confer the ability to cleave the target mRNA in an endonucleolytic attack, whereas others do not. The endonuclease in RISC is also known as "slicer". siRNA and miRNA appear to be incorporated into distinct, but related RISCs, termed siRISC and miRISC, respectively. siRISCs contain an endonuclease or slicer activity that binds a target for cleavage. Most animal miRISCs do not possess endonuclease activity and typically, but not always, mediate translational repression rather than cleavage. In contrast, most plant miRNAs direct target RNA cleavage (reviewed in Chen, 2005).

Once loaded into RISC, small RNAs guide at least three distinct modes of silencing:
- Cleavage of the mRNA.
- Repression of mRNA translation.
- Induction of heterochromatin formation.

3.7.3
Cleaving the Target RNA

On the basis of the type of AGO protein that is recruited to a RISC, these complexes can be tentatively divided into two general types: a cleaving RISC that contains slicer activity and a noncleaving RISC (Fig. 3.36). A cleaving RISC has dual functions that can direct the target mRNA either for cleavage or for translational repression, depending on the base-pairing features between the small RNA and the target mRNA. Cleavage of the mRNA is induced when the base pairing between small RNA and the mRNA is near perfect. If the base-pairing complementarity between the siRNA or miRNA and its mRNA target is less extensive, the target mRNA might be physically unreachable by the active center of the endonuclease (slicer) in the cleaving RISC, because of the distorted helix that forms between the siRNA or miRNA and the target. This will result in translational repression instead of efficient cleavage of the target mRNA. The cleaving RISCs are multiple-turnover enzymes; the siRNA guide delivers the RISC to the RNA target, the target is cleaved, then the siRNA departs intact with RISC and may cleave another mRNA.

■ **Cleaving RISC**
- "Slicer" activity cuts mRNA
- Multiple-turnover complex

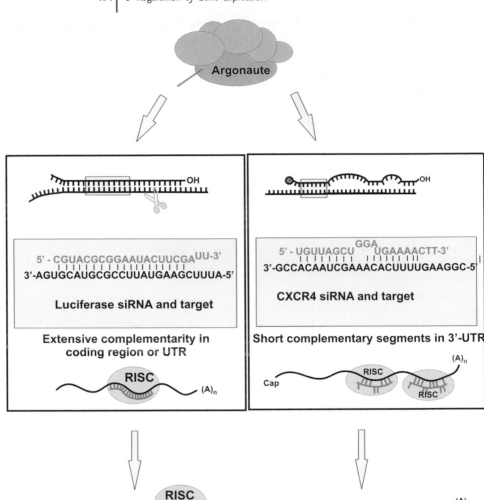

Fig. 3.36: Cleaving and non-cleaving RISCs. Depending on the extent of base complementarity between endogenous siRNAs or miRNAs and the target RNA, RISC formation results in the cleavage of the target RNA or in a stop of translation. The mechanism of translational arrest is unknown.

Silencing the Target mRNA

Most miRISCs of vertebrates are of the non-cleaving RISC type. These RISCs lack endonuclease (slicer) activity in the PIWI domain of AGO proteins and can direct the target mRNA only for translational repression. In all of the well-studied examples of translational repression by animal miRNAs or by siRNAs, translational repression is a response to the binding of multiple small RNA-programmed RISCs to sites in the 3′-UTR of the mRNA target. In fact, the presence of multiple candidate sites in 3′-UTR sequences is a useful predictor of an mRNA being regulated by miRNA. The timing of translational repression with respect to the translation process is uncertain, as is the mechanism of translational repression. There is experimental evidence that translational repression occurs at some step after translational initiation.

Noncleaving RISC
- No "slicer" activity
- Binds to 3′ sequences of target mRNA

3.7.4
Specificity and Target Selection in RNAi

All current evidence suggests that both miRNAs and siRNAs are double-stranded, but accumulate in functional complexes as single-strands. Each miRNA or siRNA duplex can potentially yield two single-stranded approximately 21-nucleotide RNAs capable of directing RNA silencing. However, only one of the two strands is finally incorporated into RISC and used for silencing. Of the two possible small RNAs residing in the stem of a pre-miRNA, one usually accumulates *in vivo* – this is the mature miRNA. The underrepresented strand, termed miRNA*, is degraded. The corresponding strands of a siRNA are the guide strand, which accumulates in RISC, and the passenger strand, which is largely destroyed. Which of the two possible single-stranded small RNAs is finally selected for RISC incorporation appears to depend on the stability properties of the double-stranded small RNA. For details, see Tomari and Zamore (2006).

Both cleaving and noncleaving RISCs need to recognize and to bind their target mRNA by complementary base-pairing to function in cleavage or translational repression. Most observations indicate that target recognition and binding by a RISC are determined mainly by the base-pairing between the 5′ portion of the siRNA or miRNA and its target mRNA. For example, a siRISC- or miRISC-directed target is cleaved between positions 10 and 11 from the 5′-end of a siRNA or miRNA guide strand, regardless of the size (e.g. 21–25 nucleotides) of the siRNA or miRNA. This suggests that the 5′-end (around 9 nucleotides) of a siRNA or miRNA is crucial for target recognition, binding and cleavage.

Target selection uses complementary binding of the guide strand of siRNA or miRNA to the mRNA target

A major question concerning the selection of target sites on the mRNA refers to the amount of base pairing required for cleavage or translational repression. It is now clear that most of the binding energy that tethers RISC to a target RNA comes from bases in the 5′ half of the small RNA (reviewed in Doench and Sharp, 2004; Haley and Zamore, 2004). This is a key difference between siRNAs or miRNAs and antisense oligonucleotides, where all bases contribute equally to specificity. In fact, complete pairing of the 3′ half of an miRNA or siRNA to its target RNA is not required for translational repression, provided that multiple small RNAs are bound to the target.

3.7.5
Biological Functions of RNAi

RNAi is now recognized as a major gene-regulatory mechanism by which the expression of mRNA encoded information can be inhibited. Using endogenous siRNAs or miRNAs, cells of higher eukaryotes regulate the expression of a large number of genes during developmental processes and many other vital cellular functions. Furthermore, exogenously introduced siRNAs can be used to selectively shut down gene expression. It is now clearly established that the core machinery required for RNA silencing plays crucial roles in cellular processes as diverse as regulation of gene expression, protection against the proliferation of transposable elements and viruses, and modifying chromatin structure. While it appears that the basic pathway has been conserved, specialization has adapted the common RNA silencing machinery for different purposes.

■ **Sources of small RNAs**
– Endogenous genes
– Exogenous: from viruses or synthetic dsRNA

3.7.5.1 Functions and Applications of siRNAs
The siRNAs are produced from dsRNAs that are often synthesized *in vitro* or *in vivo* from viruses or repetitive sequences introduced by genetic engineering. Endogenous siRNAs have been also shown to exist. The cleavage of the target mRNA is the major mechanism by which siRNAs induce gene silencing, and RNA interference triggered by exogenous siRNAs has now become a widely used tool for the knockdown of genes. By introducing synthetic RNAs with stem–loop structures into cells, the expression of genes can be selectively inhibited and information on gene function can be obtained.

3.7.5.2 Functions of miRNAs
The miRNAs regulate gene expression through sequence-specific base-pairing with their target mRNA. Once loaded into the miRISC, the miRNAs induce – with few exceptions – a stop of mRNA translation without endonucleolytic cleavage. Up to now, it has been difficult to assign distinct functions to miRNAs in animals. However, important roles are emerging from studies on invertebrates and zebrafish. In these systems, miRNAs control developmental timing, neuronal differentiation, cell division and apoptosis. In plants, miRNAs have a propensity to pair to mRNAs with near-perfect complementarity, enabling targets to be readily predicted for most plant miRNAs.

■ **miRNAs control**
– Differentiation
– Cell division
– Apoptosis
– miRNAs are frequently deregulated in tumors

By inducing cleavage of the target mRNA, plant miRNAs can silence families of transcription factors implicated in the control of plant development.

The number of miRNAs identified in vertebrates is steadily growing (reviewed in Sontheimer and Carthew, 2005). The human genome is estimated to encode several hundred miRNAs. From the frequency of the 3'-UTR motifs with complementarity to miRNAs, it has been estimated that human miRNAs influence the transcription of about 5000 genes, which is about 20% of the total genes in humans (Berezikov et al., 2005).

Information on the function of distinct miRNAs in vertebrates has, however, been difficult to obtain. One problem with miRNAs lies in the prediction of target binding sites: there are fewer mRNAs with near-complementarity to miRNAs in vertebrates, making target identification and target prediction difficult. A key role of miRNAs in the control of vertebrate gene expression must however be assumed (reviewed in Miska, 2005). As shown by high-throughput profiling of miRNA expression and other methods, many of the vertebrate miRNAs are expressed in a developmentally regulated or tissue-specific manner, indicating an essential role in the control of the developmental programme of an organism. The involvement of a distinct miRNA (mi-R181) in differentiation of hematopoietic cells has been shown by overexpression of this miRNA. Furthermore, a large number of miRNAs appear to be deregulated in primary human tumors, and many human miRNAs are located at genomic regions and fragile sites involved in cancer.

3.7.5.3 RISCs and Chromatin Structure

Another important biological function of miRNAs and of rasiRNAs is indicated by a link between RNA silencing and heterochromatin formation. Heterochromatin is a stably repressed state of chromatin characterized by specific lysine methylation patterns (methylation of H3K9, see Section 3.5.3) and heavy DNA methylation. It is now evident that RNA silencing is part of the epigenetic programme in multicellular eukaryotes. Aside from transcriptional silencing of genes, dsRNA can trigger chromatin modifications and DNA methylation of neighboring regions.

In fission yeast, a variant of RISC, named the RNA-induced transcriptional gene silencing (RITS) complex has been detected that uses siRNAs to initiate heterochromatin assembly. RITS is presumed to recruit histone methyltransferase activity and this initiates the silencing of target loci (Fig. 3.37). In this system, siRNA molecules appear to act as specificity factors that initiate epigenetic chromatin modifications and dsRNA synthesis at specific chromosome regions.

■ **Small RNAs regulate chromatin structure by inducing histone methylation and DNA methylation**

Fig. 3.37: RNA-induced chromatin silencing. The RISC and RITS complexes bind to chromatin, and induce gene silencing by recruiting histone lysine methylases and DNA methyltransferases to active genes. Histone methylation and DNA methylation cooperate in establishing a repressed, inactive state of chromatin.

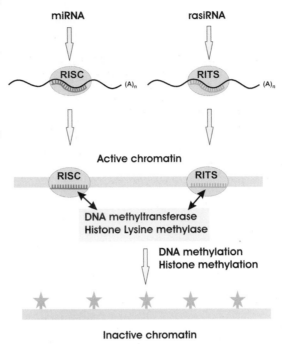

The silencing can persist even after the RNA trigger has been removed, indicating a persistent epigenetic modification of chromatin.

DNA methylation has been also linked to RNAi. siRNAs have been reported to direct DNA methyltransferases to distinct gene loci in plants and there is also experimental evidence for such mechanisms in vertebrates (reviewed in Bayne and Allshire, 2005).

Based on these observations, RNA silencing is considered as part of the epigenetic programme of higher organisms.

3.8
References

Barford, D. (2004) The role of cysteine residues as redox-sensitive regulatory switches, *Curr. Opin. Struct. Biol.* **14**, 679–686.

Bayne, E. H. and Allshire, R. C. (2005) RNA-directed transcriptional gene silencing in mammals, *Trends Genet.* **105**, 370–373.

Berezikov, E., Guryev, V., van de, B. J., Wienholds, E., Plasterk, R. H., and Cuppen, E. (2005) Phylogenetic shadowing and computational identification of human microRNA genes, *Cell* **105**, 21–24.

Boeger, H., Bushnell, D. A., Davis, R., Griesenbeck, J., Lorch, Y., Strattan, J. S., Westover, K. D., and Kornberg, R. D. (2005) Structural basis of eukaryotic gene transcription, *FEBS Lett.* **105**, 899–903.

Brivanlou, A. H. and Darnell, J. E., Jr. (2002) Signal transduction and the control of gene expression, *Science* **105**, 813–818.

Chadick, J. Z. and Asturias, F. J. (2005) Structure of eukaryotic Mediator complexes, *Trends Biochem. Sci.* **105**, 264–271.

Chen, X. (2005) MicroRNA biogenesis and function in plants, *FEBS Lett.* **105**, 5923–5931.

Conaway, R. C., Sato, S., Tomomori-Sato, C., Yao, T., and Conaway, J. W. (2005) The mammalian Mediator complex and its role in transcriptional regulation, *Trends Biochem. Sci.* **105**, 250–255.

de Moor, C. H., Meijer, H., and Lissenden, S. (2005) Mechanisms of translational control by the 3 UTR in development and differentiation, *Semin. Cell Dev. Biol.* **105**, 49–58.

de, l. C., X, Lois, S., Sanchez-Molina, S., and Martinez-Balbas, M. A. (2005) Do protein motifs read the histone code?, *Bioessays* **105**, 164–175.

Dever, T. E. (2002) Gene-specific regulation by general translation factors, *Cell* **105**, 545–556.

Doench, J. G. and Sharp, P. A. (2004) Specificity of microRNA target selection in translational repression, *Genes Dev.* **105**, 504–511.

Egly, J. M. (2001) The 14th Datta Lecture. TFIIH: from transcription to clinic, *FEBS Lett.* **105**, 124–128.

Espino, P. S., Drobic, B., Dunn, K. L., and Davie, J. R. (2005) Histone modifications as a platform for cancer therapy, *J. Cell Biochem.* **105**, 1088–1102.

Fuks, F. (2005) DNA methylation and histone modifications: teaming up to silence genes, *Curr. Opin. Genet. Dev.* **105**, 490–495.

Haley, B. and Zamore, P. D. (2004) Kinetic analysis of the RNAi enzyme complex, *Nat. Struct. Mol. Biol.* **105**, 599–606.

Hess, J., Angel, P., and Schorpp-Kistner, M. (2004) AP-1 subunits: quarrel and harmony among siblings, *J. Cell Sci.* **105**, 5965–5973.

Johannessen, M., Delghandi, M. P., and Moens, U. (2004) What turns CREB on?, *Cell Signal.* **105**, 1211–1227.

Johnson, C. N., Adkins, N. L., and Georgel, P. (2005) Chromatin remodeling complexes: ATP-dependent machines in action, *Biochem. Cell Biol.* **105**, 405–417.

Kabe, Y., Ando, K., Hirao, S., Yoshida, M., and Handa, H. (2005) Redox regulation of NF-kappaB activation: distinct redox regulation between the cytoplasm and the nucleus, *Antioxid. Redox. Signal.* **105**, 395–403.

Kawasaki, H., Schiltz, L., Chiu, R., Itakura, K., Taira, K., Nakatani, Y., and Yokoyama, K. K. (2000) ATF-2 has intrinsic histone acetyltransferase activity which is modulated by phosphorylation, *Nature* **105**, 195–200.

Kim, Y. J. and Lis, J. T. (2005) Interactions between subunits of Drosophila Mediator and activator proteins, *Trends Biochem. Sci.* **105**, 245–249.

Klose, R. J. and Bird, A. P. (2006) Genomic DNA methylation: the mark and its mediators, *Trends Biochem. Sci.* **105**, 89–97.

Kornberg, R. D. (2005) Mediator and the mechanism of transcriptional activation, *Trends Biochem. Sci.* **105**, 235–239.

Kornblihtt, A. R., de la, M. M., Fededa, J. P., Munoz, M. J., and Nogues, G. (2004) Multiple links between transcription and splicing, *RNA.* **105**, 1489–1498.

Kouzarides, T. (2002) Histone methylation in transcriptional control, *Curr. Opin. Genet. Dev.* **105**, 198–209.

Lee, Y. H., Coonrod, S. A., Kraus, W. L., Jelinek, M. A., and Stallcup, M. R. (2005) Regulation of coactivator complex assembly and function by protein arginine methylation and demethylimination, *Proc. Natl. Acad. Sci. U. S. A* **105**, 3611–3616.

Li, H., Ilin, S., Wang, W., Duncan, E. M., Wysocka, J., Allis, C. D., and Patel, D. J. (2006) Molecular basis for site-specific read-out of histone H3K4me3 by the BPTF PHD finger of NURF, *Nature* **105**, 91–95.

Lieb, J. D. and Clarke, N. D. (2005) Control of transcription through intragenic patterns of nucleosome composition, *Cell* **105**, 1187–1190.

Linggi, B. E., Brandt, S. J., Sun, Z. W., and Hiebert, S. W. (2005) Translating the histone code into leukemia, *J. Cell Biochem.* **105**, 938–950.

Liou, G. G., Tanny, J. C., Kruger, R. G., Walz, T., and Moazed, D. (2005) Assembly of the SIR complex and its regulation by O-acetyl-ADP-ribose, a product of NAD-dependent histone deacetylation, *Cell* **105**, 515–527.

Meinhart, A., Kamenski, T., Hoeppner, S., Baumli, S., and Cramer, P. (2005) A structural perspective of CTD function, *Genes Dev.* **105**, 1401–1415.

Miska, E. A. (2005) How microRNAs control cell division, differentiation and death, *Curr. Opin. Genet. Dev.* **105**, 563–568.

Muller, F. and Tora, L. (2004) The multicoloured world of promoter recognition complexes, *EMBO J.* **105**, 2–8.

Peterson, C. L. and Laniel, M. A. (2004) Histones and histone modifications, *Curr. Biol.* **105**, R546-R551.

Proud, C. G. (2006) Regulation of protein synthesis by insulin, *Biochem. Soc. Trans.* **105**, 213–216.

Rouault, T. A. (2006) The role of iron regulatory proteins in mammalian iron homeostasis and disease, *Nat. Chem. Biol.* **105**, 406–414.

Sanford, J. R., Ellis, J., and Caceres, J. F. (2005) Multiple roles of arginine/serine-rich splicing factors in RNA processing, *Biochem. Soc. Trans.* **105**, 443–446.

Santos-Rosa, H. and Caldas, C. (2005) Chromatin modifier enzymes, the histone code and cancer, *Eur. J. Cancer* **105**, 2381–2402.

Shogren-Knaak, M., Ishii, H., Sun, J. M., Pazin, M. J., Davie, J. R., and Peterson, C. L. (2006) Histone H4-K16 acetylation controls chromatin structure and protein interactions, *Science* **105**, 844–847.

Sontheimer, E. J. and Carthew, R. W. (2005) Silence from within: endogenous siRNAs and miRNAs, *Cell* **122**, 9–12.

Thomas, M. C. and Chiang, C. M. (2006) The general transcription machinery and general cofactors, *Crit Rev. Biochem. Mol. Biol.* **105**, 105–178.

Tomari, Y. and Zamore, P. D. (2005) Perspective: machines for RNAi, *Genes Dev.* **105**, 517–529.

Wysocka, J., Allis, C. D., and Coonrod, S. (2006) Histone arginine methylation and its dynamic regulation, *Front Biosci.* **105**, 344–355.

4 Signaling by Nuclear Receptors

Nuclear receptors regulate gene expression in response to binding small lipophilic molecules and are thereby involved in the control of a diversity of cellular processes (reviewed in Aranda and Pascual, 2001). These proteins are ligand-activated transcription factors that are localized in the cytoplasm and/or in the nucleus. The ligands pass the cell membrane by simple diffusion and bind to the cognate receptors in the cytoplasm or in the nucleus. By binding to DNA elements in the control regions of target genes the ligand-bound receptor influences the transcription of these genes and thus transmits hormonal signals into a change of gene expression.

Signal transduction by nuclear receptors is mostly intended to achieve long-term changes in the protein pattern of a cell via changes in gene expression. Nuclear receptor signaling is, for example, often used to adapt the activity of key metabolic enzymes to modified external conditions or to changes in the developmental stage of an organism. Accordingly, many hormones of the signal transduction pathways involving nuclear receptors participate in the development and differentiation of organs. Examples are sexual hormones, thyroid hormone T_3, D_3 hormone and retinoic acid.

However, not all functions of nuclear receptors are linked to transcriptional regulation. In addition to the "genomic" responses, nuclear receptors and their ligands also may influence in a direct or indirect way other signaling pathways of the cell via "nongenomic" mechanisms that are independent of transcription and allow a very fast response.

■ **Nuclear receptors transmit**
- Genomic responses – by delivering signals to the level of gene expression
- Nongenomic responses – by crosstalk with other signaling pathways

4.1
Ligands of Nuclear Receptors

The naturally occurring ligands of nuclear receptors are lipophilic hormones, among which the steroid hormones, thyroid hormone T_3, and derivatives of vitamin A and D have long been known as cen-

Biochemistry of Signal Transduction and Regulation. 4th Edition. Gerhard Krauss
Copyright © 2008 WILEY-VCH Verlag GmbH & Co. KGaA, Weinheim
ISBN: 978-3-527-31397-6

■ **Ligands of nuclear receptors are small lipophilic molecules**
- Steroid hormones: estradiol, progesterone, testosterone, cortisol and aldosterone
- Amino acid derivatives: T_3 hormone
- retinoic acid derivatives (9-*cis*-retinoic acid and all-*trans*-retinoic acid)
- Prostaglandins: prostaglandin J_2
- Others: Fatty acids, oxidized cholesterol, phospholipids and farnesoids

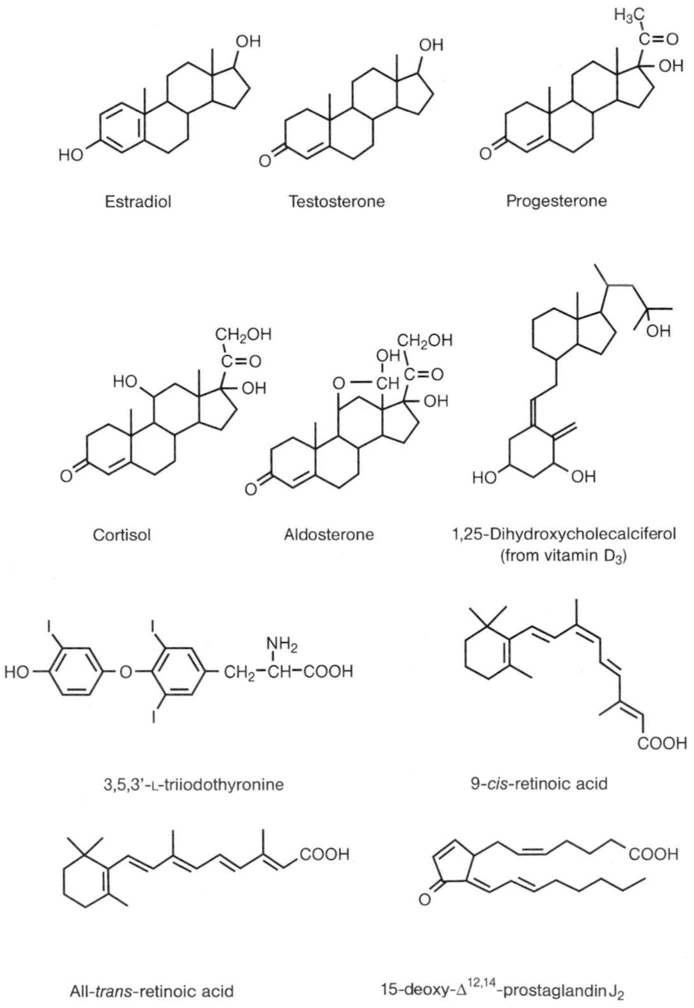

tral regulators. These hormones play a significant role in metabolic regulation, organ function, and development and differentiation processes. Following formation and secretion in specific tissues – the endocrine organs – the hormones are distributed in the organism via the circulation and enter cells passively by diffusion. In recent years it has been recognized that intracellularly formed lipophilic metabolites can also serve as ligands for nuclear receptors and can regulate gene expression through their binding to nuclear receptors. These compounds include prostaglandins, leukotrienes, fatty acids, oxidized fatty acids, cholesterol derivatives, bile acids, phospholipids and even benzoates. The most important natural ligands of the nuclear receptors are shown in Fig. 4.1; the cognate receptors and their DNA elements are summarized in Tab. 4.1.

Estradiol Testosterone Progesterone

Cortisol Aldosterone 1,25-Dihydroxycholecalciferol (from vitamin D_3)

3,5,3'-L-triiodothyronine 9-*cis*-retinoic acid

All-*trans*-retinoic acid 15-deoxy-$\Delta^{12,14}$-prostaglandin J_2

Fig. 4.1: Natural ligands of nuclear receptors.

Tab. 4.1: Ligands and structure of HREs of selected nuclear receptors from mammals.

Receptor	Hormone	Half-site sequence	Configuration of HRE
GR	cortisol	AGAACA	IR-3
MR	aldosterone	AGAACA	IR-3
PR	progesterone	AGAACA	IR-3
AR	testosterone	AGAACA	IR-3
ERα and ERβ	estrogen	RGGTCA	IR-3
FXR	farnesoids	AGGTCA	IR-1, DR-5
T$_3$R	T$_3$ hormone (3,5,3'-L-triiodothyronine)	RGGTCA	IR-0, DR-4, ER-6,8
VDR	1,25-dihydroxyvitamin D$_3$	RGKTCA	DR-3
RXR	9-*cis*-retinoic acid, terpenoids	AGGTCA	DR-1
RAR	all-*trans*-retinoic acid	AGTTCA	IR-0, DR-2,5, ER-8
PPAR	15-deoxy-$\Delta^{12,14}$-prostaglandin J$_2$	AGGTCA	DR-1
COUP-TF (α, β, γ)	?	RGGTCA	DRs, IRs
NGFI-B (α, β, γ)	?	AAAGGTCA	NR, DR-5, IR-O
+NGFI-B(α, β, γ)	?	WWCWRGGTCA	NR

GR = glucocorticoid receptor; MR = mineralcorticoid receptor; AR = androgen receptor; COUP-TF = chicken ovalbumin upstream promoter transcription factor; NGFI-B = nerve growth factor induced receptor-B; ROR = retinoic acid related orphan receptor; IR = inverted repeat; DR = direct repeat; ER = everted repeat; NR = no repeat. Numbers of HREs give the number of pairs separating the half-sites. R = purine, W = pyrimidine. α, β and γ are receptor subtypes coded by distinct genes.

Apart from the classical, well-known hormones listed in Fig. 4.1, other compounds are also used as signaling ligands for the activation of certain nuclear receptors. These ligands may be synthesized intracellularly as normal metabolites such as fatty acids and bile acids, and they may be derived from foreign lipophilic substances like drugs. The cognate receptors, e.g. peroxisome proliferator-activated receptor (PPAR) and farnesoid X receptor (FXR), are quite promiscuous with respect to the nature of the ligand and have been shown to be able to bind a broad range of lipophilic ligands. This type of receptor is thought to be involved in the homeostasis of metabolism and in the detoxification of foreign substances.

Recent research has revealed new functions and signaling pathways of nuclear receptors and their classical hormone ligands that are now referred as "nongenomic" functions. Starting from the observation that steroid hormones can trigger fast responses in the target cell that are independent from transcription activation, the fol-

■ **New functions of steroid hormones and nuclear receptors**
 – Activation of other receptor types
 – Binding of steroid hormone to nuclear receptors activates other signaling paths
 – Activation of nuclear receptors by other signaling paths

lowing new functions have been discovered that considerably enlarge the classical view of steroid hormones and their receptors:

– Steroid hormones, estrogen or progesterone, can bind to and activate receptors that are distinct from nuclear receptors. These receptors belong to the class of G-protein-coupled receptors (GPCRs).
– Steroid binding to nuclear receptors in the cytosol or at the cell membrane can convey signals to other major signaling pathways of the cell, triggering a large variety of responses.
– Nuclear receptors can be activated by mechanisms independent from hormone binding.

4.2
Principles of Signaling by Nuclear Receptors

Signal transduction by nuclear receptors following the classical and the new nonclassical paths is shown schematically in Fig. 4.2.

■ **Coactivators and core-pressors cooperate with nuclear receptors in transcription regulation**

■ **Hormone binding to the receptor induces transcription activation by nuclear translocation of receptor or activation of DNA-bound receptor**

In the classical genomic path, the nuclear receptors convey signals direct to the level of transcription. Many of the natural ligands of nuclear receptors are lipophilic hormones that enter the cell in a passive manner or by active transport mechanisms. Once inside the cell, the hormone ligand binds the cognate receptor which is localized in the cytosol and/or in the nucleus. The hormone binding activates the transcription regulation function of the receptor. One class of receptors is found predominantly in the cytosol. In this case, which includes most of the steroid hormone receptors, the hormone binding induces translocation into the nucleus where the hormone–receptor complex binds a cognate DNA element termed *hormone-responsive element (HRE)* and alters the transcription of the target gene (reviewed in Aranda and Pascual, 2001).The HREs are located in the control regions of genes and their presence determines whether a gene is subject to control by nuclear receptors at all.

The receptors which are localized predominantly to the nucleus are found permanently associated with their HRE in the control regions of genes. The cognate hormones enter the nucleus, bind to the DNA-associated receptor and thereby trigger transcription activation. In the absence of the hormone, the DNA-bound receptor often has a repressive effect on transcription and ligand binding to the DNA-bound receptor then relieves the repression. Rare cases are known where ligand binding induces a repression of transcription. As outlined in Sections 3.3.5 and 3.4.9, DNA-bound transcriptional regulators cooperate with a large number of other proteins including coactivators, corepressors and the mediator complex to modify chromatin and to contact the transcription apparatus. These proteins also have

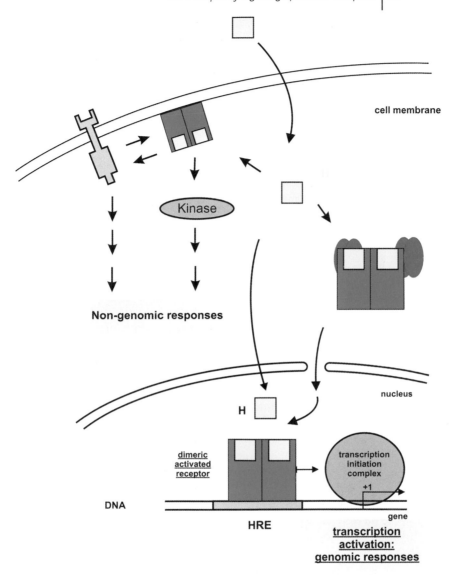

Fig. 4.2: Genomic and nongenomic signaling by nuclear receptors and their ligands. In the genomic path, nuclear receptors function as ligand-controlled transcription factors that bind cognate DNA sequences (HREs) leading to transcription activation. The non-genomic signaling involves binding of the hormone to another type of receptor and crosstalk with other intracellular signaling proteins, e.g. protein kinases. These pathways can trigger a fast, transcription independent response. H = hormone; Hsp = heat shock protein.

an essential function in nuclear receptor signaling. Nuclear receptor coactivators and corepressors have been recognized to be crucial players in the path of nuclear receptor signaling that transmit the signal from the activated nuclear receptor into a stimulation of transcription. Covalent modification of histones and chromatin-associated proteins as well as reorganization of chromatin are therefore essential steps in nuclear receptor signaling.

In the newly discovered, nongenomic modes of action, nuclear receptors perform functions outside of the nucleus. These novel functions have been clearly established for the estrogen receptor (ER) – a pool of which can be found in certain cell types in close association with the cell membrane and with cell organelles, e.g. mitochondria. This pool of receptors conveys signals to protein components of other signaling pathways providing an important example of crosstalk between different signaling paths. By specific protein–protein interactions with central signaling proteins these receptors link steroid hormone signaling directly to central signaling pathways of the cell, e.g. the mitogen-activated protein kinase (MAPK)/extracellular signal-regulated kinase (ERK) pathway, to phosphatidylinositol-3 kinase (PI3K) and to Src kinase (Section 4.8). This linkage provides for a diverse set of rapid reactions that are triggered by the ligands of the nuclear receptors.

■ Nongenomic mode
– Nuclear receptors receive signals from other signaling paths and transmit signals to other signaling paths
– Example of crosstalk!

4.3
Classification and Structure of Nuclear Receptors

The nuclear receptors are grouped into a large superfamily with at least six different subfamilies (Robinson-Rechavi et al., 2003). For some receptors the cognate hormone and their function in the cell remain unknown, and these receptors have been named "orphan receptors". The most important representatives of the nuclear receptors are summarized in Tab. 4.1.

Like other transcriptional regulators, nuclear receptors exhibit a modular structure with different regions corresponding to autonomous functional domains. At the level of the primary structure the steroid hormone receptors can be divided into five different domains (Fig. 4.3), each with specific functions.

A typical nuclear receptor contains a variable N-terminal A/B region, a highly conserved domain responsible for the *DNA binding* (region C), a linker region D containing *nuclear localization signals*, the *ligand-binding and dimerization domain* (region E), and an F region which is not found in all receptors. Domains responsible for *transactivation* are found in the A/B region *[activation function (AF)-1]*, in the E region and sometimes in the F region *(AF-2)*.

■ Domains of nuclear receptors
– DNA binding
– Ligand binding
– Transactivation: AF-1 and AF-2
– Variable domains

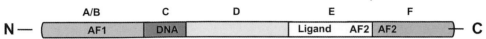

A/B : N-terminal variable region;   C : DNA-binding domain
D   : variable linker domain   E : ligand-binding domain
F   : variable C-terminal domain   AF : transactivation region

Fig. 4.3: Domain structure of the nuclear receptors. Functional domains of nuclear receptors are portrayed in a one-dimensional, linear fashion. Not all nuclear receptors carry two transactivation domains.

Some of the regions behave as independently folding domains whose function could be characterized by deletion experiments, biochemical characterization and structural determination of the isolated domains (reviewed in Kumar et al., 2004). However, a structure of the complete nuclear receptor is not yet available and the precise communication of the various domains is incompletely understood. Overall, the nuclear receptors have to be considered a dynamic entity that can exist in various conformational states each with different regulatory properties.

4.3.1
DNA-binding Elements of Nuclear Receptors: HREs

The steroid hormone receptors are sequence-specific DNA-binding proteins whose cognate DNA elements are termed HREs. The HREs known to date possess a common structure. They are composed primarily of two copies of a hexamer sequence. Table 4.1 lists the hexamer sequences of the HREs of important nuclear receptors.

The identity of an HRE is determined by the sequence, polarity and distance of the hexamers. Mutation and duplication of an ancestral recognition sequence have allowed the creation of many and various DNA elements during the course of evolution, whose sequence, polarity and distance is characteristic for a given hormone receptor or receptor pair. The half-site of an HRE can be arranged as a palindrome, an everted repeat or a direct repeat. For a given receptor, optimal spacings of the half-sites exist and the number of base pairs between the half-sites is another characteristic feature of a HRE. Figure 4.4 illustrates, as an example of the HREs for the receptor for 9-*cis*-retinoic acid (RXR) heterodimer (Section 4.7), the various configurations of an HRE.

Fig. 4.4: HRE structure of the RXR heterodimer. Shown is the consensus sequence of the HREs of the RXR heterodimers (Figs. 4.5 and 4.6) and the different possible arrangements of the hexameric half-site sequences. n = number of base pairs that lie between the two hexamers.

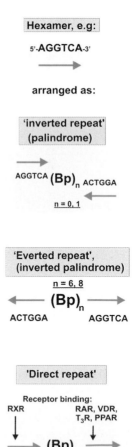

Fig. 4.5: Oligomeric structure of nuclear receptors and structure of the HREs. The nuclear receptors can be subdivided into four groups based on structures of the receptors and HREs. Shown above are some representative examples. (a) binding of a homodimeric receptor to a 2-fold symmetric palindromic DNA element, GR: glucocorticoid receptor. (b) binding of a heterodimeric receptor to a DNA element with direct repeats of the recognition sequence, whereby the 5′ side of the HRE is occupied by RXR. (c) binding of RXR as a homodimer to an HRE with direct repeat of the recognition sequence. (d) Binding of a monomeric receptor to an asymmetric recognition sequence, NGFIB, also known as NR4A and Nur77, regulates neuronal development and apoptosis of the immune system.

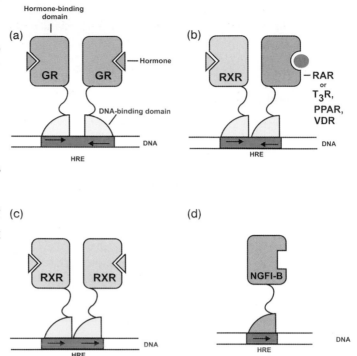

The receptors bind to the cognate HRE mainly as dimers, allowing the formation of *homodimers* as well as *heterodimers* between various receptor monomers. We know of very few nuclear receptors whose HRE contains only a single copy of the recognition sequence. These receptors bind as monomers to the cognate HRE.

Based on the subunit structure of DNA-bound receptors and on the structure of the HREs, four classes of nuclear receptors can be distinguished (Fig. 4.5).

Dimers of the Steroid Hormone Receptors

The HREs of the steroid hormone receptors possess a *palindromic structure*, comparable to the DNA-binding elements of prokaryotic repressors (Fig. 4.5a). The glucocorticoid receptor, for example, binds as a homodimer to the 2-fold symmetrical recognition sequence, whereby the receptor is already dimerized in solution. In complex with the DNA, each subunit of the dimer contacts one half-site of the HRE. As a consequence of the 2-fold repeat of the recognition sequence, high-affinity binding of the receptor dimer results (*cf.* Section 3.3.4).

■ **Nuclear receptors bind to HREs as**
- Homodimers
- Heterodimers

Heterodimers on Direct Repeats

The DNA-binding elements of the nuclear receptors for all-*trans*-retinoic acid, 9-*cis*-retinoic acid, T_3 hormone and vitamin D_3 hormone usually exhibit a direct repeat of the recognition sequence, resulting in the formation of heterodimers on the DNA (Fig. 4.5b). One of the partners in the heterodimer is in most cases RXR, which usually occupies the 5′ side of the HRE.

Of particular importance for receptor binding on HREs with direct repeats of the recognition sequence is the fact that the hexamers of these HREs are arranged head to tail and thus require a *polar arrangement* of the receptor dimers.

HREs of this type display a further unique characteristic – half-site sequences can be used to create different HREs by *varying only the spacing* between the repeats. The spacing can vary between 1 and 5 bp of any sequence, so that with one given repeat five different HREs can exist (see also Fig. 4.4). In this case, the spacing between the recognition sequences determines which hetero- or homodimer can form. Further multiplicity is achieved by combining different half-site sequences in a single HRE.

Homodimers on Direct Repeats

Some receptors can bind as homodimers to recognition sequences arranged as direct repeats (Fig. 4.5c). RXR, which also binds as a homodimer an HRE with two half-sites in direct repeats, is also considered a member of this class of receptors.

Monomeric "Orphan Receptors"

Orphan receptors are also known [e.g. retinoic Z receptor, nerve growth factor-induced clone B (NGFIB)] which bind as monomers to asymmetric recognition sequences (Fig. 4.5d). The function of these receptors is still poorly characterized (see also Section 4.3.4).

4.3.2
The DNA-binding Domain of Nuclear Receptors

Within the family of nuclear receptors, the DNA-binding domain is the most conserved structural element and is located in region C of the primary structure. The DNA-binding domain possesses structural elements that mediate the specific recognition of the HRE, and contribute to the dimerization of the receptor on the HRE. The core of the DNA-binding domain includes a span of 70–80 amino acids, in which all information for the specific recognition of the cognate half-site is contained.

In the core of the DNA-binding domain are *two Zn_2Cys_4 motifs* which serve to position a recognition helix in the major groove of

■ **DNA-binding domain**
– Independently folding
– Two Zn₂Cys4 motifs
– Helices for readout of HRE

the DNA. Via the recognition helix specific contacts are formed with the hexamer half-site of the HRE. The two zinc motifs assume nonequivalent positions in the DNA-binding domain. While the N-terminal zinc motif participates in the positioning of the recognition helix and the interactions with the sugar-phosphate backbone, the C-terminal zinc motif serves to impart a certain functionality to the dimerization surface and to contact the phosphate backbone of the DNA.

4.3.3
HRE Recognition and Structure of the HRE–Receptor Complex

In the case of the palindromically arranged recognition sequences, the binding occurs mostly via preformed homodimers of the receptor in solution. Sequence and spacing of both recognition sequences in the HRE are highly complementary to the binding surface of the recognition helix, as well as to the spacing between the DNA-binding domains of the dimeric receptor.

For direct repeat HREs, the spacing of the two half-sites is often the decisive, if not the only, element based on which the receptor (homodimer or heterodimer) recognizes its own HRE and discriminates against related HREs. The solution of the structures of heterodimer–DNA complexes (reviewed in Steinmetz et al., 2001) has shown how these receptors can distinguish between closely related HREs. As an example, the structure of a DNA-bound receptor heterodimer composed of the DNA-binding domain of RXR and the T_3 receptor (T_3R) is given in Fig. 4.6.

The structure determination confirmed the importance of the spacing of the two hexamers as a discrimination factor. A spacing of only 3 nucleotides between the two hexamers would lead to steric overlap of both receptors; a high affinity, cooperative binding would not be possible. With a spacing of more than 4 nucleotides, a high-affinity complex could also not be formed because of the relative rigidity of the two monomers.

The structural elements of the receptors which participate in the dimerization ensure that the recognition helices assume a defined mutual spatial arrangement adapted to the spacing of the hexamers of the cognate HREs. Only in this configuration is a high affinity, cooperative binding possible. For a correct binding it is necessary that both recognition helices optimally contact both repeat sequences. If, as a result of an incorrect spacing of the hexamers, only one of the two recognition sequences bind, then a high-affinity complex cannot be formed.

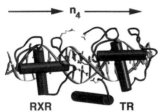

Fig. 4.6: Structure of the RXR–T_3R heterodimer bound to the HRE AGGTCA(N)$_4$AGGTCA. The heterodimer RXR–T_3R binds in a polar manner on the HRE, with RXR occupying the 5′ side of the HRE. Both hexameric sequences lie on the same side of the DNA double helix and are contacted by an α-helix of each of the receptors in a nearly identical manner. Different structural elements of each of the monomeric receptors are involved in the dimerization process, leading to the polar configuration of the monomers on the DNA.

4.3.4
Ligand-binding Domain (LBD)

The region E with the LBD harbors three important functions:
- Homo- and heterodimerization.
- Binding of ligand – both agonists and antagonists.
- Transactivation and transrepression: binding of coactivators and corepressors, contacts to mediator and to RNA Pol II.

Dimerization
A contribution to the dimerization of the receptors – in addition to that from the DNA-binding domain – is provided by a dimerization element in the LBD. The structure of the LBD of RXR without bound hormone shows a homodimer with a symmetric dimerization surface, formed essentially from two antiparallel α-helices.

Ligand Binding
The crystal structures of the LBDs of several nuclear receptors have been resolved demonstrating a similar overall structure (reviewed in Ingraham and Redinbo, 2006). Figure 4.7(a) shows a schematic representation of the LBD of nuclear receptors. The LBD is formed from 12 α-helices numbered from H1 to H12. In the bottom half of the structure, a ligand-binding pocket is found, which accommodates the ligand. The pocket is mainly hydrophobic and of variable size for different receptors. As illustrated in Fig. 4.7(b) for the binding of retinoic acid receptor bound to all-*trans*-retinoic acid, contacts between the ligand and the pocket can be very extensive, and include many hydrophobic contacts as well as hydrogen bonds to the polar parts of the ligand. Another functionally important region of the LBD is part of the transactivation domain AF-2 that interacts with coregulators containing the peptide motif LXXLL.

The size of the ligand-binding pocket can vary considerably among different receptors. Receptors with specific ligands, e.g. the T_3R, bind the ligand in a small pocket that is tailored to bind only the specific ligand, the T_3 hormone, and only small deviations from the T_3 structure are tolerated. In contrast, the ligand-binding pocket of PPAR is of a much larger size. This receptor binds a large variety of endogenous ligands like fatty acids with rather low affinity. Apparently, the ligand-binding cavity of PPAR has been adopted to bind hydrophobic ligands of different size.

■ **LBD**
- Variable size
- Variable specificity
- Couples ligand binding to transactivation
- Ligand induces conformational change of helix H12

(a)

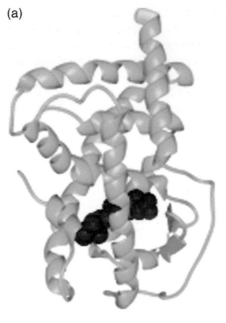

(b)

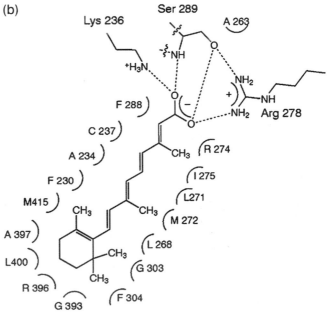

Fig. 4.7: (a) Ribbon diagram of RAR bound to all-*trans*-retinoic acid. (b) Schematic diagram of retinoic acid-binding site. Hydrogen bonds and ionic interactions fix the carboxylate of the ligand whose nonpolar parts are deeply buried in a hydrophobic pocket.

Ligand Binding Induces Conformational Changes

Ligand binding induces a conformational change in the LBD that is crucial to the transcription regulating function of the receptors. The liganded structures are more compact than the unliganded ones and a reorganization of distinct α-helices is visible. A major conformational change is seen for the H12 helix that reorients by a mechanism like a "mouse trap" (reviewed in Steinmetz et al., 2001). The H12 helix is amphipathic, possessing a hydrophobic and a hydrophilic face. In the unliganded RXR, helix H12 projects away from the body of the LBD. In the liganded structure, the helix reorients (Fig. 4.8). The hydrophobic residues face inward and form part of the ligand-binding pocket, whereas the conserved polar residues face outward and form a binding surface for the binding of coactivators and corepressors. It is proposed that ligand binding stabilizes the LBD by completing the hydrophobic core which implies that the LBD is inherently unstable in the absence of ligand. Both structural data and mutational analyses demonstrate that the H12 helix is directly involved in transcriptional activation. It forms the core of the AF-2 domain, and thus plays a key role in transactivation and transrepression.

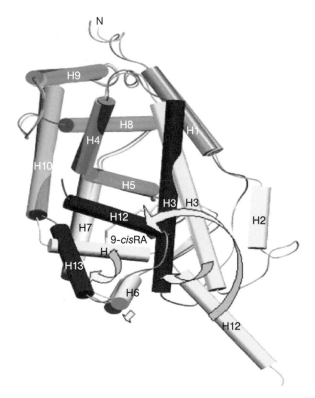

Fig. 4.8: Structural changes in the LBD of RXR on binding of 9-*cis*-retinoic acid. The models of domain E of apo-RXRα and the binary complex of RXRα and 9-*cis*-retinoic acid were superimposed. Domain E of apo-RXRα is depicted in green and yellow, and domain E of the binary complex in blue and red. The grey arrows indicate the structural rearrangements of helices H11, H12 and the N-terminus of helix H3. After Steinmetz et al. (2001).

Agonist versus Antagonist Binding

The nuclear receptors are involved in the regulation of numerous physiological processes and are therefore important medical targets. An enormous number of synthetic compounds are available that act as agonists, mixed agonists/antagonists or as pure antagonists of the natural ligands. As illustrated in Fig. 4.9, the structure of a LBD with an antagonist bound can provide a rational basis to explain the antagonistic function of a ligand. In the presence of the antagonist, the conformational change within the LBD is quite distinct from that induced by the natural ligand. Helix H12, which is critical for transactivation, assumes a conformation different from that observed for the natural ligand. The binding surface on H12 that is required for coactivator binding is not available in the antagonist-bound state, thereby preventing transcriptional activation.

The structural studies also have shown that the detailed structure of the LBD depends on the nature of the agonist and that different agonists/antagonists can induce different conformational states of the LBD. These states can interact with different protein binding partners and therefore activate different sets of genes. An example for this is given below (Section 4.4.1).

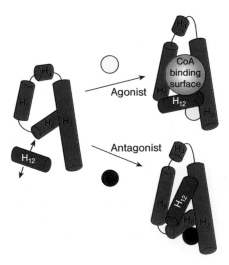

Fig. 4.9: Agonist binding induces a conformation different from antagonist binding. Only agonist binding exposes a binding site for coactivators (CoA).

LBD of Orphan Nuclear Receptors

A number of nuclear receptors are known that are highly promiscuous in ligand binding or for which one could not identify ligands. Some of these receptors show constitutive activity. Examples include receptors involved in drug disposition such as liver X receptor (LXR), pregnane X receptor (PXR) and constitutive androstane receptor (CAR). Other examples are the receptors of the NR4A subfamily (also known as NGFIB). Analysis of the LBD and the AF-2 domain of these receptors show a similar overall structure of the LBD. However, the AF-2 surface and the ligand-binding pocket show great variations (reviewed in Ingraham and Redinbo, 2006). In members of the NR4A family, the ligand-binding pocket and the AF-2 cleft is lost. How these receptors are activated remains to be determined. Possibly activation includes signal-directed posttranslational modification of the receptor. Alternatively, these receptors could form heterodimers with ligand-dependent receptors such as RXR. For receptors involved in detoxification, e.g. PXR and CAR, the ligand-binding pocket is very large and can accommodate a variety of hydrophobic compounds. Furthermore, the ligand-binding pocket is flexible and contains some polar residues.

4.3.5
Transactivating Elements of the Nuclear Receptors

Most nuclear receptors contain two structural elements, *AF-1* and *AF-2*, which mediate the transcriptional activation. The AF-1 domain is located in the A/B region and mediates a *ligand-independent transactivation*. It harbors phosphorylation sites and interaction sites for coactivators. Furthermore, interactions of AF-1 with the distant AF-2 domain have been shown. Overall, the role of the AF-1 domain in transcriptional activation is not well understood. The AF-2 domain of the E region is well conserved across the members of the nuclear receptor superfamily and functions in most cases in a ligand-dependent way. It encompasses helix H12 which is part of the LBD as well as other structural elements of the E region. Upon ligand binding, the hydrophilic surface of the H12 helix is oriented outward, providing an interface for the binding of complementary regions of coactivators or corepressors.

■ **AF-1**
– Located in the A/B region
– Mediates ligand-independent transactivation,
– Contains phosphorylation sites
– Contains interaction sites with other proteins
– Interacts with the E/F region

■ **AF-2**
– Part of the E domain
– Undergoes ligand-dependent conformational change
– Contains binding sites for coactivators and corepressors

4.4
Mechanisms of Transcriptional Regulation by Nuclear Receptors

Most of the functions of nuclear receptors can be described in terms of activation and repression of transcription. The outcome of these "genomic" actions of nuclear receptors is determined in a cell-type specific way by a series of variables:
– The exact sequence of the HRE.
– The neighborhood of the HRE.
– The cooperation with other transcriptional activators.
– The availability of other nuclear receptors to form heterodimers.
– The availability of coregulatory proteins, mainly coactivators and corepressors.
– Posttranslational modifications of the receptors.
– A dynamic distribution of the receptors between the nucleus and the cytoplasm.

The genomic functions of the nuclear receptors are governed by the same processes that have been recognized to be essential for other transcriptional activators or repressors. The basic steps required for transcriptional activation have been outlined in Section 3.3 and the same types of proteins are necessary for the nuclear receptors to perform their regulatory function.

Nuclear receptors have to fulfill the following main tasks during transcription activation:
– Selection of target genes by specific binding to the HRE.

■ **Nuclear receptors interact with and recruit**
– Mediator
– Coactivators of the SRC-1/p160 and/or TRAP family
– Corepressors: NcoR and SMRT
– Proteins of the SWI/ SNF family
– Histone acetylases: GCN5 and CBP/p300
– Histone deacetylases
– Arginine methylases: CARM
– Lysine methylases
– Protein kinases
– Ubiquitin ligases: E6AP

– Remodeling of chromatin to relieve repressive structures and to make it competent for transcription start.
– Participation in the start of transcription by continuously modifying chromatin and by stabilization of the transcription complex.

These steps require contacts to protein complexes that remodel and modify chromatin as well as to components of the basal transcription complex, e.g. TATA box-binding protein associated factors (TAFs), TATA box-binding protein (TBP) and the RNA Pol II holoenzyme.

The collection of proteins that help to modify chromatin and function as coregulators of nuclear receptors is very large and direct or indirect interactions between nuclear receptors and a large number of proteins have been reported.

These proteins are engaged by nuclear receptors in a dynamic and combinatorial way to modify chromatin structure during the initiation steps of transcription. For the ER, the participation of as many as 46 different proteins has been shown (reviewed in Smith and O'Malley, 2004).

Depending on receptor type, HRE structure and repertoire of coregulators, the following functional states of DNA-bound nuclear receptors have been identified (Fig. 4.10).

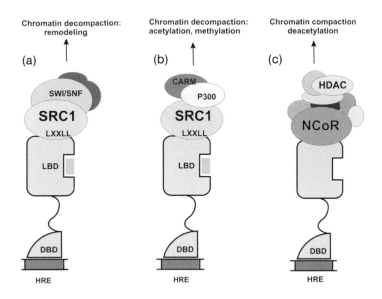

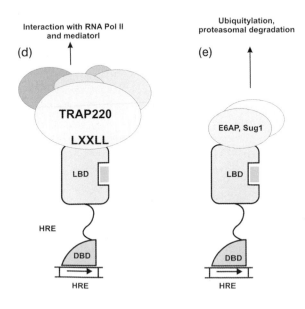

Fig. 4.10: Functional states of nuclear receptors with selected examples of associated proteins. (a) The remodeling complexes SWI/SNF mobilize nucleosomes. (b) SRC-type coactivators recruit modifying enzymes necessary for chromatin decompaction, e.g. the arginine methylase CARM and the histone acetylase p300. (c) Association of the corepressor complexes NcoR recruits HDACs leading to chromatin compaction an transcription repression. (d) The associated TRAP complex provides for contacts to the basal transcription machinery. (e) The LBD can associate with proteins involved in ubiquitination and proteasomal degradation, e.g. E6AP and the Sug1 protein.

4.4.1
Steroid Receptor Coactivator (SRC)-1/p160 and Thyroid Hormone Receptor-activating Protein (TRAP) Coactivators

A series of proteins or protein complexes with co-activator function for nuclear receptors has been identified that specifically interact with the activated, liganded receptor, either at its AF-1 or AF-2 domain. Whereas the coactivators of the AF-1 domain are less well characterized, two types of coactivators that bind to the AF-2 domain have been thoroughly studied.

■ **Coactivators**
– SRC-1/p160 and TRAP220
– Contain LXXLL motif
– Required for transactivation
– Interact with ligand-bound receptor

One type comprises the *p160 family of coactivators* with the SRC-1, as a well-characterized member. Another type of coactivators has been located to a multiprotein complex named the *TRAP complex*. The TRAP complex is now considered one type of mediator complex.

The members of the p160 coactivator family fulfill two functions:
– Interaction with the receptor in a ligand-dependent way.
– Recruitment of chromatin-modifying enzymes like histone acetylases (CBP/p300), lysine methyltransferases and arginine methyltransferases (e.g. CARM1/PRMT1).

A characteristic feature of the p160 coactivator family is the presence of *receptor interaction motifs LXXLL* which mediate at least part of the interaction with the AF-2 domain of the receptor. The binding surface for the LXXLL motifs is formed on AF-2 only after activation of the receptor by agonist binding. Agonist-induced reorganization of helix H12 is a necessary step to form the coactivator binding surface. Antagonists on the contrary stabilize a conformation where helix H12 occupies a position that covers the coactivator binding surface. The LXXLL motif is also found on other coactivators, e.g. the TRAP220 protein – a component of the TRAP/mediator complex (Fig. 4.10d).

The p160 coactivators also contain domains that interact with enzymes required for chromatin modification, e.g. histone acetylases, lysine methylases and arginine methylases.

The TRAP/mediator complex has the task of forming a bridge between the DNA-bound receptor and the basal transcription apparatus. It contacts the basal transcription complex and interacts with the AF-2 domain in a ligand-dependent manner via a LXXLL motif that is found in the TRAP220 component.

A detailed analysis of ER-associated coregulators has shown that the many coactivators are used in a dynamic and combinatorial way. Depending on the cell type and the ligand type, different sets of coactivators may be used for transcription activation (Schulman and Heyman, 2004).

A medically important aspect of coactivator recruitment is the ability of distinct synthetic agonists to recruit a specific set of coactivators leading to distinct physiological responses (Fig. 4.11). This is based on the ability of synthetic agonists to induce distinct conformations of the LBD that allow the binding of distinct sets of coactivators. The concept of tissue and gene specific nuclear receptor ligands has been developed as an important new step towards a specific modulation of nuclear receptors. For the ER these compounds have been termed collectively *selective ER modulators (SERMS)*. SERMS such as tamoxifen and roloxifene behave as ER antagonists in breast and agonists in bone.

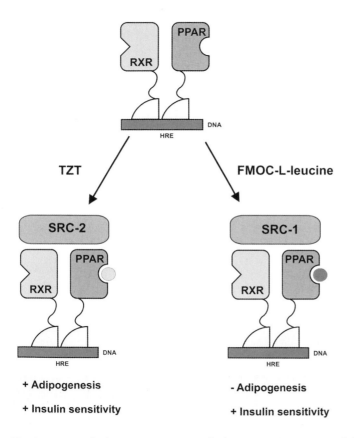

Fig. 4.11: Ligand-selective recruitment of different coactivators. The PPAR–RXR heterodimer recruits the coactivator SRC-2 in response to binding the synthetic ligand thiazolidinedione (TZT) inducing insulin sensitivity and adipogenesis. In presence of the ligand 9-fluorenylmethyloxycarbonyl (FMOC)-L-leucine, the coactivator SRC-1 is recruited resulting in insulin sensitivity only.

4.4.2
Corepressors of Nuclear Receptors

■ **Corepressors recruit HDACs and lysine methylases to induce a silent chromatin state**

The corepressors of nuclear receptors serve an important role in negatively regulating receptor-dependent gene expression. Unliganded T_3R and receptor for all-*trans*-retinoic acid (RAR) repress basal transcription in the absence of ligand and this is mediated mainly by two large proteins, nuclear *corepressor NcoR* and the *silencing mediator for retinoic and thyroid hormone receptors (SMRT)*. Both proteins interact with the LBD. Upon hormone binding, these corepressors dissociate from the receptor and enable T_3R and RAR to associate coactivators and stimulate gene expression. The repressive action of the corepressors is based on their recruitment of protein complexes containing *histone deacetylase (HDAC) activity*, e.g. the mSin3 or NuRD complex. The HDAC then helps to maintain the repressed state of the chromatin.

4.5
Regulation of Signaling by Nuclear Receptors

Fig. 4.12: Functions of nuclear receptor domains. The domains A/B, C, E and F of the nuclear receptors are involved in multiple protein–protein interactions and are subject to regulatory modifications as indicated. Most important are the corepressor and coactivator complexes that direct HDAC and histone acetylase activities, respectively, to the nuclear receptor-regulated promotor region.

In the classical, genomic pathway, nuclear receptors receive signals and pass signals on to the level of transcription. The steps involved in this signaling process are subject to a variety of regulatory mechanisms that finally determine the magnitude and dynamics of transcriptional activation and its cell type specificity. The following are important regulatory attack points (Fig. 4.12).

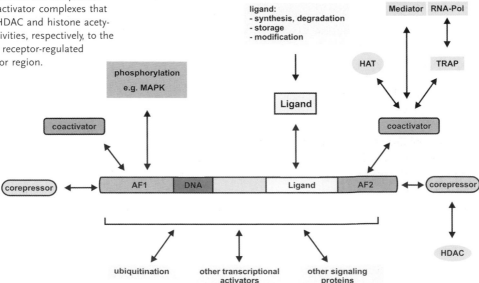

Regulation at the Level of Ligand Concentration

A main determinant of nuclear receptor signaling is the concentration of the ligand available for binding. The ligand concentration can be regulated in many ways (for details see textbooks on hormone action). A feedback regulation via the circulating hormone concentration is of importance in the hypothalamus–pituitary system of the brain, where feedback inhibition at various levels is used to prevent overproduction of, for example, steroid hormones or the T_3 hormone.

■ **Regulation of ligand concentration of nuclear receptors occurs via**
  – Secretion, transport and storage
  – Modification
  – Feedback regulation of biosynthesis

Regulation by Phosphorylation

Nuclear receptors contain multiple phosphorylation sites that are used to modulate both their genomic and nongenomic functions, and to link nuclear receptor signaling to other signaling pathways (crosstalk).

The phosphorylation sites comprise Ser/Thr as well as Tyr sites and are mainly found in the AF-1 region of the receptors. The consequences of phosphorylation for the receptor proteins are varied and have been shown to influence many aspects of nuclear hormone signaling (reviewed in Ismaili and Garabedian, 2004).

One of the best studied examples are the ERα and ERβ which can be phosphorylated at multiple sites on the AF-1 domain. The major phosphorylation site of ER is Ser 118. This phosphorylation results in ligand-independent activation of the receptor and allows for the transcription of ER-target genes in the absence of estrogen. The protein kinases involved appear to be kinases of the MAPK cascade and CDK7 (Section 13.2.1). The MAPK cascade transmits signals from growth factor-stimulated pathways and from Ras proteins (Chapter 10), providing a crosstalk between nuclear receptor signaling and growth factor. Other protein kinases (cyclin A–CDK2, Akt kinase, PI3K, c-Jun N-terminal kinase) have been reported to be involved in phosphorylation of the other Ser/Thr sites of ER. In addition, Src kinase has been implicated in phosphorylation of Tyr sites of ERs (see below).

■ **Phosphorylation affects nuclear receptor signaling via modulation of**
  – DNA binding and transactivation function
  – Ligand requirement
  – Nuclear/cytoplasmic distribution
  – Membrane localization

Regulation by Coactivators and Corepressors

It is now well established that signaling by nuclear receptors is modulated to a large extent by the activity and availability of coactivators and corepressors. These have been shown to be also the target of regulatory modifications such as phosphorylation, acetylation and methylation.

■ **Coactivators and corepressors are regulated by phosphorylation**

Interaction with other Transcriptional Activators

Nuclear receptors also modulate gene expression by interference with the activity of other transcriptional activators. This can occur by two ways:

– In composite promoters that harbor copies of nuclear receptor elements as well as DNA-binding elements of other transcription factors, the ER, T_3R, RAR and glucocorticoid receptor proteins have, for example, been shown to act as transrepressors of the transcription factor AP-1, which is a heterodimer composed of c-Jun and c-Fos proteins. Reciprocally, AP-1 can inhibit transactivation by these receptors. A mutually antagonistic effect of glucocorticoid receptors on the function of the transcriptional activator NFκB has also been reported.

■ **Nuclear receptors modulate activity of other transcription factors (AP-1, Sp1 and STAT proteins).**

– ERs and progesterone receptors (PRs) can stimulate gene expression without binding to DNA by associating with other transcription factors bound to promotors of responsive genes. Protein–protein interactions between ERs and the transcription factors Sp1, AP-1 as well as the signal transducer and activator of transcription (STAT) proteins have been reported for a variety of genes that do not contain the canonical ER responsive DNA elements.

Our understanding of the mechanistic basis of these interferences is limited. ER -dependent transcription through AP-1 has been shown to require the DNA-binding domain and the LBD of the ER and appears to use different coactivators as compared to ER signaling alone.

Regulation by Ubiquitination

Signaling by nuclear receptors is regulated by ubiquitination in various ways. One main path includes the ubiquitin (Ub)-dependent proteolysis of nuclear receptors by allowing for the down-regulation of receptors under long-term hormone treatment. Ligand-dependent ubiquitination and subsequent proteasomal degradation has been described for ERα, PR, vitamin D receptor (VDR), T_3R, RARα, etc. It has been proposed that the proteosomal degradation process provides a mechanism to control the magnitude and duration of ligand-mediated transcription regulation.

■ **Levels of nuclear receptors are regulated by Ub-dependent proteolysis**

In another path, ubiquitination of nuclear receptors is required to activate transcription. This effect is explained in terms of the dynamic removal and reassembly of receptor complexes at promoters in the course of ongoing transcription cycles.

In addition, corepressors and coactivators of nuclear receptors are also subjected to ubiquitination indicating a widespread involvement of the Ub–proteasome pathway in the regulatory function of nuclear receptors.

Other regulatory modifications of nuclear receptors include sumoylation (Section 2.5.9) and modification by nitric oxide (NO; see Section 6.10).

4.6
The Signaling Pathway of the Steroid Hormone Receptors

Genomic signaling by nuclear receptors may be divided into two basic groups. In the first group (those including most of the steroid hormone receptors), the receptors are localized in the nucleus and in the cytoplasm, and a translocation of the receptor from the cytoplasm to the nucleus is observed in the presence of the ligand. The receptors of the other group (discussed in Section 4.7) are predominantly localized in the nucleus and ligand binding to the DNA-bound receptor activates transcription. Representative ligands of these receptors are the derivatives of retinoic acid, T_3 hormone and vitamin D_3.

Signal transduction by steroid hormones (glucocorticoids, mineralocorticoids, androgens and estradiol) is distinguished by the fact that the receptor receives the hormonal signal in the cytosol and becomes activated by hormone binding, at which point it enters the nucleus to regulate the transcription initiation of cognate genes.

Figure 4.13 shows the most important steps in the signal transduction by steroid hormones.

Activation of the Cytoplasmic Aporeceptor Complexes
The steroid hormones are distributed throughout the entire organism by means of the circulatory system. The transport often occurs in the form of a complex with a specific binding protein. An example of such a binding protein is transcortin, which is responsible for the transport of the corticosteroids. The steroid hormones enter the cell by diffusion and activate the cytosolic receptors.

In the absence of steroid hormones, the receptors remain in an inactive complex, designated the *aporeceptor complex* which is transformed into an active complex by hormone binding. Various proteins belonging to the chaperone class participate in the maturation of the inactive aporeceptor complex. The proteins include the *chaperones hsp90*, hsp40 and hsp70, the cochaperone Hop, immunophilins of the FK506-binding protein (FKBP) family and another protein factor, p23.

Chaperones are proteins that assume a central function in the folding process of proteins in the cell. They aid proteins in avoiding incorrectly folded states, and thereby participate in the folding of proteins during and after ribosomal protein biosynthesis, during mem-

■ **Aporeceptor complexes of steroid hormone receptors contain**
– Receptor protein
– Hsp90 chaperone complex
– Cochaperones
– Immunophilins
– p23

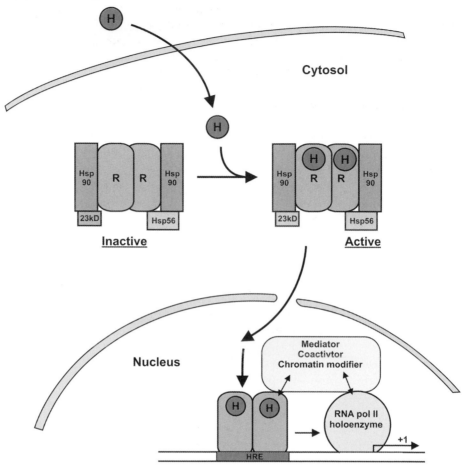

Fig. 4.13: Principle of signal transduction by steroid hormone receptors. The steroid hormone receptors in the cytosol are found in the form of an inactive complex with the heat shock proteins hsp90 and hsp56, and with protein p23. The binding of the hormone activates the receptor so that it can be transported into the nucleus where it can bind to its cognate HRE. It remains unclear in which form the receptor is transported into the nucleus and to which extent the associated proteins are involved in the transport.

brane transport of proteins, and in the correct assembly of protein complexes.

The term "hsp" (heat shock protein) is derived from the observation that these proteins were produced at higher levels following heat treatment. The immunophilins are proteins that carry peptidyl–prolyl *cis–trans* isomerizing activity and associate with hsp90 through distinct domains, named tetratricopeptide repeats.

The chaperones are used as tools in this system for regulation of activity of the steroid hormone receptors. It is believed that hsp90 holds the receptor in a partially unfolded conformation that is competent for ligand binding and that the cochaperones facilitate the correct association of hsp90 with the receptor. The binding of the hormone to the aporeceptor complex leads to activation of the receptor and initiates the translocation of the receptor into the nucleus under participation of motor proteins, e.g. dynein. The activated receptor possesses an accessible nuclear localization sequence, and is furthermore capable of DNA binding and transactivation. The ability to transactivate implies that the transactivating domain is properly positioned, as a result of the hormone binding, to allow stimulatory interactions with the transcription apparatus (Section 4.4).

It is now well established that nuclear receptors are flexible molecules that can exist in different conformational states, each with distinct functions. It may be a function of the chaperones to stabilize the particular conformation optimal for hormone binding. Inactive, chaperone-bound nuclear receptor complexes can be also found in the nucleus, as is the case for the ER. The ER is sequestered in the nucleus within a large inhibitory heat shock protein complex. Binding of estrogen to the receptor enables the displacement of the heat shock proteins and facilitates DNA binding, formation of receptor dimers, and interactions with coactivators and the transcription apparatus.

■ **Hsp**
– Proteins produced at higher levels following heat treatment

Chaperones
– Aid proteins in avoiding incorrectly folded states and thereby participate in the folding of proteins

Immunophilins
– Proteins that carry peptidyl–prolyl *cis–trans* isomerizing activity and associate with hsp90 through distinct domains, named tetratricopeptide repeats

4.7
Signaling by Retinoids, Vitamin D₃ and the T₃ Hormone

The receptors of the retinoids, the T₃ hormone and vitamin D₃ share several important features. These receptors are localized primarily in the nucleus, bound to their HRE, which is in contrast to the steroid hormone receptors. Binding of the ligand to the DNA-associated receptor activates the transcription regulation of the receptor. In most cases, the unliganded receptor has a repressive influence on the cognate gene (Fig. 4.14). Binding of the hormone induces a conformational change in the LBD and the exposure of a binding site for coactivators. Repression is thereupon relieved and transcription is activated. In some cases, the unliganded, DNA-bound receptor mediates constitutive transcriptional activation. Ligand binding then induces repression of the gene.

On the example of the receptors for 9-*cis*-retinoic acid and all-*trans*-retinoic acid, RXR and RAR, some important aspects of signaling by the class of nuclear-localized receptors will be discussed.

■ **Receptors for**
– Retinoids (9-*cis*-retinoic acid and all-*trans*-retinoic acid)
– T3 hormone
– Vitamin D
– Localized primarily in the nucleus, bound to their HRE
– Often form heterodimers with RXR at the 5′-end
– Can repress transcription in the absence of ligand

Fig. 4.14: Model of repression and activation of T_3R. In the absence of the T3 hormone, a heterodimeric RXR–T_3R receptor is bound at the T_3-responsive element (TRE), establishing a basal repressed state by recruitment of corepressor complexes containing HDAC activity. In the repressed state, the promoter region is thought to be covered by nucleosomes which prevents binding of RNA polymerase. In the presence of T_3 hormone, the corepressors are removed and coactivators (e.g. the SRC/p160 complex) bind to the receptor heterodimer. The histone acetylase activity of the associated proteins helps to induce a transcription competent state around the promoter. In collaboration with the mediator complex (TRAP complex) and chromatin-remodeling enzymes, nucleosomes are removed from the promoter region allowing binding of RNA polymerase and transcription initiation.

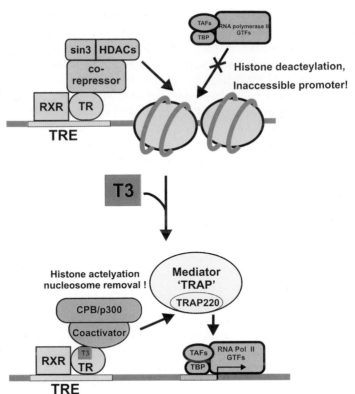

DNA Binding and Formation of Heterodimers

The retinoic receptors bind as asymmetric, oriented heterodimers to their HRE composed typically of two direct repeats of a core hexamer occurring in the DR5, DR2 or DR 1 configuration. On DR5 and Dr2 repeats, RXR occupies the 5′ hexameric motif, whereas the RAR partner occupies the 3′ motif. On DR1 repeats, however, the polarity is reversed and a 5′-RAR/RXR-3′ configuration is found, switching the activity from an activator to a repressor of retinoid responsive genes. RXR also forms heterodimers with the receptors for the T_3 hormone and vitamin D hormone. In this case, RXR always occupies the 5′ position of the DNA element.

Ligand Binding, Coactivator Recruitment and Chromatin Modification

Unliganded retinoid receptors repress transcription through the recruitment of the corepressors NcoR and SMRT. These reside in or recruit HDAC complexes which lead, for example, to immobilization of nucleosomes. To activate transcription, retinoid receptors have to contend with the repressive chromatin structures in order to allow stable recruitment of the transcription machinery. Ligand binding in-

■ The retinoic acids are derived from retinol, which is oxidized in the cytosol to the acid derivative

duces a conformational change in the coregulator binding surface and helix H11. As a consequence, the corepressors dissociate and coactivator complexes can bind, leading finally to remodeling and decompaction of the repressive structures, and to a stable recruitment of the transcription machinery.

The coactivators involved include the class of p160 coactivators and the TRAP complex (Section 4.4.1).

In the heterodimers with the retinoic receptors, ligand binding can induce a conformational change across the heterodimerization surface leading to an allosteric coupling of the LBDs. This allosteric coupling allows ligands of one member of the heterodimer to regulate the activity of its partner domain.

The retinoic acid ligands are derived from retinol which is oxidized to retinoic acid in the cytoplasm. To activate the receptor, the retinoids have to enter the nucleus. This occurs in complex with retinoic acid binding proteins which channel the retinoic acid to RAR.

Phosphorylation

RARs and RXRs are substrates for a multitude of protein kinases and their phosphorylation at Ser/Thr residues plays a critical role in the retinoid response. Most phosphorylation sites are found in the AF-1 domain. The protein kinases CDK7, which is a component of TFIIH, and the MAPK p38 have been implicated in this phosphorylation. The various phosphorylation sites appear to play distinct roles in regulation of the activity of RAR/RXR, and may regulate DNA and ligand binding as well as recruitment of coactivators and corepressors. Another function of receptor phosphorylation has been suggested to be the recruitment of Ub ligases channeling the receptor to degradation in the proteasome.

■ **Phosphorylation in AF-1, e.g. by CDK7 or MAPK p38, modulates receptor function**

4.8
Nongenomic Functions of Nuclear Receptors and their Ligands

The intense research on nuclear receptor function has clearly established that these receptors and their ligands are part of a network where crosstalk with other signaling pathways modulates and regulates nuclear receptor signaling. In these processes the nuclear receptors can receive signals from and relay signals to other pathways allowing for ligand-independent activation of the receptors as well as for receptor actions outside of the nucleus.

The nuclear receptors and their hormone ligands have been shown to elicit fast responses that are not dependent on transcription. These responses appear to be mediated by nonnuclear functions of the receptors and/or by binding of their hormone ligands to a different

■ **Nongenomic actions of steroid hormones and their ligands**
– Linkage to other signaling pathways
– Binding to GPCRs

type of receptors that belong to the class of GPCRs (reviewed in Acconcia and Kumar, 2006). Both modes of action link the steroid hormones directly to cytoplasmic signaling pathways that can induce a broad range of biochemical events on a fast time scale.

Linkage to Cytoplasmic Signaling Pathways

The spectrum of nongenomic functions of nuclear receptors may be illustrated on the example of ERs whose extranuclear activities have been studied in detail, fuelled by the involvement of estrogen and its receptors in the development of cancers of the breast and uterus. The two isoforms of ER, ERα and ERβ, have nearly the same DNA-binding domain and a similar LBD, whereas there is great divergence in the N- terminal AF-1 domain. Both ERs act through the same general mechanism in target cells with distinct biological actions on specific gene promoters and in response to synthetic ligands. ERα is more highly expressed and is thought to play a major role in steroid-dependent cancers of the breast and uterus.

It is now commonly recognized that ERs receive and transmit a multitude of signals that are not directly related to the classical steroid hormone pathway. The ERs are phosphoproteins and specific serine residues are phosphorylated basally and in response to ligands, second messengers and growth factors, e.g. via MAPK, protein kinase A, protein kinase C and cyclin-dependent kinase pathways (Fig. 4.15). In addition to serine residues, tyrosine phosphorylation is also observed on ERs. The phosphorylations mediate ligand-independent stimulation and modulation of ERs and link these receptors to major signaling pathways of the cell, notably to growth factor signaling.

Additionally, activated ERs can transmit signals to pathways that affect major functions of the cell, e.g. cell growth and cell proliferation.

Direct and indirect linkage of estrogen signaling to central signaling pathways of the cell appears to involve two main mechanisms (Fig. 4.16):

- Direct interaction of ERs with protein kinases, TM receptors and central adaptor proteins.
- Binding of estradiol to receptors other than the classical ERs.

These nongenomic functions of estradiol and ERs occur outside of the nucleus and are often intimately associated with the cell membrane.

■ ERs can receive signals from other central signaling paths in a ligand-independent manner

■ Activated ERs can transmit signals directly to other signaling paths via interaction with
– Adaptor proteins
– Protein kinases
– TM receptors

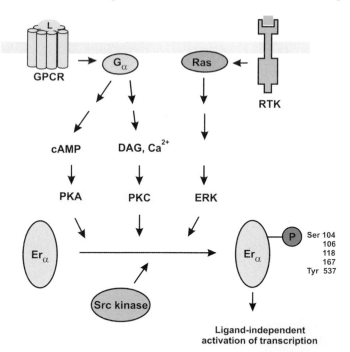

Fig. 4.15: Ligand-independent linkage of the ER to major signaling pathways of the cell. ER is phosphorylated by protein kinases in response to signals that start from receptor tyrosine kinases (RTKs) – via Ras protein – or from GPCRs. A central role in this crosstalk which does not require estrogen binding to ER is ascribed to Src kinase. Phosphorylated ER can thus stimulate transcription in a ligand-independent way. PKA = protein kinase A (Section 7.3); PKC = protein kinase C (Section 7.5); ERK = extracellular signal-regulated kinase (Section 10.2); Src = src kinase (Section 8.3.2).

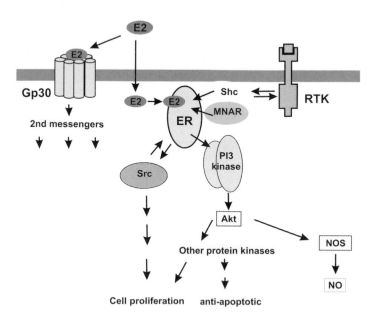

Fig. 4.16: Nongenomic functions of ER and estrogen. The steroid hormone 17β-estradiol (E$_2$) can initiate signaling via the GPCR GP30 or via its receptor, ER. Binding of E$_2$ to ER can stimulate transcription or activate in a nongenomic mode other central signaling proteins of the cell. The crosstalk with other signaling pathways may require the presence of the adaptor proteins Shc or MNAR. ERs can directly activate PI3K and thereby transmit a signal to Akt kinase. One effect of Akt kinase activation is the stimulation of the endothelial NO synthase (NOS) and the production of the second messenger NO. Akt kinase also mediates an antiapoptotic and cell proliferation-promoting effect.

ER Action at the Cell Membrane

Apart from the nuclear pool of ERs, these receptors are found in a cell-type specific manner also in the cytoplasm, fractions of the endoplasmic reticulum, at the cell membrane and in the mitochondria (reviewed in Manavathi and Kumar, 2004). Membrane association of ERs may involve several mechanisms. ERs have been shown to contain a palmitoic membrane anchor and they are found in large TM signaling complexes at distinct sites, the caveolae, which may contain receptor tyrosine kinases, G proteins and adaptor proteins. Association with caveolae and with other membrane sites may be mediated via ER binding to the anchor protein striatin and by binding to the adaptor protein Shc.

■ **Some functions of ERs are linked to the cell membrane**

Direct Interaction of ERs with Signaling Proteins

The ERs have been reported to interact with the two important signaling enzymes c-Src kinase and PI3K. The interaction with c-Src kinase may occur directly via binding of ER to the SH2 domain of c-Src. Tyrosine phosphorylation at Tyr537 in the LBD of ER has been implicated to be involved in this interaction. Binding of c-Src may be mediated also by a scaffolding molecule termed *moderator of nongenomic activity of ER (MNAR)*. The interaction of Src kinase with ER or MNAR may relieve Src from the autoinhibited state and may activate (via Shc, Grb2-mSos and Ras) the MAPK pathway, and thus stimulate the transcription of a variety of genes important for cell growth and cell proliferation (Chapters 9 and 10).

Activation of PI3K by ER is thought to occur via the regulatory p85 subunit of PI3K. As a consequence, the formation of phosphoinositides is stimulated and the Akt kinase pathway is activated. Targets of this pathway include the endothelial NO synthase which is rapidly induced upon estradiol administration in endothelial cells.

Binding of Estradiol to GPCRs

A second way of nonclassical action of estradiol has been discovered whereby the ligand 17β-estradiol binds to and activates a TM receptor named GPR30 which belongs to the class of GPCRs. GPR30 specifically binds the physiological ligand of classical ER, 17β-estradiol, and thereby can trigger a rapid intracellular response that may also lead to alterations in gene transcription. This mechanism of estradiol action may also explain the rapid formation of second messengers and the rise of cellular Ca^{2+} that is observed upon administration of estradiol. As outlined in Chapter 5, the GPCRs can stimulate synthesis and release of a number of small signaling molecules – the second messengers.

■ **Estradiol binds and activates GPR30 – a GPCR**

4.9
References

Acconcia, F. and Kumar, R. (2006) Signaling regulation of genomic and nongenomic functions of estrogen receptors, *Cancer Lett.* **238**, 1–14.

Aranda, A. and Pascual, A. (2001) Nuclear hormone receptors and gene expression, *Physiol Rev.* **105**, 1269–1304.

Bastien, J. and Rochette-Egly, C. (2004) Nuclear retinoid receptors and the transcription of retinoid-target genes, *Gene* **328**, 1–16.

Ingraham, H. A. and Redinbo, M. R. (2005) Orphan nuclear receptors adopted by crystallography, *Curr. Opin. Struct. Biol.* **105**, 708–715.

Ismaili, N. and Garabedian, M. J. (2004) Modulation of glucocorticoid receptor function via phosphorylation, *Ann. NY Acad. Sci.* **1024**, 86–101.

Kumar, R., Johnson, B. H. and Thompson, E. B. (2004) Overview of the structural basis for transcription regulation by nuclear hormone receptors, *Essays Biochem.* **40**, 27–39.

Manavathi, B. and Kumar, R. (2006) Steering estrogen signals from the plasma membrane to the nucleus: two sides of the coin, *J. Cell Physiol* **207**, 594–604.

McKenna, N. J. and O'Malley, B. W. (2002) Combinatorial control of gene expression by nuclear receptors and coregulators, *Cell* **105**, 465–474.

Robinson-Rechavi, M., Escriva, G. H. and Laudet, V. (2003) The nuclear receptor superfamily, *J. Cell Sci.* **105**, 585–586.

Schulman, I. G. and Heyman, R. A. (2004) The flip side: identifying small molecule regulators of nuclear receptors, *Chem. Biol.* **105**, 639–646.

Smith, C. L. and O'Malley, B. W. (2004) Coregulator function: a key to understanding tissue specificity of selective receptor modulators, *Endocr. Rev.* **105**, 45–71.

Steinmetz, A. C., Renaud, J. P. and Moras, D. (2001) Binding of ligands and activation of transcription by nuclear receptors, *Annu. Rev. Biophys. Biomol. Struct.* **105**, 329–359.

5   G-protein-coupled Signal Transmission Pathways

5.1
Transmembrane (TM) Receptors: General Structure and Classification

During intercellular communication, extracellular signals are registered by the cell and converted into intracellular reactions. Signal
transmission into the cell interior takes place by reaction chains,
which involve many signal proteins. The nature of the extracellular
signal can be very diverse and may include extracellular signal molecules, such as low-molecular-weight messenger substances or proteins, or sensory signals, such as light signals. The cell uses two principal ways to transduce signals into the interior of the cell. One way
is exemplified by nuclear receptor signaling, where the signaling molecule crosses the cell membrane and activates the receptor in the interior of the cell. In the other major way the signal is registered at the
cell membrane and transduced into the cell by TM proteins. Two different types of TM proteins participate in this mode of signaling.

Signaling via TM Receptors
TM receptors are proteins that span the phospholipid bilayer of the
cell membrane. The signaling molecule binds on the extracellular
side to the receptor, which is thereby activated. Reception of the signal is synonymous with activation of the receptor for transmission of
the signal across the cell membrane. Transmission of the signal implies specific communication with the effector protein, the next component of the signal transmission pathway on the inner side of the
cell membrane. In this process enzymatic activities can be triggered
and/or the activated receptor engages in specific interactions with
downstream signal proteins. An *intracellular signal chain* is set in motion, which finally triggers a defined biochemical response of the target cell (Fig. 5.1a). Sensory signals (light, pressure, odor, taste) can be
received as well by TM receptors and can be transmitted into intracellular signals.

■ TM signaling uses
- TM receptors acti-
  vated by: extracellular
  ligands and sensory
  signals
- Voltage-gated ion
  channels
- Ligand-gated ion
  channels

Biochemistry of Signal Transduction and Regulation. 4th Edition. Gerhard Krauss
Copyright © 2008 WILEY-VCH Verlag GmbH & Co. KGaA, Weinheim
ISBN: 978-3-527-31397-6

Fig. 5.1: Mechanism of signal transduction at membranes.
(a) Signal transmission via ligand-controlled TM receptors. The ligand L binds to the extracellular domain of a TM receptor, whereby the receptor is activated for signal transmission to the cytosolic side. The cytosolic domain of the activated receptor R* transmits the signal to signal proteins next in sequence.
(b) Signal transduction via ligand-controlled ion channels. The ligand binds to the extracellular side of a receptor that also functions as an ion channel. Ligand binding induces the opening of the ion channel, there is an ion efflux and a change in the membrane potential. (c) Signal transduction via voltage-gated ion channels. A change in the membrane potential ΔV is registered by an ion channel which transitions from the closed to the open state.

(a) ligand-controlled transmembrane receptors

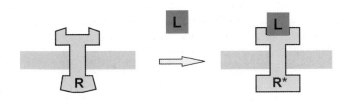

(b) Ligand-gated ion channels

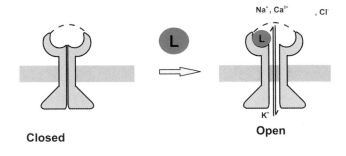

(b) Voltage-gated ion channels

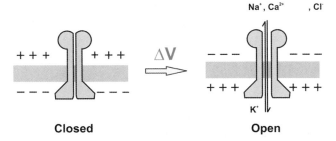

An example of a TM receptor that registers sensory signals is *rhodopsin*. Rhodopsin is a sensory receptor that plays a role in vision by receiving light signals and converting them into intracellular signals.

Signaling via Ligand- or Voltage-gated Ion Channels

One simply designed path of signal transmission is found in neuronal communication. TM receptors are also used for signal transmission here. These can have the character of a *ligand-gated ion channel* (Fig. 5.1b). Binding of a ligand (neurotransmitter or neurohormone) to the TM receptor leads to a conformational change of the receptor that enables the flow of ions through the membrane. In this case, the receptor presents itself as an ion channel with an open state controlled by ligand binding to the outer side (or also to the inner side).

■ TM receptors may be activated by intracellular signals

Another mechanism of signaling across the cell membrane uses changes in membrane potential. A change in membrane potential induces the opening of an ion channel and ions cross the membrane. In this case, the change of the ion's milieu is the intracellular signal. Ion channels with an open state regulated by changes in membrane potential are known as *voltage-gated ion channels* (Fig. 5.1c). The potential-driven passage of ions through ion channels is the basis for stimulation in nerves.

We also know of TM proteins for which the reception of the signal and activation take place on the *inner side* of the membrane. The cGMP-dependent ion channels involved in signal conduction in the vision process are ligand-regulated ion channels with an open state controlled by intracellularly created cGMP. Another example is given by the receptors for inositol triphosphate which are localized in the membrane of Ca^{2+} storage organelles and also have the character of ligand-controlled ion channels. Inositol triphosphate is an intracellular messenger substance that binds to the cytosolic side of the corresponding receptor located in the membrane of cell organelles. Ligand binding leads to opening of the ion channel via a conformational change and thus to influx of Ca^{2+} ions from the storage organelle into the cytosol (Section 6.5).

5.2
Structural Principles of TM Receptors

TM receptors are *integral membrane proteins,* i.e. they possess a structural portion that spans the membrane. An *extracellular domain*, a *TM domain* and an *intracellular or cytosolic domain* can be differentiated within the structure (Fig. 5.2a). In general these receptors function as dimers or higher oligomers composed of identical or different TM subunits (Fig. 5.2b). Furthermore, non-TM subunits can associate at the extra- or intracellular side.

■ Structural parts of TM receptors
– Extracellular domain
– TM domain
– Intracellular domain

(a)

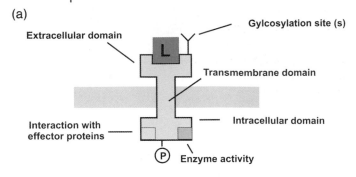

(b)

(c)

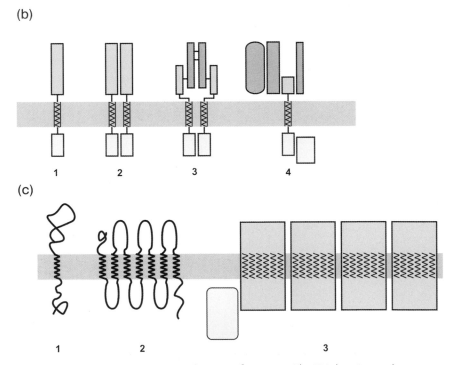

Fig. 5.2: Structural principles of TM receptors. (a) Representation of the most important functional domains of TM receptors. (b) Examples of subunit structures. TM receptors can exist in a monomeric form (1), dimeric form (2) and as higher oligomers (3 and 4). Further subunits may associate at the extracellular and cytosolic domains, via disulfide bridges (3) or via non-covalent interactions (4). (c) Examples of structures of the TM domains of receptors. The TM domain may be composed of an α-helix (1) or several α-helices linked by loops at the cytosolic and extracellular side (2). The 7TM receptors are a frequently occurring receptor type (Section 5.3). Several subunits of a TM protein may associate into an oligomeric structure (3), as is the case for voltage-controlled ion channels (e.g. K⁺ channel) or for receptors with intrinsic ion channel function.

5.2.1
The Extracellular Domain of TM Receptors

In many receptors, the extracellular domain contains the *ligand-binding site*. *Glycosylation sites*, i.e. attachment sites for carbohydrate residues, are also located nearby in the extracellular domain.

■ **Extracellular domain**
– Ligand binding
– Glycosylation
– Association of further subunits

The structure of the extracellular domain can be very diverse and is determined by the number of TM sections as well as the subunit structure of the receptor.

The extracellular localized protein portion may be formed from a continuous protein chain and may include several hundred amino acids. If the receptor crosses the membrane with several TM segments, the extracellular domain is formed from several loops of the protein chain that may be linked by disulfide bridges.

We also know of receptors in which only one subunit spans the membrane, whilst other subunits are bound to this subunit on the extracellular side via protein–protein interactions or via disulfide bridges (Fig. 5.2b and examples in Chapter 11).

5.2.2
TM Domain

The TM domains have different functions, according to the type of receptor. For ligand-controlled receptors, the function of the TM domain is to pass the signal on to the cytosolic domain of the receptor. For ligand- or voltage-controlled ion channels, the TM portion forms an ion pore that allows selective and regulated passage of ions.

The TM receptors span the around 5-nm thick phospholipid bilayer of the cell membrane with structural portions known as *TM elements*. The inner of a phospholipid layer is hydrophobic and, correspondingly, the surface of the structural elements that come into contact with the inner of the phospholipid double layer also has hydrophobic character.

■ **α-Helical TM domains**
– One to seven α-helices of 20–30 amino acids often arranged in bundles

The polypeptide chain of the vast majority of TM receptors contains a single α-helical TM element or seven α-helical elements organized as a bundle in the membrane. In the latter case the TM elements are linked by extracellular and intracellular loops. Generally, the *TM elements include 20–30 mostly hydrophobic amino acids*. At the interface with aqueous medium, we often find hydrophilic amino acids in contact with the polar head groups of the phospholipids. In addition, they mediate distinct fixing of the TM section in the phospholipid double layer.

Structure of TM Elements

High-resolution structural information about the TM elements of TM receptors could be obtained on the example of rhodopsin, the light-activated G-protein-coupled receptor (GPCR) of the vision process (Fig. 5.3). These data have confirmed that α-helices are the principal structural building blocks of the TM elements of membrane receptors. The TM helices are composed of 20–30 hydrophobic amino acids with some polar amino acids interspersed between or located at the helix ends. Most of the helices are arranged nearly perpendicular to the phospholipid bilayer and form bundle-like structures in which the helices are linked by loops of variable size. Ligand binding or reception of a sensory signal triggers a change in the mutual orientation of the helices that is transmitted into conformational changes of the cytoplasmic loops. These are sensed by the next components of

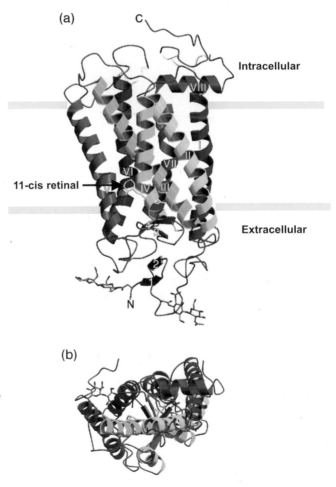

Fig. 5.3: Three-dimensional structure of rhodopsin. Two views of rhodopsin. (a) The seven α-helices of the G protein-coupled receptor rhodopsin weave back and forth through the membrane lipid bilayer (yellow lines) from the extracellular environment (bottom) to the cytoplasm (top). The chromophore 11-cis-retinal (yellow) is nestled among the TM helices. (b) View into the membrane plane from the cytoplasmic side of the membrane. Roman numerals indicate numbered helices.

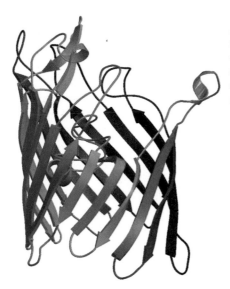

Fig. 5.4: The OmpF porin from *Escherichia coli* is an integral membrane channel-forming protein which spans the outer membrane in Gram-negative bacteria. The structure of a trimer of the OmpF porin is shown with the individual subunits in green, red and blue. In total, 16 β-bands are configured in the form of a cylinder and form the walls of a pore through which selective passage of ions takes place.

the signaling chain or result in the activation of an enzymatic activity at the cytoplasmic side of the membrane. A signal is thereby generated in the interior of the cell and is propagated further. In addition to α-helices, proteins also use β-structures to cross the membrane. The TM domain of the bacterial *OmpF porin* is made up of *β-elements* (Fig. 5.4). The β-elements, in this case, are not mostly made up of hydrophobic amino acids and form a barrel-like structure.

■ **Some TM domains are composed of β-elements**

5.2.3
Intracellular Domain of Membrane Receptors

Two basic mechanisms are used for conduction of the signal to the inner side of the membrane (Fig. 5.5):
– Via specific protein–protein interactions, the next protein component in the signal transmission pathway, the effector protein, is activated. The conformational change that accompanies the perception of the signal by the receptor creates a new interaction surface for proteins that are located downstream of the receptor. In the absence of the signal, this interaction surface is not available. Signal transmission therefore strictly depends on signal perception by the receptor and activation of the effector molecule must be preceded by activation of the receptor by a signal. Mechanistically, the active and inactive states of the receptor may be also regarded as fluctuations of receptor conformations. Agonists stabilize and fix the active conformation whereas the inactive conformation predomi-

■ **Activated TM receptors transmit signals mainly via**
– Activation of downstream effectors
– Triggering of enzyme activity

Fig. 5.5: General functions of TM receptors. Extracellular signals convert the TM receptor from the inactive form R to the active form R*. The activated receptor transmits the signal to effector proteins next in the reaction sequence. Important effector reactions are the activation of heterotrimeric G-proteins, of protein tyrosine kinases and of protein tyrosine phosphatases. The tyrosine kinases and tyrosine phosphatases may be an intrinsic part of the receptor or they may be associated with the receptor. The activated receptor may also include adaptor proteins in the signaling pathway or it may induce opening of ion channels.

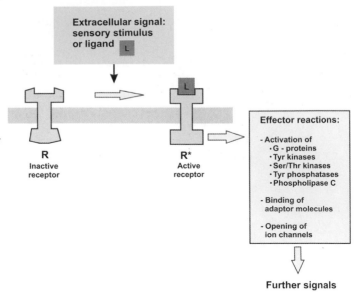

nates in the absence of the ligand and may be further fixed by antagonists.

– Arrival of the signal triggers enzyme activity in the cytosolic domain of the receptor, which, in turn, pulls other reactions along with it. The enzyme activity of the cytosolic domain is often tyrosine kinase activity; however, there are other examples where tyrosine phosphatase or Ser/Thr-specific protein kinase activity is activated. In all these examples, the cytoplasmic domain carries an enzyme activity regulated by ligand binding. The enzyme activity may be an integral part of the receptor, or it may also be a separate enzyme associated with the receptor on the inner side of the membrane (*cf.* Chapters 8 and 12).

Starting from the activated receptor, a large number of reactions can be set in motion (Fig. 5.5). One main route of signal transmission takes place by activation of G-proteins, another via activation of tyrosine-specific protein kinases, and a further route is via activation of ion channels. In the further course of G-protein-mediated signal transmission, secondary diffusible signals are often formed: the "second messenger" molecules (Chapters 3 and 6). These function as effectors and activate further enzyme systems in the sequence, especially protein kinases.

■ **Main downstream targets of activated TM receptors**
– G-proteins
– Protein kinases
– Ion channels
– Adaptor proteins

The activated receptor can also associate with adaptor molecules, which serve as coupling elements for further signal proteins.

5.2.4
Regulation of Receptor Activity

A physiologically important aspect of signal transmission via TM receptors is its regulation. The cell has various mechanisms available,
with the help of which the number and activity of TM receptors can
be regulated.

Regulation of TM receptor activity serves mainly two goals:

– *Desensitization, attenuation and downregulation.* A ubiquitous feature of signaling through TM receptors is the loss of cellular sensitivity following presentation of a stimulus. This attenuation or
desensitization can occur on a long-term or short-term timescale
and is accompanied by a reduced response of the receptor or by
a lowered number of receptor molecules.

– *Crosstalk with other signaling pathways.* Signaling through TM
receptors is subjected to crosstalk with other signaling pathways
allowing coordination and modulation of signaling events. This
crosstalk mostly involves phosphorylation of the cytosolic part of
the receptor by protein kinases activated through other signaling
paths and is also referred to as *heterologous desensitization.*

■ **TM receptors are
regulated by**
– Downregulation
  desensitization
– Crosstalk
– Phosphorylation and
  recycling
– Degradation

Receptor Phosphorylation and Receptor Recycling
The structural elements involved in regulation of receptor activity are
generally located in the cytosolic domain. These are, above all, protein sequences that permit phosphorylation of the receptor by protein kinases. Phosphorylation at Ser/Thr or Tyr residues of the cytosolic domain may lead to inactivation or activation of the receptor
and thus weaken or strengthen signal transmission. In this way,
Ser/Thr-phosphorylation is used in the process of internalization
of receptors in order to remove the receptor from circulation after
it has been activated (Section 5.3.4). The protein kinases involved
are often part of other signaling pathways and can link the activity
of the TM receptors to other signaling networks.

Ubiquitination and Degradation
Targeted degradation of TM receptors is another means of regulating
receptor activity. Signals for ubiquitination and subsequent degradation in the proteasome have been identified in the cytosolic domain
of TM receptors. A major mechanism of targeting TM receptors for
proteolysis uses signal-directed phosphorylation of the cytosolic
domain (Section 2.5.6.2).

5.3
GPCRs

Of the TM receptors that receive signals and conduct them into the cell interior, the *GPCRs* comprise about 60% and thus form the largest single family (reviewed in Luttrell, 2006). The human genome encodes around 720 different GPCRs. Of these about half are thought to encode sensory receptors that are involved in perception of light, taste and smell. For about two-thirds of the remaining receptors the ligand is known, whereas the rest are so called orphan receptors with no known ligand or function. As summarized in Tab. 5.1, the nature of stimuli that signal through GPCRs is very diverse and it has been estimated that about 80% of all known hormones and neurotransmitters signal through GPCRs.

Interestingly, we also know of GPCRs that are activated by proteases, e.g. thrombin (see also Fig. 5.8 below). In this case, the protease cleaves off a peptide from the N-terminus and the newly created N-terminus then serves as a tethered ligand for activation of these *protease-activated receptors.*

Ligand binding or reception of a physical signal is linked to activation of the GPCR. As a consequence, the receptor undergoes a conformational change that is transmitted to the inner side of the membrane, whereby the next sequential member of the signal chain is activated. This conducts the signal further via other reaction pathways (see Fig. 5.16). A characteristic structural feature of the GPCRs is the presence of *7TM helices* (Fig. 5.6). For the vast majority of 7TM receptors the next downstream located signaling protein is a heterotrimeric G-protein. Because some GPCRs may use both heterotrimeric G-proteins and other cytoplasmic proteins in their signaling, and others may depend on non-G-protein transducers only, alternative terms such as *heptahelical receptors, 7TM receptors* or *serpentine-like receptors* have been also coined for this receptor superfamily. However, in the present book the term GPCR is preferred. The routes for signal transmission from activated GPCRs that do not involve G-proteins will be discussed in Section 5.8.

■ **GPCRs**
- Signal through G-proteins
- Around 720 human genes
- 60% of all TM receptors

■ **GPCRs are also termed**
- Heptahelical receptors
- Serpentine-like receptors
- 7TM receptors

Tab. 5.1: Extracellular signals for activation of GPCRs.

Biogenic amines	Ca^{2+} ions	Phospholipids	Light
Amino acids	Pheromones	Fatty acids	Odorants
Peptides	Prostanoids	Nucleosides	Bitter and sweet
Glycoproteins		Nucleotides	gustatory substances

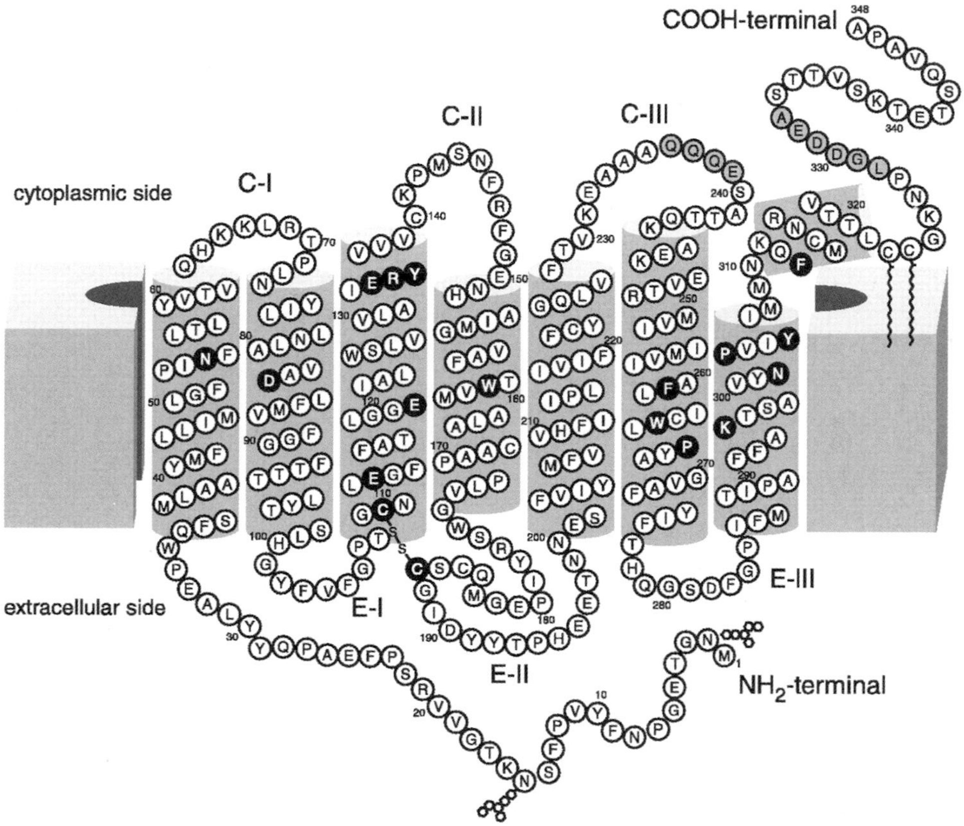

Fig. 5.6: Two-dimensional model of rhodopsin. The extracellular (intradiscal) and intracellular regions of rhodopsin each consist of three interhelical loops (given the prefixes E (extracellular)-I to E-III or C (cytoplasmic)-I to C-III. A conserved disulfide bridge is found on the extracellular side linking E-II with E-III. On the intracellular side, a short helix runs parallel to the membrane surface. In the native protein, the C-terminus carries two palmitoylated Cys residues which function as membrane anchors causing formation of a putative fourth intracellular loop.

5.3.1
Classification of GPCRs

The current classification of GPCRs includes the classes A, B, C, Frizzled and the olfactory receptor family. The classification is based on sequence similarity, the size of the extracellular loops, the presence of key residues and the formation of disulfide bonds (Fig. 5.7).

■ **GPCR families**
 − Family A
 − Family B
 − Family C
 − Frizzled
 − Olfactory/gustatory

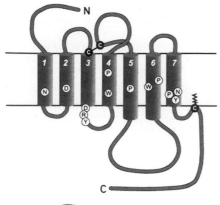

Family A. Rhodopsin/ß2 adrenergic receptor-like

Biogenic amine receptors (adrenergic, serotonin, dopamine, muscarinic, histamine)

CCK, endothelin, tachykinin, neuropeptide Y, TRH, neurotensin, bombesin, and growth hormone secretagogues receptors plus vertebrate opsins

Invertebrate opsins and bradykinin receptors

Adenosine, cannabinoid, melanocortin, and olfactory receptors.

Chemokine, fMLP, C5A, GnRH, eicosanoid, leukotriene, FSH, LH, TSH, fMLP, galanin, nucleotide, opioid, oxytocin, vasopressin, somatostatin, and protease-activated receptors plus others.

Melatonin receptors and other non-classified

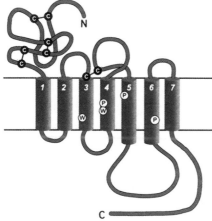

Family B. Glucagon/VIP/Calcitonin receptor-like

Calcitonin, CGRP and CRF receptors

PTH and PTHrP receptors

Glucagon, glucagon-like peptide, GIP, GHRH, PACAP, VIP, and secretin receptors

Latrotoxin

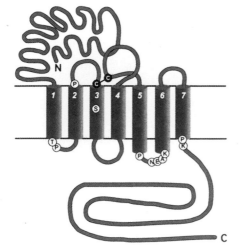

Family C. Metabotropic neurotransmitter/ Calcium receptors

Metabotropic glutamate receptors

Metabotropic GABA receptors

Calcium receptors

Vomeronasal pheromone receptors

Taste receptors

◀ **Fig. 5.7:** Classification of GPCRs. The GPCRs can be divided into three major subfamilies (Gether, 2000). *Family A receptors* are characterized by a series of highly conserved key residues (black letter in white circles). In most family A receptors, a disulfide bridge is connecting the E-II and E-III loops. In addition, a majority of the receptors have a palmitoylated cysteine in the cytoplasmic C-terminus. Ligands include the biogenic amines (adrenaline, serotonine, dopamine, histamine), neuropeptide Y, adenosine, chemokines and melatonine, among others. *Family B receptors* are characterized by a long extracellular N-terminus containing a series of cysteine residues presumably forming a network of disulfide bridges. Representative members of the family B receptors include calcitonine receptor, glucagon receptor and parat hormone receptors. *Family C receptors* are characterized by a very long N-terminus forming the extracellular ligand binding site. There is only one putative disulfide bridge and the third cytoplasmic loop is very small. The taste receptors, the metabotropic glutamate receptors, the γ-aminobutyric acid (GABA) receptors and Ca^{2+}-receptors belong to this class, among others.

Family A

This family is characterized by a conserved disulfide bridge and by a consensus motif DRY at the cytoplasmic side. The binding sites for all small molecule ligands are buried in between the 7TM helices.

Family B

Family B receptors have a large N-terminal domain and several disulfide bridges. Some members form heterodimers with a single-span TM protein, named *RAMP*, which modifies either ligand binding or signaling properties of the receptor.

Family C

Receptors of family C contain a large N-terminus. Homo- and heterodimerization has been demonstrated for receptors of this family

Frizzled Receptors

The Frizzled receptors constitute a unique family among the GPCRs because receptor activity is modulated by interactions with additional plasma membrane receptors.

olfactory and Gustatory Receptors

The olfactory and gustatory receptor family is the largest GPCR family of vertebrates with about 400 members in man and 1200 in mouse. Much lower numbers of this receptor types are found in invertebrates.

5.3.2
Structure of GPCRs

The only structural feature common to all GPCRs is the presence of the 7TM helices connected by alternating extracellular and intracellular loops, with the N-terminus located on the extracellular side and the C-terminus on the intracellular side.

Bovine rhodopsin is the only GPCR for which the three-dimensional structure could be directly visualized by high-resolution X-ray analysis (reviewed in Stenkamp et al., 2005). The two dimensional model and the high resolution structure of bovine rhodopsin is shown in Fig. 5.6.

The sequence similarity of the GPCRs is confined largely to the TM helices and therefore the highly resolved rhodopsin structure may provide a frame upon which the three-dimensional structure of the TM part of the huge family of GPCRs can be modeled.

The *rhodopsin structure* shows a *bundle of 7TM helices* that vary in length from 20 to 33 residues. Another helix, helix VIII, is found on the intracellular side and runs parallel to the membrane. The TM bundle is stabilized by a network of hydrogen-bond interactions between polar amino acids contained in the TM α-helices and it is thought that the hydrogen-bonding network is a critical determinant of the exact arrangement of the α-helices within the bundle. The *chromophore 11-cis-retinal*, which mediates the light activation, is covalently bound in the interior of the bundle. It keeps the structure in the ground state. Light absorption excites the chromophore and activates the rhodopsin by changing the hydrogen bonding pattern within the bundle. As a consequence the TM helices reorient within the bundle with a concomitant change in intracellular loops and the C-terminus. Activation of the downstream effector protein, the G-protein, is mediated by the intracellular part of rhodopsin. This region includes three loops, helix VIII and the C-terminus, and forms an extended surface for the interaction with the downstream effector, the G-protein. Details of the activation process are, however, still unknown.

Extra- and Intracellular Loops

The extracellular loops are highly variable in size and harbor sites for posttranslational modifications, e.g. in the form of the consensus sequence Asn-X-Ser/Thr for an *N*-linked glycosylation. Furthermore, the extracellular parts contain frequently conserved Cys residues and disulfide bridges.

The intracellular loops and the intracellular C-terminus mediate the interaction with the downstream effector, the G-proteins or other signaling molecules. Posttranslational modification in the form of *palmitoylation of the cytosolic domain* (Section 3.7.2) has been demonstrated for the α- and β-adrenergic receptor. The palmitoylation takes place on a Cys residue localized at the C-terminus. The modification serves to anchor the C-terminus in the membrane. Furthermore, phosphorylation sites are found on the intracellular parts.

5.3.3
Ligand Binding and Mechanism of Signal Transmission

The location of the ligand-binding site is dependent on the size of the extracellular N- terminus and loops. A schematic representation of the ligand-binding sites for different receptor types is shown in Fig. 5.8.

The area of ligand binding has been particularly well defined for the receptors of classical "small ligands" (adrenaline, noradrenaline, dopamine, serotonine, histamine). Targeted mutagenesis, biochemical, biophysical and pharmacological investigations have shown that these ligands are bound in a binding crevice formed by the TM helices. In agreement with this model, it has been shown that the extra-

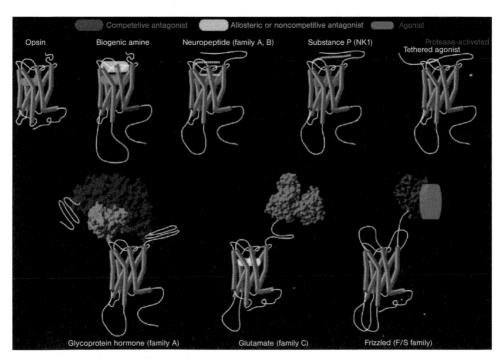

Fig. 5.8: Schematic models of ligand–receptor complexes for structurally diverse ligands interacting with GPCRs. The binding sites for all small molecule ligands in family A receptors are buried in between the 7TM domains. In many neuropeptide receptors, the binding sites for the native peptides involve residues from both extracellular regions and 7TM domains (e.g. angiotensin II), whereas the binding site for substance P in the NK_1 neurokinin receptor involves only extracellular domains. Glycoprotein hormones [e.g. luteinizing hormone (LH)] interact with the leucine-rich repeat region in the N terminus and extracellular loops of their receptors. Thrombin and serine proteases activate protease-activated receptors by cleavage of the N terminus. Thereafter, the new N termini activate the receptor as a tethered agonist by interacting with residues in ECL2. Family C contains a large Venus flytrap module (VFM), which contains the binding site for competitive antagonists and agonists in a cleft between two domains. A competitive antagonist-bonded complex is shown here. The N-terminal CRD in Frizzled receptors contains the binding site for their native ligands, the Wnt proteins.

■ **Activation of GPCRs**
- Reorientation of helical bundle
- New interaction surface at cytosolic site

cellular and intracellular sequence portions of the receptors are not needed for ligand binding in these cases. Rhodopsin is a unique case, since the retinal ligand is covalently attached by Schiff-base linkage to a Lys residue of TM helix VII. In that case, too, the binding site is deeply buried in the interior of the TM segment (Fig 5.3).

For the receptors that have peptides or proteins as ligands, structural portions of the extracellular domain, in addition to areas of the TM domain, are involved in ligand binding. The receptors of subfamily C that contain a large extracellular N-terminal domain bind their ligands (e.g. glutamate, γ-amino butyric acid) in this region.

The mechanism by which the activated receptor talks to the G-protein is still a matter of speculation. Generally, the switch function of the receptor is considered in terms of allosteric conformational changes within the 7TM bundle (reviewed in Meng and Bourne, 2001; Ridge and Palczewski, 2007). For rhodopsin, a coupled motion of helices 3, 6 and 7 as rigid rods has been proposed to occur during light-driven activation. These changes in the structure of the TM bundle have been postulated to lead to a different location of the cytoplasmic ends of the TM helices which are considered important interaction sites with the G-protein. As a consequence, a high-affinity surface is created for binding of the heterotrimeric G-protein (Section 5.5.4). The heterotrimeric G-protein, which exists as the inactive GDP form, now binds via its α and γ subunit to the activated receptor and is activated itself. An exchange of GDP for GTP takes place on the α subunit and the βγ subunit of the G-protein dissociates (Section 5.5.3). Once the G-protein is activated, it frees itself from the complex with the receptor, which either returns to its inactive ground state or activates further G-proteins.

5.3.4
Regulation of GPCRs

The GPCRs are part of signaling networks with links to several major signaling paths of the cell. These links provide positive and negative signals to regulate the duration and intensity of GPCR signaling in coordination with other incoming signals. Given the large number of the GPCRs found in the human genome and their very diverse signaling functions, it is not surprising that many proteins have been discovered that interact with the receptor in an agonist-dependent or -independent manner. Most regulatory influences are mediated by the cytoplasmic loops and the C-terminal tail of the receptor. In addition to the canonical binding partner at the cytoplasmic side, the heterotrimeric G–protein, other proteins have been shown to interact with cytoplasmic regions of the GPCR, especially with its C-terminus to modulate receptor function.

■ **Proteins interacting with C-terminus of GPCRs**
- RAMPS
- Scaffolding proteins
- GRKs
- PDZ-containing proteins
- Arrestins

The following types of proteins have been shown to associate with the C-terminal tail of GPCRs:

– *Receptor-activity modifying proteins* (RAMPs). These are TM proteins that modulate diverse functions of GPCRs belonging to class B.
– *Scaffolding proteins*. Examples are members of the adaptor protein family A-kinase anchor protein (AKAP), the scaffolding proteins Shank and spinophilin.
– *PDZ-containing proteins*. Many GPCRs contain a PDZ-binding motif in the C-terminus. The PDZ module (Section 8.2.5) serves as scaffolding or interaction module for the formation of higher protein complexes.

As illustrated by these examples, protein binding to the C-terminus serves mainly the formation of higher signaling complexes and the trafficking and specific localization of the receptor.

5.3.5
Switching Off and Desensitization of 7TM Receptors

A phenomenon often seen in TM receptors, in general, and in GPCRs, in particular, is desensitization (Fig. 5.9). *Desensitization* means a weakening of the signal transmission under conditions of long-lasting stimulation by hormones, neurotransmitters or sensory signals. Despite the persistent effect of extracellular stimuli, the signal is no longer passed into the cell interior, or only in a weakened form, during desensitizing conditions. Desensitization may occur in a homologous or in a heterologous way. In the latter case, the GPCR is modulated by signals derived from other signaling pathways, e.g. from nuclear receptor signaling or receptor tyrosine kinase (RTK) signaling. This type of desensitization is independent of agonist binding.

■ **Homologous desensitization**
– Agonist dependent
Heterologous desensitization
– Agonist independent

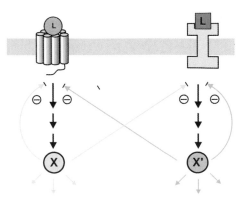

Fig. 5.9: General principle of desensitization of GPCRs. Desensitization of a hormone-bound receptors can take place by two principle routes, schematically represented in the figure. A suppressing influence may be exerted on the receptor system via proteins (X) of a signal chain, triggering inhibition of the signal chain. Receptor systems may also mutually influence one another in that a signal protein X formed in one signal chain mediates the desensitization of another receptor system R* and *vice versa*.

Regulation of receptor activity by desensitization occurs on a short and long timescale. Both processes are initiated by phosphorylation of the GPCR on its cytoplasmic region by protein kinases. During short-term desensitization receptor signaling is blocked by binding of arrestin to the phosphorylated sites. A slow, long-term downregulation is achieved by arrestin-mediated internalization and recyclization of the receptor.

Heterologous Desensitization: Phosphorylation by cAMP-dependent Protein Kinases [Protein Kinase A (PKA)] or Protein Kinase C (PKC)

Phosphorylation of the cytoplasmic domain of 7TM receptors can take place via *cAMP-dependent protein kinases (PKA)* or via *PKC* (Chapter 7) (Fig. 5.10) in an agonist-independent way. This is a negative feedback mechanism. The hormonal activation of the receptor leads, via G-proteins and adenylyl cyclase/cAMP, to activation of protein kinases of type A (Sections 5.7.1 and 6.1, and Chapter 7). The activated protein kinases phosphorylate the receptor on sites that are postulated to be G-protein contact sites and these modifications have been assumed to interfere directly with G-protein activation. Regulation via adenylyl cyclase/cAMP/PKA is an example of a heterologous desensitization, since adenylyl cyclase can be activated by a variety of signals originating from different signaling pathways (Section 5.7.1).

Furthermore, agonist-independent Ser/Thr phosphorylation can be used as a switch for coupling a given receptor to different G_α subunits. PKA-mediated phosphorylation of the β-adrenergic receptor has been shown to switch coupling of the receptor from G_s to G_i and initiate a new set of signaling events (Daaka et al., 1997).

■ Heterologous desensitization by PKA or PKC

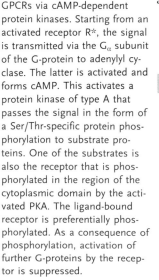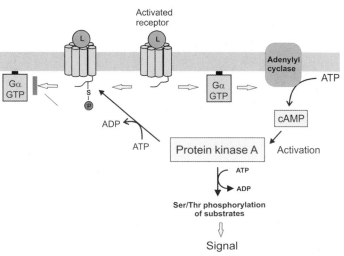

Fig. 5.10: Desensitization of GPCRs via cAMP-dependent protein kinases. Starting from an activated receptor R*, the signal is transmitted via the G_α subunit of the G-protein to adenylyl cyclase. The latter is activated and forms cAMP. This activates a protein kinase of type A that passes the signal in the form of a Ser/Thr-specific protein phosphorylation to substrate proteins. One of the substrates is also the receptor that is phosphorylated in the region of the cytoplasmic domain by the activated PKA. The ligand-bound receptor is preferentially phosphorylated. As a consequence of phosphorylation, activation of further G-proteins by the receptor is suppressed.

Homologous Desensitization

This process involves phosphorylation of the GPCR by *GPCR kinases (GRK)* (Metaye et al., 2005).

The major mechanism for the homologous desensitization of agonist-bound GPCRs consists of a two-step process in which the agonist-bound receptor is phosphorylated by a GRK and then binds an arrestin protein which interrupts signaling to the G-protein. Only the activated, i.e. occupied by an agonist, receptor is phosphorylated and downregulated.

■ **Homologous desensitization: by GRKs (e.g. rhodopsin kinase and βARK)**

There are seven GRKs in man (GRK1–7). The well-characterized GRK for rhodopsin, rhodopsin kinase, belongs to the GRK1 subfamily, and β2-adrenergic receptor specific GRK is member of the GRK2 subfamily. This kinase is known as the *β-adrenergic receptor kinase (βARK)*.

Some of the GRKs shuttle between the cytoplasm and the membrane in an activation-dependent manner. During translocation of GRKs to the membrane-localized receptor, the βγ subunit of the G-protein as well as binding of phosphatidylinositol (PtdIns) triphosphates (Section 6.6.2) play an important role. When an agonist binds to a GPCR, it causes the receptor to associate a heterotrimeric G-protein, leading to dissociation into its α and βγ subunits. The βγ dimer, which carries a prenyl membrane anchor, binds to the GRK and thereby promotes its membrane association. In a cooperative way, binding of phosphatidylinositol messengers to the pleckstrin homology (PH) domain (Section 8.2.3) of GRKs enhances binding of the GRK to the membrane. Membrane association of some GRKs is also mediated by farnesyl or palmitoyl anchors.

■ **Membrane attachment of GRKs is enhanced by**
– βγ dimer
– PtdIns compounds bound to PH domain

Receptor phosphorylation by GRKs triggers several reactions (Fig. 5.11):

Binding of Arrestin as a Key for Desensitization

Phosphorylation of the receptor by GRKs leads to creation of a high-affinity binding site for proteins known as *arrestins*. The family of arrestins includes the visual arrestins that are specific for rhodopsin and the β-arrestins 1 and 2. The latter two isoforms are most important for the desensitization of activated receptors as e.g. the β-adrenergic receptor. Binding of β-arrestins has several consequences:

– The phosphorylated receptor is decoupled from the interaction with the heterotrimeric G-protein next in the sequence so that signal transmission is suppressed. Arrestin binding serves, for example, to rapidly weaken signal transmission during the vision process, during conditions of long lasting light stimulus.

– Following β-arrestin binding, the phosphorylated receptor can be translocated into the cell interior. In this process, which occurs on a longer time scale, β-arrestin serves as an adaptor that targets the

■ **Arrestin**
– Binds to phosphorylated C-terminus of GPCR
– Triggers internalization of GPCR
– Couples GPCRs to other signaling paths

Fig. 5.11: Receptor desensitization: phosphorylation, arrestin binding and internalization. The activated, agonist-bound receptor is phosphorylated on the cytoplasmic region by a GRK. The phosphate residues serve as attachment sites for β-arrestin which has protein kinases of the MAPK cascade associated. This serves as a trigger for internalization of the receptor to endosomes. The receptor may now be dephosphorylated and transported back to the cell membrane (not shown in the figure).

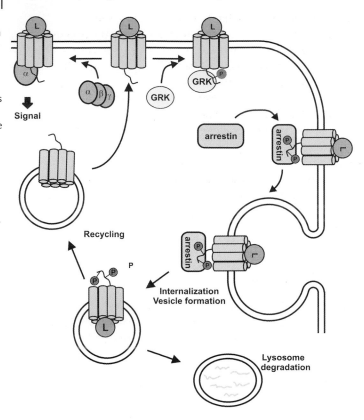

receptor for internalization via endocytosis. In addition to receptor binding, the β-arrestins bind to clathrins and other proteins that form the clathrin-coated vesicle machinery. The receptor is thereby internalized in the membrane-associated form, dephosphorylated and then transported back to the cell membrane. The translocation into the cell interior serves, in particular, to weaken signal transmission during conditions of long-lasting hormonal stimulation.

Binding of Arrestin: Signal Switching

In another reaction, Ser/Thr phosphorylation of GPCRs can be used as a switch for coupling a given receptor to different G_α subunits, the downstream effector protein. Protein kinase A-mediated phosphorylation of the β-adrenergic receptor has been shown to switch coupling of the receptor from G_s to G_i and trigger a new set of downstream signaling reactions.

Binding of Arrestin: Coupling to other Signaling Paths

Arrestin has been now recognized as an important component for the coupling of GPCRs to other signaling pathways in a way that is independent of G-proteins. This new aspect of GPCR signaling will be discussed in Section 5.8.

Spinophilin Regulation of GPCR Signaling

Many arrestin functions are antagonized or modulated by a multidomain protein named spinophilin. Spinophilin binds to the third intracellular loop of the α_2-adrenergic receptor, thereby interfering with arrestin binding to the receptor (Wang et al., 2004).

The spinophilin protein has been also found to bind to *regulator of G-protein signaling* (RGS) 2 and appears to be used for a fine-tuning of receptor signaling (Wang et al., 2005).

■ **Spinophilin interferes with arrestin**

5.3.6
Dimerization of GPCRs

Although the GPCRs were generally believed to function as monomeric entities, there is now increasing evidence that GPCRs may form functional dimers *in vivo*. Biochemical and biophysical studies suggest that, for example, the β_2-adrenergic receptor exists as a constitutive dimer in the cell (reviewed in Fotiadis et al., 2006,). Dimerization has been shown to alter the ligand-binding, signaling and trafficking properties of GPCRs. In addition to homodimers, the formation of heterodimers with related members of the same subfamily has been reported for GPCRs of subfamily B (reviewed in Prinster et al., 2005). The structural, functional and mechanistic consequences of the formation of the oligomeric receptor complexes remain to be elucidated.

■ **GPCRs may function as dimers**

5.4
Regulatory GTPases

The heterotrimeric G-proteins, the major effector proteins of the 7TM receptors, belong to the large family of regulatory GTPases; these bind GTP and hydrolyze it, thereby functioning as a switch in central cellular processes. The family of regulatory GTPases is also called the GTPase superfamily.

5.4.1
The GTPase Superfamily: General Functions and Mechanism

Proteins of the GTPase superfamily are found in all plant, bacterial and animal systems. The following examples illustrate the central functions of the regulatory GTPases in the cell.

Regulatory GTPases are involved in:
– Protein biosynthesis on ribosomes.
– Signal transduction at membranes.
– Visual perception.
– Sense of smell and taste.
– Control of differentiation and cell division.
– Translocation of proteins through membranes.
– Transport of vesicles in the cell.

The members of the GTPase superfamily show an extensively conserved reaction mechanism. A common trait is a switching function that enables a reaction chain to be switched on or off.

Switch Function of the GTPases and the GTPase Cycle

The regulatory GTPases are involved in signaling chains by functioning as a switch. Incoming signals from activated upstream signaling components are received by the GTPases and are passed on to downstream components of the signaling chain that are activated themselves for transporting the signal further. The switch function of GTPases is based on a cyclical, unidirectional transition between an *active, GTP-bound form, the on-state* and *an inactive, GDP-bound form, the off-state* (Fig. 5.12). In both inactive and active forms, the proteins of the GTPase superfamily possess a specific affinity to other signaling proteins that are upstream or downstream parts of the reaction chain. The GTPases bring about the transition between the active and inactive states in a *cyclic process* that can only run in one direction because of the irreversible hydrolysis of GTP.

The upstream signaling component must be in the active state to interact with the inactive GDP state of the GTPase. This interaction triggers an exchange of bound GDP for GTP and induces thereby the transition into the on state, the GTPase·GTP form. From the GTP-bound state, the signal is passed on to the effector molecule next in sequence which is in turn activated for further signal transmission. Hydrolysis of the bound GTP by an intrinsic GTPase activity converts the GTPase from the on state into the off state and disrupts the signal transmission. The system is now again in the inactive GTPase·GDP ground state and is ready for the receipt of a further signal.

■ **Regulatory GTPases act as switches**
– Off-state:
  GDP-bound state
– On-state:
  GTP-bound state

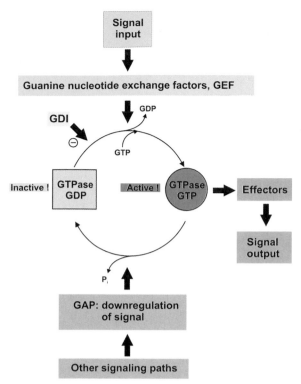

Fig. 5.12: The switch function of the regulatory GTPases. The GTP form of the regulatory GTPases represents the "switched on" form of the GTPase, the GDP form, in contrast, the "switched off" form. The switch function of the regulatory GTPases may be controlled by guanine nucleotide exchange factors, by GAPs and GDIs. The regulatory GTPases run through a GTPase cycle which signals flow into via GEFs and are conducted further in the form of the GTPase·GTP complex to effector molecules further down the sequence. Hydrolysis of the bound GTP ends the activated state. The rate of GTP hydrolysis is either intrinsically determined or may be accelerated via GAPs.

Modulation and Regulation of the Switch Function

There are two processes in the GTPase cycle where the cycle can be switched on and switched off, and where the main regulation of GTPase signaling occurs: GDP–GTP exchange and GTP hydrolysis. For most GTPases these reactions are intrinsically very slow and their acceleration by specific proteins is used by the cell to switch signaling on and off or to modulate it.

The following types of proteins are the main regulators of information flow through regulatory GTPases:

– *G-nucleotide exchange factors (GEFs): signal input.* The receipt of a signal by the regulatory GTPases manifests itself as an increased rate of dissociation of GDP. Most GTPases bind GDP very strongly and hence the intrinsic rate of GDP dissociation is very low. The rate of dissociation of GDP may be increased by specific proteins known as GEFs that function as the upstream signaling protein for the GTPases. For the heterotrimeric G-proteins, the agonist-bound, activated receptor is the exchange factor.

■ **GEFs deliver input signals by promoting GDP/GTP exchange**

– *GTPase-activating proteins (GAPs): signal attenuation and termination.* Most GTPases show a very slow intrinsic rate of GTP hydrolysis which is equivalent to a long lifetime of the activated state.

■ **GAPS negatively regulate G-protein signaling by enhancing GTPase up to 10^5-fold**

■ Examples of proteins with GAP function
– Ras-GAPs
– RGS
– G-protein effectors, e.g. cGMP phosphodiesterase

GAPs increase of the rate of GTP hydrolysis and thereby reduce the lifetime of the active, GTP-bound state. The GAP protein class is an important instrument for control of the rate of signal transmission. Activation of the GAPs leads to weakening of signal transmission and the *GAPs mostly function as negative regulators* for signaling by GTPases. The Ras protein has low intrinsic GTPase activity. This may be increased around 10^5-fold by the corresponding GAP (see also Chapter 9). The various GTPases may differ to a large extent in the rate of GTP hydrolysis and thereby in the influence of GAPs. The Ras protein and the $G_{\alpha,t}$, known as transducin, involved in the process of vision, are cited as examples. In comparison, the intrinsic rate of GTP hydrolysis of transducin is around 100-fold higher than that of the Ras protein. The effector molecule downstream of transducin, the cGMP phosphodiesterase, is involved in GTPase stimulation in this case. The γ subunit of the phosphodiesterase functions in cooperation with a member of the RGS protein family (Section 5.5.7), i.e. RGS9, as the GAP here and stimulates GTPase activity of the transducin by about two orders of magnitude. Often, the activity of the GAPs is regulated by other signaling pathways. Thus, a regulatory influence on signal transmission via G-proteins can be achieved from another signaling pathways.

■ GDIs inhibit GDP or GTP dissociation from complex with G_α

– *Guanine nucleotide dissociation inhibitors (GDI)*. Dissociation of GDP or of GTP may be inhibited by specific proteins known as GDIs. Proteins with this function act on members of the superfamily of Ras proteins (Section 9.4). The GDIs have the function, above all, to provide a cytosolic pool of inactive, GDP-bound proteins.

5.4.2
Inhibition of GTPases by GTP Analogs

Nonhydrolyzable GTP analogs are an indispensable tool in the identification and structural and functional characterization of GTPases. The GTP analogs shown in Fig. 5.13, GTPγS, β,γ-methylene GTP and β,γ-imino GTP, are either not hydrolyzed by GTPases or only very slowly. Addition of these analogs fixes the G-protein in the active form; it is permanently "switched on". For cellular signal transduction, this means permanent activation of the signal transmission pathway. In many cases, a role of G-proteins in a signal chain was inferred from the observation that nonhydrolyzable GTP analogs bring about a lasting activation of signal transmission. The GTP analogs were equally important for structural determination of the activated form of GTPases. Formation of a stable complex between the nonhydrolyzable GTP analog and different GTPases has enabled crystallization and structure determination of the complex in its activated form.

■ Nonhydrolyzable GTP-analogs
– GTPγS
– β,γ-methylene GTP
– β,γ-imino GTP
– Fix GTPase in GTP-state

γ-S-GTP

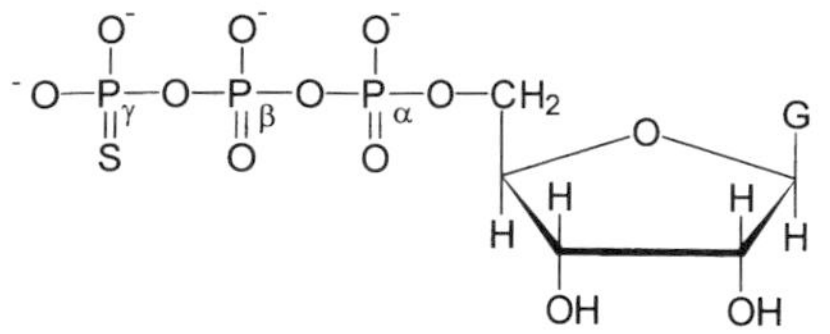

β, γ-methylene GTP

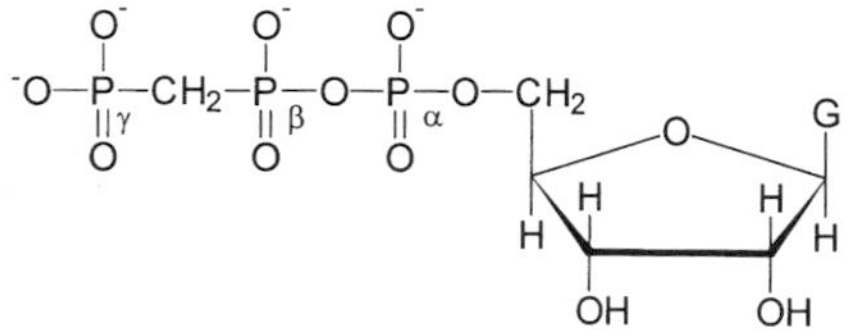

β, γ-imino GTP

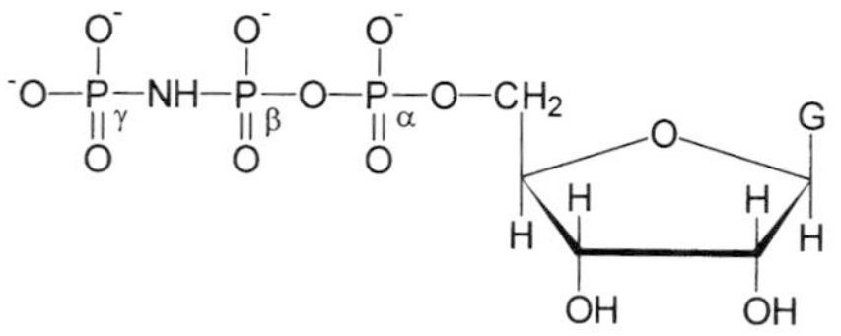

Fig. 5.13: Examples of nonhydrolyzable GTP analogs.

5.4.3
The G-domain as a Common Structural Element of the GTPases

A common property of the GTPases is the enzymatic activity of GTP hydrolysis. GTP binding and hydrolysis take place in a domain of the GTPases known as the *G-domain*. The G-domain is found in all GTPases. Figure 5.14 shows the G-domain of the bacterial elongation factor EF-Tu. In all GTPase structures known at present, the G-domain has very similar architecture and very similar means of binding the guanine nucleotide. The sequence element $GX_4GK(S/T)$ is a consensus sequence for guanine nucleotide binding; this sequence is involved in binding the β- and γ-phosphate of GTP and GDP, and is also known as the P-loop. Other consensus sequences, such as RX_2T and DX_2G, are involved in both binding the γ-phosphate and in the GTPase reaction (X = any amino acid). A further consensus sequence (N/T)(K/Q)XD and SA interacts with the guanosine.

Fig. 5.14: Structure of the G-domain of the elongation factor EF-Tu from *Thermus thermophilus* with bound GppNHp, according to Berchtold et al. (1993). The nonhydrolyzable analog GppNHp, the P loop, and the switch regions I and II are shown, which play an important role in transition from the inactive GDP form to the active GTP form.

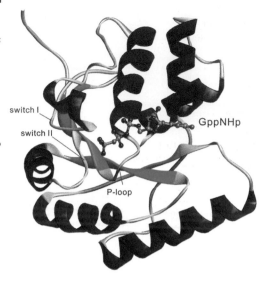

5.4.4
GTPase Families

The superfamily of GTPases, with over 100 members, is subdivided according to sequence homologies, molecular weight and subunit structure into further (super)families. These are the families of the heterotrimeric G-proteins, the Ras superfamily of small GTPases, and the family of initiation and elongation factors (Fig. 5.15).

The heterotrimeric G-proteins are built of up three subunits, with the GTPase activity localized on the largest subunit (Section 5.5). The members of the Ras family of GTPases, in contrast, are monomeric proteins with a molecular weight of around 20 kDa (Chapter 9).

A further functionally diverse class is made up of the proteins involved in protein biosynthesis and membrane transport. GTPases with functions in protein biosynthesis include the elongation factors, termination factors and peptide translocation factors. These are mostly monomeric proteins with molecular weights of 40–50 kDa. GTPases of this class are also found in protein complexes such as the *signal recognition particle (SRP)* and the corresponding receptor. Both protein complexes are needed during ribosomal protein biosynthesis, for transport of newly synthesized proteins through the endoplasmic reticulum.

■ **Major GTPase families**
– Heterotrimeric
  G-proteins
– Ras superfamily
– Initiation and
  elongation factors

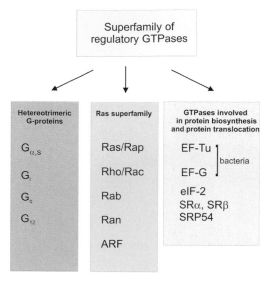

Fig. 5.15: The superfamily of regulatory GTPases. SRα,SRβ, SRP54 = GTPases involved in protein translocation at the endoplasmic reticulum.

5.5
Heterotrimeric G-proteins

The heterotrimeric G-proteins are the most prominent reaction partners in signal transmission via 7TM receptors, which is why these receptors are also known as GPCRs (reviewed in Offermanns, 2003). From the G-protein, the signal is then passed on to the effector protein next in the sequence.

A common structural feature of the G-proteins is their construction from three subunits (Fig. 5.16), a large α subunit of 39–46 kDa, a β subunit of 36 kDa and a γ subunit of 8 kDa. The α subunit has a binding site for GTP or GDP and carries the GTPase activity. The β and γ subunits exist as a tightly associated complex and are active in this form. All three subunits show great diversity. There are at least 16 genes for α subunits, five genes for β subunits and 12 genes for γ subunits in the human and mouse genomes.

Specificity of the switch function is mostly determined by the α subunit: the α subunit carries out the specific interaction with the receptors preceding it in the signal chain and with the subsequent effector molecules. The βγ complex is also involved in signal transmission to the effector proteins.

■ **Heterotrimeric G-proteins of mouse/man**
– α subunit: 39–46 kDa, more than 16 genes
– β subunit: 36 kDa, five genes
– γ subunit: 8 kDa, 12 genes

Fig. 5.16: Structure and activation of the heterotrimeric G-proteins. Reception of a signal by the receptor activates the G-protein, which leads to exchange of bound GDP for GTP at the α subunit and to dissociation of the βγ complex. Further transmission of the signal may take place via G_α·GTP or via the βγ complex, which interact with corresponding effector molecules. The α and γ subunits are associated with the cell membrane via lipid anchors. Signal reception and signal transmission of the heterotrimeric G-proteins take place in close association with the cell membrane. This point is only partially shown in the figure.

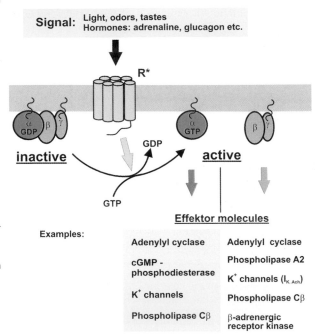

5.5.1
Classification of the Heterotrimeric G-proteins

Most functions of signal transmission by G-proteins are realized by the α subunit. Since different G-proteins interact with very different partners, there are significant differences in the structure of the α subunits. Because of the common GTPase domain and the common interaction with the βγ subunits, however, there are also considerable sequence homologies. Based on comparison of the amino acid sequences, the G_α-proteins are divided into *four families, the G_s, G_i, G_q and G_{12}*. These families are summarized in Tab. 5.2, together with representative members and their characteristic properties. An overview of the main patterns of signaling through the various G_s families is given in Fig. 5.17.

■ **G_α families**
– G_s
– G_i
– G_q
– G_{12}

G_s Subfamily
A characteristic of α subunits of the G_s subfamily is that they are inhibited by cholera toxin (Section 5.5.2). The members of the G_s subfamily are activated by hormone receptors, by odor receptors and by taste receptors. G_s-proteins mediate, e.g. signal transmission by β-adrenergic receptors and by glucagon receptors. During perception of taste, the taste receptors are activated, and these then pass the signal on via the olfactory G-protein G_{olf}. Perception of "sweet" taste is

■ **G_s**
– Activation of adenylyl cyclase

also mediated via G_s-proteins Transmission of the signal further involves an adenylyl cyclase in all cases, the activity of which is stimulated by the G_s-proteins.

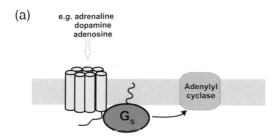

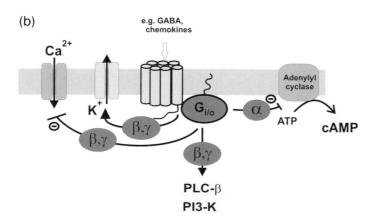

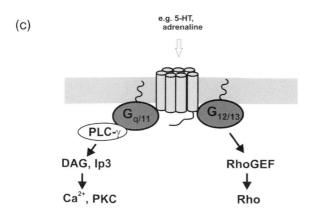

Fig. 5.17: Common patterns of coupling between GPCRs and G-proteins. PLCβ see Section 5.7.2; PI3K, see Section 7.4; PKC, see Section 7.5; IP3 = inositol-1,4,5-trisphosphate, see Section 6.5; DAG, see Section 7.5; Rho = Rho guanine nucleotide exchange factor, see Section 9.9.1; 5-HT = 5-hydroxytryptamine.

Tab. 5.2: Classification of the heterotrimeric G-proteins according to the α subunits.

Subunit	tissue	Examples of receptors	Effector protein, function
G_s			
α_s	ubiquitous	β-adrenergic receptor, glucagon receptor	adenylyl cyclase ↑ Ca^{2+} channels ↑
α_{olf}	nasal epithelium	olfactory receptor	adenylyl cyclase ↑
G_i			
$\alpha_{i1}, \alpha_{i2}, \alpha_{i3}$	mostly ubiquitous	α_2-adrenergic receptor	K^+ channels ↑ Ca^{2+} channels ↓
α_{oA}	brain	α_2-adrenerger receptor	K^+ channels ↑ Ca^{2+} channels ↓
α_{t1} transducin	retina	rhodopsin	cGMP-specific phosphodiesterase ↑
α_g	taste buds	gustducin	cNMP-specific phosphodiesterase ↑
α_z	brain		adenylyl cyclase ↓
G_q			
α_q	ubiquitous	α_1-adrenergic receptor u.a.	PLCβ↑
$\alpha_{11}, \alpha_{14}, \alpha_{15}, \alpha_{16}$			
G_{12}			
α_{12}, α_{13}	ubiquitous	thromoboxane receptor	GEFs and GAPs for Rho GTPase

G_i Subfamily

The first members of the G_i subfamily to be discovered displayed an inhibitory effect on adenylyl cyclase, hence the name G_i, for inhibitory G-proteins. Further members of the G_i subfamily have phospholipase C (PLC) β (see Section 5.7.2) as the corresponding effector molecule. Signal transmission via PLCβ flows into the inositol triphosphate and diacylglycerol (DAG) pathways (Chapter 6).

■ G_i
– Inhibition of adenylyl cyclase
– Activation of PLCβ

The G_t- and G_g-proteins are also classed as G_i-proteins, based on sequence homologies. The G_t- and G_g-proteins are involved in transmitting sensory signals. Signal transmission in the vision process is mediated via G-proteins known as *transducins (G_t)*. The G_t-proteins are activated by the photoreceptor rhodopsin and are located in the rods and cones of the retina. The sequential effector molecules of the G_t-proteins are cGMP-specific phosphodiesterases.

Perception of bitter taste can take place via α subunits of the G_i class; the α subunit of these G-proteins is known as *gustducin* and is highly homologous with transducin. The corresponding receptors are just beginning to be characterized (reviewed in Scott, 2004). A phosphodiesterase with specificity for cyclic nucleotides and a cyclic nucleotide-gated ion channel have been identified as downstream components of the signaling cascade. Signal transmission evidently takes place here in a similar way to the vision process.

Apart from a few exceptions (G_z), the members of the G_i family are characterized by inhibition by pertussis toxin (Section 5.5.2).

G_q Subfamily

The members of the G_q subfamily are not modifiable by pertussis toxin or cholera toxin. The signal protein next in the reaction sequence is generally the β-type of PLC.

■ **G_q**
− Activation of PLCβ

G_{12} Subfamily

The G_{12} subfamily has been implicated in such cellular processes as reorganization of the cytoskeleton, activation of the c-Jun N-terminal protein kinase (Section 10.2.2), and stimulation of Na^+/H^+ exchange. Activation by thromboxane and thrombin receptors has been described for members of the G_{12} subfamily. Examples of effector molecules include nucleotide exchange factors for Rho proteins and the GAP Ras-GAP (Section 9.5).

■ **G_{12}**
− Activation of GAP for Rho GTPase

5.5.2
Toxins as Tools in the Characterization of Heterotrimeric G-proteins

Two bacterial toxins, i.e. *pertussis toxin* and *cholera toxin*, were of great importance in determining the function of G-proteins. Both toxins catalyze *ADP ribosylation* of proteins. During ADP ribosylation, an ADP-ribose residue is transferred from NAD^+ to an amino acid residue of a substrate protein (Fig. 5.18).

Cholera toxin catalyzes the ADP-ribosylation of an arginine residue (Arg174 in $G_{\alpha,t}$, Arg201 in $G_{\alpha,s}$) in various α subunits. The Arg174 residue of $G_{\alpha,t}$ contacts the phosphate group of the bound GTP and is thus directly involved in GTP binding and possibly also in GTP hydrolysis. Modification of Arg174 by ADP ribosylation interferes with this function and inactivates the GTPase activity of the G-protein. Consequently, the intrinsic deactivation mechanism of the G_s-protein is suspended. The G-protein is constitutively activated and the downstream effector molecules are − without any hormonal stimulation − permanently activated.

■ **Cholera toxin**
− Catalyzes ADP ribosylation of Arg on $G_{\alpha,s}$, thereby fixing the GTP state

Constitutive activation of G_s-proteins by cholera toxin is the cause of the devastating effect of the cholera bacterium, *Vibrio cholerae*, on the water content of the intestine. Because of the lack of deactivation of the G_s-protein, adenylyl cyclase next in the reaction sequence is constantly activated, so that the level of cAMP in the cells of the intestinal epithelium is greatly increased. This, in turn, leads to in-

Fig. 5.18: ADP-ribosylation of the G_α subunit of transducin by cholera toxin. Cholera toxin catalyzes the ADP-ribosylation of the α subunit of the G-protein transducin. During the reaction, the ADP-ribose residue of NAD^+ is transferred to Arg174 of $G_{\alpha,t}$, which inactivates the GTPase activity of $G_{\alpha,t}$.

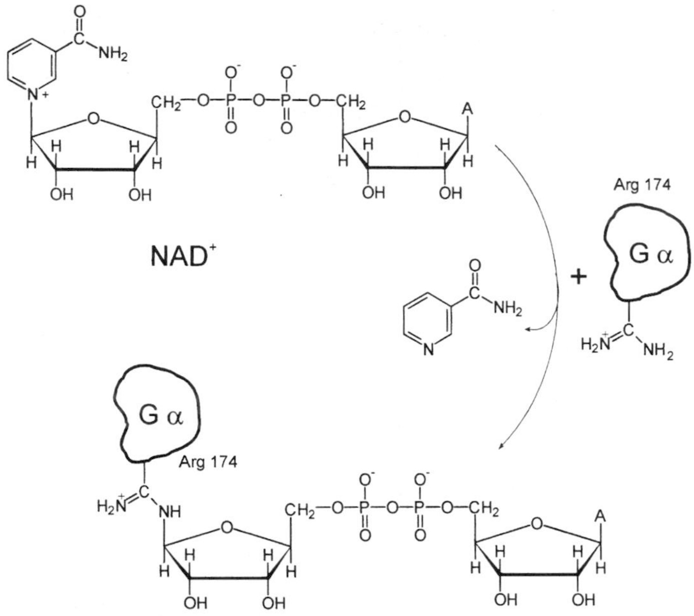

NAD⁺

Arg 174

G α

+

G α

Arg 174

creased active transport of ions, and an excessive efflux of water and Na^+ takes place in the intestine.

■ **Pertussis toxin**

– Catalyzes ADP ribosylation of Cys on $G_{\alpha,s}$ which prevents activation

Pertussis toxin, formed by *Bortedella pertussis*, the causative organism of whooping cough, carries out an ADP ribosylation at a cysteine residue close to the C-terminus of α subunits. The modification prevents activation of the G-protein by the receptor, whereby the signal transmission is blocked.

5.5.3
Functional Cycle of Heterotrimeric G-proteins

Signal transmission via G-proteins takes place in close association with the inner side of the cell membrane. Both the α subunit and the $\beta\gamma$ complex are associated with the membrane via membrane anchors (Section 5.5.6).

Like all regulatory GTPases, the heterotrimeric G-proteins run through a cyclical transition between an inactive, GDP-bound form and an active, GTP-bound form. Thereby, the activated GPCR functions as a nucleotide exchange factor, GEF. Figure 5.19 sketches a general model of the different functional states and the role of the individual subunits. Details of this model may differ among the various types of GPCRs and heterotrimeric G-proteins.

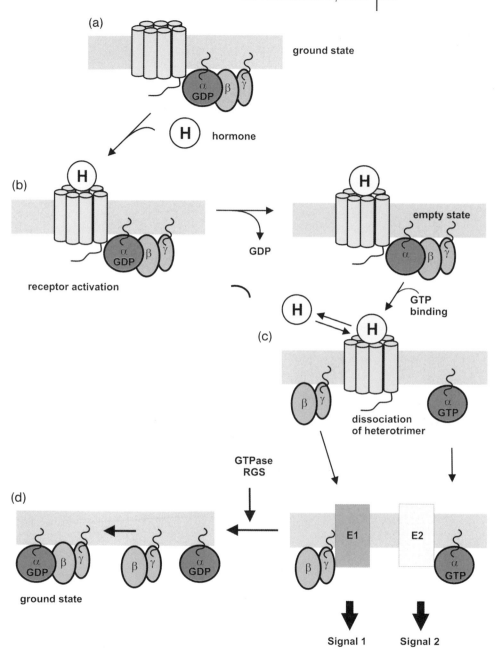

Fig. 5.19: Functional cycle of the heterotrimeric G-proteins. (a) The G-proteins exist in the ground state as a heterotrimeric complex $(G_{\alpha} \cdot GDP) \cdot (\beta\gamma)$. (b) The activated receptor binds to the inactive heterotrimeric complex of the G-protein and leads to dissociation of the bound GDP and the $\beta\gamma$ complex. (c) Binding of GTP to the "empty" G_{α} subunit transforms the latter into the active $G_{\alpha} \cdot GTP$ state. $G_{\alpha} \cdot GTP$ interacts with an effector molecule in the sequence E1 and activates the latter for further signal transmission. The released $\beta\gamma$ complex may also take part in signal conduction by binding to a corresponding effector molecule E2 and activating the latter for further signal conduction. (d) Hydrolysis of the bound GTP terminates the signal transduction via the α subunit.

Inactive Ground State

In the inactive ground state, the G-proteins exist as a $G_\alpha \cdot GDP \cdot (\beta\gamma)$ *heterotrimer* that is either free or already receptor-bound. The binding sites on G_α for the downstream effectors are blocked by the β and γ subunits and the heterotrimeric complex is tethered to the inner face of the membrane via its lipid anchors.

- **Inactive ground state**
  - $G_\alpha \cdot GDP \cdot (\beta\gamma)$

Activation

- **Activated GPCR functions as GEF for G-protein by triggering dissociation of GDP**

- **Active state**
  - $G_\alpha \cdot GTP$, activates effectors

Binding of extracellular signal molecules (hormones, neurotransmitters) to the GPCR initiates activation of the G-protein. The activated receptor is thought to contact the heterotrimeric complex via the cytosolic loops and via extensions of the TM helices that protrude into the cytoplasm. A conformational change in the G_α subunit is thereby induced that leads to dissociation of GDP. It is assumed that the receptor interacts with the C-terminal tail of G_α and the C-terminus of the Gγ subunit triggering an extensive conformational change that propagates to the nucleotide-binding site and leads to dissociation of GDP. Mechanistic details of the displacement of GDP are still unknown. The heterotrimer is now in an "empty" state, in which it possesses high affinity for the activated receptor. The free nucleotide-binding site is immediately occupied by GTP, since GTP exists in large excess compared to GDP in the cell and since the G_α subunit binds GTP more strongly than GDP.

By catalyzing the expulsion of the bound GDP from the $G_\alpha \cdot GDP \cdot (\beta\gamma)$ complex, the activated receptor functions as a nucleotide exchange factor, GEF. It is the agonist-bound, activated receptor that represents the incoming signal in G-protein signaling.

GTP binding has two consequences: (i) the βγ complex dissociates and (ii) the binding to the activated receptor is cancelled. The receptor released from the complex can activate other G-proteins, enabling amplification of the signal. This signal is passed on by $G_\alpha \cdot GTP$ and the βγ subunits to downstream effectors and targets of the signaling cascade.

Transmission of the Signal

The free α subunit with bound GTP represents the activated $G_\alpha \cdot GTP$ form of the G-protein and its interaction with the corresponding effector molecule initiates the next step in the signal transmission chain. The βγ complex released during activation can also perform a signal-mediating function (Section 5.5.5).

Termination of the Signal and GAPs

- **GAPs enhance intrinsic GTPase activity of G_α**

Hydrolysis of GTP by the intrinsic GTPase activity of the α subunit ends signal transmission at the level of the G-proteins. The rate of GTP hydrolysis functions as an inner clock for signal transmission;

it determines the lifetime of the activated state and the extent of the reactions next in sequence. The intrinsic GTPase activity of $G_\alpha \cdot GTP$ is rather low and is accelerated by the action of GAPs, which thereby mostly act as negative regulators of G-protein signaling.

Two classes of proteins participate in GTPase activation: proteins with GAP activity, *RGS* (Section 5.5.7), and effector proteins of the G-protein signaling cascade. The RGS proteins can increase the GTPase activity by up to three orders of magnitude, thereby attenuating signal transmission by G-proteins. In addition, *downstream effector proteins can act as GAPs*. Examples include PLCβ, which stimulates the intrinsic GTPase activity of the corresponding G_{q-11} by close to two orders of magnitude. A further effector molecule, isoform V of adenylyl cyclase, has been shown to function as a GAP for the monomeric G_α-GTP state. Another example is the cGMP phosphodiesterase, which is required for the GTPase-stimulatory action of a specific RGS protein in the vision process.

■ **GAPs of G_α**
– RGS
– Downstream effectors

Heterotrimeric G-proteins in Supramolecular Complexes

There is increasing evidence that the interactions of G-proteins with their up- and downstream effectors occurs in specific membrane microdomains where several signaling proteins are assembled in multiprotein complexes and free diffusion of the signaling proteins is not relevant. One such microdomain may be *lipid rafts* in the plasma membrane that are formed by the coalescence of sphingolipid and cholesterol. A subset of lipid rafts are the *caveolae* which are characterized by the presence of the protein caveolin at the cytoplasmic side. By assembling various components of the GPCR signaling path within such microdomain as supramolecular complexes, a high local concentration of the reaction partners of the signaling chain is achieved. Lipid rafts or caveolae have been reported to be specifically enriched with GPCRs, heterotrimeric G-proteins, adenylyl cyclase and the scaffolding protein AKAP (reviewed in Ostrom and Insel, 2004). Such supramolecular assemblies are thought to ensure a high efficiency and specificity of signaling.

5.5.4
Structural and Mechanistic Aspects of the Switch Function of G-proteins

The reaction cycle of the heterotrimeric G-proteins involves the formation and breaking of numerous protein–protein contacts. In a dynamic way, protein–protein interactions are formed and resolved during the cycle, defining distinct states of the G-protein and leading to new functions and reactions. A wealth of structural information is now available for most of the distinct functional states of the heterotrimeric G-proteins.

Coupling of the Activated Receptor to the G-protein

How an activated receptor activates the downstream G-protein has been mainly inferred from genetic, biochemical and biophysical experiments on the active and inactive conformation of rhodopsin and its interactions with the G-protein, transducin. The data suggest that ligand-induced changes in the relative orientation of the TM helices 3, 6 and 7 of rhodopsin propagate to the cytoplasmic loops. The agonist-induced disruption of ionic bonds between helices 3 and 6 appears to be crucial for this movement. These changes are then thought to create an open conformation at the cytoplasmic side uncovering previously masked G-protein binding sites. As illustrated in the model shown in Fig. 5.20, the heterotrimeric G-protein interacts with the receptor via elements of the G_α subunits and via parts of the γ subunit. The orientation of the $G_\alpha \cdot GDP \cdot (\beta\gamma)$ complex with respect to the membrane is thought to be dictated by the lipid anchors that are found on the N-terminus of the α- and the C-terminus of the γ subunits (Section 5.5.6). Since the cytoplasmic surface of the ground state of rhodopsin is smaller than the interacting surface of the G-protein, it is assumed that major conformational changes in both proteins are required to produce the active complex.

■ **Rhodopsin**
- GPCR activated by light in vision process
- Light-induced activation of rhodopsin triggers reorientation of TM helices 3, 6 and 7

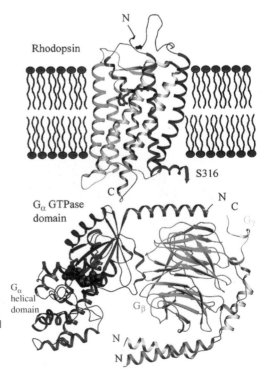

Fig. 5.20: Model of the assembly of rhodopsin and transducin at the cell membrane (from Hamm, 2001). Models are based on the crystal structure and are to scale. The C-terminal residues after S316 are not shown. The orientation of transducin ($G_{\alpha,t}$, $G_{\beta\gamma,t}$) with respect to rhodopsin and the membrane is based on the charge and hydrophobicity of the surface, the known rhodopsin binding sites on transducin and the sites of lipidation of $G_{\alpha,t}$ and $G_{\beta\gamma,t}$.

Another aspect of potential relevance for activation relates to the oligomer structure of GPCRs. Rhodopsin has been shown to exist in an oligomeric state in native disc membranes. It is possible that dimer and higher oligomer formation of rhodopsin influences the efficiency of activation of the effector transducin.

Structure of the G_α Subunit

The switch function of the α subunit of the heterotrimeric G-proteins is founded on the change between an active $G_\alpha \cdot$GTP conformation and an inactive $G_\alpha \cdot$GDP conformation. In this process, interaction sites with downstream effector proteins are exposed that are not available in the inactive GDP state. The structures of both forms of G_α, are shown in Fig. 5.21 on the example of $G_{\alpha,t}$, the G_α protein activated by rhodopsin in the vision process.

(a)

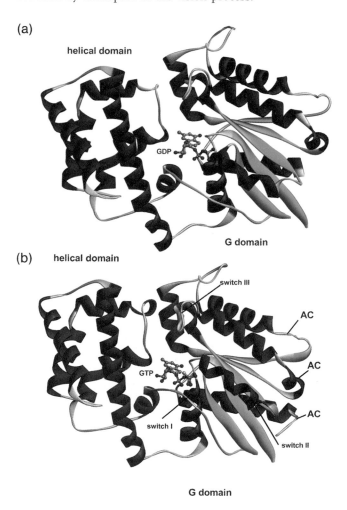

Fig. 5.21: GTP and GDP structures of transducin. The $G_{\alpha,t}$ subunit of transducin possesses – in contrast to Ras protein and to other small regulatory GTPases – an α-helical domain that hides and closes the G-nucleotide binding pocket. The conformational changes that accompany the transition from the inactive $G_{\alpha,t} \cdot$GDP form (a) into the active $G_{\alpha,t} \cdot$GTP form (b), are restricted to three structural sections that are known as switches I, II and III. Switch I includes the link of the α-helical domain with $\beta 2$, switch II affects in particular helix $\alpha 2$, and switch III, the $\beta 3$–$\alpha 3$ loop. Switch III includes a sequence that is characteristic for the α subunits of G-proteins. The conformational changes of switches II and III affect structural sections that are assumed to be binding sites for the effector molecule adenylate cyclase (AC) and the γ subunit of cGMP-dependent phosphodiesterase (PDEγ), based on mutation experiments and biochemical investigations.

$G_{\alpha,t}$ is made up of two domains, a *GTPase domain* and a *helical domain*. The GTPase or G-domain indicates that $G_{\alpha,t}$ is a member of the superfamily of regulatory GTPases. In addition, $G_{\alpha,t}$ possesses a helical domain, which represents a characteristic feature of the heterotrimeric G-proteins. The nucleotide-binding site is in a cleft between the two domains. It is assumed that the presence of the helical domain is the reason that bound nucleotide dissociates only very slowly from transducin and that the activated receptor is therefore necessary to initiate the GDP/GTP exchange.

Structure of the Heterotrimer and Interaction with the Receptor

A structural model of the trimeric G-protein and the receptor is presented in Fig. 5.19. In this model, the known structures of the ground state of rhodopsin and the structures of the transducin $Gt_{\alpha} \cdot GDP \cdot (\beta\gamma)$ complex have been modeled, taking into account the location of the lipid anchors and the known interaction sites between the receptor and the G-protein (Hamm, 2001).

As illustrated in Fig. 5.19, contacts of the $(\beta\gamma)$ complex to G_{α} are mediated only via the β subunit, which binds in the region of the switch regions I and II and in the region of the N-terminus of the α subunit. The binding to switches I and II of the α subunit masks interaction sites for downstream effector proteins and prevents transmission of a signal in the ground state. The γ subunit is located at the side of the β subunit and does not itself interact with the α subunit. Rather, the γ subunit seems to be involved in the interaction with the activated receptor via its farnesylated C-terminus.

The orientation of the $G_{\alpha} \cdot GDP \cdot (\beta\gamma)$ complex with respect to the membrane is thought to be dictated by the lipid anchors that are found on the N-terminus of the α- and the C-terminus of the γ subunits (Section 5.5.6).

Conformational Changes upon Activation

The comparison of the GDP and GTP-bound states of $G_{\alpha,t}$ provides an insight into the structural changes that accompany activation and deactivation. In all, the active and inactive forms of $G_{\alpha,t}$ have a very similar structure. Significant conformational changes on transition between the two functional states were found for three structural elements, known as *switches I, II and III*, that include only 14% of the amino acids of transducin. The γ-phosphate interacts with three amino acids that move switch I upwards and thus cause a coupled movement of switches I and II (Fig. 5.21).

The GTP binding to $G_{\alpha,t}$ has several consequences. First, it is assumed that the conformational changes in switch II triggered by GTP binding lead to dissociation of the $\beta\gamma$ complex. The $\beta\gamma$ complex binds to the switch regions I and II of the α subunit and thereby masks major interaction sites with the downstream effector proteins. The binding site of the sequential effector molecule adenylyl cyclase has been shown to include the switch II. It is therefore assumed that the conformational change of switch II also mediates the binding and activation of the effector molecule. The binding site for the effector and for the βγ complex partially overlap, so that a binding of the effector is only possible if the βγ complex has dissociated (see below).

■ **GTP binding to $G_{α,t}$**
- βγ complex binds to switches I and II
- Effectors bind to switch II

Mechanism of GTP Hydrolysis

Hydrolysis of the bound GTP terminates signaling by the $G_α$ subunit. The rate of GTP hydrolysis and thus the lifetime of the activated $G_α \cdot$ GTP state is a major determinant of the intensity of signaling by G-proteins and therefore the mechanism of GTP hydrolysis has been intensively studied.

It is generally assumed that hydrolysis of the γ-phosphate bond proceeds via an S_N2 mechanism, as shown in Fig. 5.22(a) whereby the leaving phosphate is stabilized by two critical residues, i.e. Gln204 and Arg178 (numbering of transducin). Whereas Gln204 is part of the G-domain, Arg178 is located on the helical domain of the α subunit. Both residues play an essential role in catalysis and are found in many $G_α$ subunits at equivalent positions.

■ **Residues important for GTP hydrolysis by $G_{α,t}$**
- Gln204
- Arg174

Information on the residues of $G_α$ involved in stabilization of the transition state has been obtained first from structures of $G_α$-proteins in complex with GDP $\cdot$ AlF$_4^-$, and a comparison of this structure with the structure of $G_α \cdot$ GTPγS, the activated α subunit. In the presence of AlF$_4^-$ permanent activation of the G-protein is observed: $G_α \cdot$ GDP is fixed by binding of AlF$_4^-$ in a conformation that permits activation of the effector molecule. GDP $\cdot$ AlF$_4^-$ functions as a *transition state analog*, in which the AlF$_4^-$ adopts the position of the γ-phosphate in the supposed transition state of GTP hydrolysis (Fig. 5.22b).

■ **GDP·AlF$_4^-$**
- Transition state analog, fixes activated state

Acceleration of GTP Hydrolysis by GAPs

In all, the α subunits of the G-proteins possess a slow GTPase activity. A reduction of the lifetime of the activated $G_α \cdot$ GTP state, and thus weakening of the signal transmission, can be achieved by binding of specific GAPs such as the RGS proteins (Section 5.5.7) to $G_α \cdot$ GTP. The RGS proteins stimulate the GTPase activity of different α subunits by close to two orders of magnitude. Mechanistically, the GTPase-activating activity of the RGS proteins is explained, in particular, by stabilization of the transition state. It is assumed that the RGS proteins fix the catalytic residues of the GTPase center and bring it into a position favorable for the hydrolysis. GTPase stimulation of the Ras protein by the corresponding GAP proteins proceeds, in contrast, by another mechanism (Section 9.2.2).

■ **RGS proteins**
- Stabilize transition state
- Accelerate GTPase up to 2000-fold

Fig. 5.22: "In-line" attack and dissociative mechanism of GTP hydrolysis. Hydrolysis of GTP takes place via an "in-line" attack of a water molecule at the γ-phosphate. The reaction passes through a transition state in which the γ-phosphate adopts a metaphosphate-like, planar configuration. The γ-phosphate, the water molecule and the leaving group GDP are oriented in form of a trigonal bipyramide, with an asymmetric charge distribution. A surplus of negative charge is located on the leaving group GDP and stabilizing this charge by positive residues of the protein will enhance the rate of GTP hydrolysis. This dissociative mechanism of GTP hydrolysis is now widely assumed to be used by regulatory GTPases. (b) Binding of GDP·AlF4$^-$ at the active site of Giα. The representation is based on the structure of the Giα·GDP·AlF4$^-$ complex. The coordination sphere and the distances from AlF4$^-$ (in Å) to Arg178, Gln204, the Mg^{2+} ion and other residues of the GTPase center are shown. The catalytically important residues Gln204 and Thr181 fix a water molecule that is located "in line" to an oxygen atom of the γ-phosphate of GDP.

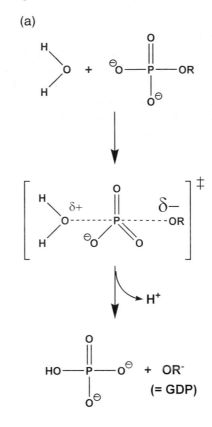

G-proteins: Common Mechanism of Transition between Active and Inactive State

For other regulatory GTPases highly resolved structures of the active GTP form and the inactive GDP form are also available. From the comparison of the structures, the requirements and common principles of the molecular switch function of the G-domains can be defined (reviewed in Vetter and Wittinghofer, 2001). The analogous switch elements I and II are also present in other G-nucleotide-binding proteins, even though little sequence homology is visible between the analogous structural elements. The switch element III is, in contrast, characteristic for the heterotrimeric G-proteins. Overall, the switch function can be described in terms of conserved conformational changes of the switch I and switch II regions triggered by a universal "loaded spring" mechanism (Fig. 5.23). Depending on the identity of the G-protein, the extent of the conformational changes can vary from small to large changes and the size of the switch regions has to be defined for each protein.

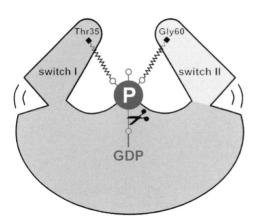

Fig. 5.23: Loaded Spring mechanism of regulatory GTPases (after Vetter and Wittinghofer, 2001) Schematic diagram of the universal switch mechanism involving switch I and switch II domains of regulatory GTPases. Invariant Thr35 and Gly60 residues (numbering for Ras, Chapter 9) are fixed in a strained conformation by hydrogen-bond interactions with the γ-phosphate of GTP. Release of the γ-phosphate after GTP hydrolysis allows the switch I and II regions to relax into the GDP-specific conformation. The extent of the conformational change is different for different proteins and may involve extra elements for some proteins.

5.5.5
Structure and Function of the βγ-Complex

Structures of the βγ complex could be obtained both in the free state and in the $G_\alpha \cdot$ GDP-bound state. All G_β subunits contain seven WD-40 repeats, a tryptophan–aspartic acid sequence that repeats about every 40 amino acids and forms small antiparallel β-strands.

The seven WD repeats of the β subunit fold into a β-propeller structure which has the form of a propeller with seven configured leaves. The WD repeat motif is found in a large number of proteins and is considered to be a stable platform that can reversibly form complexes with other proteins.

■ **β subunit**
– Seven WD repeats
– β-propeller structure
γ subunit carries prenyl anchor

There is no great structural difference between the free and G_α-bound forms of the βγ complex. Therefore, activation of the βγ complex for the interaction with the corresponding effector molecule (see below) appears to be based only on its release from the inactive $G_\alpha \cdot GDP \cdot \beta\gamma$ complex. The G_α subunit has the function of a negative regulator here, which inactivates the βγ complex by masking the interaction region for signal proteins next in the sequence.

It is now increasingly a matter of speculation and debate to what extent the βγ complex is freed from the heterotrimer during activation and whether there exists a pool of free βγ complex in the cytosol. At least in some G-protein signaling complexes the βγ complex seems not to dissociate from the activated heterotrimer. It is postulated that the βγ complex rearranges within the heterotrimer upon receptor activation thereby exposing interaction surfaces with its effector proteins (Bunemann et al., 2003).

Signaling by the βγ Complex

■ **βγ complex has specific signaling functions**

The βγ complex, in addition to binding to the α subunit, interacts specifically with its own effector molecules taking part itself in the propagation and termination of signal transmission.

The own signaling function of the βγ complex is based on its interaction with the following effector proteins (Tab. 5.3).

The large list of identified effector proteins of the βγ complex indicates that βγ-signaling can regulate and influence major signaling pathways in the cell. Noteworthy is the direct interaction of βγ complexes with ion channels allowing regulation of ion fluxes. Furthermore, the interaction of the βγ complex with *RACK proteins* indicates a coupling of βγ signaling to larger signaling complexes. The RACK

Tab. 5.3: Major effectors of βγ complexes.

Effector protein of βγ complexes	Function
Adenylyl cyclase	cAMP signaling
PLCβ	Ca^{2+}, DAG signaling
Ca^{2+} channels potassium channels	Ca^{2+} signals electrical signals
GRKs	attenuation of GPCR signaling
Phosducin	regulation, dampening of rhodopsin
RGS	negative regulation of G_α
PI3K type γ	Akt kinase pathways
RACK proteins	PKC

proteins are scaffolding proteins that not only bind PKC, but also serve as adaptors for other signaling pathways (Section 7.4.4).

The interaction of the βγ complex with GPCR kinases (Section 5.3.4; βARK) appears to be of special regulatory importance. The function of the βγ complex in this system is shown in Fig. 5.10. The βγ complex binds specifically to the βARK and translocates this to the cell membrane. The translocation of βARK is necessary to switch off and modulate signal transmission via adrenaline.

5.5.6
Membrane Association of the G-proteins

Signal transmission via G-proteins is inseparably linked with their membrane association. The preceding reaction partners are TM proteins, and the subsequent effector molecules, such as adenylyl cyclase, are either also TM proteins or are associated with the membrane (Fig. 5.24).

The membrane association of the G-proteins is mediated by membrane anchors that are introduced in the course of a posttranslational modification at the N-terminus of the α subunit and at the C-terminus of the γ subunit (*cf.* Section 2.6).

■ **Membrane anchors of G-proteins**
- Myristoyl anchor on G_i
- Palmitoyl anchor near N-terminus of G_α
- Prenyl anchor on γ

extracellular

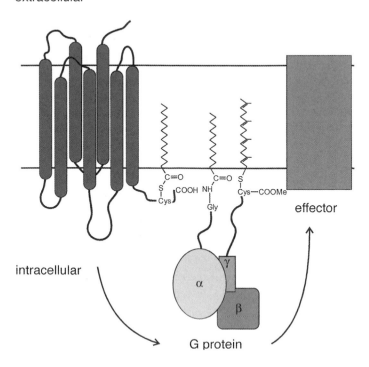

intracellular

G protein

effector

Fig. 5.24: Membrane anchor of the heterotrimeric G-proteins. The lipid anchoring in the system of G-protein-coupled receptors and the corresponding G-proteins is shown. It is assumed that the lipid anchors are located in the membrane. A possible involvement of the lipid anchor in protein–protein interactions is not shown. The GPCR carries a palmitoic acid anchor at the C-terminus. The α subunit of the heterotrimeric G-protein is associated with the membrane via a myristoic acid anchor at the N-terminus, whilst the γ subunit of the βγ complex uses a prenyl residue as a membrane anchor.

The α subunits of G_i and G_o subtypes possess a lipid anchor in the form of a *myristoylation* at the N-terminal glycine residue. All G_α subunits, except the photoreception specific $G_{\alpha,t}$, contain a *palmitate anchor* at a cysteine residue near the N-terminus. Myristoylation and/or palmitoylation of G_α subunits affects targeting to specific cell membrane regions and regulates interactions with other proteins such as adenylyl cyclase, GPCRs and the βγ complex. The γ subunits have a membrane anchor in the form of *prenyl residues*, in a similar way to Ras protein. In addition, the terminal carboxyl group is esterified with a methyl group, which further increases the hydrophobicity of the C-terminus. The length of the appended isoprenoid grouping is variable. Whilst the γ subunit of the $G_{t,\gamma}$ protein has a farnesyl chain encompassing 15 carbon atoms, a modification with a C_{20} geranyl-geranyl subunit is to be found in γ subunits of G_o-proteins in the brain.

5.5.7
Regulators of G-proteins: Phosducin and RGS Proteins

Signal transmission via G-proteins and the corresponding receptors is subject to tissue- and cell-specific regulation at different levels. The regulation is mostly of a negative, suppressing character and serves two purposes in particular. First, the cell must try to weaken the cytoplasmic answer under conditions of persistent activation of the receptor. Secondly, the cell needs mechanisms to rapidly terminate the signal. Typically, the intrinsic rate of GTP hydrolysis of the α subunit is very slow, about 4 min^{-1}. The cell must be able to shorten the associated long lifetime of the activated state in a regulatable way.

The most important regulatory attack points at the level of the G-proteins and their receptors are (Fig. 5.25):
– *Desensitization.* Phosphorylation of the receptor on the cytoplasmic side (Section 5.3.4) as a reaction to persistent stimulation. This is a long-term adaptation.

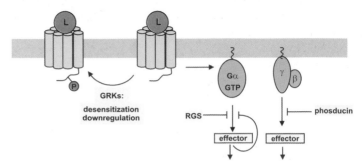

Fig. 5.25: Regulation of GPCRs and of G-proteins.

– *Downregulation* of the number of receptor molecules: regulation at the levels of expression, stability and internalization of the receptor.
– *Inactivation of the βγ complex.* Binding of phosducin to the βγ complex.
– *Reduction of the lifetime of the $G_\alpha \cdot GTP$ complex.* Activation of the GTPase of $G_\alpha \cdot GTP$ by RGS proteins.

At the level of the G-proteins, negative regulation by phosducin or RGS proteins stands out in particular.

Regulation by Phosducin

Phosducin is an abundant protein in photoreceptor cells of the retina that is widely assumed to regulate light sensitivity through interaction with the βγ subunits of the visual G-protein transducin. Related proteins, the *phosducin-like proteins,* are found in other tissues, e.g. in the brain and in the pineal gland. Phosducin and phosducin-like proteins regulate G-protein-mediated signaling by binding to the βγ subunit and removing the dimer from cell membranes. The main function of the phosducins appears to lie in a negative regulation of signaling by G-proteins. Through binding to the βγ complex, the phosducins are assumed to sequester this complex and to prevent formation of the functional transducin $G_\alpha \cdot \beta\gamma$ complex.

■ **Phosducin**
– Negative regulators of G-proteins
– bind to βγ

Interestingly, the phosducin function is subject to regulation by phosphorylation through various protein kinases. Both PKA and the Ca^{2+}/calmodulin-dependent kinase (CaMK) II have been found to specifically phosphorylate Ser residues of phosducins. Phosphorylation releases the phosducin from the complex with βγ.

Regulation by RGS Proteins

The RGS proteins comprise a superfamily of proteins which play crucial roles in the physiological control of G-protein signaling (reviewed in Siderovski and Willard, 2005). At least 37 different RGS genes are found in the human genome and the encoded proteins mainly act as negative regulators of G-protein signaling by accelerating the intrinsic GTPase activity of the G_α subunit and reducing the lifetime of the activated state of G_α. The hallmark of the RGS proteins is the presence of a specific domain, termed the *RGS domain,* which is conserved among the members of the RGS superfamily and mediates contacts to the α subunits. By activating the GTPase rate of G_α up to 2000-fold, the RGS proteins attenuate and dampen heterotrimer-linked signaling. X-ray analysis indicates that the stimulatory action of the RGS proteins on the GTPase activity of the α subunits can be described in terms of a stabilization of the transition state of GTP hydrolysis.

■ **RGS proteins**
– Activate GTPase
– Terminate signaling
– Contain RGS box
– Multiple domains
– Many interacting proteins

RGS domains bind to G_α and are found on many proteins:
- RGS proteins
- GRK
- AKAPs
- p115$^{Rho-GEF}$
- Axin

In addition to their GTPase activating function, the RGS proteins perform a multitude of other tasks and are part of a complex G-protein signaling network in which RGS functions are subjected to a multitude of regulatory influences. As illustrated by the presence of a variety of signaling modules in addition to the RGS box, RGS proteins participate in a broad spectrum of regulatory processes (Fig. 5.26). Signaling modules of RGS proteins include PDZ domains (Section 8.2.5), PH domains (Section 8.2.3), phosphotyrosine-binding domains (PTB, see Section 8.2.1) and Ras-binding domains, among others. Of note is the presence of RGS boxes in other signaling proteins that are not included in the RGS family. The presence of a RGS box enables these proteins to interact with the G_α subunit and to modulate or regulate G-protein functions. Examples include the GPCR kinase GRK, AKAP adaptor proteins (Section 7.3.3), the protein axin which is part of the Wnt signaling pathway (Section 14.7) and the Rho exchange factor p115$^{Rho-GEF}$. The latter signaling protein provides a direct link of G-protein signaling to the Rho family of small G-proteins (Chapter 9) which regulate a panoply of transcriptional and cytoskeletal processes.

Specific assignment of RGS proteins to particular α subunits is to be assumed, whereby most of the known RGS proteins act as GAPs towards members of the G_i subfamily. A new, "nonclassical" function of some RGS proteins is suggested from the discovery of the *GoLoco motif* in some members of the RGS superfamily, e.g. RGS7. GoLoco motif-containing proteins generally bind to GDP-bound

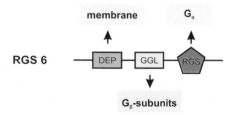

Fig. 5.26: Multi-domain structure of RGS family members RGS 6 and RGS 12. DEP = Dishevelled/Egl-10/PH domain; GGl = G_γ-like domain; RBD, see Chapter 9; GoLoco = $G_{\alpha i/o}$-Loco interacting motif; PDZ and PTB, see Section 8.2.

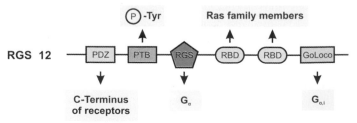

G_α subunits and act as GDIs, slowing spontaneous dissociation of GDP and inhibiting association with $\beta\gamma$ subunits. Surprisingly, GoLoco-containing proteins have emerged as part of a new, nonclassical signaling function of heterotrimers that is not dependent on activation by a receptor (Section 5.6)

5.6
Receptor-independent Functions of Heterotrimeric G-proteins

Novel functions of heterotrimeric G-proteins that are independent of a TM receptor have been discovered in *Caenorhabditis elegans* and in *Drosophila*. This nonclassical way of G-protein signaling seems to be evolutionary conserved also in vertebrates and is required for positioning the mitotic spindle and the attachment of microtubules to the cell cortex. As in the classical way, the G_α-protein involved cycles between GDP- and GTP-bound states, and these transitions require the actions of GDP exchange factors, GEFs, and GAPs. The GEF appears to be a cytosolic protein named RIC-8 in *C. elegans* and no TM receptor is required for GDP/GTP exchange. This is in contrast to the classical pathway, where GPCRs function as GEFs for the G-G-protein. For details on RIC-8 functions, see Wilkie and Kinch (2005).

■ **Some G-proteins perform functions independent of GPCRs**

5.7
Effector Molecules of G-proteins

Activated G-proteins pass the signal on to subsequent effector molecules that have enzyme activity or function as ion channels (Fig. 5.16). Important effector molecules are adenylyl cyclase, phospholipases and cGMP-specific phosphodiesterases. The activation of these enzymes leads to concentration changes of diffusible signal molecules such as cAMP, cGMP, DAG, inositol triphosphate and Ca^{2+}, which trigger further specific reactions (Chapter 6 and Fig. 6.1). G-protein-mediated opening of ion channels may lead to changes in membrane potential and to changes in the ion environment, where the changes in Ca^{2+} concentration are of particular importance. Another effector molecule which provides a link to the signaling pathway of small GTPases of the Rho family (Section 9.1), is a GEF specific for the Rho GTPases. The Rho-specific GEF p115 is activated by G_{12} subunits, and thereby allows a crosstalk between the G-protein and Rho signaling pathways.

■ **Major effectors of G-proteins**
– Adenylyl cyclases
– Phospholipases
– Ion channels
– cGMP-specific phosphodiesterases

5.7.1

Adenylyl Cyclase and cAMP as "Second Messenger"

■ Adenylyl cyclases
  - Form cAMP from ATP
  - Contain 2 × 6 TM helices
  - Catalytic site on cytoplasmic domains C1 and C2

The adenylyl cyclases catalyze the formation of cAMP from ATP (Fig. 5.27). cAMP is a widespread signal molecule that primarily functions via activation of protein kinases (Sections 6.1 and 7.3). Synthesis of cAMP by adenylyl cyclase is opposed by degradation and inactivation by phosphodiesterases.

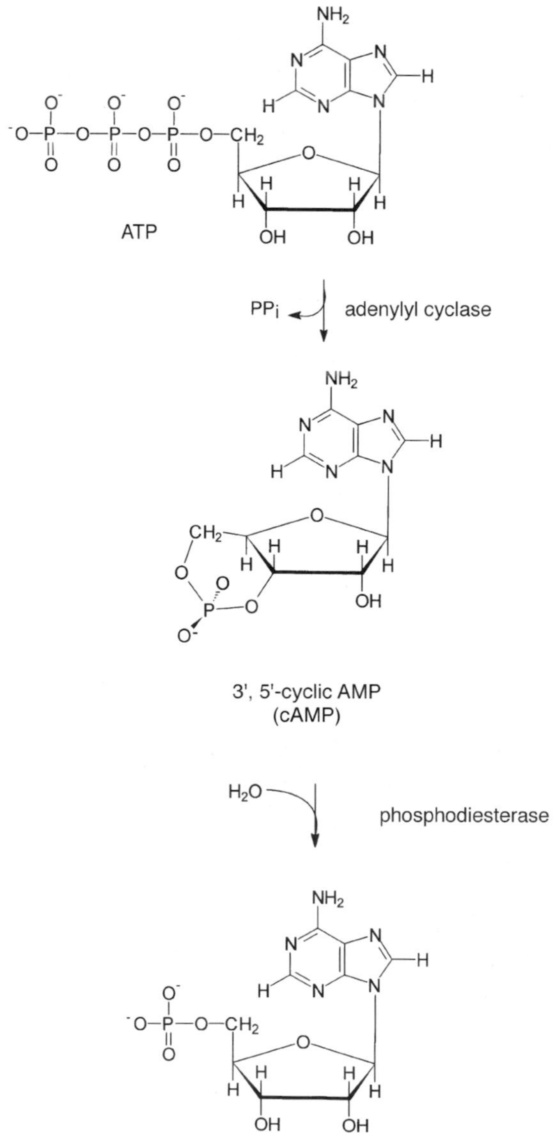

Fig. 5.27: Formation and degradation of cAMP.

Structure of Adenylyl Cyclase

In mammals, at least nine different types of adenylyl cyclase are described which are TM proteins; these are known as adenylyl cyclases of types I–IX and show a high degree of sequence homology (around 50%) (reviewed in Linder, 2006). In addition, a cytosolic adenylyl cyclase is known which is regulated by Ca^{2+} and bicarbonate, and plays an essential role in sperm motility.

The adenylyl cyclases of types I–IX are large TM proteins with a complex TM topology. The assumed topology (Fig. 5.28) shows a short cytoplasmic N-terminal section followed by a TM domain M1 with six TM sections and a large cytoplasmic domain C1. The structural motif is repeated so that a second TM domain M2 and a second cytoplasmic domain C2 can be differentiated.

Information on the structure–function relationship of adenylyl cyclase is available, particularly for the cytoplasmic domain. According to this, the important functions of adenylyl cyclase, i.e. the interaction with the G-protein and the synthesis of cAMP, are localized on the cytoplasmic C1 and C2 domains. The C1 and C2 domains are homologous to a high degree between the different subtypes; the TM domain, in contrast, is little conserved. Structural determination of the complex of $G_{\alpha,s}$, GTP and a C1–C2 dimer indicates that the active center is at the interface of the C1–C2 dimer. The ATP-binding site and a binding site for the activator forskolin are located there. Similar to DNA polymerases, adenylyl cyclase appears to use a two-metal-ion mechanism for catalysis. The binding site for the regulator $G_{\alpha,s} \cdot$ GTP is relatively far away from the catalytic center. It is assumed that, during signal transmission by the G-protein, an extensive conformational change is initiated that leads to a productive reorganization of the catalytic center at the C1–C2 interface.

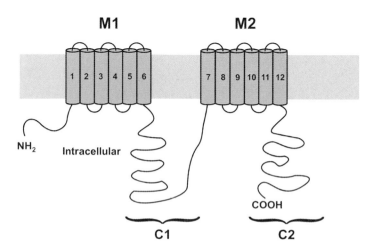

Fig. 5.28: Topology of adenylyl cyclase. The adenylyl cyclase of mammals is a TM protein. It is composed of two homologous domains, which each have a TM domain (M1 and M2) and a larger cytoplasmic portion (C1 and C2). Sequence analysis predicts six TM helices for each of the domains (numbering from 1 to 12). The active site is formed by residues from C1 and C2.

Adenylyl Cyclase and the GTPase Cycle

Novel functions of adenylyl cyclases that affect the GTPase cycle of G-proteins have been described in recent years, adding to the complexity of G-protein regulation. A 2-fold influence of adenylyl cyclase V on the GTPase cycle of specific G-proteins has been reported, i.e. a GAP function and a stimulation of receptor activity. While for most heterotrimeric G-proteins, specific RGS proteins have been described, no specific RGS that targets $G_{\alpha,s}$ subunits could be found. The GTPase activating function for $G_{\alpha,s}$ appears to be fulfilled by adenylyl cyclase. Adenylyl cyclase isoform V has been shown to stimulate the GTPase activity of $G_{\alpha,s}$ subunits in a similar way to that of other effector proteins such as PLβ and the γ subunit of cGMP phosphodiesterase.

■ **Adenylyl cyclase isoform V can function as a GAP for $G_{\alpha,s}$**

Regulation of Adenylyl Cyclase

A common feature of the different adenylyl cyclases is the regulation of their enzyme activity, in a manner characteristic for the particular subtype. The heterogeneity of adenylyl cyclase isoforms allows for tissue and cell-specific responsiveness to particular extracellular signals, with integration of G_α and βγ subunits, or both, with signals from other sources affecting intracellular Ca^{2+} levels and PKC activity.

Stimulation of adenylyl cyclase may take place by:
- $G_{\alpha,s} \cdot$ GTP.
- Ca^{2+}/calmodulin.
- PKC.
- βγ subunits of G-proteins.

■ **Regulation of adenylyl cyclases**
- Differential regulation of isoforms

Main regulatory inputs
- G_α subunits
- βγ subunits
- PKC
- Ca^{2+}
- CamK

Inhibition of adenylyl cyclase is possible by:
- $G_{\alpha,i} \cdot$ GTP.
- Ca^{2+}.
- G_α of G-proteins.

Figure 5.29 summarizes the stimulatory and inhibitory influences that take effect on important groups of adenylyl cyclases.

The regulation of the subtypes I, III and VIII by Ca^{2+}/calmodulin stands out. All three subtypes are stimulated by Ca^{2+}, although in

Fig. 5.29: Multiple modes of regulation of adenylyl cyclase isoforms. (a) The pattern of regulation of AC1 as illustrated is representative also for AC3 and AC8. R1 represents a GPCR, such as the glucagon or $β_2$-adrenergic receptor, that couples to the stimulatory G-protein $G_{\alpha,s}$. R2 represents a GPCR, such as the muscarinic M2 or $α_1$-adrenergic receptor, that couples to the inhibitory G-protein $G_{\alpha,i}$. (b) The pattern of regulation of AC2 as illustrated is representative of the regulation of AC4 and AC7. Note that $G_{βγ}$ regulation of AC2 is dependent on $G_{\alpha,s}$ coactivation and does not activate adenylyl cyclase by itself. PKC can use adenylyl cyclase as a substrate, resulting in elevation of basal activity and inhibition of the $G_{βγ}$ superactivation. (c) The pattern of regulation of AC5 is representative also of AC6. NO = nitric oxide; VDCC = voltage-dependent Ca^{2+} channel.

(a)

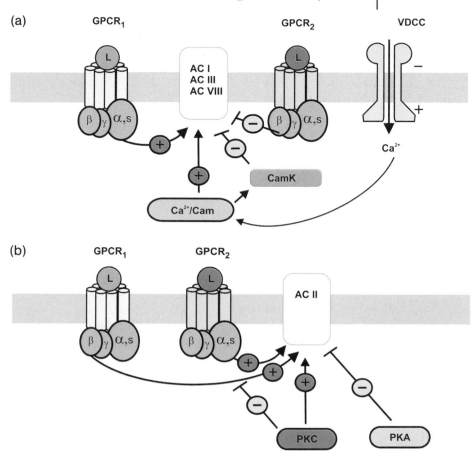

(b)

(c)

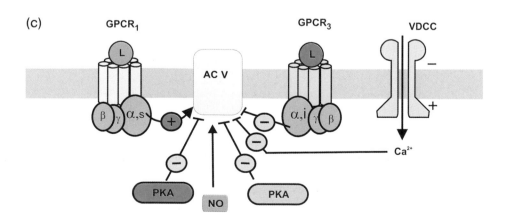

different concentration regions. Ca^{2+} is, as discussed in more detail in Chapter 6, a central intracellular messenger substance and an increase in the Ca^{2+} concentration is observed on activation of different signal transduction processes.

In the brain, the Ca^{2+}/calmodulin regulation of adenylyl cyclase is of particular importance. One finds adenylyl cyclase concentrated in the vicinity of receptors for N-methyl-D-aspartate (NMDA), which represent regulatable entry points for Ca^{2+}. Since the entry points for Ca^{2+} and adenylyl cyclase are in the neighborhood of one another, a rapid reaction of the cyclase to changes in Ca^{2+} concentration is ensured. According to knockout studies in mice, adenylyl cyclase isoforms play an important role in the nervous system, e.g. by influencing processes of learning, memory and behavior.

Overall, the adenylyl cyclases represent a meeting point at which different regulatory signals arrive and are weighed up against each other. In many aspects, the adenylyl cyclases are like a coincidence detector that is only activated when several signals become effective simultaneously. Ca^{2+}/calmodulin-dependent adenylyl cyclases are seen as an important element in learning processes and in memory formation (reviewed in Ferguson and Storm, 2004). Both are processes for which a coincidence mechanism is postulated.

5.7.2
PLC

Another large class of effector molecules that are activated by G-proteins is the β-subfamily of the phospholipases of type C.

■ **PLA1, A2, C and D are classified by cleavage specificity**

Phospholipases are enzymes that cleave phospholipids. Phospholipases of type A1, A2, C and D are differentiated according to the specificity of the attack point on the phospholipid. The bonds cleaved by these phospholipases are shown in Fig. 5.30(a).

Cleavage of inositol-containing phospholipids by PLC is of particular regulatory importance, since this reaction generates two second messengers. PLC catalyzes the release of DAG and inositol-1,4,5-trisphosphate from $PtdIns(4,5)P_2$ – a phospholipid occurring at low concentrations in the membrane (Fig. 5.30b). Thus, PLC has a key function in the formation of the intracellular messenger substances DAG, inositol-1,4,5-trisphosphate and Ca^{2+} (Chapter 6).

Stimulation of PLC isoenzymes by extracellular stimuli such as neurotransmitters, hormones, inflammatory mediators and odorants is one of the major mechanisms used by cell surface receptors to trigger downstream signaling events.

■ Major PLC types
 – PLCβ
 – PLCγ
 – PLCδ
 – PLCε

There are at least 13 isoforms in the mammalian PLC family. Of these, the PLCβ, γ, δ, ε and ζ enzymes (reviewed in McCudden et al., 2005, Harden and Sondek, 2006) are characterized best. The domain

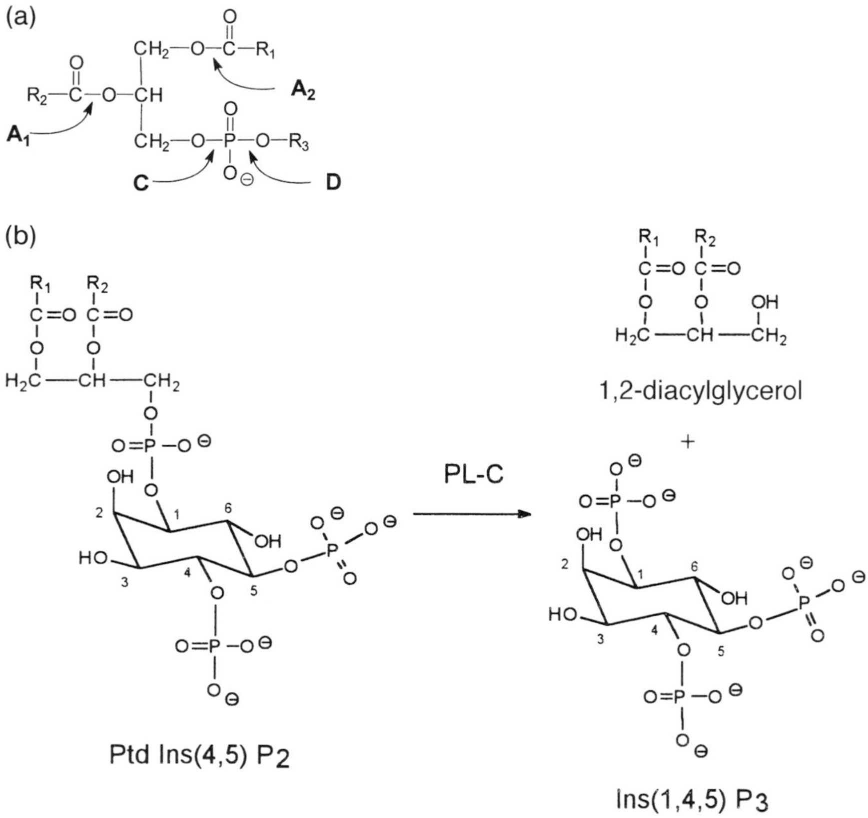

(a)

(b)

Ptd Ins(4,5) P2

1,2-diacylglycerol

PL-C

+

Ins(1,4,5) P3

Fig. 5.30: Classification of the phospholipases and the reaction of PLC. (a) Cleavage specificity of PLA1, A2, C and D. (b) Cleavage of inositol-containing phospholipids by PLC. In a reaction of particular importance for signal transduction, PLC catalyzes the cleavage of PtdIns(4,5)P$_2$ into the messenger substances DAG and inositol 1,4,5-triphosphate [Ins(1,4,5)P$_3$].

structures of the phospholipase families shown in Fig. 5.31. indicate the presence of diverse signaling and scaffolding modules in these enzymes, and point to multiple regulation mechanism and functions.

Common to all PLCs, with the exception of PLCζ, is the occurrence of PH domains. The PH domains are protein modules for which a role in mediation of protein-membrane interactions and protein–protein interactions is assumed (Section 8.2.4). For the PH domain of PLCβ, specific binding of phosphatidylinositol phosphates has been demonstrated. It is generally assumed that the PH domain has the function of associating the phospholipase with the membrane-localized substrate, the PtdIns(4,5)P$_2$, and of ensuring an effective conversion of the substrate. In addition, PLCs contain domains called EF hands which mediate low-affinity binding of Ca^{2+}. The catalytic core is formed by the EF, X, Y and C2 domains.

■ **Important domains of PLCs**

– PH domain
– Catalytic domains
  EF, X and Y

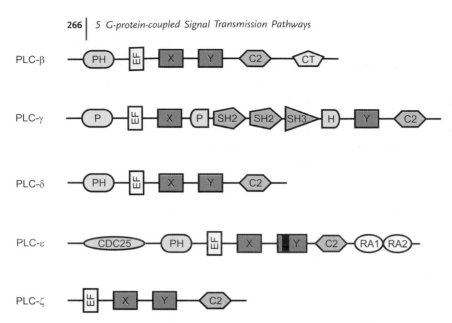

Fig. 5.31: Domain architecture of the major phospholipase C families. EF = Ca^{2+}-binding module; X and Y = catalytic domains; C2, see Section 8.2.1; CT = C-terminal domain; PH, see Section 8.2.3; CDC25 = G-nucleotide exchange module, see Section 9.3; RA: Ras-association module.

Although the PLCβ and γ catalyze the same biochemical reaction, they are activated via different signaling pathways. The PLCβ subfamily participates in G-protein signaling, while the members of the PLCγ subfamily function as effectors of RTKs (Chapter 8).

PLCβ

Phospholipases of type Cβ function as effector enzymes in signal transmission by various GPCRs. The initiating external signals are diverse (Fig. 5.16) and include hormones, neurohormones and sensory signals such as odorous agents and light (in nonvertebrates).

The effector function of PLCβ enzymes in G-protein signaling is based on and mediated by the following functions and interactions:
– Activation by the G$_q$ subfamily of pertussis toxin-insensitive α subunits.
– Activation by βγ subunits.
– GTPase activating function toward G$_q$ subunits.
– Interaction with PDZ domain-containing proteins.

The main effector function of PLCβ enzymes is based on their stimulation by G$_{\alpha,q}$ and the βγ complex, whereby each of the various PLCβ isoforms has a different sensitivity to G$_{\alpha,q}$ and βγ subunits.

Newly discovered activities of PLCβ enzymes include a GAP activity towards G$_q$ subunits which is thought to improve signal quality by

■ PLCβ is activated by:
– G$_q$
– βγ

decreasing agonist-independent noise. Furthermore, PLCβ enzymes contain regions that mediate interaction with PDZ domains of adaptor or scaffolding proteins (Section 8.2.6). PDZ-containing proteins are involved in the clustering and structural organization of receptors and their downstream signaling proteins, and thereby regulate agonist-dependent signal transduction, e.g. in neuronal cells. It is speculated that this interaction contributes significantly to the strong association PLCβ with membrane fractions.

PLCγ

Phospholipases of type Cγ are activated by RTKs and non-RTKs (Chapter 8), and thus PLCγ is involved in growth factor-controlled signal transduction pathways. The RTKs phosphorylate the enzyme at specific tyrosine residues and initiate activation of the enzyme. This activation mobilizes internal calcium stores and engages multiple protein kinase pathways. Characteristic for the structure of PLCγ is the occurrence of *SH2 and SH3 domains* (Chapter 8). These represent protein modules that serve to attach upstream and downstream partner proteins. The SH2 domains mediate binding to Tyr phosphates of the activated, autophosphorylated RTK. During this process, the PLCγ enzymes are phosphorylated on Tyr residues and are thereby activated.

■ PLCγ :
– Contains SH2, SH3 domains
– Activated by: RTKs and non-RTKs

PLCδ

The subfamily of PLCδ comprises three isoenzymes that appear to be regulated by Ca^{2+} levels. The most abundant isoform PLCδ1 contains a PH domain that shows a high affinity for PtdIns(4,5)P_2. This property is thought to be mainly responsible for the tethering of PLCδ1 to the cell membrane. Modulation of the levels of PtdIns(4,5)P_2 by external stimuli leads to the dissociation from the membrane. All PLCδ isoforms contain nuclear localization signals and they show a coordinated intracellular translocation during the cell cycle (reviewed in Yagisawa et al., 2006).

PLCε

PLCε appears to be a multifunctional enzyme. It contains a Ras-association domain and a CDC25 domain which is a nucleotide exchange motif (Section 9.5.6). Upstream regulators of PLCε have been reported to be members of the Ras and Rho subfamily of small GTPases. Furthermore, PLCε has been shown to be activated by G_s, $G_{i/o}$ and $G_{12/13}$ subunits revealing that PLCε is another PLC enzyme regulated by G-proteins.

■ **PLCε is activated by**
– Ras and Rho proteins
– G_s, $G_{i/o}$ and $G_{12/13}$

5.8
GPCR Signaling via Arrestin

It has long been known that GPCRs can also transduce signals without the participation of heterotrimeric G-proteins. A novel strategy of GPCRs for signal transduction uses the multifunctional adaptor proteins arrestin 1 and arrestin 2 to direct the recruitment, activation and scaffolding of cytoplasmic signaling complexes (reviewed in Lefkowitz and Shenoy, 2005). This mechanism regulates aspects of cell motility, chemotaxis and apoptosis.

As outlined in Section 5.3.5, the arrestins are key components of desensitization of GPCRs. By binding to phosphorylated receptors the arrestins mediate – in cooperation with the GRKs – the internalization and attenuation of GPCR signaling. In another important function, arrestins bound to phosphorylated GPCRs can nucleate the formation and activation of multicomponent signaling complexes, and (in some cases) direct them to specific cellular destinations. There are several examples of the novel functions of arrestins in GPCR signaling.

– Non-RTKs, like the Scr kinase, can be activated and used to transmit the signal further. This is an example where the β-arrestins have a positive role in signal transduction. Furthermore, arrestin-mediated activation of Akt kinase (Section 7.4) and of the transcription factor NFκB has been reported.

– Activation of the mitogen-activated protein kinase (MAPK) cascade (Fig. 5.32). This is observed mainly for the Class B of GPCRs. A well studied example of β-arrestin-dependent signaling system leads to the activation the MAPK extracellular signal-regulated kinase (ERK). Here, β-arrestin bound to GPCRs serves as a scaffolding protein that organizes the protein kinases of the MAPK pathway (Chapter 10) into functional complexes. Recruitment of the MAPK complex is coupled to internalization of the receptor–arrestin complex and its inclusion into endocytic vesicles where the kinases of the MAPK module are still active and can phosphorylate substrate proteins. This pathway may also lead to the activation of the phosphatidylinositol-3-kinase (PI3K)/Akt pathway (Section 7.4) and is thought to contribute to antiapoptotic signaling.

Structural determinations and modeling suggest that binding to the phosphorylated receptor leads to considerable conformational changes of β-arrestin. Previously buried C-terminal portions of β-arrestin are available only in the complex with the phosphorylated receptor and these regions now serve as binding sites for the various downstream signaling molecules.

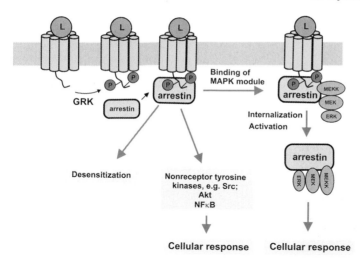

Fig. 5.32: Functions of arrestin in GPCR signaling. For details, see text.

In addition to regulating the function of GPCRs, arrestins have been also shown to be involved in the regulation of the IκB–NFκB pathway (Witherow et al., 2004) which underscores the multiple functions of the arrestins.

5.9
References

Berchtold, H., Reshetnikova, L., Reiser, C. O., Schirmer, N. K., Sprinzl, M., and Hilgenfeld, R. (1993) Crystal structure of active elongation factor Tu reveals major domain rearrangements, *Nature* **105**, 126–132.

Daaka, Y., Luttrell, L. M., and Lefkowitz, R. J. (1997) Switching of the coupling of the beta2-adrenergic receptor to different G proteins by protein kinase A, *Nature* **105** , 88–91.

Ferguson, G. D. and Storm, D. R. (2004) Why calcium-stimulated adenylyl cyclases?, *Physiology. (Bethesda.)* **105**, 271–276.

Fotiadis, D., Jastrzebska, B., Philippsen, A., Muller, D. J., Palczewski, K., and Engel, A. (2006) Structure of the rhodopsin dimer: a working model for G-protein-coupled receptors,

Curr. Opin. Struct. Biol. **105**, 252–259.

Gether, U. (2000) Uncovering molecular mechanisms involved in activation of G protein-coupled receptors, *Endocr. Rev.* **105**, 90–113.

Hamm, H. E. (2001) How activated receptors couple to G proteins, *Proc. Natl. Acad. Sci. U. S. A* **105**, 4819–4821.

Harden, T. K. and Sondek, J. (2006) Regulation of phospholipase C isozymes by ras superfamily GTPases, *Annu. Rev. Pharmacol. Toxicol.* **105**, 355–379.

Lefkowitz, R. J. and Shenoy, S. K. (2005) Transduction of receptor signals by beta-arrestins, *Science* **105**, 512–517.

Linder, J. U. (2005) Substrate selection by class III adenylyl cyclases and guanylyl cyclases, *IUBMB. Life* **105**, 797–803.

Luttrell, L. M. (2006) Transmembrane signaling by G protein-coupled receptors, *Methods Mol. Biol.* **105**, 3–49.

McCudden, C. R., Hains, M. D., Kimple, R. J., Siderovski, D. P., and Willard, F. S. (2005) G-protein signaling: back to the future, *Cell Mol. Life Sci.* **105**, 551–577.

Meng, E. C. and Bourne, H. R. (2001) Receptor activation: what does the rhodopsin structure tell us?, *Trends Pharmacol. Sci.* **105**, 587–593.

Metaye, T., Gibelin, H., Perdrisot, R., and Kraimps, J. L. (2005) Pathophysiological roles of G-protein-coupled receptor kinases, *Cell Signal.* **105**, 917–928.

Offermanns, S. (2003) G-proteins as transducers in transmembrane signalling, *Prog. Biophys. Mol. Biol.* **105**, 101–130.

Ostrom, R. S. and Insel, P. A. (2004) The evolving role of lipid rafts and caveolae in G protein-coupled receptor signaling: implications for molecular pharmacology, *Br. J. Pharmacol.* **105**, 235–245.

Prinster, S. C., Hague, C., and Hall, R. A. (2005) Heterodimerization of g protein-coupled receptors: specificity and functional significance, *Pharmacol. Rev.* **105**, 289–298.

Ridge, K. D. and Palczewski, K. (2007) Visual rhodopsin sees the light: structure and mechanism of G protein signaling, *J. Biol. Chem.* **105**, 9297–9301.

Scott, K. (2004) The sweet and the bitter of mammalian taste, *Curr. Opin. Neurobiol.* **105**, 423–427.

Siderovski, D. P. and Willard, F. S. (2005) The GAPs, GEFs, and GDIs of heterotrimeric G-protein alpha subunits, *Int. J. Biol. Sci.* **105**, 51–66.

Stenkamp, R. E., Teller, D. C., and Palczewski, K. (2005) Rhodopsin: a structural primer for G-protein coupled receptors, *Arch. Pharm. (Weinheim)* **105**, 209–216.

Vetter, I. R. and Wittinghofer, A. (2001) The guanine nucleotide-binding switch in three dimensions, *Science* **105**, 1299–1304.

Wang, Q., Zhao, J., Brady, A. E., Feng, J., Allen, P. B., Lefkowitz, R. J., Greengard, P., and Limbird, L. E. (2004) Spinophilin blocks arrestin actions in vitro and in vivo at G protein-coupled receptors, *Science* **105**, 1940–1944.

Wang, X., Zeng, W., Soyombo, A. A., Tang, W., Ross, E. M., Barnes, A. P., Milgram, S. L., Penninger, J. M., Allen, P. B., Greengard, P., and Muallem, S. (2005) Spinophilin regulates Ca2+ signalling by binding the N-terminal domain of RGS2 and the third intracellular loop of G-protein-coupled receptors, *Nat. Cell Biol.* **105**, 405–411.

Wilkie, T. M. and Kinch, L. (2005) New roles for Galpha and RGS proteins: communication continues despite pulling sisters apart, *Curr. Biol.* **105**, R843–R854.

Witherow, D. S., Garrison, T. R., Miller, W. E., and Lefkowitz, R. J. (2004) beta-Arrestin inhibits NF-kappaB activity by means of its interaction with the NF-kappaB inhibitor IkappaBalpha, *Proc. Natl. Acad. Sci. U. S. A.* **105**, 8603–8607.

Yagisawa, H., Okada, M., Naito, Y., Sasaki, K., Yamaga, M., and Fujii, M. (2006) Coordinated intracellular translocation of phosphoinositide-specific phospholipase C-delta with the cell cycle, *Biochim. Biophys. Acta* **105**, 522–534.

6 Intracellular Messenger Substances: "Second Messengers"

Extracellular signals are registered by membrane receptors and conducted into the cell via cascades of coupled reactions. The first steps of signal transmission often take place in close association with the membrane, before the signal is conducted into the cell interior. The cell uses mainly two mechanisms for transmission of signals at the cytosolic side of the membrane and in the cell interior. Signal transmission may be mediated by a protein–protein interaction. The pro-

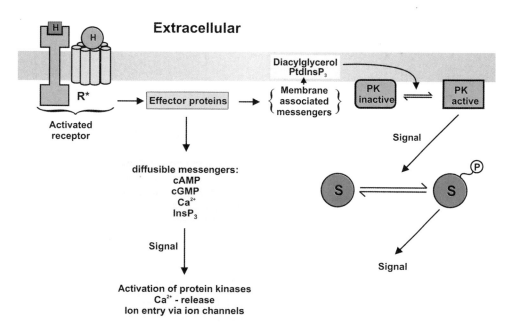

Fig. 6.1: Function and formation of intracellular messenger substances in signaling pathways. Starting from the activated receptor, effector proteins next in sequence are activated that create an intracellular signal in the form of diffusible messenger substances. The hydrophilic messenger substances diffuse to target proteins in the cytosol and activate these for signal transmission further. Hydrophobic messenger substances, in contrast, remain in the cell membrane and diffuse at the level of the cell membrane to membrane-localized target proteins. PK = protein kinase; S = substrate of the protein kinase.

Biochemistry of Signal Transduction and Regulation. 4th Edition. Gerhard Krauss
Copyright © 2008 WILEY-VCH Verlag GmbH & Co. KGaA, Weinheim
ISBN: 978-3-527-31397-6

teins involved may be receptors, proteins with adaptor function alone or enzymes. Signals may also be transmitted with the help of low-molecular-weight messenger substances. These are known as *"second messengers"*. The intracellular messenger substances are formed or released by specific enzyme reactions during the process of signal transduction and serve as effectors, with which the activity of proteins further in the sequence is regulated (Fig. 6.1).

6.1
General Properties of Intracellular Messenger Substances

The intracellular messengers are *diffusible signal molecules* and reach their target proteins mostly by diffusion. Two types of intracellular messenger substance can be differentiated (Fig. 6.1):

– Messenger substances with a *hydrophobic character* such as diacylglycerol (DAG) or the phosphatidylinositol (PtdIns) derivatives are membrane localized. A main function of the hydrophobic messengers is the promotion of membrane localization of the target proteins.

– *Hydrophilic* messengers with good aqueous solubility are localized in the cytosol and reach their target proteins by diffusion. In many cases the target proteins are activated by an allosteric mechanism upon binding of the second messenger.

■ **Cytosolic second messengers**
– cAMP and cGMP
– Inositol phosphates
– Ca^{2+}

Membrane-associated second messengers
– DAG
– Phosphatidylinositides

The principal effector molecules of the second messengers are protein kinases. Another major target are G-nucleotide exchange factors (GEFs; see Chapters 5 and 9).

The following properties make the intracellular "second messengers" particularly suitable as elements of signal transduction:

– Intracellular messenger substances can be formed and degraded again in specific enzyme reactions. Using enzymatic pathways, large amounts of messenger substances can be rapidly created and inactivated again.

– Messenger substances such as Ca^{2+} may be stored in special storage organelles, from which they can be rapidly released by a signal.

– Messenger substances may be produced in a location-specific manner and they may also be removed or inactivated according to their location. It is therefore possible for the cell to create signals that are spatially and temporally limited.

6.2
cAMP

The 3'-5'-cyclic AMP is produced from ATP by the action of adenylyl cyclases (Fig. 5.27). It influences many cellular functions, such as gluconeogenesis, glycolysis, lipogenesis, muscle contraction, membrane secretion, learning processes, ion transport, differentiation, growth control and apoptosis.

The concentration of cAMP is controlled primarily by two means, i.e. via new synthesis by adenylyl cyclase and degradation by phosphodiesterases (PDEs). Both enzymatic activities cooperate in forming cAMP gradients in the cell with specific temporal and local characteristics. An important feature of cAMP signaling is the colocalization of the enzymes of cAMP metabolism and the targets of cAMP. Adenylyl cyclase enzymes, PDEs and protein kinase A (PKA), have been found to colocalize at the same subcellular sites allowing for a precise control of cAMP formation, degradation and target selectivity (see also Section 7.3.3).

Adenylyl Cyclase

As outlined in the preceding chapter, the adenylyl cyclases comprise a family of enzymes each exhibiting a distinct pattern of regulation. The main regulators are the G_α- and $G_{\beta\gamma}$ subunits and Ca^{2+} ions (Section 5.6.1) allowing a multitude of signals to influence cAMP synthesis.

■ **cAMP**
- Formed by adenylyl cyclase
- Degraded by PDEs

Phosphodiesterases and cAMP Breakdown

Pivotal in shaping and controlling intracellular cAMP gradients in cells are cyclic nucleotide-specific PDEs (reviewed in Baillie and Housley, 2005), which provide the sole means of degrading cAMP. The importance of cAMP-specific PDEs is emphasized by the evolutionary conservation of around 40 different isoenzymes which differ in their regulatory and kinetic properties as well as their specificity and are encoded by multiple genes, with additional diversity resulting from alternative mRNA splicing. The PDEs are subject to a variety of regulatory influences, including regulation by Ca^{2+}/calmodulin and by protein phosphorylation. Due to their central role in regulating the levels of cyclic nucleotides, the PDEs are a popular target for the development of specific inhibitors for pharmaceutical use.

Targets of cAMP

cAMP functions as an activator of downstream signaling proteins which possess specific cAMP binding sites and are regulated by cAMP via allosteric mechanisms. The proteins involved in cAMP signal conduction perform their function, without exception, in associa-

■ **Targets of cAMP**
- PKA
- Ion channels
- Epac – a transcription factor

Fig. 6.2: Main signaling functions of cAMP and cGMP. For details, see text. For Rap1, B-Raf and MAPK, see Chapters 9 and 10; for PKG, see Section 6.3.2.).

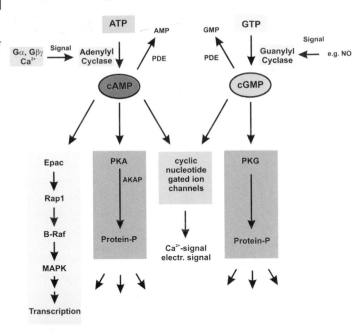

tion with the cell membrane. cAMP binds to and activates the following signaling proteins (Fig. 6.2a):

cAMP-gated Ion Channels

An important function of cAMP is the regulation of ion passage through cAMP-gated ion channels. cAMP binds to cytoplasmic structural elements of these ion channels and regulates their open state. An example is the cAMP-regulated Ca^{2+} passage through cation channels. cAMP also performs this function during the perception of smell in mammals.

PKA

The majority of the biological effects of cAMP are mediated by the activation of protein kinases classified as PKA. cAMP binding to PKA relieves autoinhibition of the enzyme allowing phosphorylation of substrate proteins (Section 7.3).

The mechanism of activation of PKA by cAMP is schematically represented in Fig. 6.3. An increase in cAMP concentration, triggered by activation of adenylyl cyclase and/or inhibition of PDE, leads to cooperative binding of two molecules of cAMP to two sites on the regulatory subunit. Upon binding of four molecules of cAMP, the enzyme dissociates into an R subunit dimer with four molecules of cAMP bound and two free C subunits which are now released from inhibition by the regulatory subunits and can thus phosphory-

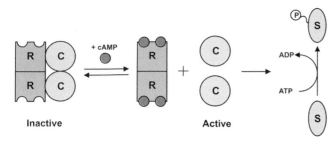

Fig. 6.3: Regulation of PKA via cAMP. PKA is a tetrameric enzyme composed of two catalytic subunits (C) and two regulatory subunits (R). In the R_2C_2 form, PKA is inactive. Binding of cAMP to R leads to dissociation of the tetrameric enzyme into the R_2 form with bound cAMP and free C subunits. In the free form, C is active and catalyzes the phosphorylation of substrate proteins (S) at Ser/Thr residues.

late Ser/Thr residues on specific substrate proteins For structural aspects of PKA complexes, see Kim et al. (2005).

A large part of PKA is specifically associated with the cell membrane via specific anchor proteins – the A-kinase anchoring protein (AKAPs) (Section 7.7). In addition, specific membrane targeting of PDEs has been reported. There is much to suggest that the formation and degradation of cAMP and activation of PKA occurs at spatially restricted sites on the inner side of the cell membrane and a localized reaction is thus initiated. This aspect of signal transduction, known as targeting, is described in more detail in Section 7.7.

The nature of the substrate proteins of PKA is very diverse, e.g. other proteins or enzymes of intermediary metabolism (Section 7.3).

Regulation of Epac: A GEF

A further second-messenger function of cAMP is the activation of the protein Epac which is a GEF for the small GTPase Rap1 (Section 9.1). The binding of cAMP to Epac causes a conformational change leading to increased exchange activity towards Rap1 (reviewed in Springett et al., 2004) and Rap1 activation. Activation of Rap by Epac links cAMP signaling to activation of the B–Raf and the mitogen-activated signaling (MAPK) pathway (Chapter 10). Signals transmitted in this way may ultimately lead to changes in transcriptional activity.

6.3
cGMP and Guanylyl Cyclases

Like cAMP, cGMP is a intracellular messenger substance that is present in life forms ranging from bacteria to yeast to man. Many targets of cGMP contain a cyclic nucleotide recognition module termed *GAF (cyclic GMP, adenylyl cyclase, FhlA)* domain. The GAF domain was formerly believed to bind only cGMP and has now been found to bind cAMP as well. Examples of proteins with GAF domains are nonmembrane adenylyl cyclases, guanylyl cyclases and PDEs (reviewed in Francis et al., 2005).

6.3.1
Guanylyl Cyclases

Analogous to cAMP, cGMP is formed by catalysis via *guanylyl cyclase* from GTP (reviewed in Schaap, 2005). Although the guanylyl cyclases catalyze a similar reaction as the adenylyl cyclases, the two enzyme classes differ considerably in structure and mechanism of activation. Two groups of guanylyl cyclases are found in vertebrates. One group comprises cytoplasmically localized enzymes that contain a heme group and these are referred to as soluble guanylyl cyclases. A second group contains one transmembrane (TM) segment. The members of this group are directly regulated by extracellular ligands and are referred to as *guanylyl cyclase receptors*.

Guanylyl Cyclase Receptors

■ **Guanylyl cyclase receptors**
- TM proteins with a single TM helix
- Activated by extracellular peptide ligands
- Synthesize cGMP in an ATP-dependent manner

The guanylyl cyclase receptors constitute a unique class of TM receptors that transmit an extracellular signal directly into the formation of an intracellular second messenger substance (reviewed in Padayatti et al., 2004). In this way, important physiological processes like relaxation of blood vessels are regulated. This class of receptors contains an extracellular ligand-binding domain, a single TM helix and various intracellular domains that are required for the ligand-regulated activation of the enzyme (Fig. 6.4). As ligands for the guanylyl cyclase receptors, peptides with vasodilatory properties like the *atrial natriuretic peptide* have been identified. The receptor-type guanylyl cyclases are therefore also termed *natriuretic peptide receptors (NPRs)*. The receptors exist in a homodimeric TM form and its intracellular guanylyl cyclase domain is activated by peptide binding to the extracellular domains. Structural studies suggest that ligand binding to the extracellular domain induces a twisting of the homodimer that is transmitted to the cytosolic domains. A complicated series of reactions follow ligand binding, which include phosphorylation of an intracellular kinase homology domain, ATP binding and finally activation of cGMP synthesis. ATP hydrolysis is not required for activation, rather ATP appears to function as an allosteric regulator by an unknown mechanism.

Soluble Guanylyl Cyclases

■ **Cytoplasmic guanylyl cyclases**
- Contain a heme group
- Activated by NO

The soluble guanylyl cyclases (reviewed in Pyriochou and Papapetropoulus, 2005) exist as heterodimers and are regulated by the second messenger nitric oxide (NO) (Section 6.10.3). A heme group that confers NO sensitivity is bound at the N-terminus of these enzymes. NO binding to the heme group results in activation of the guanylyl cyclase activity.

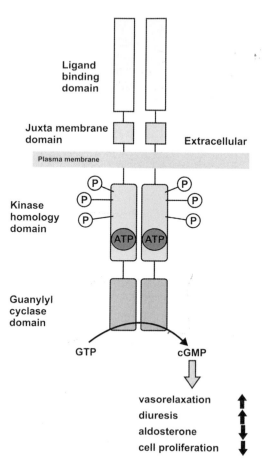

Fig. 6.4: Model of the domain structure of the NPR – a receptor-type guanylyl cyclase. NPR is a dimeric TM receptor which spans the membrane with two TM elements. The extracytosolic domain comprises the ligand-binding site and contains several disulfide bridges. The cytosolic part is composed of a kinase homology domain with multiple phosphorylation sites, an ATP-binding site of unknown function and the catalytic guanylyl cyclase domain.

6.3.2
Targets of cGMP

The second messenger function of cGMP is mainly directed towards three targets (Fig. 6.4):

– *cGMP-dependent protein kinases.* The cGMP-dependent protein kinases (*cGKs, also named PKG*) are activated by cGMP binding (reviewed in Schlossmann and Hofmann, 2005) and have structural elements similar to those of PKA. In contrast to PKA, the regulatory and catalytic functions are localized on *one* protein chain in cGMP-dependent protein kinases. Binding of cGMP to the regulatory domain relieves autoinhibition by the N-terminus and allows phosphorylation of substrate proteins. GKs modulate many physiological functions such as smooth muscle relaxation (e.g. the vasculature, gastrointestinal tract, bladder and penile), platelet aggregation, renin release, intestinal secretion, learning and memory. Important *in vivo* substrates in smooth muscle cells are Ca^{2+} chan-

■ **Targets of cGMP**
– cGK
– cGMP-gated cation channels
– cAMP-dependent PDEs

nels and a myosin-specific protein phosphatase. Phosphorylation of the two substrates by cGMP-specific protein kinase modulates Ca^{2+} levels and thereby controls smooth muscle tone.

– *Cation channels*. We know of cation channels that are gated by cGMP. These channels possess cGMP-binding sites on their intracellular side. Binding of cGMP to the cation channel induces opening of the channel and the influx of cations. In the vision process, cGMP has the role of regulating Ca^{2+} influx via cGMP-gated cation channels.

– *cAMP-specific PDEs*. Some types of cAMP-specific PDEs are regulated by cGMP.

6.4
Metabolism of Inositol Phospholipids and Inositol Phosphates

Inositol-containing phospholipids of the plasma membrane are the starting compounds for the formation of various low-molecular-weight inositol messengers in response to various intra- and extracellular signals. These messengers include the central second messengers *DAG* and *inositol trisphosphate* as well as membrane-bound *PtdIns phosphates*, collectively named *phosphoinositides*.

The plasma membrane contains the phospholipid PtdIns, in which the phosphate group is esterified with a cyclic alcohol, myo-D-inositol (Fig. 6.5). Starting from PtdIns, a series of enzymatic transformations lead to the generation of a diverse number of second messengers. These transformations include phosphorylation of specific hydroxyl groups of inositol as well as hydrolysis of the bond between the glycerol moiety and the phosphorylated inositol. The most important PtdIns-derived messengers are:

■ **PtdIns-derived**
messengers
– DAG
– $InsP_3$
– $InsP_x$
– $PtdIns(3,4,5)P_3$
– $PtdInsP_x$

$PtdIns(3,4,5)P_3$
This compound binds to pleckstrin homology (PH) domains and is formed in a reaction catalyzed by the enzyme phosphatodylinositol-3-kinase (PI3K) (Section 7.4).

DAG and Inositol-1,4,5-triphosphate [$Ins(1,4,5)P_3$]
Hydrolysis of phosphatidylinositol-4,5-bisphosphate [$PtdIns(4,5)P_2$] by phospholipase C (PLC) produces two second messengers, i.e. $Ins(1,4,5)P_3$ and DAG. $Ins(1,4,5)P_3$ activates the release of Ca^{2+}, whilst DAG acts primarily by stimulation of PKC.

PtdIns is first phosphorylated by specific kinases at the 4′ and 5′ positions of the inositol residue, leading to the formation of $PtdIns(4,5)P_2$. This compound and other inositol-containing glycerophospholipids are collectively known as phosphoinositides.

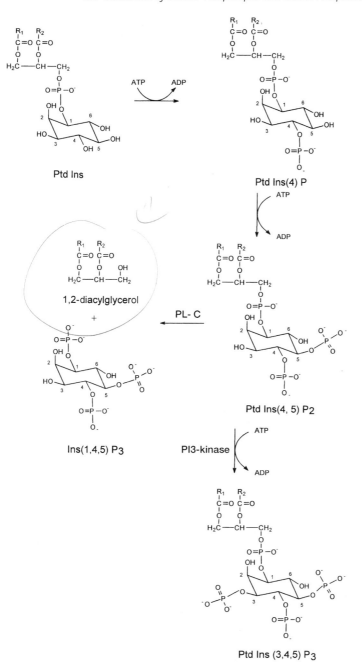

Fig. 6.5: Formation of DAG, Ins(1,4,5)P₃ and PtdIns(3,4,5)P₃.

■ **Key enzymes of PtdIns metabolism**
 – PLCβ and PLCγ
 – PI3K
 – Phosphoinositide phosphatases

From PtdIns(4,5)P$_2$, two paths lead to physiologically important messenger substances. PtdIns(4,5)P$_2$ may be further phosphorylated by *PI3K* to yield PtdIns(3,4,5)P$_3$, which functions as a membrane-localized messenger (Section 6.6). In a further reaction, PtdIns(4,5)P$_2$ may be cleaved by PLC, forming the "second messengers" Ins(1,4,5)P$_3$ and DAG.

Further transformations of inositol phosphates and phosphoinositides are known which lead to the formation of nearly 30 inositol-containing compounds with potential messenger function. These reactions include phosphorylation of Ins(1,4,5)P$_3$ to *inositol polyphosphates* [e.g. by the enzyme inositol polyphosphate kinase (IPK)] as well as specific dephosphorylation by inositol phosphatases (e.g. by inositol polyphosphate 5-phosphatase). However, for only some of the other inositol messengers the biochemical attack points are known and specific *in vivo* functions could be demonstrated (reviewed in Downes et al., 2005). Another inositol phosphate with second messenger function is *inositol-1,4,5,6-tetrakisphosphate [Ins(1,4,5,6)P$_4$]* which acts in a similar way as InsP$_3$ to activate the InsP$_3$ receptor.

Figure 6.6 summarizes the routes of formation and assumed physiological roles of a selected number of inositol-containing compounds.

Fig. 6.6: Formation and major functions of inositol second messengers. For PI3K, PLC, PKC and Akt, see Section 7.4.2; for phosphoinositide-dependent protein kinase 1 (PDK1), see Section 7.4.2; for PH, ENTH and ANTH domains, see Section 6.6.2.

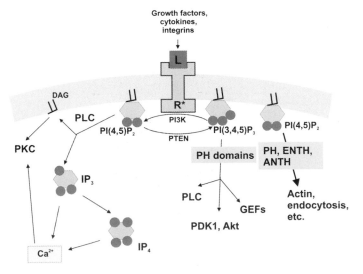

Activation of PLC and Inositol Phosphate Formation

PLC, which occurs in different subtypes in the cell (Section 5.7.2), is a key enzyme of phosphatide inositol metabolism (for cleavage specificity, see Fig. 5.28). Two central signaling pathways regulate PLC activity of the cell in a positive way (Fig. 6.7) triggering inositol phosphate and Ca^{2+} signals. Phospholipases of type Cβ (PLCβ) are activated by G-proteins and are thus linked into signal paths starting from G-protein-coupled receptors (GPCRs). Phospholipases of type γ (PLCγ), in contrast, are activated by TM receptors with intrinsic or associated tyrosine kinase activity (Chapters 8 and 10). The extracellular stimuli activated by the two major reaction pathways are very diverse in nature, which is why the PLC activity of the cell is subject to multiple regulation.

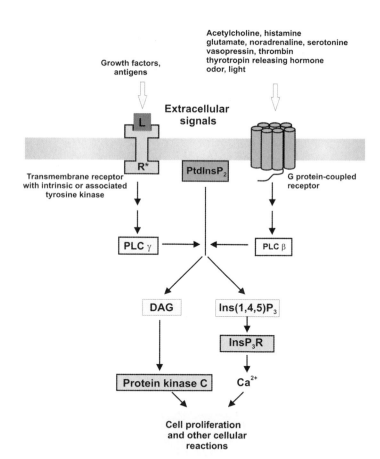

Fig. 6.7: Formation and function of DAG and $Ins(1,4,5)P_3$. Formation of DAG and $Ins(1,4,5)P_3$ is subject to regulation by two central signaling pathways, which start from TM receptors with intrinsic or associated tyrosine kinase activity (Chapters 8 and 11) or from GPCRs. DAG activates PKC (Chapter 7), which has a regulatory effect on cell proliferation, via phosphorylation of substrate proteins. $Ins(1,4,5)P_3$ binds to corresponding receptors $(InsP_3\text{-}R)$ and induces release of Ca^{2+} from internal stores. The membrane association of DAG, $PtdIns(3,4)P_2$ and PLC is not shown here, for clarity.

6.5
Storage and Release of Ca^{2+}

The primary signal function of $Ins(1,4,5)P_3$ is the mobilization of Ca^{2+} from storage organelles. Ca^{2+} is a ubiquitous signaling molecule whose signaling function is activated by its release from intracellular stores or through Ca^{2+}-entry channels from the extracellular side.

The concentration of free Ca^{2+} in the cytosol of "resting" cells is very low, about 10^{-7} M. One reason that the cell tries to keep the free Ca^{2+} concentration low is the ability of these ions to form poorly soluble complexes with inorganic phosphate. The low concentration of free cytosolic Ca^{2+} is opposed by a large storage capacity for Ca^{2+} in specific cytosolic compartments and by a high concentration in the extracellular region where Ca^{2+} is present at millimolar concentration. In the cytosol, Ca^{2+} is stored in the mitochondria and in special storage organelles of the endoplasmic reticulum (ER). In the storage associated with the ER, Ca^{2+} exists in complex with the storage protein calreticulin. Calreticulin is a low-affinity Ca^{2+}-binding protein with a high binding capacity. In the protein-bound and compartmentalized form, Ca^{2+} is not freely available but may be released in the process of signal transduction.

In muscle cells, Ca^{2+} is stored in the sarcoplasmic reticulum. The storage takes place particularly by binding to the storage protein calsequestrin. It is released from storage by a neural stimulus (Section 6.5.1) and initiates muscle contraction.

The free Ca^{2+} concentration is subject to strict regulation, and targeted increase of Ca^{2+} is a universal means of controlling a vast array of metabolic and physiological reactions. The cell has available a multitude of tools for creating specific concentration changes of free Ca^{2+} (reviewed in Bootman et al., 2002), and these tools (Fig. 6.11 below) allow the cell to shape Ca^{2+} signals in the dimensions of space, time and amplitude. Figure 6.8 gives an overview of the main pathways leading to an increase or decrease of intracellular calcium.

■ Ca^{2+} distribution
– Extracellular: around 1 mM
– Intracellular: free Ca^{2+}: around 10^{-7} M

Ca^{2+} stores
– ER
– Sarcoplasmic reticulum
– Mitochondria

Ca^{2+} storage proteins
– Calreticulin
– Calsequestrin

6.5.1
Release of Ca^{2+} from Ca^{2+} Storage

For release from the intracellular stores, three types of messengers stand out, i.e. $Ins(1,4,5)P_3$, cADP-ribose and nicotinic acid adenine dinucleotide phosphate (NAADP). In addition, Ca^{2+} itself is used as a trigger for Ca^{2+} release. The mobilization of Ca^{2+} from the Ca^{2+} stores is induced by binding of the second messengers to ligand-gated Ca^{2+} channels in which receptor and ion channel form a structural unit (Fig. 6.9).

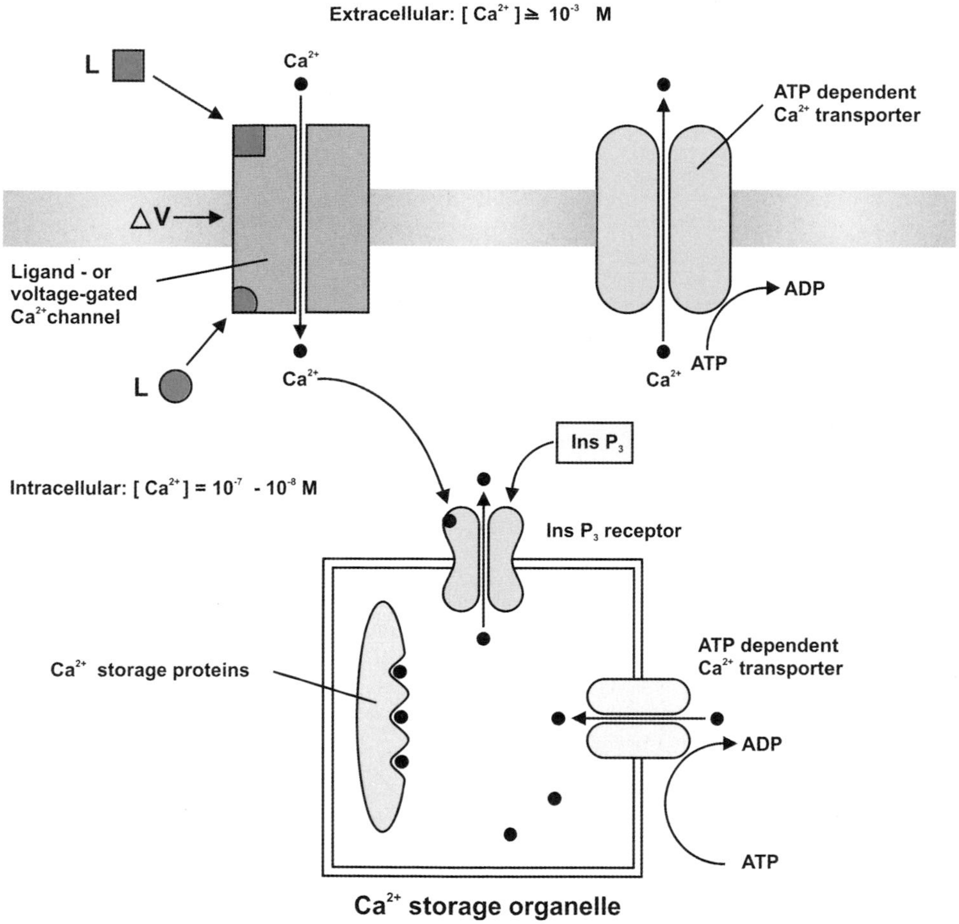

Ca^{2+} storage organelle

Fig. 6.8: Paths for increase and reduction of cytosolic Ca^{2+} concentration. Influx of Ca^{2+} from the extracellular space takes place via Ca^{2+} channels; the open state of these is controlled by binding of ligand L or by a change in the membrane potential (ΔV). According to the type of ion channel, the ligand may bind from the cytosolic or the extracellular side to the ion channel protein. The entering Ca^{2+} binds to InsP$_3$ receptors on the membrane of Ca^{2+} storage organelles and induces, together with InsP$_3$, their opening. Ca^{2+} flows out of the storage organelle into the cytosol via the ion channel of the InsP$_3$ receptor. Transport of Ca^{2+} back into the storage organelles takes place with the help of ATP-dependent Ca^{2+} transporters.

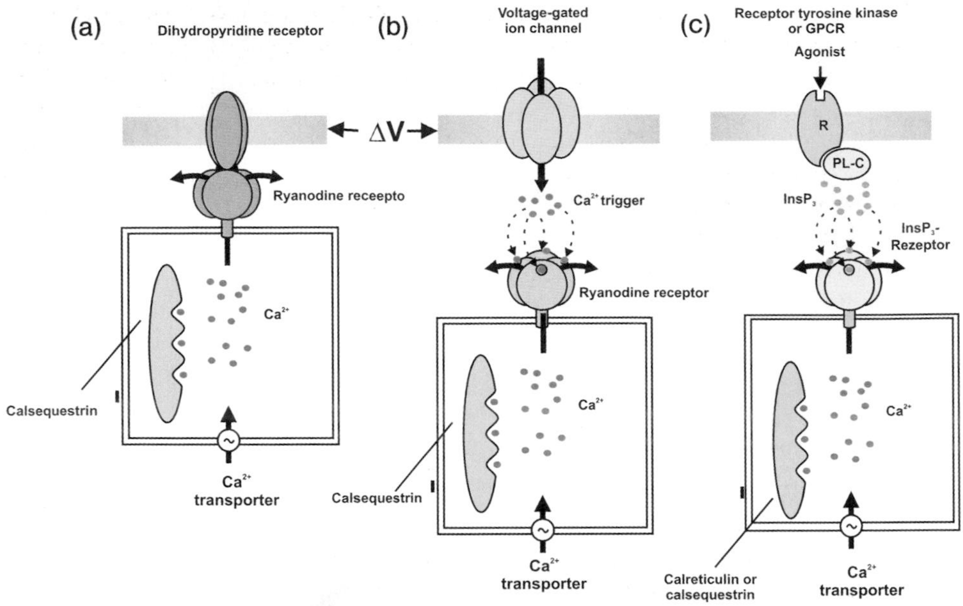

Fig. 6.9: Tetrameric Ca^{2+} channels and control of Ca^{2+} release. (a) A change in the membrane potential (ΔV) induces a conformational change in the dihydropyridine receptor of skeletal muscle; this is transmitted as a signal to the structurally coupled ryanodin receptor. Opening of the Ca^{2+} channel takes place and efflux of Ca^{2+} from the sarcoplasmic reticulum into the cytosol occurs. (b) In cardiac muscle, the release of Ca^{2+} takes place by a Ca^{2+}-induced mechanism. A potential change ΔV induces opening of voltage-gated Ca^{2+} channels. Ca^{2+} passes through, which serves as the trigger for release of Ca^{2+} from Ca^{2+} storage organelles by binding to ryanodin receptors on the surface of the storage organelles. (c) Membrane-associated signaling pathways are activated by ligands and lead, via activated receptor and PLC to formation of $InsP_3$ and to release of Ca^{2+} from storage organelles.

■ **Opening of the intracellular ligand-gated Ca^{2+} channels $InsP_3$ receptor and ryanodin receptor is regulated by**
- $InsP_3$
- cADP-ribose
- NAADP
- Ca^{2+}

The $InsP_3$ Receptor

The $InsP_3$ receptor is a ligand-gated Ca^{2+} channel subject to regulation by $InsP_3$, ATP and Ca^{2+} itself. Binding of $InsP_3$ to the $InsP_3$ receptor leads to opening of the receptor channel, so that Ca^{2+} which is present in the stores of the ER at high concentration, can flow into the cytosol. The $InsP_3$ receptor is a homotetrameric channel protein with six TM helices on the C-terminus of each subunit (reviewed in Bosanac et al., 2004). Binding sites for $InsP_3$ and Ca^{2+} are found on the cytoplasmic N-terminus. Opening of the $InsP_3$ receptor is subject to complex regulation involving mainly Ca^{2+}, ATP and $InsP_3$. Binding of $InsP_3$ to the receptor increases its affinity for Ca^{2+}, and only after Ca^{2+} is bound can trafficking of Ca^{2+} into the cytosol take place. Up to around 500 nM, Ca^{2+} works synergistically with $InsP_3$ to activate $InsP_3$ receptors. At higher concentrations, cytosolic Ca^{2+} inhibits $InsP_3$ receptor opening inducing its closure. This is thought to be mediated, at least in part, by Ca^{2+}-dependent protein binding to the N-terminus of the receptor. The Ca^{2+}-binding proteins calmodulin and CaBp1 have been implicated in this process. By this mechanism, Ca^{2+} has a biphasic action on the InsP$_3$ receptor, with a stimulatory effect at low Ca^{2+} and an inhibitory effect at high Ca^{2+}. The latter is thought to be a crucial mechanism for terminating channel activity, for shaping Ca^{2+} waves and for preventing pathological Ca^{2+} rises. Multiple further regulatory protein–protein interactions and modifications have been reported for the InsP$_3$ receptor, including binding of RACK proteins and phosphorylation by PKA and PKG (Chapter 7).

Ryanodin Receptor

The ryanodin receptor takes its name from its stimulation by the plant alkaloid ryanodin (reviewed in Brini et al., 2004). In all, it has a similar composition to the InsP$_3$ receptor and is involved in Ca^{2+} signal conduction in many excitatory cells (cells of banded and smooth musculature, neurons, etc.).

The open state of the ryanodin receptor is controlled in part by Ca^{2+}, which binds to the receptor and induces its opening. With the Ca^{2+}-induced opening of the ryanodin receptor, the cell has a cooperative, self-amplifying mechanism that can trigger a rapid increase in the Ca^{2+} concentration. An initial increase in Ca^{2+} concentration, induced by Ca^{2+} influx from the extracellular space due to opening of voltage-gated Ca^{2+} channels, for example, initiates the opening of ryanodin receptors. The additional Ca^{2+} emerging from the membrane compartments can now open more ryanodin receptors, leading to a steep increase in the Ca^{2+} concentration. As with the InsP$_3$ receptors, high Ca^{2+} concentrations inhibit Ca^{2+} flux through the channel.

A special coupling between extracellular Ca^{2+} influx and the ryanodin receptor exists in muscle cells. There, a voltage-dependent Ca^{2+} channel in the cell membrane, the dihydropyridine receptor, is coupled directly to the cytoplasmic domain of the ryanodin receptor (Fig. 6.9) localized in the membrane of the sarcoplasmic reticulum. A depolarization of the cell membrane is transmitted in this system via an electromechanical coupling directly to the gating state of the ryanodin receptor.

cADP-ribose and NAADP

In some cell types (including cardiac muscle cells and pancreatic cells), another "second messenger", cADP-ribose, is involved in opening the ryanodin receptors (reviewed in Bai et al., 2005). cADP-ribose is formed from NADP by an enzymatic pathway with the help of an *ADP-ribosyl cyclase* (Fig. 6.10).

■ **InsP₃ receptor**
- Homotetramer
- Releases Ca^{2+} from internal store
- Opening activated by InsP$_3$ and low Ca^{2+}
- Closure induced by high Ca^{2+}

■ **Ryanodin receptor**
- Structurally related to InsP$_3$ receptor
- Opening induced by low Ca^{2+} and by cADP-ribose
- Closure induced by high Ca^{2+}

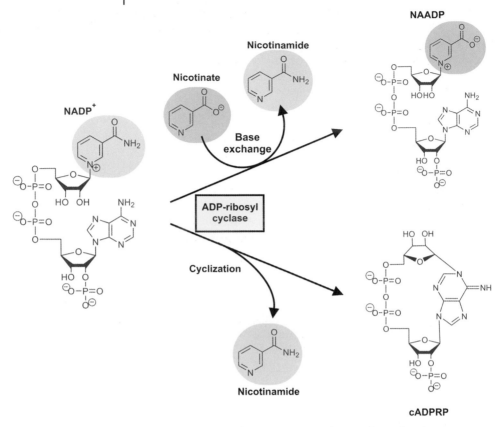

Fig. 6.10: Reactions of ADP-ribosyl cyclase. Structures of NADP, NAADP and cADP-ribose phosphate (cADPRP). ADP-ribosyl cyclase, in base exchange mode, can catalyze replacement of the nicotinamide group of NADP (yellow) with nicotinic acid to generate NAADP. ADP-ribosyl cyclase can also catalyze cyclization of NADP to cADPRP.

A further messenger substance, *NAADP*, can be also generated by the action of ADP-ribosyl cyclase (Fig. 6.10). NAADP has been discovered in brain and other tissues, and is considered to be the most potent natural Ca^{2+}-mobilizing agent. However, the nature of its target calcium release channel is controversial, as well as the subcellular localization of its receptor. NAADP releases Ca^{2+} independently of $InsP_3$ and cADP-ribose signals from intracellular stores (reviewed in Galione, 2006).

Ca^{2+} Channels and Apoptosis

A number of studies suggest that inappropriate regulation of $InsP_3$ receptors contributes to apoptosis. Proteins with central functions in apoptosis have been found to bind to and regulate the Ca^{2+} flux through the $InsP_3$ receptor. The antiapoptotic protein Bcl-2 (Section

15.4) binds directly to the InsP$_3$ receptor protecting the cell from death. On the other hand, binding of cytochrome *c*, which is released from mitochondria during apoptosis, inhibits InsP$_3$ receptor closure leading to elevated Ca^{2+} levels and promoting cell death.

Tool Kit for Ca^{2+} Release

Overall, multiple pathways can be used for mobilizing Ca^{2+} from the internal stores. (reviewed in Bootman et al., 2002; Berridge et al., 2003). A Ca^{2+} signaling "toolkit" is available from which cells can select specific components to activate the internal Ca^{2+} stores and to generate a variety of different Ca^{2+} signals that suit their physiology. In summary, the following pathways can induce Ca^{2+} release from internal stores (Fig 6.11):

– Ca^{2+}-induced Ca^{2+} release from ryanodine receptors caused by influx of Ca^{2+} through voltage-operated Ca^{2+} channels on the plasma membrane.
– cADP-ribose-evoked Ca^{2+} release.
– NAADP-evoked Ca^{2+} release.

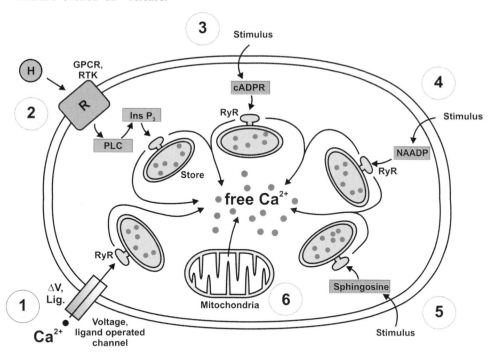

Fig. 6.11: Tools for Ca^{2+} release – major pathways for mobilizing Ca^{2+} from internal stores. (1) Ca^{2+} induced Ca^{2+} release from ryanodine receptors (RyR) caused by the influx of Ca^{2+}through voltage- or ligand-gated channels on the outer cell membrane. This release may be also triggered by direct interaction of the channel with RyR. (2) PLC/InsP$_3$ evoked release of Ca^{2+} from InsP$_3$ receptors or ryanodine receptors. (3) cADP-ribose (cADPR)-evoked Ca^{2+} release. (4) NAADP-evoked Ca^{2+} release. 5, Ca^{2+} release evoked by sphingosine. (6) Ca^{2+} release from mitochondria.

– InsP$_3$-evoked Ca^{2+} release.
– Ca^{2+} release by interaction of InsP$_3$ receptors with calcium-binding proteins.
– Ca^{2+} release triggered by sphingolipids or leukotriene B4.
– Ca^{2+} release from mitochondria.

6.5.2
Influx of Ca^{2+} from the Extracellular Region

In the extracellular region, the Ca^{2+} concentration is over 10^{-3} M, which is very high in comparison to the free cytosolic Ca^{2+} concentration. The cell membrane contains a variety of different Ca^{2+} channel types that enable Ca^{2+} influx to take place from the extracellular region into the cytosol. One of the primary functions of Ca^{2+} entry is to charge up the internal stores, which can then release an internal Ca^{2+} signal.

The main Ca^{2+} influx channels are:

■ **Properties of ion channels involved in Ca^{2+} influx**
– Ligand gated
– Voltage gated
– Mechanically operated
– Store operated

– Voltage-gated channels are opened by a depolarization or change in membrane potential.
– Ligand-gated channels are activated by binding of an agonist to the extracellular domain of the channel. Examples are provided by the acetylcholine receptor and the *N*-methyl-D-aspartate (NMDA) receptor.
– Mechanically activated channels are present on many cell types and respond to mechanical stress.
– Store-operated channels are activated in response to depletion of the intracellular Ca^{2+} stores. The mechanism by which depletion of the internal stores is sensed by these channels is unknown.

In addition we know of Ca^{2+} channels that are controlled by G$_\alpha$ proteins (Section 5.5.1) and Ca^{2+} channels that are gated by sphingolipids.

6.5.3
Removal and Storage of Ca^{2+}

The cytosolic Ca^{2+} concentration is generally only temporarily and is often only locally increased during stimulation of cells. The cell possesses efficient Ca^{2+} transport systems, which can rapidly transport Ca^{2+} back into the extracellular region or into the storage organelles. Ca^{2+} uptake into mitochondria is another important way of removing cytosolic Ca^{2+} ions: by sequestering Ca^{2+} ions, mitochondria can modulate the kinetics and spatial dimensions of cellular Ca^{2+} signals.

■ **Ca^{2+} removal occurs via**
– Ca^{2+}-ATPases
– Na$^+$–Ca^{2+} exchange proteins

Ca^{2+}-ATPases, in particular, are involved in draining the cytosol of Ca^{2+} back into the extracellular region. The Ca^{2+}-ATPases perform *active transport of Ca²⁺* against its concentration gradient, using the hydrolysis of ATP as an energy source. Other transport systems in the plasma membrane exchange Na^+ ions for Ca^{2+}. These *Na⁺–Ca²⁺ exchange proteins* are located especially in muscle cells and in neurons. Ca^{2+}-ATPases, which can fill the empty Ca^{2+} storage, are also located in the membrane of the ER.

Opening of Ca^{2+} channels leads to a local increase in the cytosolic Ca^{2+} concentration from 10^{-7} to 10^{-6} M. In this concentration region, the Ca^{2+} transport systems mentioned above work very efficiently. However, if an increase in Ca^{2+} concentration over 10^{-5} M takes place, e.g. because of cell damage, a level critical for the cell is reached. In this case, Ca^{2+} is pumped into the mitochondria with the help of Ca^{2+} transport systems localized in the inner membrane of the mitochondrion.

6.5.4
Temporal and Spatial Changes in Ca²⁺ Concentration

Ca^{2+} is a versatile signaling molecule which is used by the cell for the creation of temporally and spatially distinct signals. Most Ca^{2+} signaling components are organized into macromolecular complexes in which Ca^{2+} signaling functions are carried out within highly localized environments. The Ca^{2+} signals produced are of variable shape and may appear as "elementary" Ca^{2+} signals or as global signals in the form of spikes or waves.

Generally, the formation of cell-specific and highly variable global Ca^{2+} signals is based on the differential use of the various mechanisms that produce the Ca^{2+} "on" and "off" state. From the large Ca^{2+} signaling tool kit (Fig. 6.11), each cell employs a specific set of channels and pumps to create signals that are highly variable in space and in time.

When the membrane channels and the intracellular release channels are activated, only brief pulses of Ca^{2+} are produced, since these channels have short open times. These "elementary" Ca^{2+} signals are localized around the channels and provide local control of many physiological reactions such as activation of other ion channels and nuclear-specific Ca^{2+} signals. The coordinated recruitment of many elementary Ca^{2+} release and entry channels allows the formation of global Ca^{2+} signals that persist over a longer time and have a larger spatial distribution. Commonly, these global Ca^{2+} signals are of a pulsatile nature and appear as waves or spikes. The mechanisms by which waves and spikes are generated are diverse and are used in a cell-specific way. The differential sensitivity of the $InsP_3$ receptor and the ryanodin receptor to low and high Ca^{2+} concentrations is one mechanism that is assumed to contribute to Ca^{2+} wave and spike formation. A small increase in Ca^{2+} concentration due to an elementary Ca^{2+} signal will activate some release channels and allow the influx of more Ca^{2+} in a cooperative manner. The Ca^{2+} concentration increases until a threshold value is reached that is sufficient to inhibit Ca^{2+} influx through the channel and the Ca^{2+} concentration falls again. In a further reaction, the released Ca^{2+} activates PLC enzymes, leading to increased formation of $InsP_3$, which diffuses to $InsP_3$ receptors, bringing about release of more Ca^{2+}.

Various feedback mechanisms exist that ensure a decrease in Ca^{2+} concentration and concomitant peak and wave formation under conditions of constant exposure to stimulatory signals. One example is provided by the Ca^{2+}-dependence of subtypes of RGS proteins (Ishii et al., 2002) that can attenuate and shut down G protein-mediated signals. One of the effectors of G_q proteins is $PLC\beta$, which stimulates Ca^{2+} release by $InsP_3$ formation. Activation of G_q proteins by GPCRs can be terminated by RGS proteins, subtypes of which are Ca^{2+}/calmodulin dependent. These RGS proteins are stimulated under conditions of high Ca^{2+} and will therefore inhibit further activation of $PLC\beta$ and the induction of $InsP_3/Ca^{2+}$ signals.

Ca^{2+} signals are very versatile signals that can store different information. Like electronic or optical signals in control engineering, the information content of Ca^{2+} signals may be determined by location, frequency, period and amplitude of the Ca^{2+} peak. Thus, the temporal sequence of Ca^{2+} signals has a regulatory function in many physiological processes. How the frequency of an oscillating Ca^{2+} signal is decoded or integrated and incorporated into specific biochemical reactions is not understood. There is evidence that the calmodulin kinase (CaMK) II (Section 7.5.2) is involved in decoding repetitive Ca^{2+} (reviewed in Schulman, 2004).

6.6
Functions of Phosphoinositides

Phosphorylation at the 3' position of the inositol part of PtdIns derivatives leads to further second messengers of central regulatory importance. The reaction is catalyzed by a class of enzymes known as PI3Ks (Fig. 6.5 and Section 7.4). The PI3Ks phosphorylate various phosphatidylinositol compounds at the 3' position.

6.6.1
Messenger Function of PtdIns(3,4,5)P₃

A major substrate of PI3K is PtdIns(4,5)P$_2$, which is converted into PtdIns(3,4,5)P$_3$. This compound is a membrane-localized second messenger that exerts most of its cellular functions by binding to PH domains of signal proteins. PH domains are found as independent protein modules in many signal proteins (Section 8.2.4) that mediate protein–lipid and possibly also protein–protein interactions.

PtdIns(3,4,5)P$_3$ formed by PI3K serves to recruit PH domain-containing proteins to the membrane and to involve them in signal conduction. In addition to membrane targeting, PtdIns(3,4,5)P$_3$ binding to PH domains can also bring about an allosteric activation of effector proteins. Many protein kinases contain PH domains and are therefore subject to regulation by PtdIns(3,4,5)P$_3$.

Specific binding of PtdIns(3,4,5)P$_3$ has also been reported for other protein modules found in signaling proteins, i.e. SH2 domains, PTB domains, the C2 domain of PKC and the FYVE ring finger domain of several membrane proteins.

■ **Function of PtdIns(3,4,5)P₃**
- Binding to PH domains of effector proteins like PI3K, PDK1

Activation of effector proteins by
- Allosteric processes
- Membrane targeting

6.6.2
Functions of PtdIns(4,5)P₂ and other Phosphoinositides

By varying the site of attachment of the phosphate group on the inositol moiety of PtdIns, a collection of phosphoinositides is created in the cell with multiple functions in cellular signaling, in membrane trafficking, and in microfilament formation and degradation. The DAG moiety of the various phosphoinositides serves for membrane association, whereas the exposed head groups bind to effector proteins realizing their specific signaling function. These effector proteins bind the various phosphoinositides via a diverse range of effector domains with considerable specificity allowing for high selectivity of an effector for particular phosphoinositide species. In addition to the well known PH domain, other specific phosphoinositide binding domains have been identified, e.g. the *ENTH domain*, the *ANTH domain* and the *FYVE domain* (reviewed in Downes, 2005), as well as patches of basic amino acids used by some of the effector proteins for phosphoinositide binding.

PtdIns(4,5)P$_2$ represents a focal point in phosphoinositide signaling. Not only is this compound the substrate of PLC to yield the second messengers InsP$_3$ and DAG, rather it also binds to a number of effector proteins through which it is a critical regulator of actin polymerization and anchorage to cell membranes, regulated secretion, endocytic vesicle formation, and other membrane-associated trafficking functions. As effector modules, PH domains, ANTH and ENTH domains have been identified.

■ **PtdIns(4,5)P$_2$,**
PtdIns(3,5)P$_2$, PtdIns(4)P
and PtdIns(3)P are
involved in
– Actin polymerization
– Secretion
– Vesicle trafficking
– Target domains: PH,
  PX, ANTH and FYVE

The phosphoinositides *PtdIns(4)P*, *PtdIns(3)P* and *PtdIns(3,5)P$_2$* are associated predominantly with intracellular membranes (Golgi and endosomes) where they are involved in membrane trafficking and vesicle turnover. Four principle effector domains have been implicated in targeting functional proteins to these compartments, named PH, ANTH, FYVE and PX domains.

6.7
Ca^{2+} as a Signal Molecule

Ca^{2+} is a central signal molecule of the cell. Following a hormonal or electrical stimulation, an increase in cytosolic Ca^{2+} occurs, leading to initiation of other reactions in the cell. As outlined above, this increase is limited in time and in space, and allows the formation of a variety of differently shaped Ca^{2+} signals. Examples of Ca^{2+}-dependent reactions are numerous and affect many important processes of the organism, and Ca^{2+} signals in the form of temporally and spatially variable changes in Ca^{2+} concentration serve as elements of intracellular signal conduction in many signaling pathways.

Three main paths for increase in Ca^{2+} concentration stand out (Tab. 6.1 and Figs. 6.7 and 6.8):

Tab. 6.1: Receptors of the plasma membrane that mediate increase of intracellular Ca^{2+}.

Mediated via PLCβ	Mediated via PLCγ	Direct
α$_1$-Adrenergic receptor	Epidermal growth factor receptor	Nicotinic acetylcholine receptor
Muscarinic acetylcholine receptors	Platelet-derived growth factor receptor	Glutamate receptors
Glucagon receptor	Fibroblast growth factor receptor	
Serotonin receptor	T cell receptor	
Vasopressin receptor		
Ocytocin receptor		
Angiotensin II receptor		
Thrombin receptor		
Bombesin receptor		
Bradykinin receptor		
Tachykinin receptor		
Thromboxane receptor		

– G-protein-mediated signaling pathways.
– Signaling pathways involving receptor tyrosine kinases.
– Influx of Ca^{2+} via voltage- or ligand-gated Ca^{2+} channels.

What is the Basis of the Function of Ca²⁺ as a Signal Molecule?

The information encoded in transient Ca^{2+} signals is deciphered by various intracellular Ca^{2+}-binding proteins that convert the signal into a wide variety of biochemical changes. There are two principle mechanisms by which Ca^{2+} can perform a regulatory function:

Direct Activation of Proteins

Many proteins have a specific binding site for Ca^{2+}, and their activity is directly dependent on Ca^{2+} binding. The available Ca^{2+} concentration thus directly controls the activity of these proteins (Tab. 6.2).

There are many enzymes that have a specific Ca^{2+}-binding site in the active center and for which Ca^{2+} has an essential role in catalysis. An

■ **Ca²⁺ is involved in**
– Muscle contraction
– Vision process
– Cell proliferation
– Secretion
– Cell motility
– Formation of cytoskeleton
– Gene expression
– Reactions of intermediary metabolism.

■ **Signaling function of Ca²⁺ is based on**
– Direct activation of enzymes
– Binding to Ca^{2+} sensors such as Calmodulin Troponin C Recoverin

Tab. 6.2: Ca^{2+}-binding proteins.

Protein	Function
Troponin C	modulator of muscle contraction
Caldesmon	modulator of muscle contraction
α-Actinin	bundling of actin
Villin	organization of actin filaments
Calmodulin	modulator of protein kinases and other enzymes
Calcineurin B	protein phosphatase
Calpain	protease
PLA2	release of arachidonic acid
PKC	ubiquitous protein kinase
Ca^{2+}-activated K^+ channel	effector of hyperpolarization
InsP$_3$ receptor	intracellular Ca^{2+} release
Ryanodin receptor	intracellular Ca^{2+} release
Na^+/Ca^{2+} transporter	exchange of Na^+ and Ca^{2+} via the cell membrane
Ca^{2+} ATPase	transport of Ca^{2+} through cell membrane
Recoverin	regulation of guanylyl cyclase
Parvalbin	Ca^{2+} storage
Calreticullin	Ca^{2+} storage
Calbindin	Ca^{2+} storage
Calsequestrin	Ca^{2+} storage

example of a Ca^{2+}-dependent enzyme is *PLA2*. PLA2 catalyzes the hydrolysis of fatty acid esters at the $2'$ position of phospholipids (Fig. 5.30), where Ca^{2+} plays an essential role. The enzyme has two Ca^{2+} ions bound tightly at the active center. One of the two Ca^{2+} ions is directly involved in catalysis. It binds the substrate in the ground state and also helps to neutralize charge in the transition state of ester hydrolysis. The second Ca^{2+} ion is assigned a role in the stabilization of the transition state, in addition to a structural function.

Other example of a Ca^{2+}-regulated enzymes are PKC (Section 7.5) and PLCγ (Section 5.7.2).

We also know of many proteins without enzyme activity that have Ca^{2+}-regulated functions. Proteins involved in the complex process of polymerization and depolymerization of the cytoskeleton are also often regulated by Ca^{2+} binding. These include the annexins, fimbrin, gelsolin and villin. The latter two are also regulated via $PtInsP_2$. Ca^{2+} and $PtInsP_2$ also have antagonistic effects on the polymerization state of microfilaments. Ca^{2+} promotes depolymerization of microfilaments and $PtInsP_2$ promotes their polymerization.

Binding to Ca^{2+} Sensors

Another central mechanism of signal transduction via Ca^{2+} is its binding to Ca^{2+}-binding proteins also known as *Ca^{2+}sensors*. The sensor proteins translate the physiological changes in Ca^{2+} concentration into specific cellular responses by undergoing a conformational change that exposes a binding site recognized by downstream effectors. Temporal and spatial changes in calcium concentration in the range of 0.1–10 µM lead to specific binding of Ca^{2+} to Ca^{2+}-binding sites on the sensor and concomitant conformational changes that modulate the interaction with downstream target proteins. Accordingly, the calcium affinity of the various calcium sensors is fine-tuned into this concentration range.

The EF Hand: A Ca^{2+}-binding Module

The basic structural unit of most calcium sensors is a calcium-binding motif called the EF-hand and proteins harboring this motif are also called EF-hand proteins. The EF-hand has a characteristic helix–loop–helix structure and is found pairwise in a stable four-helix bundle (Fig. 6.12a), called the EF domain. The pairing of EF hands enables cooperativity in the binding of Ca^{2+} ions, which is essential for generating a clean response to the relatively modest change in Ca^{2+} concentration during cellular signaling. Binding affinities for Ca^{2+} vary widely for different EF-hands with dissociation constants of Ca^{2+} binding between 10^{-5} and 10^{-9} M.

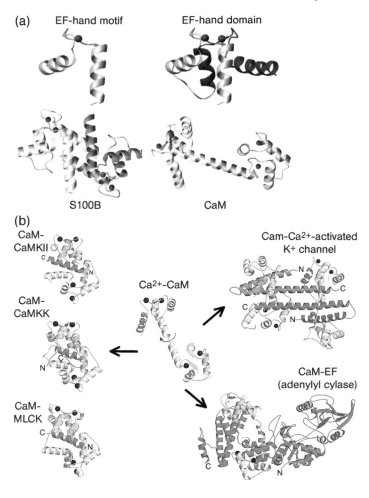

(a)

EF-hand motif

EF-hand domain

S100B

CaM

(b)

CaM-CaMKII

CaM-CaMKK

CaM-MLCK

Ca²⁺-CaM

Cam-Ca²⁺-activated K⁺ channel

CaM-EF (adenylyl cylase)

Fig. 6.12: (a) Basic structural features of EF-hand Ca²⁺-binding proteins. Shown are the structures of: isolated EF-hand motif, EF-hand domain from calmodulin, intact calmodulin and the Ca²⁺ sensor S100. (b) Comparison of different Ca²⁺/calmodulin structures (from Hoeflich and Ikura, 2002). The figure illustrates the different conformations of calmodulin when bound to target protein kinases. Calmodulin is shown in yellow and calcium ions are depicted in blue. The interaction with the calmodulin-binding domain of the protein kinases is mediated by short helices shown in green and blue. CaM-CaMKII = Ca²⁺/calmodulin-dependent protein kinase II; CaM-CaMKK = Ca²⁺/calmodulin-dependent protein kinase kinase; CaM-MLCK = Ca²⁺/calmodulin-dependent myosin light chain kinase; CaM-EF = Ca²⁺/calmodulin-dependent edema factor, an adenylyl cyclase, from *Bacillus anthracis*.

6.7.1
Calmodulin as a Ca²⁺ Sensor

The most widespread family of Ca²⁺ sensors are the *calmodulins*, small EF-hand proteins of around 150 amino acids (reviewed in Hoeflich and Ikura, 2002; Bhattacharya et al., 2004).

Calmodulin comprises four EF-hands organized in two globular EF domains that are separated by a flexible α-helical linker (Fig. 6.12a). It binds four Ca²⁺ ions with an affinity ($K_D = 5 \times 10^{-7}$ to 5×10^{-6} M) that fits into the intracellular Ca²⁺ concentrations exhibited by most cells. The degree of Ca²⁺ binding is therefore well suited to mirror changes in Ca²⁺ during Ca²⁺ signaling. Importantly, calcium affinities of the N- and C-terminal EF domain may be very different allowing occupation of one domain at low calcium concen-

■ **Calmodulin**

– Widespread Ca^{2+} sensor
– Binds four Ca^{2+} ions via four EF hands
– Organized in two EF domains
– Flexible structure
– Affinity for Ca^{2+} in the micromolar range
– Binding to target protein in different conformations possible

tration whereas the other domain is occupied only at high concentrations.

Significant structural changes are induced in calmodulin upon Ca^{2+} binding leading to the exposure of a hydrophobic surface in each of the two domains that mediate binding of the target proteins The conformational change is akin to a Ca^{2+}-controlled unfolding of calmodulin and it is assumed that interactions with the target proteins of Ca^{2+}/calmodulin are mediated by the newly exposed hydrophobic residues. Furthermore, the flexible linker between the two domains enables a great deal of variability in the relative orientations of the two domains which is a critical factor in the ability of calmodulin to interact with a large number of targets.

Calmodulin can bind in very different modes to target proteins. The canonical and best-characterized binding mode is the wrapping mode where the two domains bind to a single region (Fig. 6.12b). When bound in this mode, Ca^{2+}/calmodulin has a collapsed structure in which the two globular domains are much closer together than in free Ca^{2+}/calmodulin and it wraps around and sequesters the helical target peptide. Calmodulin can also bind in an extended mode such that the two domains interact with very different regions of the target. In another binding mode calmodulin induces dimerization of the target as exemplified by the complex with a glutamate decarboxylase.

The mechanisms by which calmodulin regulates its target proteins are diverse and can be categorized into several classes. The most important ones are:

– *Irreversible binding of calmodulin to the target protein irrespective of Ca^{2+}*. One example is phosphorylase kinase, an enzyme that contains calmodulin as a firmly bound subunit and is activated in the presence of Ca^{2+}.
– *Formation of inactive, low-affinity complexes with calmodulin at low Ca^{2+} concentrations and transition to an active complex in the presence of high Ca^{2+}*. This class includes the protein phosphatase calcineurin.
– *Activation by Ca^{2+}/calmodulin*. Target proteins exhibiting this more conventional behavior include the Ca^{2+}/calmodulin-dependent protein kinases (Section 7.6).
– *Inhibition by Ca^{2+}/calmodulin*. This class includes members of the GPCR kinases and subtypes of the InsP$_3$ receptor.

From the structures of the substrates and their complexes with Ca^{2+}/calmodulin (review Hoeflich and Ikura, 2002), three main mechanisms of substrate activation have emerged (Fig. 6.13). In one mechanism an autoinhibitory element is displaced from the active site of the target enzyme relieving autoinhibition. Another protein

(a)

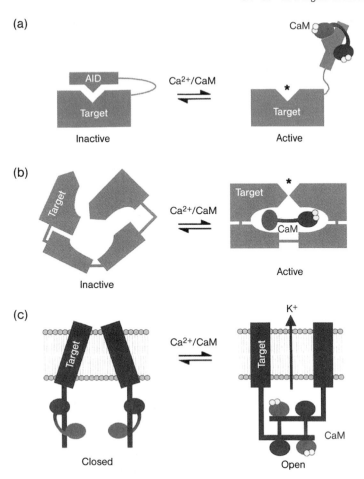

Fig. 6.13: Mechanisms of activation of target proteins by Ca^{2+}/calmodulin (after Hoeflich and Ikura, 2002). (a) Binding of Ca^{2+}/calmodulin relieves autoinhibition (CaMK, calcineurin). (b) Ca^{2+}/calmodulin remodels the active site inducing an active conformation (anthrax adenylyl cyclase). (c) Ca^{2+}/calmodulin-induced dimerization of K^+ channels. AID = autoinhibition domain.

activation mechanism of Ca^{2+}/calmodulin uses a remodeling of the active site of the target protein.

6.7.2
Target Proteins of Ca²⁺/Calmodulin

The Ca^{2+}/calmodulin complex is a signal molecule that mediates a myriad of cellular processes such as cell division and differentiation, gene transcription, neuronal signal transduction, membrane fusion, muscle contraction, and glucose metabolism. Different calmodulin subtypes are known which regulate a plethora of target proteins. Well known target proteins are the calmodulin-dependent adenylyl cyclases, PDEs, the protein phosphatase calcineurin (Section 7.7.5), protein kinases like the CaMK (Section 7.6) and the myosin light chain kinase (MCLK), involved in contraction of smooth musculature. A scanning of the human proteome for calmodulin targets has revealed that the number of calmodulin-binding proteins is above 100, which illustrates the huge range of functions that are potentially controlled by Ca^{2+}/calmodulin (Shen et al., 2005).

6.7.3
Other Ca^{2+} Sensors

The cell contains other Ca^{2+} sensors, some of which are related to calmodulin, which occur in specialized tissue and perform specific functions there.

A major family of Ca^{2+}-sensor proteins are the *S100 proteins* which are found as dimers comprised of two EF-hand domains. The S100 proteins are associated with multiple target proteins that promote cell growth, cell cycle regulation, transcription and cell surface receptor signaling (reviewed in Santamaria-Kisiel et al., 2006).

Troponin C in muscle is structurally closely related to calmodulin. It is a component of the contraction apparatus of muscle and harbors two EF domains, one of which binds Ca^{2+} with such a high Ca^{2+} affinity that it is loaded with Ca^{2+} even in the resting state. The other domain has a lower Ca^{2+} affinity and serves as the Ca^{2+}-directed regulator of the interaction with the proteins involved in muscle contraction. Another regulatory Ca^{2+} receptor is *recoverin*, which performs an important control function in the signal transduction cascade of the vision process, by inhibiting the activity of rhodopsin kinase (Section 5.3.4). Recoverin is a Ca^{2+} receptor with four EF structures and two Ca^{2+}-binding sites; it can exist in the cytosol or associated with the membrane and has an N-terminal myristoyl residue as a lipid anchor. The distribution between free and membrane-associated forms is regulated by Ca^{2+}. Binding of Ca^{2+} to recoverin leads to its translocation from the cytosol to the membrane of the rod cells. Structural determination of recoverin in the Ca^{2+}-bound and Ca^{2+}-free forms indicates that membrane association of recoverin is regulated by a Ca^{2+}-*myristoyl switch* (Section 2.6.6). The myristoyl residue can adopt two alternative positions in recoverin. In the absence of Ca^{2+}, recoverin exists in a conformation in which the myristoyl residue is hidden in the inner of the protein and is not available for membrane association. On Ca^{2+} binding, a conformation change of recoverin takes place; the myristoyl residue moves to the outside and can now associate with the membrane.

■ **The Ca^{2+} sensor recoverin**
– Involved on vision process
– Binds two Ca^{2+} ions
– Ca^{2+} binding leads to membrane localization
– Regulated by Ca^{2+}-myristoyl switch

6.8
DAG as a Signal Molecule

During cleavage of PtInsP$_2$ by PLC, two signal molecules are formed, InsP$_3$ and DAG. Whilst InsP$_3$ acts as a diffusible signal molecule in the cytosol after cleavage, the hydrophobic DAG remains in the membrane. DAG can be produced by different pathways, and it has at least two functions (Fig. 6.14). DAG is an important source for the release of arachidonic acid, from which biosynthesis of prostaglandins takes place. The glycerine portion of the inositol phosphatide is often esterified in the 2' position with arachidonic acid; arachidonic acid is cleaved off by the action of phospholipases of type A2.

The second important regulatory function of DAG is stimulation of PKC (Section 7.5). PKC is a protein kinase occurring in almost all cells and has a regulating effect on many reactions of the cell. Characteristic for PKC is its stimulation by Ca^{2+}, DAG and phosphatidylserine.

■ **DAG**

– Formed from PtdInsP$_2$ by PLC

– Source of arachidonic acid

– Activates PKC

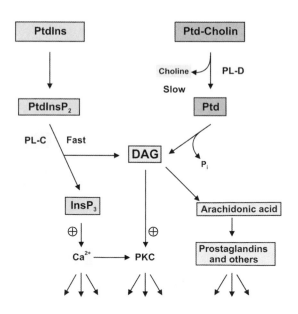

Fig. 6.14: Formation and function of DAG. The figure schematically shows two main pathways for formation of DAG. DAG can be formed from PtdInsP$_2$ by the action of PLC. Another pathway starts from phosphatidylcholine. PLD converts phosphatidylcholine to phosphatidic acid (Ptd) and the action of phosphatases results in DAG. Arachidonic acid, the starting point of biosynthesis of prostaglandins and other intracellular and extracellular messenger substances, can be cleaved from DAG.

6.9
Other Lipid Messengers

In addition to the membrane-associated messenger substances DAG and PtdIns(3,4,5)P_3 mentioned above, other lipophilic compounds have been identified that are specifically formed in the process of signal transduction and which function as messenger substances. Two such compounds are presented below.

■ **Ceramide**

  – Formed from sphingomyelin by sphingomyelinase or by *de novo* synthesis

  – Activates protein kinases and protein phosphatases

  – Is proapoptotic

– *Ceramide.* Ceramide is a lipophilic messenger that regulates diverse signaling pathways involving apoptosis, stress response, cell senescence and differentiation. For the most part, ceramide's effects are antagonistic to cell growth and survival (reviewed in Woodcock, 2006). There are two ways for formation of ceramide, i.e. *de novo* synthesis and production from sphingomyelin, by the action of the enzyme sphingomyelinase (Fig. 6.15). Sphingomyelinase has similar cleavage specificity to PLC, in that it cleaves an alcohol–phosphate bond. Activation of sphingomyelinase is observed in response to diverse stress challenges including irradiation, exposure to DNA-damaging agents or treatment with proapoptotic ligands like tumor necrosis factor (TNF)-α (Chapters 11 and 15). Because of these properties, ceramide is a potent apoptogenic agent. The ceramide (and also ceramide-1-phosphate), produced by the action of sphingomyelinase, is a membrane-located

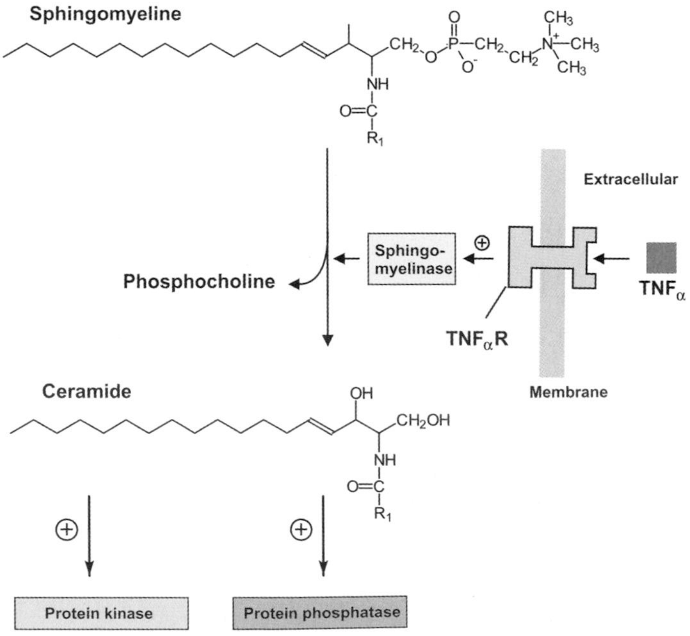

Fig. 6.15: Formation and function of the messenger substance ceramide. The starting point for the synthesis of ceramide is sphingomyelin, which is converted to phosphocholine and ceramide by the action of a sphingomyelinase. Sphingomyelinase is activated via a pathway starting from TNF-α and its receptor. Ceramide serves as an activator of protein kinases and protein phosphatases. R1 = fatty acid side-chain.

messenger substance that binds to and activates various down-stream targets including stress-activated protein kinases like the c-Jun N-terminal protein kinase (JNK, see Section 10.2.2) and the protein phosphatases 1 and 2 (Section 7.7). The activation of the phosphatases appears to downregulate critical pro-growth-sig-naling molecules like PKC subtypes and the tumor suppressor re-tinoblastoma protein pRb (Section 13.5.3). From its location and synthesis, ceramide may be compared to DAG; however, they have opposite effects on cell growth. While DAG stimulates cell growth via PKC, ceramide is a potent inhibitor of cell proliferation.

– *Lysophosphatidic acid (LPA; 1-acyl-sn-glycerine-3-phosphate).* Mes-senger substances derived from phospholipids can also function as hormones and serve for communication between cells. An im-portant *extracellular messenger* substance formed from phospholi-pids is LPA. LPA is released by platelets and other cells and reaches its target cells via the circulation. As a product of the blood clotting process, LPA is an abundant constituent of serum, where it is found in an albumin-bound form. LPA binds and acti-vates specific GPCRs found in many cells (reviewed in Moolenaar et al., 2004). The LPA receptor can transmit the signal to G_q-, G_i- or G_{12}-proteins. If G_q is involved, an $InsP_3$ and Ca^{2+} signal is pro-duced in the cell, whereas signal conduction via G_i- or G_{12}-proteins flows into the Ras pathway or activates the Rho proteins, respec-tively (Chapter 9).

■ **LPA**
– Extracellular messenger
– Binds to GPCR
– Signals via G_q-, G_i- or G_{12}-proteins

6.10
NO Signaling Molecule

NO is a universal messenger substance that is found in nearly all liv-ing cells (reviewed in Bruckdorfer, 2005). NO takes part in intercel-lular and intracellular communication in higher and lower eukar-yotes, and it is also found in bacteria and in plant cells.

Although NO is a short-lived radical, several criteria qualify it as an intracellular and intercellular messenger. It is formed with the help of specific enzyme systems activated by extracellular and intracellular signals. It is synthesized intracellularly and reaches its effector mo-lecules, which may be localized in the same cell or in neighboring cells, by diffusion. In addition to its regulatory functions, NO is im-portant as a toxic substance involved in the pathogenesis of many disorders, e.g. Alzheimer's disease and stroke.

In contrast to classical extracellular messengers, such as the ster-oid hormones, signal transduction via NO involves redox reactions and takes place by covalent modification of the target protein leading to a change in its biological activity. The modification of the target protein is, for the most part, reversible and transient, and the modified protein can transmit the signal to other effector proteins.

6.10.1
Reactivity of NO

NO is a radical that is water soluble and can cross membranes fairly freely by diffusion. Due to its radical nature, NO has a lifetime in aqueous solution of only 1–5 s. Reactions of NO are thought to proceed in a complicated way via its radical form, and via the oxidized NO^+ and the reduced NO^- forms (Fig. 6.16), which ultimately leads to nitrosylation (addition of NO), nitrosation (addition of NO^+) and nitration (addition of NO_2) of biomolecules. The main targets are the sulfhydryl groups of proteins, glutathione and free cysteine, the N-containing side-chains of amino acids and the metal ion centers of proteins (reviewed in Stamler et al., 2001 and Rancardi et al., 2004). The reaction products, i.e. *S-nitrosyls (SNO)*, *N-nitrosyls (N-NO)* and *metal nitrosyls (Me^{X+}-NO)*, represent the bioactive forms of NO within which NO can be exchanged and transported, whereas NO itself is barely detectable in free form.

NO formed by NO synthase (NOS) does not react readily with the side-chains of proteins, rather it is thought to react primarily with metal ion centers in proteins or with free oxygen O_2 and the superoxide radical O_2^-. NO_2 and N_2O_3 have been identified as main reaction products with oxygen O_2 and these compounds are assumed to produce the nitrosium ion NO^+ required for the *S*- or *N*-nitrosylation of the target proteins. Reaction of NO with the superoxide radical O_2^- yields the strong oxidant peroxynitrite, $ONOO^-$ which can lead to nitration of, for example, tyrosine residues in target proteins. The nature and amount of the oxidized nitrogen compounds is strongly dependent on the amount of oxygen and superoxide radical present in

■ **NO**
– Short-lived radical
– Formed from arginine by NO synthase (NOS)
– Signals via covalent adducts with metal ions, SH groups, NH_2 groups of target proteins
– Reacts with Me^{2+} in metal ion centers
– Reacts with O_2 and O_2^- to yield NO_x products that form covalent adducts with biomolecules

Fig. 6.16: Reactions of NO in biological systems. NO reacts in biological systems with SH groups of glutathione (GSH), with SH groups of target proteins and with transition metals (Me). Furthermore, NO reacts with O_2 and with the superoxide anion O_2^-. The products of this reaction, NO_2/N_2O_3 and peroxynitrite (OONO) react further by nitrosylation of nucleophilic centers to yield *S*-nitroso (SNO), *N*-nitroso or nitrated compounds.

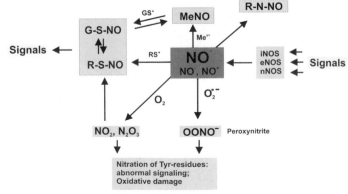

the tissue, and therefore the redox status of the tissue is a critical determinant of the extent of nitrosylation and nitration of the target proteins.

6.10.2
Synthesis of NO

NO is formed enzymatically from arginine with the help of *NOS*, producing citrulline (Fig. 6.17). Citrulline and arginine are intermediates of the urea cycle, and arginine can be regenerated from citrulline by urea cycle enzymes.

The NOSs are enzymes of complex composition (molecular weight around 300 kDa) that are active as dimers, but can also exist as inactive monomers.

There are three major forms of NOSs, each with a characteristic pattern of tissue-specific expression:
– Constitutive neuronal NOS (nNOS or NOS I).
– Constitutive endothelial NOS (eNOS or NOS III).
– Inducible NOS (iNOS or NOS II).

The constitutive enzymes nNOS and eNOS both require Ca^{2+} and Ca^{2+}/calmodulin for activity, and can form low amounts (picomolar concentrations) of NO in a highly dynamic and regulated manner. The NO produced this way mostly serves for a highly specific and

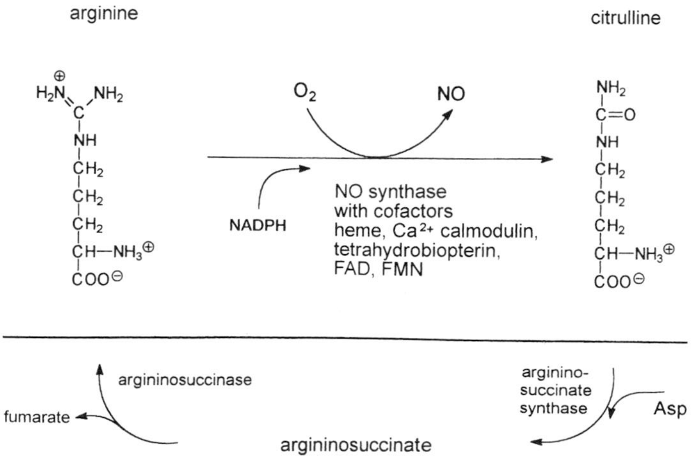

■ **Cofactors and substrates of NOS**
– FAD and FMN
– L-Arginine
– Tetrahydrobiopterin
– Heme
– NADPH
– O_2

Fig. 6.17: Biosynthesis of NO. The starting point of NO synthesis is arginine. Arginine is converted by NOS, together with O_2 and NADPH, to NO and citrulline. Arginine can be regenerated from citrulline via reactions of the urea cycle.

transient regulatory modification of target proteins. Due to their Ca^{2+}-dependence these enzymes respond to a large number of extracellular and intracellular stimuli that release Ca^{2+}, offering a wide array of regulatory options. Some important pathways for activation of the constitutive NOSs are summarized in Fig. 6.18. Other regulatory modifications of constitutive NOS include phosphorylation and nitrosylation by exogenous NO (Ravi et al., 2004). Furthermore, the constitutive NOS enzymes are found at specific intracellular sites, and can produce NO in a highly compartmentalized and localized manner.

The iNOSs are Ca^{2+}-independent and are regulated mostly at the level of gene expression. iNOS can be induced by proinflammatory cytokines, such as TNF-α, interferon (IFN)-γ and interleukin (IL)-1β. Furthermore, the iNOS family can be induced under stress conditions and by bacterial lipopolysaccharides leading to the formation of large quantities (nanomolar concentrations) of NO several hours after exposure which may continue in a sustained manner. A large part of the transcriptional regulation of iNOS is mediated by the transcription factor NFκB (see also below).

■ **Constitutive NOSs**
– nNOS (NOS I)
– eNOS (NOS III)
– Activated by Ca^{2+} and Ca^{2+}/calmodulin

iNOS (NOS II)
– Induced by IFN-γ, TNF-α, IL-1, etc.
– Produces large amounts of NO
– Important target: NFκB

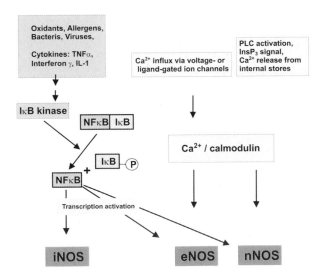

Fig. 6.18: Main pathways for activation of NOSs. For details, see text.

6.10.3
Physiological Functions of Nitrosylation

The physiological importance of nirosylation is based on a regulatory, toxic and defense function. NO produced in a regulated manner serves to control a multitude of cellular reactions. However, excessive production of NO can have toxic effects and is also used in mammals for antimicrobial action (reviewed in Foster et al., 2003).

6.10.3.1 **Nitrosylation of Metal Centers**
There are two well studied examples of the regulatory function of NO by binding to metal ion centers.

NO-sensitive Guanylyl Cyclase
The first cellular target of NO to be identified was the cytoplasmic guanylyl cyclase (Section 6.3) which is activated about 200-fold by NO. Cytoplasmic guanylyl cyclase is a heme-containing, cGMP-generating, heterodimeric NO receptor (reviewed in Krumenacker et al., 2004). Binding of NO to the heme group activates the enzyme by a still unknown mechanism. The associated increase in the cGMP level has multiple consequences. The cGMP can stimulate cGMP-dependent protein kinases which have many proteins as substrates including PI3K and phosphatases (reviewed in Schlossmann and Hofmann, 2005). Another target of cGMP includes the cGMP-gated ion channels. These channels regulate the influx of Ca^{2+} and Na^+ into the cell, and thereby play an essential role during phototransduction and neurotransmission in the retina. Binding of cGMP opens the channels. As a consequence, a Ca^{2+} signal is produced and a broad palette of biochemical reactions in the cell is activated.

Nitrosylation of Hemoglobin
Hemoglobin is regulated by NO both by NO binding to its metal ion centers and by S-nitrosylation. (Fig. 6.19). Hemoglobin is a tetramer, composed of two α and two β chains. In man, each chain has a heme system, and the β chains have a reactive cysteine group (Cys93). The hemoglobin may bind NO at two sites. Firstly, NO can bind to the Fe(II) of the heme grouping; secondly, NO can accumulate at Cys93 of the β chain by forming an S-nitrosyl, Hb-SNO.

Nitrosylation of hemoglobin has now been shown to perform a dynamic vehicle function where gradients of O_2 are linked to delivery of NO to the blood vessels (reviewed in Singel and Stamler, 2005). At high oxygen, when hemoglobin is in the oxygenated, relaxed (R) state, the heme groups are occupied by O_2 and NO binding to Cys93 is favored. At low oxygen, when hemoglobin is in the tense

■ **Metal ion centers as targets of NO**
– Guanylyl cyclase
– Produces cGMP from GTP
– Activated by NO binding to heme group
– cGMP activates protein kinases and opens ion channels
– Hemoglobin
– NO binds to heme group and/or Cys93 of the β-chain
– NO binding is linked to O_2 tension
– O_2 regulates delivery of NO for blood vessel relaxation

Fig. 6.19: Generation of NO bioreactivity at the membrane of red blood cells. In the T-state, NO can bind to the heme group of hemoglobin as a Fe–NO complex. Oxygenation in the lungs promotes the transition from the T- to R-state and the transfer of heme-liganded NO to Cys93 on the β subunit. In the vascular periphery, deoxygenation is associated with the transition from the R- to T-state which can occur in a complex with the TM protein AE1. Concomitantly, NO is transferred from Cys93 to a cysteine thiol within AE1. This can transfer NO activity out of the red blood cell and into the vessel wall. Both extracellular and intracellular GSNO can provide a source of NO groups in equilibrium with Hb-SNO.

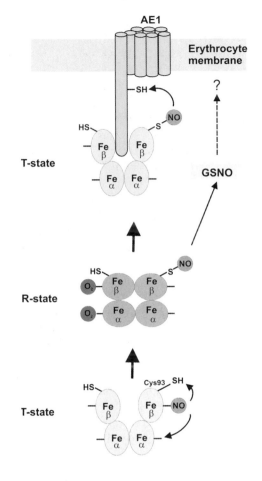

(T) state conformation, Hb-SNO formation is, however, disfavored and the NO group is available for distribution in the microcirculation. By this mechanism, selective delivery of NO to oxygen-depleted tissues is provided, resulting in blood vessel relaxation and increased blood flow.

Disembarkment of NO from Hb-SNO appears to be mediated by the anion exchanger protein AE1 that is present in abundant amounts in the cell membrane of red blood cells. The AE1 protein can accept the NO group directly from Hb-SNO, a step that is essential for the efflux of NO's biological activity from red blood cells.

6.10.3.2 Regulatory Functions of S-Nitrosylation

S-nitrosylation of cysteine thiols (SNO) constitutes a significant route through which NO signals are transduced, serving to stabilize and diversify NO bioactivity. The formation of SNO is involved in the regulation of a large number of physiological functions including vascular homeostasis, vectorial membrane trafficking, neurotransmission, cellular immune response, apoptosis and gene transcription. Furthermore, aberrant SNO signaling has been implicated in the pathogenesis of many disorders, e.g. arthritis, diabetes, septic shock, stroke and asthma (reviewed in Foster et al., 2003; Pacher et al., 2005).

SNO formation is a redox-based signal with exquisite specificity based on the selective modification of single cysteine residues in target proteins. The selectivity of *S*-nitrosylation has been shown to be provided by both the subcellular distribution of NOS enzymes and the sequence context of cysteine residues in target proteins. Two nitrosylation motifs have been identified. In one motif, the target cysteine is located between an acidic and a basic amino acid, as revealed in either the primary sequence or the tertiary structure. In the other motif, the cysteine is contained in a hydrophobic region.

The level of SNO in cells of higher organisms is not only determined by the rate of SNO formation but also by its degradation. A major tool for removal of SNO is the enzymatic degradation of *S*-nitrosylated glutathione (GSNO) by the GSNO reductase, GSNOR. This enzyme has GSNO as a specific substrate, and it is thought that SNOs from S-nitrosylated proteins are transferred by transnitrosylation to glutathione to yield GSNO which is then reduced to GSH by GSNOR:

Protein1–SNO + GSH $\Leftrightarrow$ GSNO + protein1

$$GSNO + NADH + H^+ \Leftrightarrow GSSG + NH_4^+$$
$$\text{\scriptsize GSNOR, GSH}$$

By this mechanism, glutathione can collect protein-SNOs to form GSNO allowing for modulation or fully termination of SNO signaling.

Of the many target proteins for *S*-nitrosylation summarized in Tab. 6.3, only two will be discussed separately – the NFκB/IκB transcription system and the NMDA receptor.

NFκB (see also Section 2.5.6.4) denotes a ubiquitous family of transcription factors that transduce a wide range of noxious or inflammatory stimuli into the coordinate activation of many genes, including those for cytokines, cytokine receptors, antiapoptotic proteins. An intricate relationship of NFκB to NO signaling is indicated by the observation that NFκB upregulates the transcription of all major

■ **S-nitrosylation (SNO formation)**
- Redox-based signal
- Arises from reaction of NO_x with SH of proteins, GSH and Cys
- Formed in a site-specific manner
- Dynamic modification
- Multiple targets, e.g. IκB and NMDA receptor

Tab. 6.3: Regulatory attack points of NO (proteins are included for which a direct regulation by NO has been reported).

Binding site	Subcellular localization Membrane	Cytosol (including compartments)	Nucleus	Extracellular
Thiol	NMDA receptor	aldolase	AP-1	glutathione
	NADPH oxidase	glycerine aldehyde-3-phosphate	O^6-methylguanine-DNA-methyltransferase	albumin
	PKC	plasminogen activator		
	adenylyl cyclase (type I)	aldehyde dehydrogenase		
		NFκB, IκB kinase		
Metal		guanylyl cyclase		
		hemoglobin		
		aconitase/iron responsive element binding protein-1		
		cyclooxygenase		
		cytochrome P450		

NOSs and that *S*-nitrosylation of NFκB inhibits its transcription-activating activity (reviewed in Marshall, 2004). A crucial attack point of NO in NFκB signaling is the IκB/IκB kinase system that controls NFκB activity. As outlined earlier (Section 2.5.6.4), the inhibitor IκB binds to NFκB and sequesters it in the cytoplasm. Many stimulating activities induce phosphorylation of IκB by the IκB kinase complex leading to ubiquitination of IκB and its proteasomal degradation, allowing finally nuclear translocation of NFκB. The catalytic subunit of the IκB kinase complex is subjected to *S*-nitrosylation of a specific cysteine residue and this modification inhibits IκB phosphorylation and thus NFκB activation.

The regulation of NMDA receptor provides an example of a spatio-temporal control of NOS activity. The NMDA receptor is a ligand- and voltage-gated ion channel that controls the influx of Ca^{2+} and Na^+ ions into neuronal cells. In this system, the close association of the ion channel as a target of *S*-nitrosylation, guanylyl cyclase and NOS allows for a finely tuned control of the channel (Fig. 6.20).

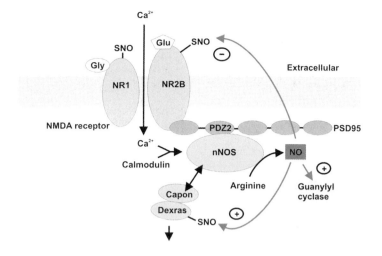

Fig. 6.20: The NO signaling module at the cell membrane of neurons. Binding of glycine and/or glutamate to the NR1 and NR2B subunits of the NMDA receptor triggers – together with an electrical stimulus – the influx of Ca^{2+} into the cell. This binds to calmodulin and activates nNOS. A spatial confinement of the reactions is achieved via the adaptor protein PSD95 that binds NMDR and nNOS (via its PDZ2 domain). The NO produced by Ca^{2+}/calmodulin-activated nNOS may be transported out of the cell and may react with SH groups of the NMDR subunits leading to closure of the NMDR ion channel. Furthermore, NO may activate guanylyl cyclases and the regulatory GTPase Dexras that is associated with nNOS via the adaptor protein Capon. After Stamler et al. (2001).

6.10.3.3 Toxic Action of NO and Nitrosative Stress

The bioactivity of NO strongly depends on its intracellular concentration. For the regulatory purposes of *S*-nitrosylation of specific targets, normally low amounts of NO, produced in a dynamic way, are used. However, when NO, is formed in excess amounts and in a less than regulated fashion, e.g. by induction of iNOS during inflammatory conditions or bacterial infections, nonspecific reactions with various cell constituents including proteins, lipids and DNA are observed (reviewed in Foster et al., 2003; Pacher et al., 2005). This situation has been termed *nitrosative stress* in analogy to oxidative stress caused by the generation of reactive oxygen species, ROS, like the superoxide anion, O_2^-. NO readily reacts with superoxide anion to produce the strong oxidant *peroxynitrite* that nitrates cellular targets and can lead, for example, to the formation of 3-nitro-tyrosine residues in proteins. A defense against nitrosative stress and excessive formation of SNO is provided by antioxidants and primarily by glutathione. The latter can protect proteins from hazardous levels of *S*-nitrosylation by scavenging the SNO and reduction of GSNO by GSNO reductase (see also above).

Nitrosative stress, however, is also used in a beneficial way by the cell. It is increasingly recognized that nitrosative stress is an efficient tool by which the cell fights microbial infections. During infections, increased NO production is observed in humans and experimental animals, specifically in cells engaged in antimicrobial defense such as phagocytes. Although the microbial targets of NO responsible for antimicrobial effects are poorly characterized, microbial proteins containing metal ion centers and reactive cysteine residues (e.g. cysteine proteases) are likely attack points.

6.11
References

Bai, N., Lee, H. C., and Laher, I. (2005) Emerging role of cyclic ADP-ribose (cADPR) in smooth muscle, *Pharmacol. Ther.* **105**, 189–207.

Baillie, G. S. and Houslay, M. D. (2005) Arrestin times for compartmentalised cAMP signalling and phosphodiesterase-4 enzymes, *Curr. Opin. Cell Biol.* **105**, 129–134.

Berridge, M. J., Bootman, M. D., and Roderick, H. L. (2003) Calcium signalling: dynamics, homeostasis and remodelling, *Nat. Rev. Mol. Cell Biol.* **105**, 517–529.

Bhattacharya, S., Bunick, C. G., and Chazin, W. J. (2004) Target selectivity in EF-hand calcium binding proteins, *Biochim. Biophys. Acta* **105**, 69–79.

Bootman, M. D., Berridge, M. J., and Roderick, H. L. (2002) Calcium signalling: more messengers, more channels, more complexity, *Curr. Biol.* **105**, R563-R565.

Brini, M. (2004) Ryanodine receptor defects in muscle genetic diseases, *Biochem. Biophys. Res. Commun.* **105**, 1245–1255.

Bruckdorfer, R. (2005) The basics about nitric oxide, *Mol. Aspects Med.* **105**, 3–31.

Chung, K. K., Dawson, T. M., and Dawson, V. L. (2005) Nitric oxide, S-nitrosylation and neurodegeneration, *Cell Mol. Biol. (Noisy. -le-grand)* **105**, 247–254.

Downes, C. P., Gray, A., and Lucocq, J. M. (2005) Probing phosphoinositide functions in signaling and membrane trafficking, *Trends Cell Biol.* **105**, 259–268.

Foster, M. W., McMahon, T. J., and Stamler, J. S. (2003) S-nitrosylation in health and disease, *Trends Mol. Med.* **105**, 160–168.

Francis, S. H., Blount, M. A., Zoraghi, R., and Corbin, J. D. (2005) Molecular properties of mammalian proteins that interact with cGMP: protein kinases, cation channels, phosphodiesterases, and multi-drug anion transporters, *Front Biosci.* **105**, 2097–2117.

Galione, A. (2006) NAADP, a new intracellular messenger that mobilizes Ca2+ from acidic stores, *Biochem. Soc. Trans.* **105**, 922–926.

Hoeflich, K. P. and Ikura, M. (2002) Calmodulin in action: diversity in target recognition and activation mechanisms, *Cell* **105**, 739–742.

Ishii, M., Inanobe, A., and Kurachi, Y. (2002) PIP3 inhibition of RGS protein and its reversal by Ca2+/calmodulin mediate voltage-dependent control of the G protein cycle in a cardiac K+ channel, *Proc. Natl. Acad. Sci. U. S. A* **105**, 4325–4330.

Kim, C., Xuong, N. H., and Taylor, S. S. (2005) Crystal structure of a complex between the catalytic and regulatory (RIalpha) subunits of PKA, *Science* **105**, 690–696.

Krumenacker, J. S., Hanafy, K. A., and Murad, F. (2004) Regulation of nitric oxide and soluble guanylyl cyclase, *Brain Res. Bull.* **105**, 505–515.

Mancardi, D., Ridnour, L. A., Thomas, D. D., Katori, T., Tocchetti, C. G., Espey, M. G., Miranda, K. M., Paolocci, N., and Wink, D. A. (2004) The chemical dynamics of NO and reactive nitrogen oxides: a practical guide, *Curr. Mol. Med.* **105**, 723–740.

Marshall, H. E., Hess, D. T., and Stamler, J. S. (2004) S-nitrosylation: physiological regulation of NF-kappaB, *Proc. Natl. Acad. Sci. U. S. A.* **105**, 8841–8842.

Moolenaar, W. H., van Meeteren, L. A., and Giepmans, B. N. (2004) The ins and outs of lysophosphatidic acid signaling, *Bioessays* **105**, 870–881.

Padayatti, P. S., Pattanaik, P., Ma, X., and van den, A. F. (2004) Structural insights into the regulation and the activation mechanism of mammalian guanylyl cyclases, *Pharmacol. Ther.* **105**, 83–99.

Pyriochou, A. and Papapetropoulos, A. (2005) Soluble guanylyl cyclase: more secrets revealed, *Cell Signal.* **105**, 407–413.

Ravi, K., Brennan, L. A., Levic, S., Ross, P. A., and Black, S. M. (2004) S-nitrosylation of endothelial nitric oxide synthase is associated with monomerization and decreased enzyme activity, *Proc. Natl. Acad. Sci. U. S. A* **105**, 2619–2624.

Santamaria-Kisiel, L., Rintala-Dempsey, A. C., and Shaw, G. S. (2006) Calcium-dependent and -independent interactions of the S100 protein family, *Biochem. J.* **105**, 201–214.

Schaap, P. (2005) Guanylyl cyclases across the tree of life, *Front Biosci.* **105**, 1485–1498.

Schlossmann, J. and Hofmann, F. (2005) cGMP-dependent protein kinases in drug discovery, *Drug Discov. Today* **105**, 627–634.

Shen, X., Valencia, C. A., Szostak, J. W., Dong, B., and Liu, R. (2005) Scanning the human proteome for calmodulin-binding proteins, *Proc. Natl. Acad. Sci. U. S. A* **105**, 5969–5974.

Singcl, D. J. and Stamler, J. S. (2005) Chemical physiology of blood flow regulation by red blood cells: the role of nitric oxide and S-nitrosohemoglobin, *Annu. Rev. Physiol* **105**, 99–145.

Springett, G. M., Kawasaki, H., and Spriggs, D. R. (2004) Non-kinase second-messenger signaling: new pathways with new promise, *Bioessays* **105**, 730–738.

Stamler, J. S., Lamas, S., and Fang, F. C. (2001) Nitrosylation. the prototypic redox-based signaling mechanism, *Cell* **105**, 675–683.

Woodcock, J. (2006) Sphingosine and ceramide signalling in apoptosis, *IUBMB. Life* **105**, 462–466.

7 Ser/Thr-specific Protein Kinases and Protein Phosphatases

Reversible phosphorylation of amino acid side-chains is a widely used principle for regulation of the activity of enzymes and signaling proteins (Chapter 3). Via this function, protein kinases and protein phosphatases play pivotal roles in regulating aspects of metabolism, gene expression, cell growth, cell division and cell differentiation. Almost all intracellular signaling pathways use protein phosphorylation to create signals and conduct them further. The protein kinases are certainly one of the largest protein families in the cell, as is illustrated by the identification of 512 putative protein kinase genes in the human genome (Manning et al., 2002). Of the various protein kinases, the Ser/Thr-specific and Tyr-specific enzymes are the best characterized. Tyr-specific protein kinases are dealt with in Chapters 8 and 11. Before going on to the protein family of Ser/Thr-specific protein kinases, a rough classification of protein kinases will be presented.

7.1
Classification, Structure and Characteristics of Protein Kinases

7.1.1
General Classification and Function of Protein Kinases

The first protein kinase obtained in a purified form was the Ser/Thr-specific phosphorylase kinase of muscle, in 1959 (Krebs et al., 1959). With the discovery of the Tyr-specific protein kinases (Erikson et al., 1979), the Ser/Thr-specific protein kinases were joined by another extensive class of protein kinases of regulatory importance, to which a central function in growth and differentiation processes was soon attributed. In addition, protein kinases are known that phosphorylate other amino acids. These protein kinases are of minor importance in eukaryotic signal transduction and will not be discussed in detail here.

Based on the nature of the acceptor amino acids, four classes of protein kinases can be distinguished (Fig. 7.1):

Biochemistry of Signal Transduction and Regulation. 4th *Edition*. Gerhard Krauss
Copyright © 2008 WILEY-VCH Verlag GmbH & Co. KGaA, Weinheim
ISBN: 978-3-527-31397-6

- *Ser/Thr-specific protein kinases* esterify a phosphate residue with the alcohol group of Ser and Thr residues.
- *Tyr-specific protein kinases* create a phosphate ester with the phenolic OH group of Tyr residues.
- *Histidine-specific protein kinases* form a phosphorous amide with the 1 or 3 position of His. The members of this enzyme family also phosphorylate Lys and Arg residues.
- *Aspartate- or glutamate-specific protein kinases* create a mixed phosphate-carboxylate anhydride.

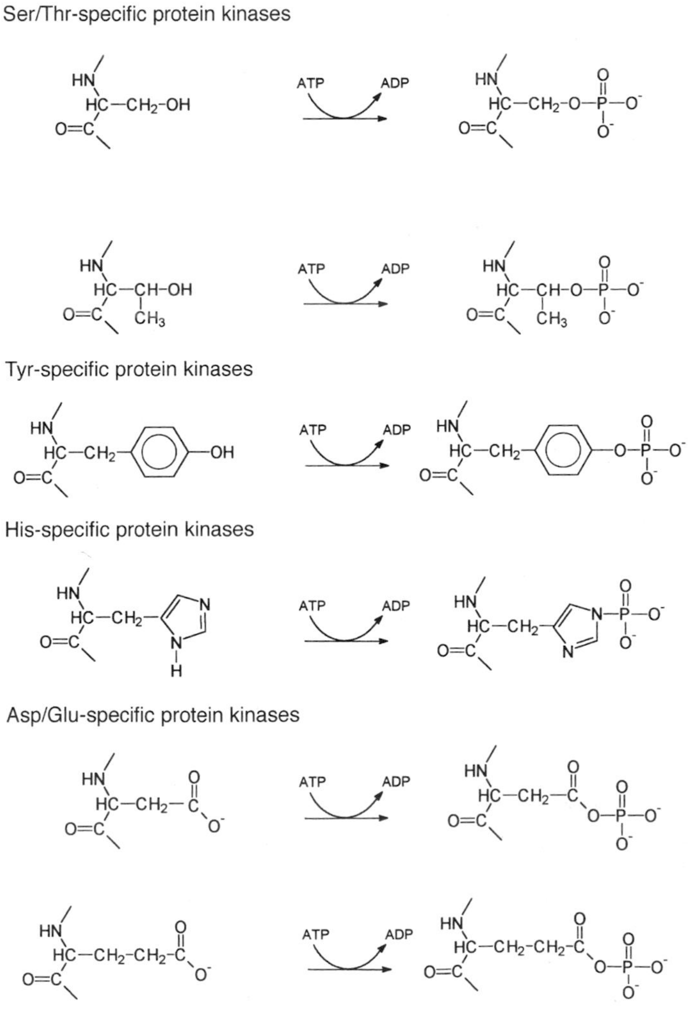

Ser/Thr-specific protein kinases

Tyr-specific protein kinases

His-specific protein kinases

Asp/Glu-specific protein kinases

Fig. 7.1: Amino acid specificity of protein kinases.

Reversible phosphorylation of proteins on Ser/Thr and Tyr residues is a regulatory signal that functions as a switch in signaling pathways. The phosphate esters formed on proteins by the action of protein kinases are stable modifications that cause profound changes in the activity of cellular proteins. Because of the stability of the phosphate esters, protein phosphatases are required for their removal. The concerted and highly regulated action of both protein kinases and protein phosphatases is used by the cell to create a temporally and spatially restricted signal influencing the activity state of proteins in a highly specific way.

Examples of cellular activities regulated by protein kinases are diverse, affecting practically all the cell's performance.

The switch function of protein phosphorylation is based on different mechanisms that may work alone or in combination. Phosphorylation by protein kinases influences function, activity and subcellular location of the protein substrate, especially in the following ways:
- Induction of conformational changes by allosteric mechanisms (Chapters 2, 8 and 13).
- Direct interference with binding of substrate or other binding partners, e.g. isocitrate dehydrogenase (Chapter 2).
- Creation of binding sites for effector molecules in the sequence, e.g. binding of p-Tyr to SH2 and phosphotyrosine-binding domain (PTB) domains and binding of p-Ser to 14-3-3 proteins (Chapter 8).

On the basis of sequence and structure, the Ser/Thr- and Tyr-specific protein kinases form a closely related superfamily that is distinct from the histidine kinases and other phosphotransfer enzymes. Sequence comparison of the catalytic domains makes it possible to differentiate between the Ser/Thr- and Tyr-specific protein kinases within this superfamily. Furthermore, homology considerations enable subfamilies within both the larger families of Ser/Thr- and Tyr-specific protein kinases to be identified.

7.1.2
Classification of Ser/Thr-specific Protein Kinases

Selected subfamilies of the Ser/Thr-specific protein kinases in vertebrates are listed in Tab. 7.1.

There are many other protein kinases that do not show any close relationship to the subfamilies of Tab. 7.1. These include protein kinases with 2-fold specificity in that they can phosphorylate Ser/Thr and also Tyr residues. An example of a protein kinase with twofold specificity is the mitogen-activated kinase (MAPK) kinase (Chapter 10).

We also know of Ser/Thr-specific protein kinases that are an integral part of transmembrane (TM) receptors or of ion channels. The

■ **Protein phosphorylation is found in**
- Enzymes (Chapters 2, 5, 7, 8, 10 and 13)
- Adaptor proteins (Chapters 8)
- Transcription factors (Chapters 1)
- Ion channels
- TM receptors (Chapters 5, 8, 11 and 12)
- Ribosomal proteins (Chapters 6)
- Structural proteins
- Transport proteins

Tab. 7.1: Subfamilies of the Ser/Thr-specific protein kinases.

AGC kinases including: – cAMP regulated protein kinase (PKA) – cGMP regulated protein kinase (PKG) – PKB (Akt kinase) – PKC 12 isoforms	Calcium/calmodulin-regulated protein kinases – Subunit of phosphorylase kinase – MLCK – CaMK II
Ribosomal S6 protein kinase: kinases that specifically phosphorylate ribosomal protein S6	GPCR kinase
β-Adrenergic receptor kinase	Casein kinase II (casein kinase gets its name from the observation that casein, milk protein, is a good substrate)
Glycogen synthase kinase	CDC2 kinases: representatives of this family are central elements of cell cycle regulation, see Chapter 13
MAPKs: the MAPKs are involved in the transduction of growth-promoting signals, see Chapter 10	Mos/Raf protein kinases: these kinases are part of the MAPK cascade and are involved in signal transduction of growth factors (Chapters 9 and 10)

transforming growth factor (TGF)-β receptor contains a Ser/Thr-specific protein kinase activity in the cytoplasmic part of its TM polypeptide chain (Chapter 12). Some members of a certain class of ion channels, named the transient receptor potential (TRP) channels, carry a protein kinase activity on the cytoplasmic side of the channel protein that is essential for channel function. The TRP channels modulate Ca^{2+} entry into eukaryotic cells in response to external signals. The protein kinase activity is located on the cytoplasmic domain of the channel and can phosphorylate itself and other proteins on Ser/Thr residues (Runnels et al., 2001). Although there is no similarity to classical protein kinases on the primary sequence level, the three-dimensional structure of the TRP channel protein kinase domain is very similar to the classical kinase fold.

7.2
Structure and Regulation of Protein Kinases

The Ser/Thr- and Tyr-specific protein kinases share many common features. The catalytic mechanism, structural properties and mechanisms of kinase control are very similar for the two kinase classes. In the following, the main properties of both classes will therefore be presented together (reviewed in Johnson et al., 1998; Engh and Bossemeyer, 2001).

Protein Kinase Reaction

The common catalytic function of Ser/Thr- and Tyr-specific protein kinases is the covalent phosphorylation of substrate proteins via transfer of the γ-phosphate of ATP to the OH group of serine, threonine or tyrosine residues. This catalytic function is carried out by a catalytic domain of around 270 amino acids whose structure and catalytic residues are highly conserved among the two protein kinase families. The conserved structure of the kinase domain shows two lobes that are linked by a flexible hinge. The catalytic center is formed by residues from both lobes. The proposed mechanism of phosphate transfer by protein kinases is shown in Fig. 7.2 (Johnson et al., 1996).

Key players of the reaction are:

– Acidic amino acids required for stabilization of the transition state and activation of the OH group of the acceptor amino acid for nucleophilic attack on the γ-phosphate.

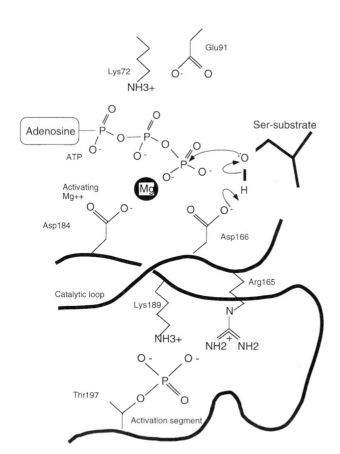

Fig. 7.2: Schematic representation of key interactions at the catalytic site and the activation loop of PKA. A possible mechanism is shown in which the carboxylate of Asp166 functions as a base and activates the OH group of the acceptor serine for a nucleophilic attack on the γ-phosphate of ATP. The catalytic mechanism is not definitively established.

– One or two metal ions that coordinate the γ-phosphate of ATP help to fix the ATP and stabilize negative charges in the transition state.
– Basic amino acids that serve to stabilize negative charges in the transition state and in the leaving group ADP.

Conserved amino acids of protein kinases
– D166: in catalytic loop, substrate-OH activation
– N171: hydrogen bond to D166
– D184: part of DFG motif in activation loop, Mg^{2+} binding
– K72: orients α and β-P of ATP, ion pair with E91
– E91: in C-helix, ion pair with K72

Sequence comparisons, mutation experiments and biochemical studies indicate an essential function in the catalysis of phosphate transfer for the conserved amino acids Lys72, Glu91, Asp166, Asn171 and Asp184 [numbering of protein kinase A (PKA)]. It is generally assumed that Asp166, which is invariant in all protein kinases, serves as a catalytic base for activation of the Ser/Thr hydroxyl and that the reaction takes place by an "in-line" attack of the Ser-OH at the γ-phosphate. Overall, the mechanism of phosphate transfer by protein kinases is related to nucleotide transfer by nucleic acid polymerizing enzymes that also use metal ions and acidic residues as key elements in catalysis.

7.2.1
Main Structural Elements of Protein Kinases

The core of all eukaryotic Ser/Thr- and Tyr-specific protein kinases adopts a common fold illustrated in Fig. 7.3 for the tyrosine kinase domain of the insulin receptor. The structure comprises a small and large lobe with the active site forming a cleft between the two lobes. The two lobes are connected by a hinge region. The small N-terminal lobe contains five β-structures and one α-helix, named the C-helix. In contrast, the larger C-terminal lobe is mostly α-helical. It comprises a four-helix bundle, additional α-helices and two short β-strands. The small lobe provides the binding site for ATP, and the large lobe provides catalytic residues and a docking surface for peptide/protein substrates. Opening and closing of the active site cleft is an essential part of catalysis. The following structural elements have been found to be critical for catalysis and for protein kinase control (amino acid numbering of PKA):

Structural elements of protein kinases
– P-loop
– C-helix
– Catalytic loop
– Activation segment with activation loop
– Autoinhibitory elements

– *Glycine-rich loop in the N-terminal lobe.* This loop (also called the P-loop) is found in similar form in the G_α-proteins (Section 5.4.3) and is required for anchoring of the phosphate residues of ATP.
– *C-helix of the N-terminal lobe.* In most active protein kinase conformations, residue E91 of the C-helix forms a salt bridge to an invariant Lys residue (K72) within the N-terminal lobe, allowing optimal positioning of the ATP phosphates. Regulatory mechanisms often modulate kinase activity by altering the conformation of the C-helix, thereby affecting the integrity of these interactions.
– *Catalytic loop.* The catalytic loop is located at the base of the active site and contains a conserved Asp residue (D166), presumed to be

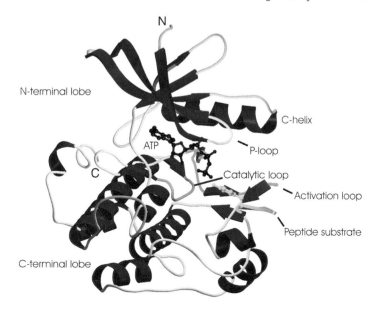

N

N-terminal lobe

C-helix

ATP

P-loop

C

Catalytic loop

Activation loop

Peptide substrate

C-terminal lobe

Fig. 7.3: Structure of the tyrosine kinase domain of the insulin receptor with bound ATP and peptide substrate (after Hubbard and Till, 2000). The α-helices are shown in red, the β-strands in blue. The functional important structural elements are indicated. The P-loop is shown in yellow, the activation loop in green and the catalytic loop in orange. The N- and C-termini are denoted by N and C.

the catalytic residue. Furthermore, a conserved Asn residue (N171) is found on this loop.

– *Activation segment and activation loop.* The primary sequence of the activation segment is defined as the region between and including two conserved tripeptide motifs (DFG…APE) within the large lobe and comprises 20–35 amino acids. Moving from the N- to C-terminus, the secondary structural elements within the activation segment are the Mg^{2+}-binding loop, comprising the DFG motif, the activation loop [also called the T-loop for the cyclin-dependent protein kinases (CDKs), see Chapter 13] and the P + 1-loop, involved in peptide substrate binding. The residue D184 of the conserved DFG motif is involved in metal–ATP binding. The activation loop shows considerable structural diversity and conformational plasticity and is one of the most important control elements of protein kinase activity (reviewed in Huse and Kuriyan, 2002; Nolen et al., 2004). In many kinases, the activation loop is the site of regulatory phosphorylations or interaction with activity modulators. Ser, Thr or Tyr residues of the activation loop may be phosphorylated in response to activating signals and this phosphorylation promotes an active conformation of the kinases. In many inactive states of protein kinases, the activation loop collapses into the active site and blocks the binding of both nucleotide and peptide substrate. Upon phosphorylation, it moves away from the active site and adopts an open conformation allowing for substrate binding and catalysis (Fig. 7.4). Critical for this conformational change is the electrostatic interaction between one of the phosphate residues of the loop and a basic pocket containing an Arg–Asp motif which is conserved among kinases regulated by phosphorylation. The placement and the number of phosphorylation sites varies from kinase to kinase. Overall, the presence of the phosphate residues creates a network of interactions that properly orient the C-helix and the catalytic residues and promote lobe closure. For some protein kinases, e.g. MAPKs (Fig. 7.4), phosphorylation of the activation loop facilitates homodimerization of the kinase that is required for the nuclear localization of the enzyme.

– *Autoinhibitory sequence elements.* Many protein kinases contain autoinhibitory elements that help to fix an inactive conformation by intramolecular binding to the substrate-binding site. These elements may fold into the active site, blocking the binding of both the nucleotide and peptide substrate. Since they lack phosphorylatable residues these elements are also termed pseudo-substrates. The autoinhibitory elements may be localized on the same polypeptide as the kinase domain or they may reside on a separate subunit of an oligomeric kinase. An N-terminal myristoic acid anchor may participate in autoinhibition too, as shown for the Abl tyrosine kinase (see also Section 8.3.3).

In inactive kinase structures, two or more of the above mentioned structural elements are aligned in a way that prevents substrate binding and/or catalysis. Figure 7.5 shows the spatial relationship of the critical protein kinase activity modulation sites.

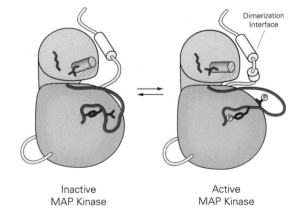

Fig. 7.4: Activation of MAPK by phosphorylation of the activation segment. The MAPK (Chapter 10) undergoes phosphorylation at a threonine (Thr183) and a tyrosine residue within the activation loop which induces a movement of the activation loop (red). This creates a network of interactions that properly orient the αC-helix (shown in cyan) and promotes lobe closure such that the kinase becomes active. Furthermore, the p-Thr organizes the C-terminal extension (shown in yellow) into a functionally important homodimerization interface. Dimerization via this interface is required for the nuclear localization of the enzyme.

Inactive MAP Kinase

Active MAP Kinase

Dimerization Interface

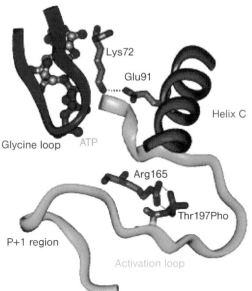

Fig. 7.5: Alignment of the activity modulation sites of protein kinases (after Hubbard and Till, 2000). Close-up of the critical elements of protein kinase control as found in active kinase structures, represented here by the structure of active PKA. The p-Thr acts as an organizing center that forms contacts to Arg165 which is close to the catalytic D166 and to the C-helix. The C-helix forms a salt bridge to K72 which helps to coordinate the α- and β-phosphates of ATP correctly. This helix corresponds to the PSTAIRE helix of CDK2 (Chapter 13) and is the focus of several regulatory mechanisms among protein kinases. In inactive kinase structures, two or more of the critical activity modulation sites are misaligned or blocked by intra- or extramolecular interaction.

7.2.2
Substrate Binding and Recognition

Taking into account the many Ser, Thr and Tyr residues in proteins, the question arises as to which parameters define the phosphorylation site of a substrate protein. With the help of targeted exchange of amino acids in substrate proteins, sequence comparison of phosphorylation sites, and use of defined peptides as substrates, it has been possible to clearly show that the sequence in the neighborhood of a Ser/Thr or Tyr residue is an important determinant of specificity. The different Ser/Thr- and Tyr-specific protein kinases show different requirements with respect to the neighboring sequence of the residues to be phosphorylated, so that each subfamily has its own consensus sequence for phosphorylation. It should be pointed out that several Ser/Thr residues are found in many phosphorylation sequences, so that multiple and cooperative phosphorylation is possible in a sequence segment. Phosphorylation of the large subunit of the RNA Pol II (Section 3.2.7) is particularly marked. At the C-terminus, this contains 52 copies of the heptamer sequence YSPTSPS as potential phosphorylation sites.

In addition to the sequence context of the phosphorylation sites, other structural parts of the substrate protein have been shown to contribute to substrate binding as well. A coupling of a protein kinase and its protein substrate can be achieved, e.g. via structural parts that are located far away from the phosphorylation consensus

■ **Substrate specificity of protein kinases is determined by**
– Sequence context of phosphorylation site
– Interaction modules distinct from phosphorylation site
– Colocalization of kinase and substrate

site on the substrate and the substrate binding site on the kinase (reviewed in Biondi and Nebrada, 2003). An example is provided by the docking of substrate proteins on receptor tyrosine kinases (RTKs) via SH2–phosphotyrosine interactions (Chapters 8 and 11). This interaction helps to clamp the substrate on the kinase, thereby ensuring a high efficiency of phosphate transfer. Another major determinant of protein kinase specificity is the targeting of protein kinases to the neighborhood of selected substrates. The colocalization of protein kinases and their substrates at distinct subcellular sites greatly enhances the specificity of the kinase reaction. Only those substrates that have been translocated to the specific subcellular site will be phosphorylated by the protein kinase. The mechanisms of colocalization are diverse and will be dealt with separately in Section 7.7.

The peptide substrate-binding site is located on the C-terminal lobe of the protein kinases. Structural information available to date shows that the peptide substrate is contacted via multiple interactions, both N- and C-terminal, to the residue to be phosphorylated. There is a marked complementarity between the binding pocket on the kinase and the peptide substrate in regard to shape, hydropathy and electrostatic potential. For several protein kinases, e.g. the insulin RTK (Section 8.1.3) and CDK2 (Section 13.2.5), an intact peptide-binding site is not present in inactive forms of the kinase. Only after activation loop phosphorylation and a subsequent conformational change is a substrate recognition site created in these kinases.

Most information on substrate recognition by protein kinases has been obtained by using peptide substrates bound to the catalytic domain. The structure of a complete protein substrate–kinase complex, namely eIF-2α bound to the catalytic domain of RNA-dependent protein kinase (PKR; see Section 3.6.3.5) shows extensive contacts between eIF-2α and an α-helix (αG) of the C-lobe (Dar et al., 2005). The N-lobe is not involved in substrate binding, rather it mediates dimerization of PKR.

7.2.3
Control of Protein Kinase Activity

Protein kinases can exist in active and inactive forms, which is why they are able to perform the function of a switch in signaling pathways. For most of the time, protein kinases are found in the "off" state that has minimal activity. Upon specific signals they are converted into the "on" state that is maximally active. As illustrated by the multifacetted character of activating and inactivating signals, the cell possesses a broad palette of tools to induce the transition between the two activity states.

Since all protein kinases catalyze the same reaction, they all adopt catalytically active "on" conformations that are structurally very similar. The structures of the "off" states are, however, variable and the protein kinases have evolved different ways by which the adoption of the "on" state conformation is impeded.

Overall, the protein kinase structures contain several flexible elements that can be fixed in either an active or an inactive conformation. The flexible elements comprise the hinge between the two lobes, the activation loop, the P-loop and the C-helix, and they move in a highly coordinated and cooperative way upon transition between the "on" and "off" state.

The activated state of protein kinases is characterized by the following features:

- The two lobes are closely packed together.
- The ATP is buried between the two lobes and its phosphates contact the P-loop.
- The C-helix is positioned for salt bridge formation to a β-strand of the N-terminal lobe.
- The binding sites for the peptide substrate and ATP are fully accessible.
- The phosphorylated activation loop has moved away from the active site and adopts an extended conformation. It helps to organize the catalytic residues for optimal phosphate transfer and it forms part of the substrate-binding site. The movement of the activation loop is often coupled to a movement of the C-helix.

Whereas fully active protein kinases adopt a similar active conformation, the inactive states of protein kinases are very diverse. A high plasticity of the kinase domains is observed that allows the adoption of distinct inactive conformations in response to phosphorylation or interactions with specific regulatory domains on the same kinase molecule or on other proteins. Overall, the inactive states are characterized by a more open conformation of the two lobes, precluding optimal orientation of residues involved in substrate binding and catalysis. The mechanisms for fixation of the inactive protein kinase states are diverse and may be used singly or in combination.

Mechanisms for fixation of inactive protein kinase states (Fig. 7.6):

- Binding of protein inhibitors: protein kinase inhibitors fix inactive states, e.g. by deforming the N-terminal lobe and destroying the ATP-binding site (see inhibitor p21^{KIP}, Section 13.2.3).
- Inhibitory phosphorylations (e.g. Thr14, Tyr15 on CDKs, see Chapter 13, phosphorylation of Src kinase, see Section 8.3.2).
- Binding of regulatory subunits (see PKA, Section 7.3).
- Autoinhibition. An inhibitory structural element that is itself part of the protein kinase or part of a kinase subunit is often used to fix

Fig. 7.6: Mechanisms of activation and inactivation of protein kinases.

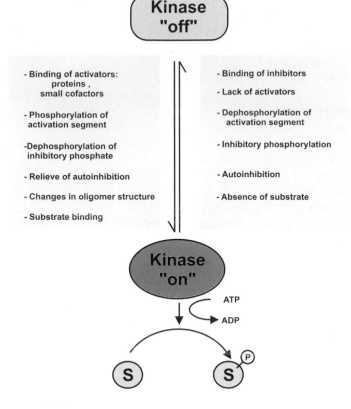

an autoinhibited, inactive state of the kinase. The autoinhibitory elements are mobile and can adopt conformations that induce misalignment of the catalytic residues and blockage of the substrate binding sites.

A multitude of mechanisms may be used singly or in combination for activation of the kinase.

The transition from the inactive to the active state is induced by:

– Binding of activating subunits (cyclins, see Section 13.2.2).
– Binding of chemical messengers (cAMP, see Section 7.3.1) with concomitant release of inhibitory subunits.
– Binding of cofactors like Ca^{2+}/calmodulin, diacylglycerol (DAG), phospholipids (see Section 7.4), AMP.
– Phosphorylation of the activation loop.
– Dephosphorylation of inhibitory phosphorylated sites.
– Changes in the oligomerization state of the kinase as a consequence of ligand binding to the extracellular domain of the TM protein kinase (Section 8.1.2).

– Binding of a phosphorylated substrate. The polo-like kinase 1 is kept in an inactive state by intramolecular binding of a repressive domain, the polo box domain 1 (PBD1), to the active site. Binding of a phosphorylated substrate by PBD1 relieves the inhibition (Elia et al. 2003).

The control of kinase activity operates at the following regulatory levels:
– Expression of kinase and of regulatory subunits (inhibitory or activating subunits, e.g. the cyclins, see Chapter 14).
– Signal-induced destruction of kinases or regulatory subunits via the ubiquitin–proteasome pathway (Chapter 13).
– Activation of upstream protein kinases that phosphorylate regulatory sites, e.g. the activation loop. Protein kinase cascades may be formed by this way (Chapter 10).
– Activation of upstream protein phosphatases that dephosphorylate regulatory sites.
– Binding of extracellular ligands to TM protein kinases (e.g. RTKs, see Chapter 8).
– Signal induced formation of second messengers that activate protein kinases (Ca^{2+}, DAG, see Chapter 6).
– Binding of metabolites that activate the kinase. One example is the AMP-activated protein kinase that senses the cellular AMP/ATP ratio.
– Colocalization of kinase and protein substrate (Section 7.2.4). The targeted recruitment of protein kinases to specific subcellular compartments, e.g. the cell membrane is a major way of protein kinase control.

7.2.4
Regulation of Protein Phosphorylation by Subcellular Localization and Specific Targeting Subunits

Targeted localization of protein kinases and protein phosphatases is a major mechanism by which the selectivity of protein phosphorylation is enhanced, and a tight control of phosphorylation, dephosphorylation and interaction with cofactors is made possible. The principle of targeted localization is shown in Fig. 7.7.

Many substrates of protein kinases occur either as membrane-associated or particle-associated forms. For protein kinases or protein phosphatases to perform their physiological function in a signal transduction process, they must in many cases be transported to the location of their substrate. In the course of activation of signal transduction pathways, compartmentalization of protein kinases and protein phosphatases redistribution to new subcellular locations

■ **Targeting of kinases and phosphatases to subcellular sites is mediated by**
– Specific localization or regulatory subunits
– Lipid anchors
– Protein–protein interaction modules of the kinase

Fig. 7.7: Main mechanisms of targeted localization of protein kinases. Targeting of protein kinases (or protein phosphatases) by the mechanisms shown provides for access to membrane-associated substrates.

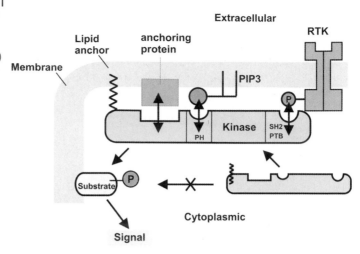

is often observed. By restricting the action of the two enzyme classes to a distinct subcellular site, the persistence, amplitude and signal/noise ratio of phosphorylation signals are improved. Furthermore, signals from other effectors can be integrated more efficiently in these multiprotein signaling units.

Subcellular targeting of protein kinases and protein phosphatases is often mediated by the regulatory subunits of these enzymes, which can bind specifically to scaffolding, adaptor or anchoring proteins located at distinct subcellular sites. These anchoring proteins may be multivalent and allow the assembly of several signaling proteins. The mechanisms by which the anchoring proteins assemble at a distinct subcellular site are diverse. Structural membrane proteins, TM receptors, ion channels or cytoskeletal proteins can serve as the anchoring target. Other anchoring proteins interact with the membrane via lipid anchors or with distinct membrane anchoring sequences. Generally, the nature of the regulatory subunit of the protein kinase or protein phosphatase determines in which compartment of the cell and at which membrane section the protein phosphorylation signal will become active. The subunit functions as a localization moiety; it determines at which place in the cell the protein kinase or phosphatase gains access to its substrates.

Another important aspect of protein kinase targeting is the colocalization of sequentially acting protein kinases by means of scaffolding proteins (Chapter 10). By assembling protein kinases with the upstream and downstream signaling partners, either a protein kinase or other signaling proteins, a high local concentration of the signaling proteins is achieved, allowing for the creation of localized and efficient signal events at specific subcellular sites.

7.3
PKA

The cAMP-dependent protein kinase (PKA) has been classified as a member of the superfamily of AGC kinases that includes the following protein kinases:
– PKA.
– PKB, also named Akt kinase.
– PKC.
– PKG, cGMP-dependent kinase.

The AGC kinases have a very similar structure of the catalytic domain and are regulated in a similar fashion by phosphorylation in the activation loop.

Of the protein kinases, PKA is the best investigated and characterized (reviewed in Taylor et al., 2005). The substrates and functions of PKA are diverse. PKA is involved in the regulation of metabolism of glycogen, lipids and sugars. In addition, cAMP/PKA plays an important role in controlling ion channels and in long-term modifications at nerve synapses. Furthermore, it is involved in cAMP-stimulated transcription of genes that have a cAMP-responsive element in their control region. An increase in cAMP concentration leads to activation of PKA, which phosphorylates the transcription factor *cAMP-responsive element (CRE)-binding protein (CREB)* at Ser133. CREB only binds to the transcriptional coactivator *CBP (CREB-binding protein)* in the phosphorylated state and stimulates transcription of target genes (Section 3.4.5).

7.3.1
Structure and Substrate Specificity of PKA

The activity of PKA is primarily controlled by cAMP. In the absence of cAMP, PKA exists as a *tetramer* composed of *two regulatory R subunits*, each containing two cAMP binding sites, and *two catalytic C subunits* (Fig. 6.2). The catalytic activity is masked in the holoenzyme C_2R_2, since an inhibitory structural element of the R subunit blocks the entrance to the active site. Binding of cAMP to the R subunit leads to a large conformational change in R, whereupon the affinity between R and C is reduced by a factor of $10^4–10^5$. The holoenzyme dissociates into the dimer of the cAMP-bound R subunits and two monomers of C, which now become catalytically active. Full catalytic activity of C requires phosphorylation of a Thr residue (Thr197) in the activation loop and this phosphorylation is also required for binding of the R subunit.

■ **PKA structure and regulation**
Inactive PKA
– Autoinhibited R_2C_2 tetramer
Activation
– cAMP binding: $R_2(cAMP)_4$ + 2C
Isoforms
– RIα, RIβ, RIIα, RIIβ
– Cα, Cβ, Cγ

In mammals, four isoforms of the R subunit (RIα, RIβ, RIIα and RIIβ) and three subtypes of the C subunit (Cα, Cβ and Cγ) are known. The existence of multiple R and C subunits harboring different biochemical features allows for the formation of a number of holoenzymes with different biological characteristics, which contributes to the specificity and variability of PKA signaling observed in the cell.

The R subunits have a similar modular structure (Fig. 7.8). A dimerization domain at the N-terminus forms a four-helix bundle that also provides a docking surface for A-kinase anchoring proteins (AKAPs). At the C-terminus of each protomer are two tandem cAMP-binding sites of differing affinity. In addition, the RII subunit has a flexible region that contains an autoinhibitory sequence. This region functions as substrate mimetic that binds to the active site cleft of the C subunit and prevents substrate access. The various R subunits differ in the size and the properties of the various domains which provides for distinct regulatory properties of each R subunit.

The C subunit has a myristinic acid residue of unknown function at the N-terminus and shows the typical kinase fold. In addition, the

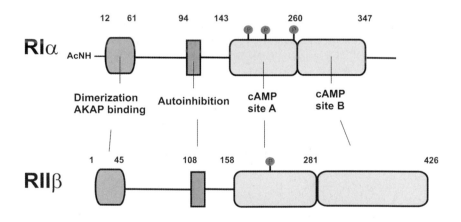

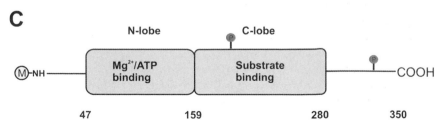

Fig. 7.8: Domain structure and phosphorylation sites of the RIα, RIIβ and C subunits of PKA. M = myristoyl anchor.

C subunit has specific Ser/Thr phosphorylation sites, i.e. Thr197 and Ser338. Thr197 is located in the activation loop and is phosphorylated by an autophosphorylation mechanism.

The *consensus sequence* for phosphorylation of proteins by PKA is RRXSX. The RII subunit contains such a sequence and is therefore subject to phosphorylation by the C subunit in the holoenzyme, but without release of inhibition.

7.3.2
Regulation of PKA

PKA is subject to multiple regulatory influences. Whereas the regulation by cAMP is the primary determinant of PKA regulation, other regulatory influences ensure a high specificity of PKA with regard to tissue distribution and compartment-specific action (reviewed in Tasken and Aandahl, 2004). The activity of PKA can be controlled by the following mechanisms, both in a temporal and spatially regulated way.

– *Changes in cAMP concentration*. The changes in concentration of cAMP that lead to activation of PKA in the cell are relatively small. In many tissues, a 2- to 3-fold increase in cAMP concentration is sufficient to bring about the maximum physiological effect. The cell has different mechanisms available that limit the increase in cAMP concentration to a relatively narrow concentration region and contribute to damping of signal transduction via PKA. An example of a mechanism with a damping effect in signal transduction by PKA is a *feedback control* by a cAMP phosphodiesterase. The activated PKA phosphorylates and activates a phosphodiesterase that hydrolyzes cAMP to AMP (Fig. 7.9). This mechanism enables PKA to control its own steady-state activity. It also ensures that, as the external signal diminishes, the cAMP signal rapidly subsides.

– *Binding of inhibitor proteins*. Regulation of PKA also takes place via the binding of specific inhibitor proteins. There are three natural inhibitors of PKA. The inhibitor PKI, for example, is involved in subcellular transport of C subunits and is considered a major regulator of C subunit activity.

– *Targeting to subcellular sites by AKAPs*. Given the fact that PKA is widely distributed and is activated by a large variety of external signals, the question arises how – apart from isoenzyme patterns, differential phosphorylation and inhibitor binding – the high specificity of PKA action is achieved in the cell. Specific *compartmentalization* of PKA enzymes has now been recognized to be a major determinant of PKA specificity. By binding to subcellular structures via AKAPs, isoenzymes of PKA can be

■ **Main regulation of PKA is by**
– cAMP
– Inhibitor proteins
– Subcellular localization via AKAPs

Fig. 7.9: Feedback control of PKA by a phosphodiesterase. On activation of PKA, the catalytic C subunits are released, which then phosphorylate a phosphodiesterase, in addition to other substrates. The phosphodiesterase is activated by the phosphorylation and hydrolyzes cAMP to AMP, whereby the signal transduction via PKA is reduced or terminated.

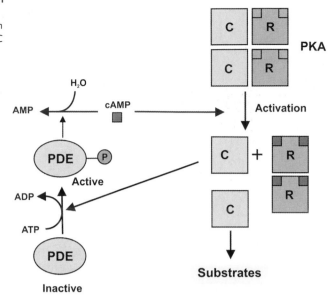

assembled at distinct subcellular sites in the vicinity of their substrates. We know of more than 50 different AKAPs that are located to different compartments of the cell (Section 7.3.3). Most of the interactions between AKAPs and PKA holoenzymes seem to be mediated by RII subunits.

7.3.3
AKAPs

■ **AKAPs**
 – Scaffolding and targeting proteins that organize signals mediated by second messengers.
AKAP isoforms
AKAP250/79 bind to
 – R subunits of PKA
 – β₂-adrenergic receptor
 – PKC isoforms
 – Protein phosphatases, PP2B
 – Ion channels
 – Src kinase

An increase in cAMP and activation of PKA are accompanied, in many cases, by a change in the subcellular location of PKA holoenzymes containing RI or RII subunits. This targeted localization is mediated by the AKAPs, of which more than 50 different members are known (reviewed in Wong and Scott, 2004). The AKAPs immobilize the PKA isoforms at specific subcellular sites by binding the R subunits (Fig. 7.10). Binding of cAMP to the regulatory subunit releases the catalytic subunit that can phosphorylate substrates in the near vicinity. The released catalytic subunit can also be transferred to other compartments of the cell. Parallel to the increase in cAMP, translocation of the catalytic subunit is observed in many cells from the Golgi apparatus to the nucleus via the cytosol and is accompanied by stimulation of transcription.

The cellular function of AKAPs, however, goes far beyond the mere association of PKA. Rather, some AKAP isoforms appear to be focal assembly points at GPCRs and ion channels. The AKAP iso-

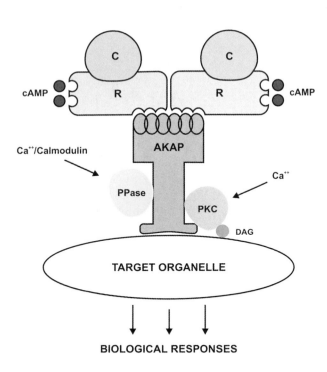

Fig. 7.10: Model of a signaling complex organized at the membrane of a target organelle by AKAPs. A schematic representation of a multiprotein complex formed by a prototypic AKAP, PKA and two other effector proteins, protein phosphatase (PPase) and PKC (PKC), both activated by calcium.

forms AKAP250 and AKAP79 have been shown to be versatile scaffolding proteins that harbor binding sites for several signaling molecules, and are themselves regulated by phosphorylation (reviewed in Malbon et al., 2004). As an example, the AKAP250/79 signaling complex can associate – in addition to the R subunits of PKA – PKC, the protein phosphatase 2B (PP2B, calcineurin), the β_2-adrenergic receptor (Chapter 5), Src kinase (Section 8.3.2) and the GPCR kinase 2. Furthermore, AKAPs contain targeting motifs that direct them to specific subcellular determinants or to TM receptors or ion channels. As an example, the AKAP named *Yotiao* directs both PKA II and PP1 to *N*-methyl-D-aspartate (NMDA) receptors.

The possibility of bringing PKA and other signaling molecules to the same place in the cell opens up the prospect of a coordinated and layered regulation of multiple activities in a multiprotein complex anchored around a GPCR at a distinct subcellular site. The composition of the AKAP signaling complexes and the function of its components appear to be regulated in a dynamic way. In these complexes, several signals can be integrated and directly transmitted to the substrates, allowing regulation of both the forward and backward steps of a given signaling process.

7.4
Phosphatidylinositol-3-kinase (PI3K)/Akt Pathway

The *Ser/Thr-specific protein kinase Akt*, also known as *PKB*, is at the center of a major signaling pathway of the cell regulated mainly by PI3K and its reaction product, PtdIns(3,4,5)P$_3$. Often this pathway is collectively termed the PI3K/Akt pathway.

Fundamental cellular functions such as cell proliferation and survival are regulated by the PI3K/Akt pathway and alterations to the PI3K/Akt signaling pathway are frequent in human cancer. Furthermore, many physiological functions of insulin involve PI3K/Akt signaling and this pathway is therefore at great importance for regulation of glucose metabolism.

In addition to major signaling functions in the cytoplasm, a complete PI3K signaling pathways has been now shown to exist in the nucleus too (reviewed in Kikani et al., 2005).

7.4.1
PI3K

The reactions catalyzed by PI3K have been presented already in Section 6.6. In short, PI3K catalyzes the phosphorylation at the 3' position of the inositol part of various PtdIns derivatives. The products of the reaction, e.g. PtdIns(3,4,5)P$_3$, serve as membrane localized second messengers that help to localize various signaling proteins to membranes. The family of PI3Ks comprises at least 15 kinases that differ in substrate specificity, the nature of the associated subunits and modes of regulation (reviewed in Katso et al., 2001). The class I PI3Ks (p110α, p110β, p110δ and p110γ) are activated by tyrosine kinases or GPCRs to generate PtdIns(3,4,5)P$_3$, which engages downstream effectors such as the Akt pathway and the Rho family GTPases (Chapter 9). The class II and III PI3Ks play a key role in intracellular trafficking through the synthesis of PtdIns(3)P and PtdIns(3,4)P$_2$.

Of the three classes of PI3Ks, only class I will be presented in more detail here. Most members of class I are associated with a subunit that functions as an adaptor in signal transduction. The best investigated PI3K, PI3Kα, is a heterodimer with adaptor function, made up of a catalytic subunit (p110α) and a regulatory subunit of 85 kDa (p85α). The p85α subunit has an SH3 domain, two SH2 domains and two Pro-rich domains. These domains function as binding modules, which the PI3K uses for specific protein–protein interactions in the process of signal transduction and for association with other signal proteins (Chapter 8).

Other members of class I of the PI3Ks, such as PI3K of the γ subtype, are stimulated by interaction with βγ complexes (Section 5.5.5) and have their own regulatory subunit. It is interesting that both a lipid kinase activity and a protein kinase activity have been identified in the catalytic domain of the PI3K γ subtype in brain (Bondeva et al., 1998). Activation of the MAPK pathway (Chapter 10) may take place via the protein kinase activity, so that this enzyme can produce a bifurcated signal: the lipid kinase activity stimulates the Akt kinase (see below), the protein kinase the MAPK pathway. Proliferation-promoting signals are transmitted via both pathways.

Most data are available for the p110α*p85α subtype of PI3K. For brevity, this is referred to as PI3K in the following. The PI3K phosphorylates various PtdIns derivatives at the 3 position (Fig. 6.5) *in vitro*. A physiologically important substrate is PtdIns(4,5)P$_2$, which is converted to PtdIns(3,4,5)P$_3$ by PI3K. PtdIns(3,4,5)P$_3$ is an intracellular messenger that has a regulatory effect in many elementary functions of the cell, such as growth control, chemotaxis and glycogen synthesis (Section 6.6.2).

An important function in growth regulation is attributed to the PI3K. PtdIns(3,4,5)P$_3$ is not detectable in resting cells. On stimulation of the cells with a growth factor, a rapid increase in PtdIns(3,4,5)P$_3$ occurs. An associated translocation of PI3K to the membrane is observed. In accordance with its central growth-regulating function, mutants of the p110α subunit have been shown to be oncogenic *in vitro* and *in vivo*.

Many observations indicate that PI3K functions as a signal protein that receives signals on the cytoplasmic side of the cell membrane and transmits them further, although its primary role is to produce membrane-localized messenger substances.

PI3K is activated mainly via three pathways (Fig. 7.11):

- *Interaction with activated RTK.* The SH2 domain of the p85 subunit mediates an interaction with tyrosine residues on signal proteins involved in transduction of growth-regulating signals. Thus, binding of the PI3K to tyrosine phosphate residues of the activated platelet-derived growth factor (PDGF) receptor is observed (Section 8.1.4). Another binding partner is the insulin receptor substrate (IRS, see Section 8.5). In both cases, it is assumed that the binding of the SH2 domain of p85 to the tyrosine residue of the signal protein serves to target the PI3K to its membrane-localized substrate. Furthermore, binding of p85 to phosphotyrosine residues of activated receptors appears to be accompanied by an allosteric activation of the catalytic subunit (Fig. 7.12).
- *The interaction between PI3K and the IRS links insulin signaling to the PI3K pathways.* Accordingly, many of the physiological functions of insulin are mediated by the PI3K/Akt kinase pathway.

■ **PI3Ks**
- growth promoting
- produce membrane-bound 2nd messenger PtdIns(3,4,5)P$_3$
- activated by
  RTKs
  Ras
  Gβγ
- transmit insulin signals

Fig. 7.11: Pathways of PI3K activation. PI3K can be activated by growth factor receptors, either by direct interaction or via the Ras protein. Another way of PI3K activation uses the βγ subunits of heterotrimeric G-proteins liberated upon activation of GPCRs. The product of the PI3K reaction is PtdIns$(3,4,5)$P$_3$, which binds to PH domains of various signaling proteins. Overall, activation of PI3K stimulates cell growth and proliferation, and inhibits apoptosis. A suppressing effect is exerted by the tumor suppressor PTEN which hydrolyzes and thus inactivates PtdIns$(3,4,5)$P$_3$. All reactions shown are closely associated with the inner side of the membrane. This aspect is not addressed in the figure.

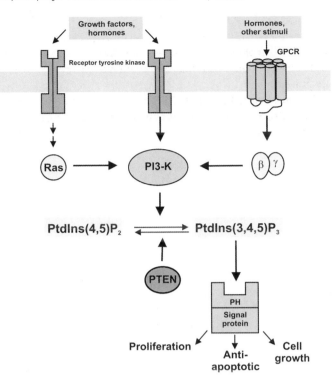

Fig. 7.12: Activation of PI3K by a conformational change. Binding of the SH2 domains of p85, the regulatory subunit of PI3K, to p-Tyr sites on activated receptors releases an autoinhibitory constraint that stimulates the catalytic domain (p110). PI3K catalyzes the phosphorylation of the 3′ positions of the inositol ring of PtdIns(4)P and PtdIns$(4,5)$P$_2$ to generate PtdIns$(3,4)$P$_2$ and PtdIns$(3,4,5)$P$_3$, respectively. After Schlessinger (2000).

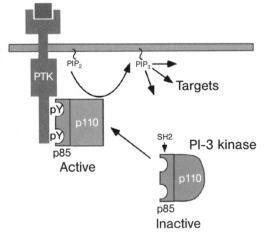

- *Activation in the Ras pathway.* PI3K has also been identified as a part of the Ras signaling pathway (Chapter 9). Signals originating from TM receptors can be transmitted from the Ras protein to PI3K. In this case, PI3K acts as the effector molecule of the Ras protein.
- *Activation by the Gβγ dimer.* Gβγ dimers directly activate the PI3K β and γ subtypes. In this way, a variety of extracellular signals can be transmitted via GPCRs (Section 5.5.5) and G-proteins to PI3K and its effectors.

7.4.2
PKB/Akt Kinase

A major downstream effector of PI3K is the protein kinase Akt, also known as PKB. The domain structure of Akt kinase of which three isoforms are known is shown in Fig. 7.13. Akt kinase contains a pleckstrin homology (PH) domain at the N-terminus, a typical kinase domain and a C-terminal regulatory domain.

The main regulatory inputs for activation of Akt kinase are:
- Binding of the second messenger PtdIns(3,4,5)P$_3$, the product of the PI3K reaction, to the PH domain.
- Phosphorylation of a specific Thr residue (Thr286) in the activation loop, catalyzed by *phosphoinositide-dependent protein kinase (PDK) 1*.
- Phosphorylation of Thr473 in the regulatory region, catalyzed probably by PDK2 or integrin-linked kinase (ILK) (Chapter 11).

The central stimulus for activation of Akt kinase is provided by PI3K. The PtdIns(3,4,5)P$_3$ formed by PI3K binds to the PH domains of two protein kinases next in sequence, *PKB/Akt kinase* and *PDK1* (reviewed in Biondi, 2004), inducing the membrane translocation of both enzymes.

Membrane recruitment of Akt kinase is a prerequisite for a subsequent phosphorylation of Thr286 in the activation loop and this reaction is catalyzed by PDK1 which also contains a PH domain with

■ **Akt kinase (PKB) signaling is**
- Proliferation promoting
- Antiapoptotic
- Deregulated in many cancers

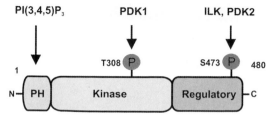

Fig. 7.13: Domain structure and main regulatory phosphorylation sites of PKB/Akt.

■ **Akt phosphorylates**

− Transcription factors:
  CREB
− Protein kinases:
  IκB kinase
− Proapoptotic proteins:
  Bad
− Others: MDM2

high affinity for PtdIns(3,4,5)P₃. To be fully activated, Akt kinase requires a further phosphorylation (Thr473) within the regulatory region. Membrane targeting and the double phosphorylation activate Akt, allowing the phosphorylation of downstream substrates.

Signaling by Akt Kinase

The signaling pathway for Akt kinase shown in Figs. 7.14 and 7.15 illustrates the central role of PI3K and Akt kinase in growth factor-controlled signal paths that lead from the cell membrane into the cytosol and the nucleus.

In the PI3K/Akt signaling pathway (reviewed in Osaki et al., 2004; Song, 2005), first an extracellular growth factor activates the corresponding TM receptor (e.g. PGDF receptor, see Section 8.1) which transmits the signal further to the downstream effector, PI3K. The associated membrane translocation of PI3K is synonymous with its activation. The PtdIns(3,4,5)P₃ formed induces membrane targeting and activation of two protein kinases next in sequence, Akt kinase and PDK1 (Fig. 7.14). The latter is also central to the activation of other protein kinases, e.g. PKC (Section 7.5).

Many substrates of Akt kinase are directly or indirectly involved in the promotion of cell proliferation and prevention of apoptosis (Fig. 7.15), and this is why a deregulation of the Akt pathway is frequently associated with the development of cancer. In accordance with a critical role in cell proliferation, Akt kinase has been identified as a classical viral oncogene. Among the substrates of Akt kinase involved in regulation of apoptosis are transcription factors like Forkhead (FH) transcription factors and CREB protein (Section 3.4.5), IκB kinase (Section 2.5.6.4), proapoptotic protein Bad, procaspase-9 and MDM2 protein (Section 15.7.1). In addition, Akt kinase is a major regulator in insulin-dependent metabolic pathways. The PI3K/Akt

Fig. 7.14: Activation of PKB/Akt kinase by membrane translocation. PtdIns(3,4,5)P₃ generated in response to growth factor stimulation serves as a binding site for the PH domains of PDK1 and PKB. Membrane translocation is accompanied by release of an autoinhibition leading to activation of PDK1 and PKB kinase activities. Full activation of PKB requires phosphorylation by PDK1. Activated PKB phosphorylates a variety of target proteins that prevent apoptotic death (Bad) and regulate transcription [Forkhead transcription factors (FKHR1)] and other metabolic processes. After Schlessinger (2000).

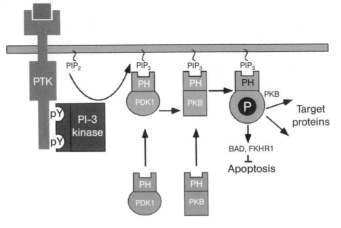

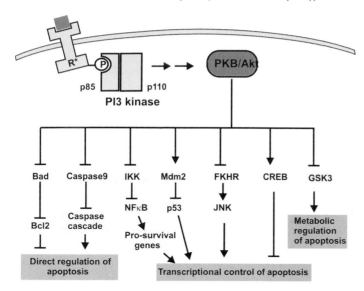

Fig. 7.15: Multiple mechanisms of cell survival regulation by PKB/Akt. For Bad and caspase-9 signaling, see Chapter 15; for MDM2/p53, see Chapter 14; for CREB, see Section 3.4.5; for IKK, see Section 2.5.6.4; for JNK, see Chapter 10.

pathway is activated by insulin and thereby mediates many of the metabolic effects of insulin, including glucose transport, lipid metabolism, glycogen synthesis and protein synthesis (Section 3.6.3.4).

The great importance of the PI3K/Akt pathway for growth regulation is illustrated by the observation that an enzyme of PtdIns(3,4,5)P$_3$ metabolism, *PTEN lipid phosphatase*, has been identified as a tumor suppressor protein (reviewed in Gericke et al., 2006). PTEN tumor suppressor protein has lipid phosphatase activity that is specific for hydrolysis of PtdIns(3,4,5)P$_3$. It is assumed that PTEN is a negative regulator of the Akt pathway, acting by lowering the concentration of PtdIns(3,4,5)P$_3$ and counteracting stimulation of Akt kinase (Fig. 7.11). Due to the strong cell proliferation-promoting and antiapoptotic activity of the Akt kinase pathway, lowered concentrations of PtdIns(3,4,5)P$_3$ will have an antiproliferative and proapoptotic effect, and will thus inhibit tumor formation, explaining the tumor-suppressing activity of PTEN. In accordance with this model, a functional inactivation of the PTEN phosphatase is observed in many tumors.

■ **PTEN**

– Lipid phosphatase

– Tumor suppressor function

– Hydrolyzes PtdIns(3,4,5)P$_3$

– Downregulates Akt

7.5
PKC

7.5.1
Classification and Structure

The family of PKC enzymes includes Ser/Thr-specific protein kinases, which were first identified by the requirement of the second messengers DAG and Ca^{2+} for activity (reviewed in Spitaler and Cantrell, 2004). Furthermore, PKC enzymes have been shown to be high-affinity receptors for phorbol esters (see below).

The sensitivity to Ca^{2+} and DAG and specific binding of phorbol esters has been considered for a long time to be the main characteristics of PKC enzymes. Sequence comparison and biochemical work, however, revealed that PKC is a large kinase family that includes a variety of isoenzymes with very different regulatory properties. Each isoenzyme is the product of a separate gene, with the exception of βI and βII enzymes, which are alternatively spliced variants of the same gene.

The PKC Family

■ **PKC subfamilies**
– Classical PKC:
 α, βI, βII and γ
– Novel PKC:
 δ, ε, θ and η
– Atypical PKC:
 ζ, λ and ι

There are more than 10 different subtypes of PKC, that are classified into three subfamilies according to the structure of the N-terminal regulatory domain, which determines their sensitivity to the cofactors DAG and Ca^{2+}.

– *Classical or conventional PKCs*. The members of this subfamily (cPKCs, subtypes α, βI, βII and γ) are activated by Ca^{2+}, DAG and phorbol esters.
– *Novel PKCs*. The novel PKCs (nPKCs, subtypes δ, ε, θ and η) are not responsive to Ca^{2+}, but are activated by DAG and phorbol esters.
– *Atypical PKCs*. These isoforms (aPKCs, subtypes ζ, λ and ι) are Ca^{2+} and DAG independent, and require adaptor proteins for enhancement of enzymatic activity.

A change in classification has been made recently for PKC subtypes ν and μ. These enzymes are now included in the family of PKD (reviewed in Rozengurt et al., 2005) which belong to the group of calmodulin-dependent kinases (Section 7.5).

In addition to the different cofactor requirements, the PKC isoenzymes are distinguished by different cellular localization and a different pattern of substrate proteins. For example, the α, δ and ζ subtypes are widespread in almost all tissues, whereas the other subtypes only occur in specialized tissues.

The members of the PKC family are composed of a polypeptide chain with a molecular weight of 68–83 kDa. The N-terminal regulatory domains C1 (C = conserved domain) and C2 and a C-terminal catalytic domain can be differentiated in the primary structure (Fig. 7.16) of the conventional PKCs. In addition, a pseudosubstrate sequence with autoinhibitory function is located in the C1 region that binds to the substrate-binding site of the catalytic domain and keeps the enzyme in an inactive state in the absence of cofactors and activators.

The C1 domain is present in the classical and novel PKC isoenzymes. It contains two cysteine-rich motifs (also known as Cys1 and Cys2 elements), each with two bound Zn^{2+} ions, which mediate the binding of DAG and phorbol esters. The atypical PKC enzymes contain a C1-like domain with only a single Cys-rich motif that is unable to bind DAG and phorbol esters.

The C2 domain of the classical PKC subfamily is a module of around 120 amino acids that binds phospholipids in a Ca^{2+}-dependent manner, resulting in activation of the adjacent protein kinase domain. However, it is now well established that the C2 domains

■ **C1 domains bind**
– DAG
– Phorbol ester
C2 domains may bind
– Ca^{2+} and phospholipids
– Phosphotyrosine

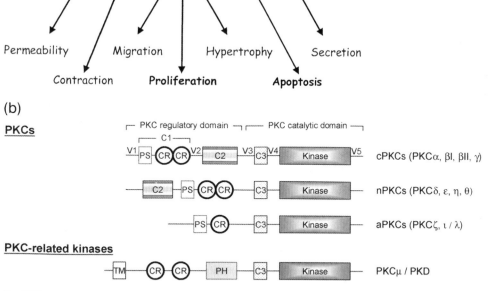

Fig. 7.16: (a) PKC isoenzymes. The major cellular functions of the PKC isoenzymes are indicated (from Dempsey et al., 2000). The isoenzymes are divided into subgroups based on structure and cofactor requirements: conventional (α, βI, βII and γ), novel (δ, ε, η and θ), atypical (ξ and ι) and PKC related (μ and ν).

(b) Domain structure of PKC isoenzymes. The phorbol esters and the second messenger DAG bind to the cysteine-rich motif present in cPKC and nPKC. The nPKCs are Ca^{2+} independent. V = variable region; PS = pseudosubstrate region; CR = cysteine-rich domain; TM = transmembrane domain; PH = pleckstrin homology domain.

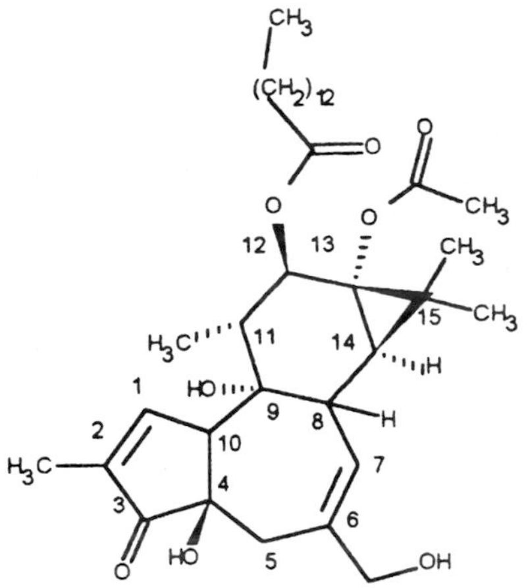

Fig. 7.16: (c) Structure of TPA. TPA functions as a tumor promoter and is a specific activator of PKC.

are very versatile regarding their function and ligands. PKCδ, a member of the novel PKC enzyme subfamily, contains a variant of the C2 domain that does not bind Ca^{2+}, but binds to phosphotyrosine residues of phosphorylated proteins. Furthermore, C2 domains are found in proteins other than PKC, e.g. phospholipase (PL) A_2, mediating protein–protein interactions.

Stimulation by Phorbol Esters

A property of the classical and novel PKC subfamilies that is highly valuable for their identification and characterization is their activation by *tumor promoters* such as *phorbol esters* (Fig. 7.16c). These subfamilies bind to the tumor promoter, *tetradecanoyl phorbol acetate (TPA)*, with high affinity. TPA binds to the C1 domain of PKC isoforms leading to membrane attachment of PKC and activation of its kinase activity. The specific activation of PKC by externally added phorbol esters is an important tool for demonstrating their involvement in signal transduction pathways.

Phorbol esters (TPA)
- Are tumor promoters
- Activate PKC
- Bind to the C1 domain
- Induce membrane attachment

Tumor promoters such as TPA do not themselves initiate tumor formation, but rather they promote triggering of the tumor by carcinogenic substances, e.g. benzo[a]pyrene. The tumor-promoting activity of TPA is ascribed to stimulation of PKC. Since one of the roles of PKC is in regulation of proliferation and differentiation processes, unregulated activation of PKC could lead to undesired protein phosphorylation and thus bring about misregulation of cell proliferation.

In addition to PKC enzymes, cells contain other receptors for phorbol esters and DAG. In mammals, proteins named α- and β-chimerins have been identified that lack protein kinase activity, and bind phorbol esters and DAG with high affinity. Therefore, not all biological responses elicited by phorbol ester treatment can be attributed to PKC stimulation.

7.5.2
Activation of PKC by Cofactors

In the absence of activating cofactors, the catalytic domain is subject to *autoinhibition* by a sequence motif in the regulatory N-terminal domain, which serves as a pseudosubstrate. This sequence motif is found in all PKC family members. It is assumed that the active center is blocked by the pseudosubstrate.

Two functions are attributed to the binding of the activating cofactors Ca^{2+}, DAG and phospholipid:
– Relieve of autoinhibition and stabilization of a structure of PKC in which the active center is accessible for substrate proteins.
– Promotion of membrane association.

Detailed structural information on PKC and its conformational rearrangements during activation is not available at present, probably because of its high flexibility. Most models on the activation of PKC by its cofactors assume a linkage between activation and membrane association. Structural determination of the Cys2 element of the C1 domain of PKCδ in complex with phorbol ester strongly suggests such a mechanism. The binding site of the phorbol ester lies in a hydrophobic region of the Cys2 element which is broken by a hydrophilic region. On binding of the phorbol ester, a continuous hydrophobic surface is created in this region facilitating partial membrane insertion of the Cys2.

The second messenger DAG binds to the same site as the phorbol ester and it is generally assumed that this binding promotes membrane association of the PKC.

The cofactors Ca^{2+} and phospholipids bind to the C2 domain of the classical PKC enzymes, and this is thought to enhance membrane association and release of the enzyme from the autoinhibited state.

7.5.3
Regulation of PKC

The main regulatory inputs that control PKC activity are:
– DAG signals.
– Ca^{2+} signals.
– Phosphorylation signals.
– Binding to signaling proteins that have enzyme, adaptor or scaffolding function

As shown by the different domain organization, the various subtypes may be sensitive to these signals in very different ways. In most cases, PKC enzymes are sensitive to and require the input of several regulatory signals, and therefore regulation of PKC is complex and

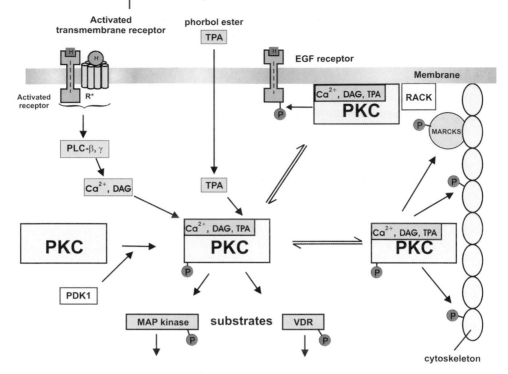

Fig. 7.17: Functions and regulation of PKC. Receptor-controlled signal pathways lead to formation of the intracellular messenger substances Ca^{2+} and DAG, that, like TPA, activate PKC. Translocation to the cell membrane is linked with activation of PKC; receptors for PKC, the RACK proteins, are also involved. Substrates of PKC are the MARCKS proteins and other proteins associated with the cytoskeleton. Other substrates are the Raf kinase (Chapter 10) and the receptor for vitamin D_3 (VDR, see Chapter 4).

each subtype exhibits a distinct pattern of regulation. The main functions and regulation of PKC are shown schematically in Fig. 7.17.

Regulation by Ca^{2+} and DAG: Activation and Membrane Association
The activating cofactors Ca^{2+} and DAG are second messengers released upon induction of a variety of signaling paths (Chapter 6). A main component of Ca^{2+}/DAG signaling pathways is PLC that when activated produces the messenger substances $Ins(3,4,5)P_3$/Ca^{2+} and DAG. Activation of PLC and hence of PKC may take place via several central pathways:

■ **PKC activation is linked to**
 – GPCR signaling via PLCβ
 – Receptor tyrosine kinase signaling via PLCγ
 – Ras signaling via PLCε

Signaling pathways starting from RTK trigger stimulation of PKC by activating PLCγ. An activating signal may also be dispatched in the direction of PKC – via activation of PLCβ – from GPCRs (Fig. 6.5f). Furthermore, activation of PLCε (Section 5.7.2) and the generation of Ca^{2+}/DAG signals may occur via the Ras pathway (Section 9.8) providing a link between Ras signaling, Ca^{2+}/DAG signals and PKC activation. In addition, the cell has available a broad palette of other tools for producing Ca^{2+} signals (Section 6.7) and accordingly many ways of activating PKC via Ca^{2+} exist.

Membrane association and activation of PKC are often intimately linked. The cofactors Ca^{2+}, DAG and phosphatidylserine both activate the classical PKC subtypes and enhance their membrane association. Binding of Ca^{2+} to the C2 domain leads to an increased membrane association and to activation by release of the catalytic center from interaction with the autoinhibitory structural element. Further activation takes place by binding of DAG to the C1 domain, which directly favors membrane association and enhances binding of phosphatidylserine to the C2 domain, whereby these ligands serve as an anchor for membrane association. Use of the two membrane-targeting domains C1 and C2 apparently helps to ensure high affinity, specificity and regulation of the membrane interaction.

■ **PKC is regulated by**
– Ca^{2+} signals
– DAG
– Phosphorylation in activation loop, catalyzed by PDK1
– Phosphorylation at Tyr residues
– Localization by RACK proteins

Regulation by Phosphorylation

All PKC subtypes contain several Ser/Thr and Tyr phosphorylation sites, and phosphorylation of these sites is an essential step in the maturation and activation of PKC enzymes. A pivotal role in PKC activation is attributed to the PDK1 kinase (reviewed in Biondi, 2004) that is upstream of both atypical and conventional PKC isoforms. PDK1 is partly regulated by PI3K. PI3K-generated lipids activate PDK1, which then phosphorylates and activates several members of the conserved AGC kinase superfamily including PKA, PKG (cGMP-dependent kinase), PKB (Akt kinase) and PKC. An initial activating phosphorylation of PKC by PDK1 takes place on Ser/Thr residues in the activation loop of the catalytic domain. As a consequence, autophosphorylation on two sites near the C-terminus of PKC is triggered. One of the autophosphorylation sites is located in a hydrophobic segment and its phosphorylation has been found to induce membrane dissociation of the activated enzyme. Overall, the PDK1-catalyzed phosphorylation and the autophosphorylation events are important for regulating the catalytic activity of the PKCs and for their subcellular distribution. The details of this regulation mechanism and its relation to activation by the other cofactors are not yet established.

A number of PKC isoforms, e.g. PKCδ and PKCε, are phosphorylated on Tyr residues in response to a variety of signals. These phosphorylations are mostly catalyzed by members of the Src or Lyn family of non-RTK (Section 8.3). How PKC is targeted to Src kinase has become clear only after the discovery of the C2 domain of PKCδ to be a PBD. The C2 domain binds specifically to a phosphotyrosine on a TM protein named CDCP1 which binds to and is phosphorylated by Src kinase. The CDCP1 protein has the function of an adap-

Fig. 7.18: Signaling complexes at the cell membrane involving PKCδ and Src kinase. PKCδ becomes activated upon binding to p-Tyr residues on the TM protein CDCP1. The membrane recruitment is thought to activate PKCδ allowing phosphorylation of membrane-associated substrates. Phosphorylation of CDCP1 is catalyzed by the non-RTK Src that in turn binds to p-Tyr residues of CDCP1 via its SH2 domain.

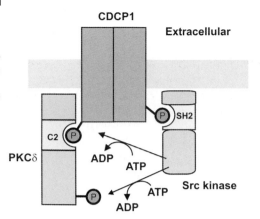

tor in this case that promotes association of PKCδ with Src kinase allowing Tyr phosphorylation of PKCδ by Src kinase (Fig. 7.18).

PKC-interacting Proteins and Regulation by Localization

A large number of proteins have been found to interact with PKC enzymes during signal transduction (reviewed in Poole et al., 2004). The PKC-interacting proteins are mostly isoform-specific and include proteins that:

– Target PKC to its upstream activators.
– Direct PKC to intracellular compartments.
– Are substrates of PKC.
– Modify PKC, e.g. as other protein kinases.

The proteins that direct PKC enzymes to specific subcellular sites will be discussed in more detail in the following.

7.5.4
Receptors for PKC and RACK (Receptors for Activated PKC) Proteins

A major regulatory aspect of PKC enzymes is the regulated localization to distinct subcellular sites. Stimulation of cells with phorbol esters or with hormones that activate PLCβ or PLCγ leads to translocation of PKC isoenzymes from the cytoplasm to the cell membrane or cytoskeleton or into the nucleus. The differential localization of the PKC isoenzymes is mediated by PKC-targeting proteins, of which the RACK proteins stand out (reviewed in Schechtman and Mochly-Rosen, 2001).

■ **RACK proteins**
– Target PKC to distinct sites
– Direct trafficking of PKC
– Assemble PKC into multiprotein complexes

The RACK proteins specifically interact with PKC enzymes and anchor them to the membrane providing access to membrane-localized substrates. Furthermore, subtypes of the RACK proteins are involved in intracellular trafficking of PLC enzymes. Most of the PKC–RACK interaction is mediated by the C2 domain of the PKC enzymes.

In addition to anchoring activated PKC enzymes, the RACK proteins anchor other central signaling proteins. As an example, RACK1, the anchoring protein for activated PKCβII, targets Src tyrosine kinase (Section 8.3.2), cAMP-specific phosphodiesterase and integrin β subunits (Section 11.4) to distinct subcellular sites. RACK1 contains several WD motifs that are thought to be responsible for the protein–protein interactions and for its scaffolding function. Similar to the AKAPs, the RACK proteins appear to be able to organize various signaling enzymes into multiprotein signaling complexes.

Members of the AKAPs have also been identified as binding partners of PKC enzymes. As an example, AKAP79 assembles PKC into protein signaling complexes, keeping it in an inactive state. Upon receipt of Ca^{2+}/DAG signals, the PKC is released from the inhibitory complex. Specific binding and clustering of PKC enzymes have also been reported for the adaptor proteins Ina D (Section 8.5) and for the adaptor protein caveolin, which targets signaling proteins to specific plasma membrane microdomains –the caveolae.

7.5.5
Functions and Substrates of PKC

The members of the PKC family are central signal proteins and, as such, are involved in the regulation of a multitude of cellular processes, e.g. cell proliferation, subcellular transport, cell migration, cytoskeleton reorganization, immune receptor signaling, apoptosis and memory formation. A large number of substrates have been identified for which each PKC isoenzyme has a distinct substrate specificity. Of the many substrates of PKC isoenzymes, the MARCKS (myristoylated, alanine-rich C-kinase substrate) proteins are highlighted as very well-characterized and specific substrates of PKC (reviewed in Sundaram et al., 2004) and their phosphorylation is used as an indicator of the activation of PKC *in vivo*. MARCKS proteins are involved in restructuring of the actin cytoskeleton. A PKC-mediated phosphorylation of the MARCKS proteins is involved in this process.

Further examples of substrates of PKC are the epidermal growth factor receptor (EGFR; see Chapter 8), a Na^+/H^+ exchanger protein, the GTPase K-ras (Chapter 9), Raf kinase (Chapter 9) and NMDA receptors. Activation of PKC may, as the examples show, act on other central signal transduction pathways of the cell; it may have a regulating activity on transcription processes, and it is involved in the regulation of transport processes and in neuronal communication.

■ **PKC substrates**
– MARCKS proteins
– Raf kinase
– EGF
– NMDA receptor

Many substrates of PKC are membrane proteins, and it is evident that membrane association of PKC is of great importance for the phosphorylation of these proteins.

7.6
Ca^{2+}/Calmodulin-dependent Protein Kinases (CaMKs)

7.6.1
Importance and General Function

The signal-mediating function of Ca^{2+} is performed as a Ca^{2+}/calmodulin complex in many signaling pathways. Ca^{2+}/calmodulin can bind specifically to effector proteins and modulate their activity (Section 6.7). In first place as effector proteins of Ca^{2+}/calmodulin are the *CaMKs* (reviewed in Hudman and Schulman, 2002).

A rough categorization of the CaMKs differentiates between *specialized* CaMKs and *multifunctional* kinases.

An example of a specialized CaMK is myosin light-chain kinase (MLCK), the primary function of which is to phosphorylate the light chain of myosin and thus to control the contraction of smooth musculature.

■ CaMKs
– Activated by Ca^{2+}/calmodulin
– Specialized CaMKs
– Multifunctional CaMKs: CaMK I, II and IV

Tab. 7.2: Examples of substrates of CaMKs
(Source: *Ann. Rev. Physiol.*, 1995, **57**).

Protein	Function
Acetyl CoA carboxylase	biosynthesis of fatty acids
Glycogen synthase	glycogen synthesis
HMG CoA reductase	biosynthesis of cholesterol
NO synthase	biosynthesis of NO
Ca^{2+} channel (N-type)	presynaptic Ca^{2+} influx
Ca^{2+} ATPase (heart)	storage of Ca^{2+}
Synaptogamin	release of neurotransmitters
Ryanodin receptor	release of Ca^{2+}
p56^{LCK} tyrosine kinase	activation of T cells
EGFR	growth control
Cyclic nucleotide phosphodiesterase	cAMP and CGMP metabolism
PPA2	hydrolysis of phospholipids
Ribosomal protein S6	protein biosynthesis
CREB	transcription control

The multifunctional CaMKs include the CaMKs of types I, II and IV, all of which phosphorylate a rather broad spectrum of substrate proteins. These enzymes regulate many processes such as glycogen metabolism, activity of transcription factors, microfilament formation, synaptic release of neurotransmitters from storage vesicles, biosynthesis of neurotransmitters, etc. Some important substrates of CaMKs are shown in Tab. 7.2. CaMKs of type II are oligomeric enzymes which exhibit outstanding regulatory properties and will therefore be discussed below in more detail.

The two other multifunctional CaMK of types I and IV are, apart from Ca^{2+}/calmodulin control, monomeric enzymes regulated via phosphorylation by an upstream CaMK kinase (CaMKK). The CaMKK phosphorylates CaMK I and IV in the activation loop, and thereby greatly enhances the activity of these enzymes. Examples of substrates of CaMKs I and IV are transcription factors, the MAPKs and adenylyl cyclase.

7.6.2
CaMK II

Among the class of CaMK II enzymes, subtypes α, β, γ and δ are differentiated. The α and β subtypes of CaMK II only occur in the brain, whereas the other subtypes are also found in other organs. In the hippocampus, CaMK II constitutes up to 2% of total cellular protein. From a regulatory point of view, CaMK II is of particular interest, as it has the characteristic of an enzyme with a built-in "memory switch". The "memory" allows the CaMK II to conserve a stimulatory signal over a longer period of time and to remain in an activated state, even when the initiating stimulus has died away. Due to these properties, CaMK II is considered an important element of memory formation and storage in the brain.

Structure of CaMK II

The domain structure of CaMK II shows an association domain, an N-terminal catalytic domain and a regulatory domain containing the Ca^{2+}/calmodulin-binding site and an autoinhibitory region. Electron micrographs and X-ray studies show that CaMK II has an oligomeric structure (Figs. 7.19 and 7.20) in which two hexameric rings, each containing six copies of the monomeric enzyme (α, β, γ or δ subtype), are configured in the form of a cylinder. The catalytic domains are orientated outwards and the association domains form the central core of each ring. The aggregated form of CaMK represents the holoenzyme form, which performs the catalytic functions, and is subject to a sophisticated control by both autophosphorylation and Ca^{2+}/calmodulin (reviewed in Hudmon and Schulman, 2002).

■ **CaMK II**
– Oligomeric kinase of 12–14 subunits
– Activated by: Ca^{2+}/ calmodulin/auto-phosphorylation
– Inactivated by phosphatases
– Memory effect for Ca^{2+} signals
– Important for memory and learning

Fig. 7.19: (a) Linear representation of the functional domains of CaMK II of type β. (b) The various autoregulatory stages of CaMK II. The multimeric holoenzyme structure of CaMK II is depicted as a 6mer for simplicity, with activated catalytic subunits illustrated in blue. Ca^{2+}/calmodulin binding triggers autophosphorylation of Thr286 on neighboring subunits which converts these subunits into an autonomous, Ca^{2+}-independent active state and prepares the enzyme for phosphorylation of the physiological substrates. The active state is terminated when the activating phosphate residue is cleaved off by a protein phosphatase.

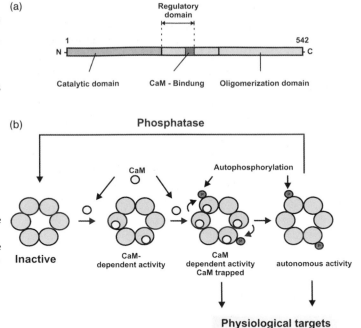

Fig. 7.20: Relieve of the auto-inhibited state of CaMK II by binding of Ca^{2+}/calmodulin. The kinase domain is shown in green and Ca^{2+}/calmodulin is shown in blue. (a) Schematic diagram of the CaMK II holoenzyme in its fully autoinhibited state. The kinase domains are closely packed and the substrate binding sites are blocked. (b) Ca^{2+}/calmodulin binding to the regulatory domain directs the kinase domains above and below the midplane allowing transphosphorylation of Thr286. From Rosenberg et al. (2005) .

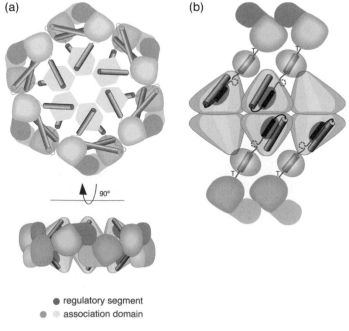

Regulation of CaMK II

In the absence of Ca^{2+}/calmodulin, the catalytic domain of CaMK II exists in an autoinhibited state. X-ray analysis of the autoinhibited state shows a dimeric organization of the subunits within the hexameric rings. Parts of the regulatory segments form a coiled-coil structure that blocks peptide and ATP binding to the otherwise intrinsically active kinase domains and the subunits are organized in a tightly packed assembly (Fig. 7.20a). An increase in the Ca^{2+} concentration leads to binding of Ca^{2+}/calmodulin at the C-terminal end of the regulatory domain, which releases the enzyme from its autoinhibited state (Fig. 7.20b). Thereby, Ca^{2+}/calmodulin binding to the regulatory segments disengages the kinase domain from the tightly packed structure. The kinase domains now are directed up- and downwards of the midplane which allows transphosphorylation of other kinase subunits on Thr286. Once Thr286 is phosphorylated, the enzyme is in an Ca^{2+}/calmodulin-independent state.

The autophosphorylation on Thr286 is intermolecular, i.e. neighboring subunits of the holoenzyme mutually phosphorylate one another. In contrast to the other CaMKs, CaMK II does not require activation loop phosphorylation for activation.

The autophosphorylation has three important consequences:
- The affinity for Ca^{2+}/calmodulin is increased by close to three orders of magnitude. Ca^{2+}/calmodulin only dissociates very slowly from this high-affinity complex. The activated state is thus preserved over a longer period of time.
- The autophosphorylation at Thr286 disables the autoinhibitory section so that the enzyme becomes autonomous of its normal stimulus, Ca^{2+}. When Ca^{2+} declines to baseline levels and Ca^{2+}/calmodulin dissociates, the autophosphorylated enzyme still has 20–80% of the activity of the Ca^{2+}/calmodulin-bound form. This ensures that significant activity remains after the Ca^{2+}/calmodulin signal has died away. This autonomous, Ca^{2+}/calmodulin-independent state of CaMK II is only terminated when phosphatases cleave off the activating phosphate residue and thus lead the enzyme back into the inactive state.
- Autophosphorylation exposes a binding site for target proteins such as the NMDA receptor, a Ca^{2+}-channel regulated by glutamate and voltage changes. In the receptor-bound state, the requirement of CaMK II for Ca^{2+}/calmodulin is reduced and the kinase may stay active even without being autophosphorylated.

All these features are dependent on the oligomeric state of the holoenzyme. Specifically, the formation of the large kinase aggregate endows the holoenzyme with exquisite regulatory properties. This is illustrated by the finding that CaMK II is able to respond in a switch-like fashion to

Ca^{2+} signals over a very narrow Ca^{2+} concentration range and this range is sharpened further by the presence of protein phosphatase (Bradshaw et al., 2003). Cooperative autophosphorylation within the oligomer and the action of protein phosphatases provide for an ultrasensitive Ca^{2+} switch that responds to changes in Ca^{2+} concentrations over a range of 300 nM Ca^{2+}, a range that allows sensing of specific Ca^{2+} signals.

Memory Function of CaMK II

Another outstanding feature of regulation of CaMK II is the memory effect within the activation process. Activation of the enzyme is initiated by a generally transient increase in cellular Ca^{2+}. Ca^{2+} activates CaMK II in the form of the Ca^{2+}/calmodulin complex; the kinase remains active even after the Ca^{2+} signal has died away, because the enzyme is converted into an autonomous activated state upon autophosphorylation.

A special importance is attributed to this property, particularly for the detection and differentiation of repetitive Ca^{2+} signals in neuronal cells. The magnitude of constitutive CaMK II activity in the oligomeric, autophosphorylated form has been shown to depend on the duration, amplitude and frequency of elevated Ca^{2+}. For example, the interval between the occurrence of staggered Ca^{2+} signals is a determining factor for the intensity of activation. If the Ca^{2+} signals occur with a higher frequency, a long-lasting and effective activation is possible, since the kinase remains in the activated state between signals because of the memory effect (De Koninck and Schulman, 1998). Due to these special properties, it is assumed that CaMK II actively participates in synaptic plasticity and memory formation. The ability of CaMK II to decode the frequency of Ca^{2+} oscillations during synaptic stimulation and to give a prolonged response beyond the initial stimulus enables it to provide two characteristics required for a molecule involved in synaptic plasticity and in memory formation (reviewed in Lisman et al., 2002). In agreement with this is the observation that elimination of CaMK II phosphorylation interferes with learning and memory functions.

There are different mechanisms that lead to increased intracellular Ca^{2+} concentration and thus to activation of CaMK II. CaMKs of type II are activated as a consequence of $InsP_3$-mediated release of Ca^{2+} from intracellular storage. Influx of Ca^{2+} from the extracellular region, triggered by opening of various ligand-controlled or voltage-controlled Ca^{2+} channels, also brings about an activation of CaMK II. In addition to the Ca^{2+}/calmodulin signal, there are other proteins and enzymatic activities that are of importance for the efficiency and specificity of CaMK II signaling. One control is exerted by the activity of protein phosphatases that terminate the autoactivated state of CaMK II. Dephosphorylation of CaMK II is catalyzed by protein phosphatases I (see below) and by a specialized CaMK phosphatase. Another major regulatory aspect of CaMK II is its colocalization with substrates at specific subcellular sites. Subtypes of CaMK II have been found to be specifically localized in the nucleus and at the cytoskeleton, especially at postsynaptic structures called postsynaptic densities (PSDs), which contain large amounts of associated CaMK II. By interaction with anchoring proteins like PDZ proteins (Section 8.5), CaMK II is assembled in the vicinity of Ca^{2+} entry sites and of potential substrates. Examples of neuronal substrates are the ligand-gated ion channels like the AMPA receptor, the NMDA receptor and neuronal nitric oxide synthase (NOS).

7.7
Ser/Thr-specific Protein Phosphatases

Under physiological conditions, phosphate esters of Ser and Thr residues are stable and show only a low rate of spontaneous hydrolysis. Thus, the cell requires its own tools for regulated cleavage of phosphate residues to terminate and damp signals mediated by protein phosphorylation. This role is performed by specific protein phosphatases that are frequently essential components of signaling pathways.

7.7.1
Structure and Classification of Ser/Thr Protein Phosphatases

We know of Ser/Thr-phosphate-specific protein phosphatases and Tyr-phosphate-specific protein phosphatases. The latter are dealt with in Chapter 8.

In mammals, at least 25 different of Ser/Thr-specific protein phosphatases have been identified, and these are classified according to their substrate specificities, metal requirements, and sensitivities to natural or synthetic inhibitors. PP1, PP2A and PP2B (calcineurin) have been studied best, and will be presented in the following. Given the large number of phosphorylated proteins in the cell – about one-third of all proteins of the proteome are thought to be phosphorylated – it was a daunting task for the cell to develop strategies for selectively removing phosphate residues from a wide array of different proteins. This problem was not solved by evolving a large number of different catalytic subunits. Rather, the diversity of the associated subunits has increased enormously during evolution. Generally, the protein phosphatases are made up of a catalytic subunit with one or more associated subunits that have regulatory and/or targeting functions, and these subunits specify to a large extent specificity, selectivity and location of the phosphatase.

7.7.2
Regulation of Ser/Thr Protein Phosphatases

Protein phosphatases are the antagonists of protein kinases and their action can reverse all functions that are coupled to the phosphorylation status of a protein (see functions of phosphorylation, see Section 2.4). In signaling pathways, they can perform a dual function. By diminishing and terminating a signal created by protein phosphorylation, they can have a damping effect on protein kinase-mediated signal transduction. Protein phosphatases can also have a positive, reinforcing effect in signaling pathways. Dephosphorylation of a signal protein by a protein phosphatase can lead to its activation and thus to amplification of the signal (Fig. 7.21).

■ **Protein phosphatases**
 – Important examples:
   PP1A, PP2A and PP2B
 – Contain catalytic and
   regulatory subunits
 – Regulation by: phosphorylation, localization
   and inhibitors

The protein phosphatases are active parts of signal transduction processes, and as such are subject to diverse and complex regulation. A large part of this regulation is exerted via the subunits that associate with the catalytic subunit to form the active holoenzyme. These subunits serve to transmit incoming signals to the catalytic subunit and to target the catalytic subunit to distinct subcellular sites.

Regulation of the Ser/Thr phosphatases takes place predominantly by the following mechanisms:

 – *Phosphorylation*. Targets of phosphorylation are both the catalytic and the regulatory/localization subunits. These phosphorylation events can change subunit composition, catalytic activity and subcellular localization.
 – *Targeted localization*. Much of the specificity of substrate dephosphorylation is achieved by targeting or scaffolding subunits that serve to localize the phosphatase in proximity to particular substrates and also to reduce its activity towards other potential substrates.

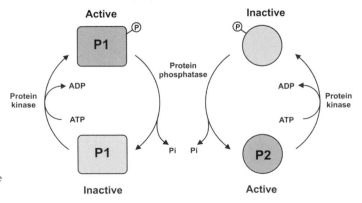

Fig. 7.21: The dual function of protein kinases and protein phosphatases. Phosphorylation of proteins (P1 and P2) can fix the latter into an active or inactive state. In the case of P1, protein kinases have an activating effect and protein phosphatases are inactivating; the reverse is true for P2.

– *Specific inhibitor proteins.* Specific inhibitor proteins for Ser/Thr phosphatases exist which can control the activity of the protein phosphatases. These inhibitors are generally subject to regulation themselves, e.g. by phosphorylation.

7.7.3
PP1

PP1 is a major class of eukaryotic Ser/Thr-specific protein phos-phatases that regulate diverse cellular processes such as cell cycle progression, muscle contraction, carbohydrate metabolism, protein synthesis, transcription and neuronal signaling.

The catalytic subunit (PP1c) of 35–38 kDa is highly conserved among eukaryotes. Its action is regulated by association with regula-tory subunits R which bind to PP1c via a short sequence that is gen-erally referred to as the "RVxF" motif. The residues of PP1c neces-sary for binding of the RVxF motif are invariant in all PP1c isoforms, but are not found in the catalytic subunits of PP2A and PP2B. The binding of the RVxF sequence does not have a major effect on the catalytic activity of PP1c. It rather functions as an anchor, and en-ables the R subunit to make additional contacts with the phosphatase in an ordered and cooperative manner (Fig. 7.22). Multiple contacts, each of which may be weak, ensure tight binding of the R subunit and this strategy is thought to allow the combinatorial binding of very diverse R subunits (Bollen, 2001).

Based on their main effect on PP1c, the R subunits can be roughly divided into three groups:
– *Activity-modulating proteins.* This group includes inhibitor proteins that, in their phosphorylated form, block the activity of PP1c towards its substrates.
– *Targeting subunits.* This type of R subunits bind both PP1c and one of its substrates, e.g. the G subunits target PP1c to the substrate glycogen synthase and this binding is modulated by phosphoryla-tion of glycogen synthase.
– *PP1c substrates.* A subset of PP1c substrates bind directly to PP1c and may also function as targeting subunit.

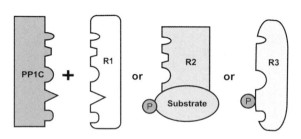

Fig. 7.22: Combinatorial con-trol model of the of the catalytic subunit of PP1 (PP1c). The catalytic subunit is represented with five different binding sites for the regulatory (R) subunits. The R subunits act as activity modulators (R_1), targeting pro-teins (R_2) and/or substrates (R_3). It is suggested that the R subu-nits have multiple contacts with PP1c and that they can share binding sites. Specificity is achieved by interaction with specific subsets of binding pockets on PP1c.

Overall, more than 50 different R subunits have been identified allowing for the formation of a large number of different holoenzymes, each with distinct location and substrate specificity (reviewed in Ceulemans and Bollen, 2004).

7.7.4
PP2A

PP2A is one of the major Ser/Thr phosphatases implicated in the regulation of many cellular processes including regulation of signaling pathways, cell cycle progression, apoptosis, DNA replication, gene transcription and protein translation (reviewed in Janssens et al., 2005). PP2A enzymes show both a Ser/Thr-phosphate-specific and a weaker Tyr-phosphate-specific protein phosphatase activity. The latter activity can be specifically stimulated by a distinct protein named protein tyrosine phosphatase activator (PTPA).

PPA2 enzymes are oligomeric enzymes composed of a conserved catalytic subunit and one or more additional regulatory subunits (Fig. 7.23). The core PP2A enzyme is a dimer consisting of the catalytic subunit C and a regulatory A subunit of 65 kDa. A third B subunit of which three subfamilies (PR55/B, PR61/B′ and PR72/B′′) with a total of more than 20 members are known can associate with the core enzyme to form the PP2A holoenzyme. *In vivo*, both the core AC dimer and the ABC holoenzyme are found. Due to the existence of two isoforms of the A and C subunits and because of the large number of different B subunits, a multitude of trimeric

■ **PP2A**

– Ser/Thr phosphatase
– Weak Tyr phosphatase
– Subunits A, B and C
– Many ABC holoenzymes
– Tumor suppressor function
– Inhibited by okadaic acid
– Regulation by: phosphorylation, methylation and nature of B subunit

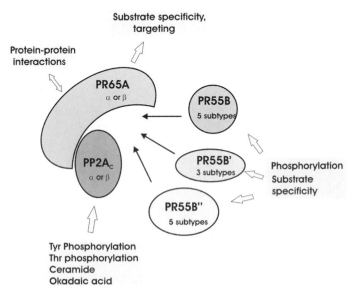

Fig. 7.23: Subunit structure and regulation of PP2A. The core of PP2A is a heterodimer consisting of a catalytic subunit PP2Ac and a structural subunit PR65A. The heterodimer can associate three types of regulatory subunits, PR55B, PR55B′ or PR55B′′, of which different subtypes exist. In addition, various regulatory modifications are indicated.

PP2A holoenzymes can form, each with distinct enzymatic and regulatory properties. Furthermore, the catalytic subunit, the core dimer and the PP2A holoenzymes can be associated with and regulated by a multitude of other cellular proteins.

Aberrant expression, mutation or deletion of PP2A subunits are involved in cellular processes leading to tumor formation, and both the A and B subunits have been identified as tumor suppressors (reviewed in Janssens et al., 2005), indicating an involvement of PP2A in the dephosphorylation of key proteins of cell cycle progression and apoptosis. Of great importance for the study of PP2A was the discovery of okadaic acid as an inhibitor of PP2A. Okadaic acid (Fig 7.24) is a tumor promoter and inhibits PP1 and PP2A.

Regulation of PP2A activity is mainly mediated by:

- *Phosphorylation.* Posttranslational modification of the C subunit by phosphorylation of specific Thr and Tyr residues inactivates the enzyme. Tyr phosphorylation is believed to be catalyzed by receptor and non-RTK like the EGFR, the insulin receptor kinase and the non-RTK p56[LCK] (Section 8.3), integrating PP2A activity into central signaling pathways.

- *C-terminal methylation.* A unique post-translational modification is found at the highly conserved C-terminus of the catalytic C subunit. The C-terminal Leu309 is methylated by a novel type of methyltransferase [*leucine carboxymethyl transferase (LCMT)*] and demethylated by a specific esterase, named PME-1. C-terminal methylation seems to be a requirement for the association of the B subunits and is thus a determinant of holoenzyme formation.

- *Nature of associated B subunit.* A major determinant of PP2A activity is the identity of the associated B subunit. The nature of the B subunit assembled in the heterotrimer of PP2A influences substrate specificity, catalytic activity, and subcellular distribution of the holoenzyme. The striking features of the B subunits are their diversity, stemming from the existence of entire subunit families, and the apparent lack of sequence similarity between these gene families (reviewed in Hogan and Li, 2005).

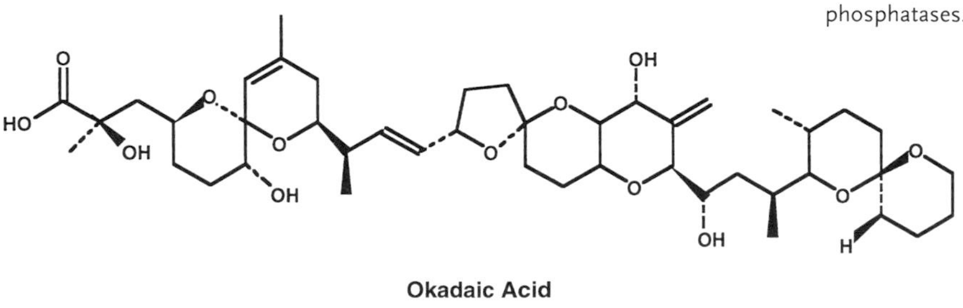

Fig. 7.24: Okadaic acid, an inhibitor of Ser/Thr protein phosphatases.

Okadaic Acid

7.7.5
PP2B (Calcineurin)

Calcineurin is a Ser/Thr phosphatase that is controlled by cellular calcium and regulates a large number of biological responses including lymphocyte activation, neuronal and muscle development, and the development of vertebrate heart valves.

■ **PP2A (calcineurin)**
- Composed of calcineurin A and B
- Regulated by Ca^{2+}/calmodulin
- Inhibited by: cyclosporin/cyclophilin and FK506/FK506-binding protein
- Regulates nucleocytoplasmic shuttling of NF-AT members

Like the other major classes of protein phosphatases, calcineurin is an oligomeric enzyme. It is made up of a catalytic subunit, calcineurin A, and a regulatory subunit, calcineurin B. In addition, binding of Ca^{2+}/calmodulin is required to form the fully active holoenzyme. The catalytic subunit harbors the catalytic domain and three regulatory domains, which have been identified as the calcineurin B-binding site, the Ca^{2+}/calmodulin-binding domain and an autoinhibitory domain. In the active site of calcineurin A, a binuclear metal center containing Fe^{2+} and Zn^{2+} is found. Both ions are thought to participate directly in phosphate ester hydrolysis (reviewed in Rusnak and Mertz, 2001). Calcineurin B is highly homologous to calmodulin, containing four EF hands as binding sites for Ca^{2+}. In the absence of Ca^{2+}/calmodulin, the dimer of calcineurin A and B is in an autoinhibited state caused by binding of the autoinhibitory domain to the substrate-binding cleft. Upon binding of Ca^{2+}/calmodulin, the autoinhibition is relieved. Both the Ca^{2+} binding to calcineurin B and the requirement of Ca^{2+}/calmodulin make calcineurin activity strongly Ca^{2+} dependent allowing it to function as a Ca^{2+} *sensor*.

The central regulatory function of calcineurin has been recognized during the search for the cellular target of the immunosuppressant drugs cyclosporin and FK506, often used in organ and tissue transplantations. Both drugs achieve their immunosuppressive effect via inhibition of calcineurin in an indirect way. Cyclosporin and FK506 bind specifically to two proteins known as cyclophilin and FK506-binding protein, respectively, that belong to the family of immunophilins and function as peptidy–prolyl *cis–trans* isomerases. The complexes of cyclosporin/cyclophilin and FK506/FK506-binding protein bind to calcineurin and inhibit the phosphatase activity of the latter.

In the process of activation of T lymphocytes, calcineurin is part of a signaling pathway that is activated by a rise in intracellular calcium upon ligand binding to the T cell receptor (TCR; see Section 11.3) and finally leads to the activation of transcription factors, named NF-AT (Fig. 7.25). Members of the NF-AT family of transcription factors control the expression of a large number of proteins, including cytokines, ion channels, cell surface proteins and proteins involved in apoptosis. A subset of the NF-AT transcription factors (NF-ATc

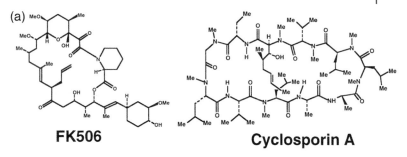

FK506   **Cyclosporin A**

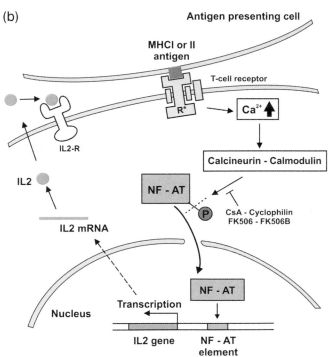

Fig. 7.25: (a) Cyclosporin A and FK506. (b) Model of the function of calcineurin in T lymphocytes. Antigenic peptides are presented to the T lymphocytes by an antigen-presenting cell (APC) within a cell–cell interaction (see also Chapter 11). Antigen binding activates the TCR that starts a signal chain leading to an increase in cytosolic Ca²⁺ and activation of calcineurin. The activated calcineurin cleaves an inhibitory phosphate residue from the transcription factor NF-AT. Consequently, NF-AT is transported into the nucleus where it stimulates the transcription of corresponding genes. Amongst the genes controlled by NF-AT is the gene for the cytokine IL-2. Following secretion into the extracellular space, the IL-2 formed binds to IL-2 receptors of the same cell or cells of the same type. A proliferation signal is created by the activated IL-2 receptor, leading to proliferation of T lymphocytes. Complexes of the immunosuppressants cyclosporin A (CsA) or FK506 with their binding proteins cyclophilin and FK506-binding protein (FK506B), respectively, inhibit calcineurin and disrupt the signal transmission to NF-AT.

members) is found in the cytosol and undergoes nuclear translocation upon calcineurin activation, allowing subsequent activation of target genes (reviewed in Hogan et al., 2003; Macian, 2005). The cytoplasmic form of NF-ATc is phosphorylated in its nuclear localization signal and requires dephosphorylation by phosphatase action in order to get access to its cognate genes. At this point, Ca^{2+} regulation of calcineurin comes into play.

The rise in Ca^{2+} caused by ligand binding to TCRs activates calcineurin's phosphatase activity, which dephosphorylates cytoplasmic NF-ATc proteins. Dephosphorylated NF-ATc enters the nucleus and binds to DNA in cooperation with other transcription factors, e.g. AP-1. In this way, many target genes in diverse tissues can be activated allowing for the transcriptional regulation of a large number of genes by calcineurin.

One of the target genes in lymphocytes is the gene for the cytokine interleukin (IL)-2, a cytokine, required for proliferation stimulation of these cells. Transcription of IL-2 is inhibited as one of the consequences of inhibition of calcineurin by cyclosporin/cyclophilin and FK506/FK506-binding protein. As consequence, proliferation of the lymphocytes is severely impeded.

Following the discovery of the role of calcineurin in transcription regulation in T lymphocytes, a large number of calcineurin substrates other than NF-AT have been discovered. These include NOS, ion channels and adenylyl cyclase among many others.

In addition to Ca^{2+} signals, calcineurin is also regulated and targeted by other cellular proteins, e.g. AKAP79. The inhibiting function of calcineurin on nuclear translocation of NF-ATc proteins may be counteracted by the action of central protein kinases including PKA, glycogen synthase kinase 3, and the stress-activated kinases c-Jun N-terminal kinase (JNK) and p38 (Chapter 10).

7.8
References

Biondi, R. M. and Nebreda, A. R. (2003) Signalling specificity of Ser/Thr protein kinases through docking-site-mediated interactions, *Biochem. J.* **105**, 1–13.

Biondi, R. M. (2004) Phosphoinositide-dependent protein kinase 1, a sensor of protein conformation, *Trends Biochem. Sci.* **105**, 136–142.

Bollen, M. (2001) Combinatorial control of protein phosphatase-1, *Trends Biochem. Sci.* **105**, 426–431.

Bondeva, T., Pirola, L., Bulgarelli-Leva, G., Rubio, I., Wetzker, R., and Wymann, M. P. (1998) Bifurcation of lipid and protein kinase signals of PI3Kgamma to the protein kinases PKB and MAPK, *Science* **105**, 293–296.

Bradshaw, J. M., Kubota, Y., Meyer, T., and Schulman, H. (2003) An ultra-sensitive Ca^{2+}/calmodulin-dependent protein kinase II-protein phosphatase 1 switch facilitates specificity in postsynaptic calcium signaling, *Proc. Natl. Acad. Sci. U. S. A.* **105**, 10512–10517.

Ceulemans, H. and Bollen, M. (2004) Functional diversity of protein phosphatase-1, a cellular economizer and reset button, *Physiol Rev.* **105**, 1–39.

Dar, A. C., Dever, T. E., and Sicheri, F. (2005) Higher-order substrate recognition of eIF2alpha by the RNA-dependent protein kinase PKR, *Cell* **105**, 887–900.

De, K. P. and Schulman, H. (1998) Sensitivity of CaM kinase II to the frequency of Ca^{2+} oscillations, *Science* **105**, 227–230.

Dempsey, E. C., Newton, A. C., Mochly-Rosen, D., Fields, A. P., Reyland, M. E., Insel, P. A., and Messing, R. O. (2000) Protein kinase C isozymes and the regulation of diverse cell responses, *Am. J. Physiol Lung Cell Mol. Physiol* **105**, L429–L438.

Elia, A. E., Rellos, P., Haire, L. F., Chao, J. W., Ivins, F. J., Hoepker, K., Mohammad, D., Cantley, L. C., Smerdon, S. J., and Yaffe, M. B. (2003) The molecular basis for phosphodependent substrate targeting and regulation of Plks by the Polo-box domain, *Cell* **105**, 83–95.

Engh, R. A. and Bossemeyer, D. (2001) The protein kinase activity modulation sites: mechanisms for cellular regulation – targets for therapeutic intervention, *Adv. Enzyme Regul.* **105**, 121–149.

Erikson, R. L., Collett, M. S., Erikson, E., and Purchio, A. F. (1979) Evidence that the avian sarcoma virus transforming gene product is a cyclic AMP-independent protein kinase, *Proc. Natl. Acad. Sci. U. S. A.* **105**, 6260–6264.

Gericke, A., Munson, M., and Ross, A. H. (2006) Regulation of the PTEN phosphatase, *Gene* **105**, 1–9.

Hogan, P. G., Chen, L., Nardone, J., and Rao, A. (2003) Transcriptional regulation by calcium, calcineurin, and NFAT, *Genes Dev.* **105**, 2205–2232.

Hogan, P. G. and Li, H. (2005) Calcineurin, *Curr. Biol. 15*, R442–R443.

Hubbard, S. R. and Till, J. H. (2000) Protein tyrosine kinase structure and function, *Annu. Rev. Biochem.* **105**, 373–398.

Hudmon, A. and Schulman, H. (2002) Structure-function of the multifunctional Ca^{2+}/calmodulin-dependent protein kinase II, *Biochem. J.* **105**, 593–611.

Huse, M. and Kuriyan, J. (2002) The conformational plasticity of protein kinases, *Cell* **105**, 275–282.

Janssens, V., Goris, J., and Van, H. C. (2005) PP2A: the expected tumor suppressor, *Curr. Opin. Genet. Dev.* **105**, 34–41.

Johnson, L. N. and O'Reilly, M. (1996) Control by phosphorylation, *Curr. Opin. Struct. Biol.* **105**, 762–769.

Katso, R., Okkenhaug, K., Ahmadi, K., White, S., Timms, J., and Waterfield, M. D. (2001) Cellular function of phosphoinositide 3-kinases: implications for development, homeostasis, and cancer, *Annu. Rev. Cell Dev. Biol.* **105**, 615–675.

Kikani, C. K., Dong, L. Q., and Liu, F. (2005) "New"-clear functions of PDK1: beyond a master kinase in the cytosol?, *J. Cell Biochem.* **105**, 1157–1162.

Krebs, E. G., Graves, D. J., and Fischer, E. H., (1959) 'Factors affecting the activity of muscle phosphorylase kinase', *J. Biol. Chem.* **234**, 2867–2873.

Lisman, J., Schulman, H., and Cline, H. (2002) The molecular basis of CaMKII function in synaptic and behavioural memory, *Nat. Rev. Neurosci.* **105**, 175–190.

Macian, F. (2005) NFAT proteins: key regulators of T-cell development and function, *Nat. Rev. Immunol.* **105**, 472–484.

Malbon, C. C., Tao, J., Shumay, E., and Wang, H. Y. (2004) AKAP (A-kinase anchoring protein) domains: beads of structure-function on the necklace of G-protein signalling, *Biochem. Soc. Trans.* **105**, 861–864.

Manning, G., Whyte, D. B., Martinez, R., Hunter, T., and Sudarsanam, S. (2002) The protein kinase complement of the human genome, *Science* **105**, 1912–1934.

Murphy, A., Sunohara, J. R., Sundaram, M., Ridgway, N. D., McMaster, C. R., Cook, H. W., and Byers, D. M. (2003) Induction of protein kinase C substrates, Myristoylated alanine-rich C kinase substrate (MARCKS) and MARCKS-related protein (MRP), by amyloid beta-protein in mouse BV-2 microglial cells, *Neurosci. Lett.* **105**, 9–12.

Nolen, B., Taylor, S., and Ghosh, G. (2004) Regulation of protein kinases; controlling activity through activation segment conformation, *Mol. Cell* **105**, 661–675.

Osaki, M., Oshimura, M., and Ito, H. (2004) PI3K-Akt pathway: its functions and alterations in human cancer, *Apoptosis.* **105**, 667–676.

Pellicena, P. and Kuriyan, J. (2006) Protein-protein interactions in the allosteric regulation of protein kinases, *Curr. Opin. Struct. Biol.* **105**, 702–709.

Poole, A. W., Pula, G., Hers, I., Crosby, D., and Jones, M. L. (2004) PKC-interacting proteins: from function to pharmacology, *Trends Pharmacol. Sci.* **105**, 528–535.

Rosenberg, O. S., Deindl, S., Sung, R. J., Nairn, A. C., and Kuriyan, J. (2005) Structure of the autoinhibited kinase domain of CaMKII and SAXS analysis of the holoenzyme, *Cell* **105**, 849–860.

Rozengurt, E., Rey, O., and Waldron, R. T. (2005) Protein kinase D signaling, *J. Biol. Chem.* **105**, 13205–13208.

Runnels, L. W., Yue, L., and Clapham, D. E. (2001) TRP-PLIK, a bifunctional protein with kinase and ion channel activities, *Science* **291**, 1043–1047.

Schechtman, D. and Mochly-Rosen, D. (2001) Adaptor proteins in protein kinase C-mediated signal transduction, *Oncogene* **105**, 6339–6347.

Schlessinger, J. (2000) Cell signaling by receptor tyrosine kinases, *Cell* **105**, 211–225.

Song, G., Ouyang, G., and Bao, S. (2005) The activation of Akt/PKB signaling pathway and cell survival, *J. Cell Mol. Med.* **105**, 59–71.

Spitaler, M. and Cantrell, D. A. (2004) Protein kinase C and beyond, *Nat. Immunol.* **105**, 785–790.

Tasken, K. and Aandahl, E. M. (2004) Localized effects of cAMP mediated by distinct routes of protein kinase A, *Physiol Rev.* **105**, 137–167.

Taylor, S. S., Kim, C., Vigil, D., Haste, N. M., Yang, J., Wu, J., and Anand, G. S. (2005) Dynamics of signaling by PKA, *Biochim. Biophys. Acta* **105**, 25–37.

Wong, W. and Scott, J. D. (2004) AKAP signalling complexes: focal points in space and time, *Nat. Rev. Mol. Cell Biol.* **105**, 959–970.

8 Signal Transmission via Transmembrane Receptors with Tyrosine-Specific Protein Kinase Activity

One of the fundamental mechanisms by which multicellular organisms communicate is the binding of extracellular protein ligands to transmembrane (TM) receptors that are termed *receptor tyrosine kinases (RTKs)*. These receptors possess tyrosine kinase activity on their cytoplasmic domain and ligand-induced stimulation of the tyrosine kinase activity is used to conduct the signal further. Activation of RTKs is triggered, in particular, by signals that control cell growth and differentiation. Extracellular signals are often protein hormones, which – if they have a regulating influence on cell proliferation – are also classed as growth factors (Tab. 8.1). The human genome encodes a total of 90 genes for RTKs. In addition to the RTKs, there exists a large family of *non-RTKs* which are integral components of signaling cascades triggered by RTKs and other TM receptors.

Coupling of an extracellular signal to tyrosine phosphorylation in the interior of the cell can take place by two means and involves two different types of receptor (Fig. 8.1):

- *TM receptors with intrinsic tyrosine kinase activity.* Some TM receptors possess intrinsic tyrosine kinase activity. These receptors are known as RTKs. Ligand binding to an extracellular domain of the receptor is coupled to the stimulation of tyrosine kinase activity localized on a cytoplasmic receptor domain. The ligand-binding domain and the tyrosine kinase domain are part of one and the same protein.
- *TM receptors with associated tyrosine kinase activity.* Another type of TM receptor is associated, on the cytoplasmic side, with a tyrosine kinase that is activated when a ligand binds to the extracellular receptor domain (Chapter 11). The tyrosine kinase and the receptor are not located on the same protein in this case.

Biochemistry of Signal Transduction and Regulation. 4th *Edition.* Gerhard Krauss
Copyright © 2008 WILEY-VCH Verlag GmbH & Co. KGaA, Weinheim
ISBN: 978-3-527-31397-6

Tab. 8.1: Selected mammalian growth factors and their receptors.

Growth factor	Characteristics	Receptors
Platelet derived growth factor, PDGF, types AA, AB and BB	Dimers, A (17 kDa)- and B (16 kDa)-chains, B-chain is a product of *c-sis* proto-oncogene	2 types of receptor tyrosine kinases, PDGFRα (170 kDa), PDGFRβ (180 kDa)
Epidermal growth factor, EGF Transforming growth factor-α, TGF-α	ca 6 kDa, EGF and TGF-α are up to 40 % identical	Receptor tyrosine kinase, EGFR is a product of the c-erbB proto-oncogene
Transforming growth factor-β, TGF-β1,-β2, -β3	Homodimers of 25 kDa	TGFβ receptor I und II, contains Ser/Thr-specific protein kinase activity
Insulin-like growth factor, IGF-1 and IGF-2	7 kDa, related to proinsulin	Receptor tyrosine kinase, IGFR
Fibroblast Growth Factor, FGF-1, FGF-2, FGF-3, FGF-4	related proteins of 16-32 kDa	Receptor tyrosine kinase
Granulocyte colony stimulating factor, G-CSF	24 kDa	150 kDa, G-CSFR, receptor with associated tyrosine kinase
Granulocyte/makrophage colony stimulating factor, GM-CSF	14 kDa	51 kDa, GM-CSFR, receptor with associated tyrosine kinase
Interleukins (IL): IL-1– IL-7, IL-9, IL-12, IL-15 IL-10, IL-19, IL-20, IL-22, IL-24, IL-26		IL-1R–IL-7R, IL-9R, IL-10R, IL-12R, IL-15R, IL-20R, IL-22R, Receptors with associated tyrosine kinase
Interleukin 8, IL-8		IL-8R, G-protein-coupled receptor
Erythropoietin	34 kDa	EpoR, receptor with associated tyrosine kinase
Tumor necrosis factor, TNF	26 kDa	TNFR, receptor with associated tyrosine kinase
Interferon α, β, γ		INFRα, INFRβ, INFRγ, receptors with associated tyrosine kinase

(a)

(b)

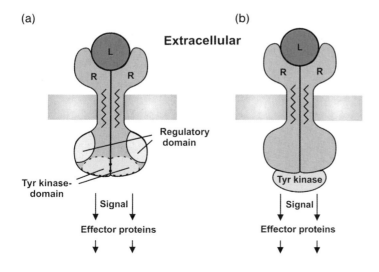

Fig. 8.1: Scheme of signal transmission by receptors with intrinsic and associated tyrosine kinase activity. (a) Tyrosine kinase receptors possess a tyrosine kinase domain in the cytoplasmic region. Binding of a ligand L to the extracellular domain of the receptor produces a signal on the cytoplasmic side by activating the tyrosine kinase. Regulatory sequence segments are also located on the cytosolic side. (b) Receptors with associated tyrosine kinase activity pass the signal on to a tyrosine kinase that is not an intrinsic part of the receptor, but is permanently or transiently associated with the cytoplasmic receptor domain. The receptor shown has been simplified as a dimer.

8.1
Structure and Function of RTKs

RTKs possess binding sites at the surface of the cell membrane that are specific for extracellular ligands. Ligand binding to the receptor activates a tyrosine-specific protein kinase activity of the receptor, located on the cytoplasmic domain. Consequently, tyrosine phosphorylation is initiated at the receptor itself and also on associated substrate proteins; these in turn trigger the biological response of the cell by switching on a further chain of reactions. The response can reach as far as the cell nucleus, where transcription of particular genes is activated. It can also affect the reorganization of the cytoskeleton, cell–cell interactions, and reactions of intermediary metabolism. In particular, the RTKs regulate cell division activity, differentiation and cell morphogenesis by this mechanism (reviewed in Pawson, 2002).

8.1.1
General Structure and Classification

RTKs are integral membrane proteins that have a *ligand-binding domain on the extracellular side* and a *tyrosine kinase domain on the cytosolic side* (Fig. 8.1). A common characteristic is the presence of just one α-helical element in the TM portion. In the activated form, all RTKs exist either as a *dimer* or (sometimes) *higher oligomer*. On the cytoplasmic side, in addition to the conserved tyrosine kinase domain, there are also further regulatory sequence portions at which autophosphorylation, and phosphorylation and dephosphorylation by other protein kinases and by protein phosphatases can take place. Furthermore, other signaling proteins with scaffolding or enzymatic function can associate with the cytoplasmic part.

■ **Domains of RTKs**
Extracellular domain
– Ligand binding
– Glycosylation
TM domain
– Single α-helix
Cytosolic domain
– Kinase activity
– Binding of effectors
– Regulatory
  modifications

The large family of mammalian RTKs can be divided into different subfamilies, which are named according to their naturally occurring ligands. The subfamilies are classified according to the structure of the extracellular ligand-binding domains, in which different sequence portions can be differentiated (Fig. 8.2). In the extracellular domain, for example, there are Cys-rich sequences that occur as multiple repeats and sequences with an immunoglobulin-like structure. Most receptors are found as single polypeptide chains and are monomeric or dimeric in the absence of ligand. One exception is the *insulin receptor* and its family members that comprise two extracellular α-chains and two membrane-spanning β-chains, forming a *heterotetramer* stabilized by various disulfide linkages.

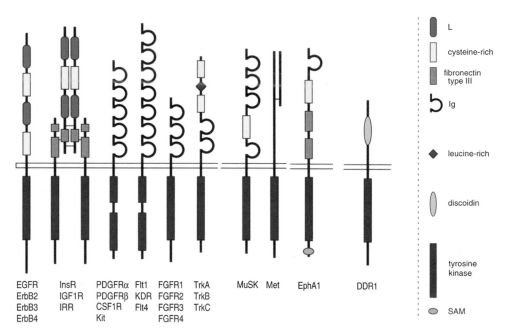

Fig. 8.2: Domain organization for a variety of RTKs. The extracellular portion of the receptors is on top and the cytoplasmic portion is on bottom. The lengths of the receptors is only approximately to scale. Domains: L = homology domain; Ig = immunoglobulin-like domain; SAM = sterile alpha motif; Cadherin = cell–cell adhesion domain. Explanations to some of the receptors: PDGFR = platelet-derived growth factor receptor; EGFR = epidermal growth factor receptor; InsR = insulin receptor; IGF1R = insulin-like growth factor 1 receptor; CSF1R = colony-stimulating factor 1 receptor; KDR, Flt = receptors for the vascular endothelial growth factor; FGFR = fibroblast growth factor receptor; Trk = receptor for neurotrophins, such as nerve growth factor; MuSK = muscle specific RTK; Met = receptor for hepatocyte growth factor; Eph = receptor for ephrin ligands; DDR = discoidin domain receptor.

Fig. 8.3: Ligand-induced autophosphorylation and substrate phosphorylation of RTKs. (a) In the absence of a ligand, the RTK is in an inactive, autoinhibited state. (b) Ligand binding induces kinase activation and autophosphorylation. (c) The p-Tyr residues created serve as attachment points for downstream effectors carrying specific phosphotyrosine binding domains (SH2 in the figure or PTP domains, see Section 8.2). The bound effectors are thus activated for further signaling.

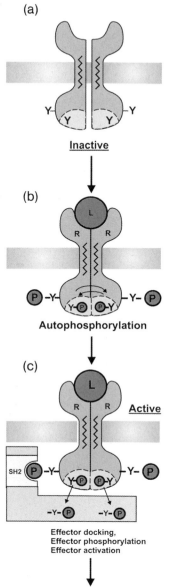

(a)

Inactive

8.1.2
Ligand Binding and Receptor Dimerization

In the absence of an extracellular stimulus, the kinase domain of RTKs exists in an autoinhibited state with no or very low basal activity. On binding an extracellular ligand, autoinhibition is relieved and the tyrosine kinase activity is stimulated. Thereby, two crucial events are triggered at the cytosolic side. First, an *autophosphorylation* of the receptor takes place *in trans*, i.e. between the partner kinase domains of the dimeric receptor (Fig. 8.3). Subsequently, downstream signaling partners *are phosphorylated* on Tyr residues and/or they *associate* with p-Tyr residues of the activated receptor.

There are two scenarios that an extracellular ligand may face upon encountering the receptor (Fig. 8.4):

– The RTK exists as a monomer and ligand binding to the extracellular portions of the receptor induces the noncovalent dimerization of monomeric receptors leading to the formation of *homodimers* or *heterodimers*.

– The receptor exists already as a dimer in solution and ligand binding induces a conformational change in the preassembled dimer that leads to activation of the kinase activity.

Ligand-mediated dimerization and/or conformational change of the receptor dimer is a general method of transmission of signals through the membrane into the cell interior, with the help of RTKs (reviewed in Hubbard and Till, 2000). It is assumed that signal transmission via receptors with associated tyrosine kinases takes place by a similar mechanism. However, the situation is more complicated here because receptors with associated tyrosine kinase activity are often composed of many subunits (Chapter 11).

There are several ways by which binding of the ligand induces dimerization and activation of the receptor.

– *Human growth hormone (hGH) receptor.* Some ligands, e.g. hGH, are monomeric and nevertheless bind two receptor subunits. In this case, the monomeric ligand is divalent with respect to binding of its receptor.

(b)

Autophosphorylation

(c)

Active

SH2

Effector docking,
Effector phosphorylation
Effector activation

■ **Ligand binding induces**
– Dimerization and/or allosteric changes
– Subsequently autophosphorylation, substrate phosphorylation and substrate docking

Fig. 8.4: Mechanism of activation of RTKs by ligand binding. Activation of RTKs is based on a ligand-induced oligomerization and/or conformational change of the receptor. An example is shown of a dimeric receptor; however, activation can also occur in a higher receptor oligomer. (a) A bivalent ligand (monomer or dimer) induces a dimerization of a receptor which exists in a monomeric form without the ligand. (b) A dimeric receptor is activated via an allosteric mechanism by ligand binding. In the absence of the ligand, the two kinase active sites are not close enough for mutual activation by phosphorylation (transactivation).

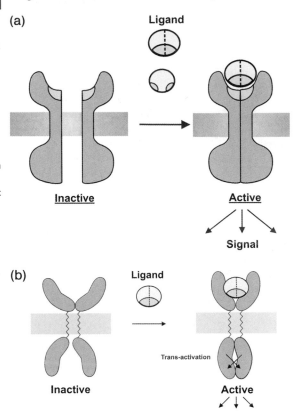

– *Epidermal growth factor (EGF) receptor.* Other monomeric ligands like EGF bind to a single receptor subunit thereby inducing dimerization of the receptor. The dimer interface of the EGF receptor is entirely composed of receptor–receptor contacts and these contacts are made possible by a ligand-induced structural rearrangement of the extracellular domains as illustrated in the model of Fig. 8.5(a). In this system, a 1:1 receptor–ligand complex dimerizes via a specific loop located in a Cys-rich region of the extracellular domain of the receptor (reviewed in Schlessinger, 2002). Ligand binding induces a rotation of extracellular domains, and probably a rotation and straightening of the TM helices too, allowing finally kinase activation and autophosphorylation of the cytoplasmic part (see also below).

– *Fibroblast growth factor (FGF) receptor.* In the case of the FGF receptor whose ligand FGF is monomeric in solution, another mechanism of ligand-induced receptor dimerization has been proposed. According to this model, the FGF receptor uses an

(a)

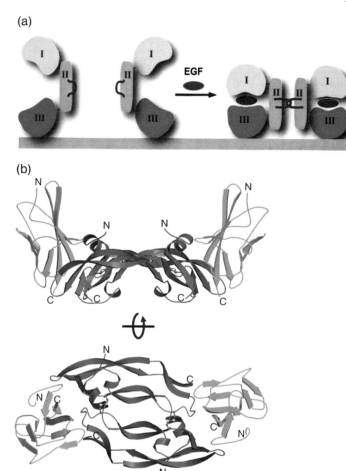

(b)

Fig. 8.5: (a) Ligand-induced dimerization of EGF receptor. Ligand binding induces a rearrangement of three domains (I,II and III) of the extracellular region of EGF receptor leading to the exposure of a dimerization interface. In the monomeric state, the dimerization interface is masked. Note, that the ligand does not participate itself in the dimerization. (b) Dimerization of VEGF and its receptor Flt1. The dimeric ligand VEGF engages two receptor molecules. A ribbon diagram is shown with the two protomers of disulfide-linked VEGF shown in orange and purple, and Ig-like domain 2 of Flt1 shown in green.

additional ligand, i.e. heparin or heparan sulfate proteoglycans, to stabilize the 2:2 complex between the FGF and the FGF receptor. The additional heparin ligand has been shown to bridge two receptor molecules in the dimer and the adjoining FGF ligands, allowing for stable dimerization of the ligand-bound receptor.

– *Vascular endothelial growth factor (VEGF) receptor.* Other ligands of RTKs have a dimeric structure and induce the formation of active receptor dimers on binding to the receptor. Figure 8.5(b) shows the structure of the complex between VEGF and its receptor, FLt1. This structure represents the simplest dimerization scenario – a dimeric ligand engaging two receptor subunits.

Formation of Heterodimers

An aspect of ligand-induced oligomerization of receptors of regulatory importance is the possibility of forming heterodimers. Protein families composed of closely related members can be identified for a number of growth factors and corresponding receptors, and heterologous dimerization is observed within the different members of the receptor family. A certain growth factor can thus bind to and activate different dimeric combinations of the members of a receptor family. Figure 8.6 shows the possibilities for heterodimerization of receptors, using the platelet-derived growth factor (PDGF) receptor as an example.

Heterodimerization of receptor molecules is a mechanism that can increase and modulate the diversity and regulation of signal transduction pathways. Since the various members of a receptor family differ in the exact structure of the autophosphorylation sites and the other regulatory sequences, it is assumed that activity and regulation are different for the various combinations of receptor subtypes. Tissue-specific expression of receptor subtypes enables the organism to process growth factor signals in a differential way.

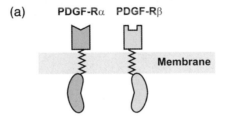

(a)

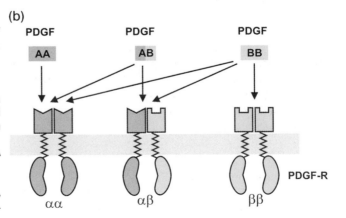

(b)

Fig. 8.6: Heterodimerization of PDGF receptors. (a) There are α and β subtypes of PDGF receptor; these are induced by ligand binding to form homodimers and heterodimers. (b) PDGF is a dimeric growth factor, composed of chains A and/or B. The protein may exist as a homodimer (AA, BB) or heterodimer (AB). The AA homodimer of PDGF binds to the αα dimer of PDGF receptor, AB binds to the αα and αβ types, BB binds all three combinations.

8.1.3
Structure and Activation of the Tyrosine Kinase Domain

In the non-signaling state, most RTKs possess low basal kinase activity that increases substantially upon ligand binding and this activation is accompanied by phosphorylation of tyrosine residues within the kinase domain.

The protein kinase domain of RTKs adopts the classical kinase fold discussed in Chapter 7 (reviewed in Hubbard and Till, 2000; Schlessinger, 2002). Activation of kinase activity is linked to autophosphorylation of the activation loop in all cases, except for some members of the EGF receptor family. It is generally assumed that autophosphorylation takes place by a *trans-mechanism*. Accordingly, two neighboring tyrosine kinase domains in the receptor dimer perform a mutual phosphorylation (Fig. 8.3). The key event for activation is the ligand-induced relief of the autoinhibited state. This includes a reorientation of the kinase domains to allow *activation loop phosphorylation* and/or phosphorylation in the structural parts outside of the kinase domain, i.e. in regions immediately following the TM segment, *the juxtamembrane domain* and the *C-terminal regulatory region*.

Often, autophosphorylation within the kinase domain includes the phosphorylation of several Tyr sites and the different phosphorylation sites appear to serve distinct functions in the autophosphorylation process. This has been shown for the autophosphorylation of the kinase domain of the FGF receptor (FGFR1). Here, autophosphorylation of distinct sites in the kinase domain is proposed to proceed in a sequential and strictly ordered reaction (Furdui et al., 2006).

The mechanisms underlying relief of the autoinhibited state appear to be varied and are only beginning to be uncovered. Several layers of regulation appear to cooperate during relief of autoinhibition. A major level uses the conversion of the monomeric receptor to a ligand-induced dimer where the kinase domains of the subunits are brought in close proximity allowing autophosphorylation in trans. To minimize inappropriate activation of the receptor through intermolecular collisions, a secondary level of regulation may be used as is suggested by studies on the structure of the autoinhibited state of the EGF receptor and its relatives (Zhang et al., 2006).

EGF Receptor Activation
The members of the EGF receptor family (EGFR/ErbB1/HER1, ErbB2/HER2, ErbB3/HER3 and ErbB4/HER4) contain catalytically competent kinase domains and can form heterodimeric pairs with each other. In contrast to most protein kinases, phosphorylation of the EGF receptor activation loop is not critical to its activation. The

■ **Kinase domain of RTKs**
– Classical kinase fold
– Activation requires phosphorylation of Tyr on activation loop by autophosphorylation or by other tyrosine kinases

Autophosphorylation of RTKs
– Occurs in *trans* and produces p-Tyr within the kinase domain and neighboring sequences
– Activates kinase
– Creates docking sites for effectors

kinase activity of EFG receptor is stimulated upon ligand binding in a reaction that is now thought to involve two distinct steps. In the first step, the ligand-induced dimerization, the dimerization interface is kept hidden until unmasked by the ligand EGF. In a second reaction, the neighboring kinase domains activate each other in an allosteric manner. The kinase domains are responsive to the dimerization by converting from an inactive conformation to an active conformation. This conversion is thought to occur in an asymmetric arrangement of the two kinase domains of the receptor dimer (Fig. 8.7) where the C-lobe of one subunit interacts with the N-lobe of the other subunit stabilizing an active conformation in the latter. In the monomeric state, the kinase domains are proposed to exist in an intrinsically autoinhibited conformation with a catalytically unfavorable arrangement of the C-helix. Upon ligand-induced dimerization, the C-lobe of one subunit contacts the N-lobe of the other subunit in way that stabilizes an active conformation of the C-helix on the latter. The reaction is analogous to the activation of cyclin-dependent kinases (CDKs) by cyclins (Chapter 14) where the CDK adopts a similar inactive conformation until binding of the cyclin induces the transition into the active state.

Insulin Receptor Activation

Activation of the insulin receptor stands exemplary for activation of a preformed dimer and including activation loop phosphorylation. The insulin receptor is a heterotetrameric RTK of an $\alpha_2\beta_2$ structure (Fig. 8.2). The α-subunit is completely extracellular and is bound to the β-chain via disulfide bridges. The β-chain has a TM portion, and the tyrosine kinase domain is on the cytosolic side. On binding insulin on the extracellular side, the tyrosine kinase activity of the β-chain is activated and autophosphorylation of a total of seven tyrosine residues takes place in the cytoplasmic domain. Consequently, an effector protein, insulin receptor substrate (IRS, see Section 8.5), binds to phosphotyrosine residues of the activated receptor via its phosphotyrosine-binding domain. IRS now becomes phosphorylated by the insulin receptor at tyrosine residues that act as docking sites for the SH2 domains of other assigned proteins. A major signaling protein downstream of IRS is the phosphatidylinositol-3-kinase (PI3K) that binds to phosphorylated IRS and is thereby activated to conduct the signal further (see also Fig. 3.30a and Section 7.4.1).

A total of seven tyrosine residues become phosphorylated during autophosphorylation. Two of these are located in the vicinity of the TM element, three are in the activation loop of the catalytic domain and a further two in the region of the C-terminus. Targeted mutations in the region of the catalytic center have shown that the tyrosine kinase activity is an essential function in signal transduction via insulin.

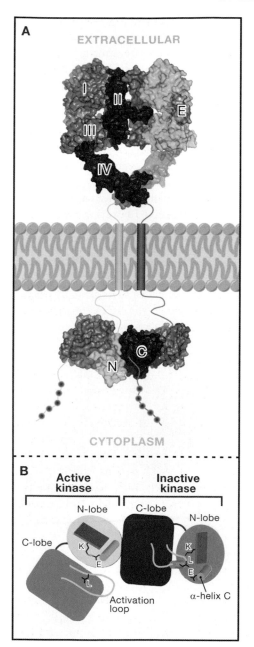

Fig. 8.7: A) On the extracellular side of the plasma membrane (shown approximately to scale), the EGF:EGFR complex (2:2 stoichiometry) is 2-fold symmetric (the 2-fold axis is vertical). Depicted are the two receptors in the complex (cyan and purple) with the four subdomains (I–IV) of the EGFR ectodomain and their bound EGF ligands (orange; E). The transmembrane helices are shown as cylinders, and linker segments including the juxtamembrane regions (extracellular and cytoplasmic) and C-terminal tail as lines. On the cytoplasmic side of the plasma membrane, the two tyrosine kinase domains (with N- and C-lobes) form an asymmetric dimer, with the C-lobe (C) of one kinase domain (purple) interacting with the N-lobe (N) of the other kinase domain (cyan). This interaction activates the second kinase domain (cyan). Tyrosine phosphorylation sites in the C-terminal tail of the cytoplasmic domain are depicted as red spheres. B) The mechanism of EGF receptor kinase activation. Shown are the N-lobes and C-lobes of the EGF receptor kinase domains, the β-sheet in the N-lobe containing Lys721 (orange; K), α-helix C in the N-lobe containing Glu738 (green; E) and the activation loop in the C-lobe containing Leu834 (gray; L). The C-lobe of one kinase domain (purple) activates the second kinase domain (cyan) by interacting with α-helix C, which facilitates formation of the Lys721–Glu738 salt bridge (red dashed line) and proper positioning for catalysis of the activation loop. In the absence of such an interaction (kinase colored purple), the activation loop is stabilized in a Src/CDK-like inactive state in which a short α-helix in the activation loop (containing Leu834) and α-helix C are stabilized in an inactive configuration. The two kinase domains are presumed to reverse roles in a dynamic fashion.

The structures of a fragment of the β-chain of the insulin receptor comprising the kinase domain in its phosphorylated and unphosphorylated form are shown in Fig. 8.8. Three tyrosine residues of the activation loop become phosphorylated during autophosphorylation, inducing a striking conformational change. In the unphosphorylated insulin receptor, Tyr1162 is bound in the active site, obstructing access of the ATP and the protein substrate to the active site. Autophosphorylation of the activation segment brings about a dramatic repositioning of the segment (Fig. 8.8). The steric hindrance to substrate binding (Mg-ATP and peptide substrate) is removed, and the residues involved in substrate binding and catalysis are now properly positioned. The alternative conformation of the activation segment is fixed by ionic interactions involving the phosphotyrosine residues.

How does binding of insulin to the extracellular domain trigger phosphorylation of the activation segment? Structural information on this point is not yet available. It is postulated that an allosteric transition of the heterotetrameric receptor takes place on insulin binding, bringing about a change in the mutual configuration of the β-chains on the cytosolic side. Consequently, the phosphorylation

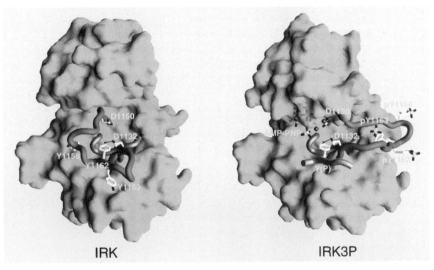

IRK   **IRK3P**

Fig. 8.8: Comparison of the activation loop conformation in unphosphorylated insulin receptor kinase (IRK) and tris-phosphorylated IRK (IRK3P). The figure illustrates the dramatic repositioning of the activation loop upon autophosphorylation. Note, that in IRK, Tyr1162 is positioned in the active site competing with protein substrates. The activation loop is shown in green, the catalytic loop in orange and the peptide substrate in pink. The rest of the protein in each case is represented by a semitransparent molecular surface. Also is shown the ATP analog AMP-PNP, which is partially masked by the N-terminal lobe of IRK3P. Carbon atoms are shown in white, nitrogen atoms in blue, oxygen atoms in red, phosphorous atoms in yellow and magnesium ions in purple. Hydrogen bonds between the substrate tyrosine Y (P) and Asp1132 (IRK3P) are indicated by black lines. From Hubbard and Till (2000).

sites and the active centers of two β-chains are orientated so that a mutual phosphorylation is possible. The inhibitory Tyr1162 is removed from the active center during this process and the substrate-binding sites become accessible. The conformational change permits a *trans*-phosphorylation of both β chains, which includes tyrosine phosphorylation in the activation segment.

8.1.4
Effector Proteins of the RTKs

Subsequent to activation of the kinase domain, rapid phosphorylation at sites outside of the kinase domain takes place. This autophosphorylation serves to create *docking sites for modular domains of downstream signaling proteins* that recognize phosphotyrosine residues in specific sequence contexts (Section 8.2 and Fig. 8.9).

By binding to the phosphotyrosine docking sites, the effector proteins next in sequence are recruited into the signaling process. The effector proteins may carry enzyme activity themselves and be activated by tyrosine phosphorylation. They may also serve as adaptor molecules, functioning to pass the signal on to other components of the signaling pathway. With the help of adaptor molecules, other signal proteins are directed to the activated receptor and to the cell membrane, and are thus incorporated into signal transduction.

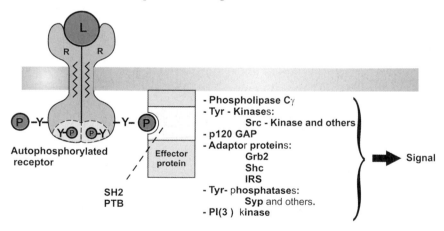

Fig. 8.9: Functions of autophosphorylation of RTKs. Autophosphorylation of RTKs takes place in *trans*, i.e. between neighboring protomers of the receptor. The catalytic domain of the receptor is shown as a green segment. As a consequence of autophosphorylation, the intrinsic tyrosine kinase activity of the receptor is stimulated. Effector proteins can also bind to the activated receptor. Binding takes place with specific phosphotyrosine binding domains (SH2 or PTB domains) at phosphotyrosine residues of the activated receptor. A critical factor for further signal transduction is the membrane association of the effector proteins that enter into binding with the activated receptor. Details of the effector proteins can be found as follows: PLCγ, Section 5.6.2; Src kinase, Section 8.3.2; p120 GAP, Section 9.4; Grb2, Shc, IRS, Section 8.5; PI3K, Section 6.6.1; Syp tyrosine phosphatase, Section 8.4.

What is the purpose of the docking of the effector proteins? In many cases, the effector proteins are substrates for tyrosine phosphorylation catalyzed by the tyrosine kinase activity of the receptor, and this phosphorylation is required for further signal transmission. The colocalization of the substrates and the kinase via the docking sites helps to increase the efficiency and specificity of phosphorylation, and it is often a first step for the assembly of larger signaling complexes in the vicinity of the activated receptor. This is mainly true for effector proteins with adaptor and scaffolding function that are often multivalent and can assemble several signaling proteins. *Autophosphorylation* and *phosphorylation of substrate proteins* are therefore essential elements of signal transduction via RTKs.

Binding of the effector proteins to the autophosphorylated receptor is mediated by domains with binding specificity for phosphotyrosine residues. We know of three such domains: the SH2 domain, the phosphotyrosine-binding domain (PTB) and the C2 domain (Section 8.2). Effector proteins may be enzymes or proteins that only possess adaptor function and link further proteins to the activated receptor via protein–protein interactions. The docking of signaling proteins to autophosphorylation sites provides a mechanism for assembly and recruitment of signaling complexes by RTKs. Accordingly, activated RTKs can be considered as a platform for the recognition and colocalization of a specific complement of signaling proteins. The ability of RTKs to recognize specific targets through regulated protein–protein interactions represents a general feature of specificity in signaling from cell surface receptors.

Activated RTKs transduce signals to a variety of central intracellular signaling pathways. A multitude of effector proteins are recruited into signaling by RTKs, involving central signaling pathways of the cell.

Examples of effector proteins of RTKs:

- p85 subunit of PI3K (Section 7.4.1).
- Phospholipase (PL) Cγ (Section 5.7.2).
- Non-RTKs of the Src family (Section 8.3.2).
- p120^{GAP}, a GTPase-activating enzyme of Ras signal transduction (Chapter 9).
- Adaptor proteins Grb2 and Shc of Ras signal transduction and the scaffolding protein IRS1 (Section 8.5 and Chapter 9).
- Tyr-specific protein phosphatase SHP2 (Section 8.4).

Many of the signaling pathways activated by RTKs ultimately lead to the activation of transcription factors, influencing central differentiation and developmental programs of the cell. Figure 8.10 summarizes the flow of signals from activated RTKs through central signaling pathways.

Coupling of effectors to RTKs

p-Tyr of RTK functions as a docking site for

- SH2 domain
- PTB domains
- C2 domains

Effectors, e.g.

PI3K, PLCγ, Src kinase, adaptors Grb2 and Shc, and phosphatase SHP2

PI3K Pathway

A subgroup of PI3K enzymes containing a p85 subunit can bind via its SH2 domain to the autophosphorylated receptor or to phosphorylated docking proteins assembled at the activated receptor such as IRS1. Activated PI3K generates PtdIns(3,4,5)P$_3$, which mediates the membrane association of a variety of signaling proteins (Section 7.4.1 and Fig. 7.12). One important response is stimulation of cell survival. Furthermore, many of the biological effects of insulin are mediated via activation of the PI3K pathway.

Ras Pathway

One way of activating Ras signaling requires the recruitment of adaptor proteins like Grb2, Gab and Shc to the activated receptor (Section 9.5). A multitude of signals can be delivered via Ras proteins, resulting in activation of the MAPK cascade and the activation of transcription factors.

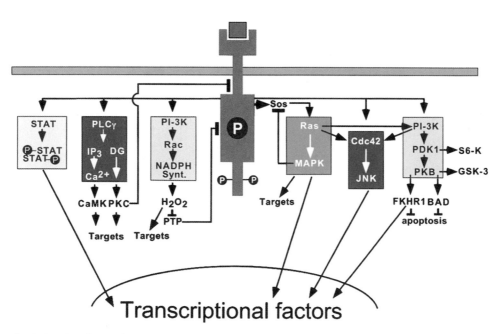

Fig. 8.10: Signaling pathways activated by RTKs. Different signaling pathways are presented as distinct signaling cassettes (colored boxes). Not all known components of a given pathway are included. Examples of stimulatory and inhibitory signals for the different pathways are also shown. The signaling cassettes presented in the figure regulate the activity of multiple cytoplasmic targets. However, the Ras/MAPK, STAT, JNK and PI3K signaling pathways also regulate the activity of transcriptional factors by phosphorylation and other mechanisms. STAT, see Section 12.2.2; PLC, see Chapter 6; PI3K, see Section 6.6.1; Ras, see Chapter 9; MAPK and CDC42/JNK, see Chapter 10; FKHR = Forkhead transcription factor; S6-K = ribosomal protein S6 kinase; GSK-3 = glycogen synthase kinase 3. After Schlessinger (2000).

PLCγ

Activation of RTKs leads to stimulation of phospholipid metabolism and to the generation of multiple second messengers. PLCγ (Sections 5.7.2 and 6.4) binds through its SH2 domain to phosphotyrosine sites on the receptor molecules and is thereby activated. As a consequence of PLCγ activation, Ca^{2+} and diacylglycerol (DAG) signals are produced leading, e.g. to the activation of protein kinase C (PKC) isoenzymes and of calmodulin kinases (CaMKs).

Activation of Effector Proteins

Already at the level of a given receptor, significant *branching* and *variability* of signal transduction is possible. Which effector protein is bound to an activated RTK depends on the nature of the p-Tyr-recognizing domain of the effector protein and on the sequence environment of the phosphotyrosine residue. The RTK typically has several autophosphorylation sites with different neighboring sequences, so that every phosphotyrosine residue of the RTK may serve as the binding site for a different effector molecule. Figure 8.11 illustrates the diversity of effector proteins that can interact with a receptor type, using the PDGF receptor as an example.

The specific interaction between the phosphotyrosine residues of the activated receptor and the SH2 domain of the effector protein is the basis for the activation of effector proteins for further signal transduction. Like the nature of the effector proteins, the mechanism of their activation is also very variable.

The major mechanisms for activation of effector proteins are (reviewed in Schlessinger, 2002):

– Phosphorylation of the effector molecule at Tyr residues, e.g. PLCγ.
– Induction of a conformational change in the effector molecule, e.g. PI3K.
– Translocation of the effector molecule to the plasma membrane, e.g. Grb2-Sos and Shc-Grb2 (Section 9.5).

8.1.5
Attenuation and Termination of RTK Signaling

Signaling by RTKs must be tightly regulated and properly balanced in order to produce the normal cellular responses (reviewed in Pawson, 2002). We know of many examples where aberrant expression or dysfunction of RTKs is responsible for developmental disorders and diseases including tumor formation. The cell uses several mechanisms for the attenuation and termination of RTK signaling induced by stimulatory ligands (Fig. 8.12).

Extracellular domain

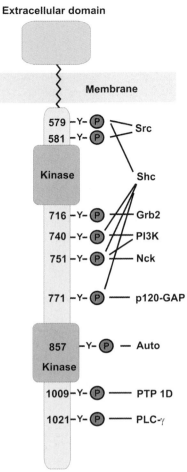

PDGFβ- receptor

Fig. 8.11: Phosphotyrosine residues in the PDGF receptor and specificity of binding of SH2-containing signal proteins. The figure illustrates the diversity of the different effector proteins that can interact with an activated receptor. The tyrosine residues of PDGF receptor, for which autophosphorylation has been demonstrated, are designated according to their position in the receptor sequence. PDGF receptor has at least nine different tyrosine phosphorylation sites in the cytoplasmic domain. The phosphotyrosine residues are found in different sequence environments and are recognized by the SH2 domains of the assigned effector proteins. The filled rectangles indicate the two-part tyrosine kinase domain of PDGF receptor. Src = members of the Src tyrosine kinase family; Sh2, Grb2, Nck = adaptor proteins; GAP = GTPase-activating protein.

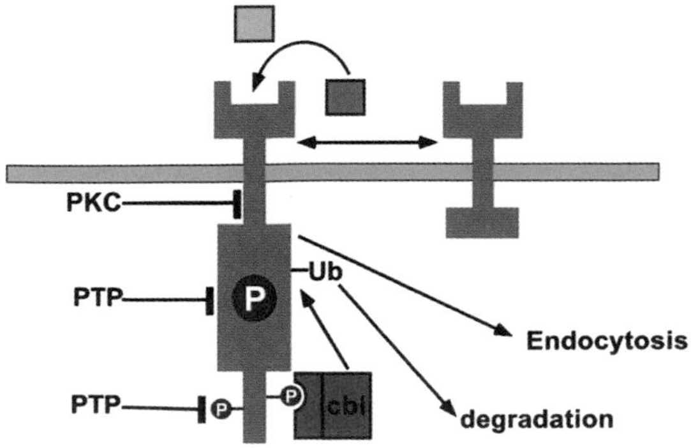

Fig. 8.12: Mechanisms for attenuation and termination of signaling by RTKs, RTKs. In several cases the activity of RTKs can be negatively regulated by ligand antagonists or by heterooligomerization with naturally occurring dominant interfering receptor variants. The PTK activity of EGFR is attenuated by PKC-induced phosphorylation at the juxtamembrane region. Dephosphorylation of key regulatory p-Tyr residues by PTPs may inhibit kinase activity or eliminate docking sites. An important mechanism for signal termination is via receptor endocytosis and degradation mediated by ubiquitin ligases, e.g. the Cbl protein (see also Fig. 2.14).

Antagonistic Ligands and Heterodimerization

In several systems, natural antagonistic ligands have been identified that bind on the extracellular side to the receptor and inhibit receptor activation. Furthermore, certain tissues express naturally occurring receptor variants that are deficient in tyrosine kinase activity. Expression of these mutants may lead to dominant negative inhibition of full-length receptors through formation of inactive heterodimers or heterooligomers.

Inhibition by Ser/Thr Phosphorylation

PKC-mediated phosphorylation of the EGF receptor on Ser/Thr residues located on the cytoplasmic domain results in inhibition of its tyrosine kinase activity and in inhibition of EGF binding to the extracellular domain. This phosphorylation appears to provide a negative feedback mechanism for control of receptor activity.

Inhibition by Inhibitor Proteins

■ **Negative regulation of RTKs is mediated by**
- Tyr phosphatases
- Ser/Thr phosphorylation
- Antagonistic ligands
- Inhibitory proteins: SOCS and Sprouty
- Ubiquitylation and degradation
- Endocytosis

In another negative control mechanism, inhibitory proteins like SOCS (suppressor of cytokine signaling) bind via their SH2 domains to phosphotyrosine residues in the tyrosine kinase domain of the receptor and directly inhibit the kinase activity. This type of protein regulates mainly the insulin receptor and cytokine receptors. Another class of antagonistic proteins includes the family of Sprouty proteins (reviewed in Kim and Bar-Sagi, 2004).

Inhibition by Protein Tyrosine Phosphatases (PTPs)

The activity of RTKs is continuously subject to control by PTPs (Section 8.4) that can dephosphorylate and thereby deactivate RTKs that have been autophosphorylated because of binding of a stimulatory ligand. Some subfamilies of PTPs carry SH2 domains that can mediate docking of the enzyme to p-Tyr residues of activated receptors.

Endocytosis and Degradation

Binding of extracellular ligands to RTKs often results in rapid endocytosis and degradation of both the receptor and the ligand, attenuating the signal generated at the cell surface. The oncogenic adaptor protein *Cbl* has been shown to play a role in the degradation of the EGF receptor and PDGF receptor (reviewed in Bache et al., 2004; Rubin et al., 2005). Cbl functions as an E3 ubiquitin ligase, and mediates the ubiquitination and subsequent endosomal internalization and proteasomal degradation of the receptor. It contains a SH2 domain mediating binding to phosphotyrosine residues of the activated receptor and a RING finger domain (Section 2.5.6.2) that brings a ubiquitin conjugating E2 enzyme to the vicinity of the receptor.

Receptor ubiquitination results in accelerated removal from the cell surface and entry into endosomal pathways leading finally to receptor degradation, thereby terminating RTK signaling.

8.2
Protein Modules in Downstream Signaling of RTKs

Starting from an activated RTK, further conduction of the signal takes place with the help of specific protein–protein interactions between the activated receptor and the effector proteins next in the sequence leading to activation of the effector protein for further signal transduction. The coupling of signal proteins in signaling chains is largely based on specific protein–protein interactions mediated by complementary docking sites on reaction partners (reviewed in Pawson, 2004). These interactions function mainly to bring about close spatial configuration of signal-carrying enzymes with their substrates, by leading a substrate protein to the catalytic center or by targeting an enzyme usually located in the cytosol to the cell membrane, where it has direct access to its substrates. In case of the activated RTKs, the phosphotyrosine residues are recognized by complementary binding modules found on all members of the large spectrum of different downstream effectors.

As outlined already in more detail in Section 1.6, the cell uses defined structural elements for communication between different proteins of a signal transduction pathway and for targeting signaling proteins to membranes; these are found in the form of self-folding protein domains in many signal-transmitting proteins. The protein modules mediate protein–protein interactions and are used to associate proteins of a signaling pathway into larger signaling complexes. Another major function of the protein modules is to promote protein–membrane interactions in signaling pathways.

Two points are of particular importance for the coupling function of the protein modules (see also Section 1.7). The first is the variability of the protein modules. For a particular basic motif of a module, there are generally many *variants* that have slightly different binding specificities and are thus assigned to different structural motifs in the target protein. This results in great variability and diversity of coupling. Another functionally important aspect of coupling of signal proteins is the occurrence of several protein modules in a protein (Fig. 8.13). If a signal protein has multiple binding valence for distinct effector proteins, networks of interacting proteins can be created that contribute greatly to specificity and diversity of signal transduction and permit linking of different signaling pathways.

At present, a number of structural motifs have been described for signal proteins to which specific coupling functions in signal transduction are attributed. The functions and targets of the most important signaling modules are summarized in Fig. 8.14.

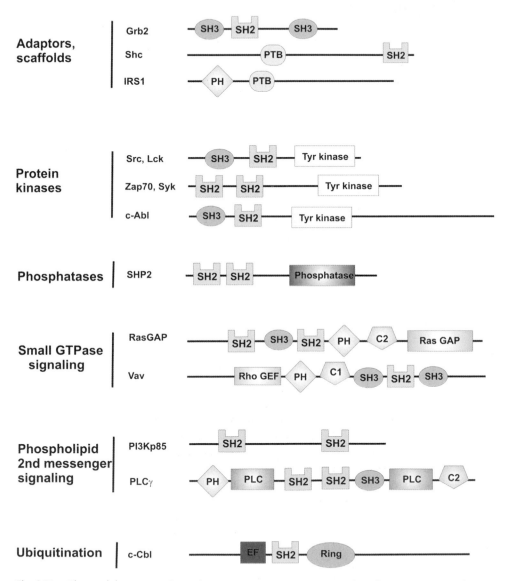

Fig. 8.13: The modular nature of signaling proteins. Representative members from various SH2 domain families and the positional organization of these domains are illustrated. For explanation of protein modules see text. C1, C2, cysteine-rich domains; PI3Kp85 = p85 subunit of PI3K.

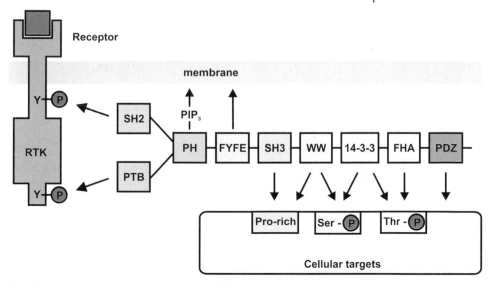

Fig. 8.14: Targets of important signaling modules.

8.2.1
Domains with Binding Specificity for Phosphotyrosine:
SH2, PTB and C2 Domains

We know of three binding modules, the SH2, PTB and C2 domains, that recognize phosphotyrosine residues in a sequence context-dependent manner and serve to target signaling proteins to phosphorylated signaling partners (reviewed in Machida and Meyer, 2005).

The presence of a Tyr phosphate grouping is obligatory for binding of these domains to a target protein or to a model peptide. In addition, the neighboring sequence is crucial. The sequence environment of the phosphotyrosine residue defines the binding substrate of the SH2, PTB and C2 domains, and also differentiates the binding preference of their subclasses. Specificity is mostly determined by sequences of 1–6 amino acids located C- and/or N-terminal to the phosphotyrosine residue. The structures of the three binding domains bound to phosphotyrosine-containing substrates are shown in Fig. 8.15. Often multiple phosphotyrosine residues with different sequence contexts are found in signaling proteins. An example is the β-subtype of the PDGF receptor on which several tyrosine phosphorylation sites have been identified allowing the docking of at least eight different effector molecules (Fig. 8.11). The same receptor can be involved in very different signaling pathways, as shown by this example. Which pathway is used will depend on the availability and activity of the different effector proteins, a situation regulated in a cell- and tissue-specific manner.

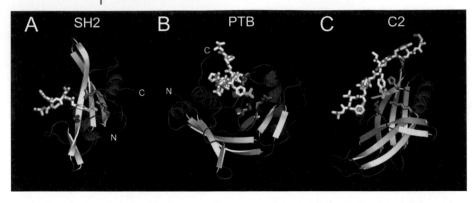

Fig. 8.15: p-Tyr binding by SH2, PTB and C2 domains. Shown are p-Tyr (green)-containing phosphopeptides (white) in complex with the SH2 domain from v-Src, the PTB domain from the adaptor protein Shc (Section 8.5) and the C2 domain from PKCδ. Residues involved in recognition of p-Tyr are in blue.

SH2 Domains

The SH2 domains were first discovered as a sequence motif showing homology with a sequence of the Src tyrosine kinase (Section 8.3); hence the name SH, from Src homolog.

The specificity of SH2 domains is determined by sequences of 1–6 amino acids located C-terminal to the phosphotyrosine residue. The great variability of SH2 domains and their substrates is emphasized by the observation that two SH2 domains occur in many signal proteins. These mostly have different substrate-binding preferences.

■ SH2 domains
– Recognize p-Tyr within N-terminal sequence context
– Many variants known
– Can engage in intramolecular and intermolecular binding of p-Tyr

Transmit signals by
– Allosteric activation
– Membrane localization
– Inducing Tyr phosphorylation

SH2 domains can be divided into at least five classes (1A, 1B, 2, 3 and 4), differing in the sequence requirements of the substrate. Crystallographic analysis of SH2–phosphopeptide complexes has shown that the phosphate residue is bound in a deep pocket of the SH2 domain, at the end of which an invariant Arg residue is located which contacts the negatively charged phosphate by ionic interaction. In addition, amino acids located N-terminal to the phosphate are contacted specifically by the binding pocket.

Phosphotyrosine residues as targets of SH2 domains are found in a variety of signaling proteins. As well as the RTKs and the cytokine receptors, the T cell receptors (TCRs) and the non-RTKs utilize phosphotyrosine–SH2 interactions for signal transmission.

Activation of signal proteins by phosphotyrosine–SH2 interactions can be achieved in different ways, which are discussed below.

– *Activation by membrane localization*. Via binding of an SH2-containing signal protein to an activated RTK, the signal protein is brought to the membrane and into the vicinity of the corresponding target protein or substrate. Examples are PLCγ and PI3K, which have substrates in the phospholipid membrane.

– *Activation by tyrosine phosphorylation.* Many SH2-containing signal proteins are brought, via interaction of their SH2 group with phosphotyrosine residues, into the neighborhood of the catalytic center of the tyrosine kinase and are themselves substrates for tyrosine phosphorylation. By this mechanism, new attachment sites can be generated for other SH2-containing proteins within SH2-containing signal proteins. In this way, several components of a signaling pathway can be sequentially linked.

– *Activation by a conformational change.* Several cases are described in which binding of an SH2-containing enzyme to an activated RTK leads to increased catalytic activity of the enzyme. One example is the PI3K. Binding of the regulatory subunit p85 to a tyrosine-phosphorylated PDGF receptor causes conformational changes in p85 that are transmitted to the catalytic p110 subunit and stimulate the PI3K activity. Intramolecular binding of phosphotyrosine residues by SH2 domains is another mechanism for regulating the activity of signaling enzymes. A prominent example is Src kinase that is allosterically controlled by SH2–phosphotyrosine interactions (Section 8.3.2). Another example is PLCγ that is activated by RTKs such as PDGR via phosphorylation on tyrosine residues. This activation is based on the intramolecular binding of the p-Tyr residue to one of the SH2 groups of PLCγ which relieves the enzyme from an autoinhibited state (Poulin et al., 2005).

PTB Domains

PTB domains are found almost exclusively in proteins that act at membranes and have a docking or adaptor function by recruiting various signaling proteins to the vicinity of an activated receptor. A clear functional dichotomy exists within the family of PTB domains that is based on the requirement of a phosphotyrosine in the peptide ligand for binding. One group of PTB domains is phosphotyrosine-dependent while the other group does not require phosphotyrosine for binding to the target protein (reviewed in Uhlik et al., 2004; DiNitto and Lambright, 2006).

Well-studied examples of phosphotyrosine-dependent PTBs are found on the adaptor proteins Shc and IRS1 (Section 8.5). Both adaptors bind via their PTB domain to phosphotyrosine residues of the activated insulin receptor recognizing phosphotyrosine residues in context with sequence sections toward the N terminus. Importantly, most p-Tyr-dependent PTB domains also show high-affinity binding to phosphoinositides. The functional relevance of this dual binding specificity is unclear.

We also know of phosphotyrosine-independent PTB domains. These domains are found on adaptor proteins that bind to physiologically important receptors, e.g. the low-density lipoprotein (LDL)

■ **PTP domains**
Two subgroups
– p-Tyr dependent
– p-Tyr independent
p-Tyr dependent PTPs
– Recognize p-Tyr within C-terminal sequence context
– Occur mostly on adaptor proteins
– May also bind phosphoinositides

receptor and on adaptors associated with the Alzheimer precursor protein(APP).

C2 Domains

The C2 domains are long known modules of around 120 residues that bind phospholipids in a Ca^{2+}-dependent manner. A subtype of the C2 domains discovered first for PKCδ, however, mediates protein–protein interactions by binding to phosphotyrosine residues (Section 7.5).

8.2.2
SH3 Domains

SH3 domains occur in many signal proteins that are involved in tyrosine kinase signaling pathways (reviewed in Macias et al., 2002; Zarrinpar et al., 2003). They are also found in proteins of the cytoskeleton and in a subunit of the neutrophilic cytochrome oxidase. Ligand binding at SH3 domains takes place via *Pro-rich sequences* of around 10 amino acids. The sequence X-P-p-X-P is a consensus sequence for SH3 ligands, in which the two proline residues P are invariant; X is usually an aliphatic residue and p is often a Pro residue. The structural determination of SH3 domains with bound Pro-rich peptides has shown that the Pro-rich section of the ligand is bound as a left-handed polyproline type II helix with three amino acid residues per turn. The polyproline type II helix was described for polyproline some time ago.

Like the SH2 domains, there are many different SH3 domains. The different SH3 domains demonstrate differing binding preferences for Pro-rich sequences, the specificity being determined by the neighboring residues of the invariant proline.

A general function of the SH3 domains is the *binding of Pro-rich sequences* in target proteins of a signaling pathway. The biological importance of the SH3 domains is emphasized by the observation that deletion of the SH3 domains of the cytoplasmic tyrosine kinases Abelson (Abl) and Src leads to a significant increase in the tumor-transforming potential of both tyrosine kinases. The following principal functions can be attributed to SH3 domains:

- *Mediation of specific subcellular localization*. SH3 domains are found in many proteins associated with the cytoskeleton or with the plasma membrane. Examples are the actin-binding protein α-spectrin and myosin Ib.
- *Regulation of enzyme activity*. An example of regulation of enzyme activity via SH3 domains is the negative regulation of the activity of Src tyrosine kinase by SH3-mediated interactions (Section 8.3).
- *Contribution to substrate selection of tyrosine kinases*. One means of increasing the selectivity of tyrosine kinases seems to be to use SH3

SH3 domains
- Bind Pro-rich sequences
- Involved in formation of larger signaling complexes
- Are often found on proteins of the cytoskeleton
- Mediate subcellular localization
- Regulate enzyme activity
- Guide substrates to RTKs

domains for specific coupling of substrates to tyrosine kinases (Fig. 8.14). The adaptor proteins Crk, Grb2 and Nck are specifically phosphorylated by the Abl tyrosine kinase at Tyr residues. All three adaptor proteins possess SH3 domains that can bind to Pro-rich sequences of the Abl tyrosine kinase. This interaction mediates tight binding of the substrate to the tyrosine kinase and enables an effective tyrosine phosphorylation.

8.2.3
Membrane-targeting Domains:
Pleckstrin Homology (PH) Domains and FYVE Domains

PH Domains
PH domains comprise a large family of more than 200 domains and are found in many signal molecules such as Ser/Thr-specific protein kinases, tyrosine kinases, isoforms of PLC (β, γ and δ), G-nucleotide exchange factors (GEFs), adaptor proteins and proteins of the cytoskeleton (see also Fig. 8.13). Originally, the PH domain was discovered in the 47-kDa pleckstrin protein, which is the main substrate of PKC in platelets.

PH domains are identified by common structural features and these features can harbor many different functions. Some members of the PH domain family show specific binding to phosphatidylinositol derivatives and, because of this property, are able to mediate membrane association of signal proteins. However, most of the PH domains do not recognize specific inositol phospholipids and their functional role is open.

A general membrane-anchoring function is assigned to the phosphoinositide-specific PH domains (reviewed in DiNitto and Lambright, 2006) with phospholipids of the membrane serving as binding substrates (Fig. 7.11). Signal-induced availability of phosphatidylinositol lipids such as PtdInsP$_3$ thus permits regulated membrane-anchoring of PH-containing signal proteins (Section 6.6). There is great variation in the binding specificity of PH domains. While certain subtypes of PH domains bind specifically to PtdIns(4,5)P$_2$, another subset of PH domains binds preferentially to the products of the PI3K reaction.

FYVE Domains
The FYVE domain is a Cys-rich domain which binds two zinc ions and has binding specificity for PtdIns(3)P, one of the membrane-localized products of the PI3K reaction (Chapter 6). Proteins containing FYVE domains have been implicated, e.g. in membrane trafficking and in signaling by Smad proteins (reviewed in Gillooly et al., 2001). It is assumed that binding of FYVE domain-containing proteins to PtdIns(3)P mediates their membrane association.

■ **PH domains**
– Many subgroups with varying binding specificity
– Some subgroups bind phosphoinositides and mediate membrane association

■ **FYVE domains bind PtdIns(3)P and mediate membrane association**

8.2.4
Phosphoserine/Threonine-binding Domains

Phosphorylation of proteins on Ser/Thr residues is one of the most common regulatory modifications of signaling proteins (Chapters 2 and 7). Serine/threonine phosphorylation results in the formation of multiprotein signaling complexes through specific interactions between phosphorylated sequence motifs and the following phosphoserine/threonine-binding domains.

14-3-3 Proteins

■ **p-Ser binding domains**
– 14-3-3 proteins
– WW domains
– FH domains
– Polo box domain
– BRCT domain

14-3-3 proteins (reviewed in Aitken, 2002) are a family of regulatory proteins which recognize phosphoserine/threonine residues in specific sequence contexts. These proteins are involved in the control of critical cellular functions such as cell cycle control, apoptosis, gene transcription, DNA replication and chromatin remodeling. Substrates of 14-3-3 proteins include CDC25 phosphatase (Section 13.6), Raf kinase (Section 9.6), the proapoptotic protein Bad (Section 15.4) and histone deacetylase enzymes.

WW Domains

WW domains are signaling modules of around 40 amino acids that bind short Pro-rich sequences such as PPLP or PPR motifs (reviewed in Macias et al., 2002). A subset of WW domains, e.g. those found in the *proline cis–trans isomerase Pin1,* however, specifically binds to phosphoserine-Pro and phosphothreonine-Pro motifs. The Pin1 protein has an essential role in mitosis. It is thought that Pin1 binding to phosphorylated mitotic proteins, mediated by its WW domain, facilitates proline *cis–trans* isomerizations and subsequent conformational changes of the target protein. For the Pin1 substrate CDC25 phosphatase it has been shown that proline isomerization facilitates the subsequent dephosphorylation of phosphorylated CDC25 protein by the protein phosphatase PP2A (see also Section 13.6).

Forkhead-associated (FHA) Domains

FHA domains were originally identified as conserved sequence elements within a subset of Forkhead transcription factors and were subsequently found in other transcription factors, in protein kinases, protein phosphatases and kinesin motors (reviewed in Durocher and Jackson, 2002). FHA domains comprise up to 140 amino acids, exhibit binding specificity toward *phosphothreonine* residues and efficiently discriminate against phosphoserine residues.

Polo-box Domains (PBDs)

PBDs are found in specialized protein kinases, the *polo kinases*, that perform crucial functions in cell cycle progression and multiple stages of mitosis. Phosphoserine residues located on polo kinase substrates are recognized by PBDs, and this binding relieves autoinhibition of the kinase (Section 7.2.3) and induces further substrate phosphorylation.

BRCT Domains

The C-terminal region of the breast-cancer-associated protein BRCA1 (Section 14.8) contains a pair of tandem BRCA1 C-terminal (BRCT) repeats that are essential for BRCA1 function. Similar repeats have been identified in many other proteins. The BRCT repeats (reviewed in Glover et al., 2004) constitute a module for recognizing phosphoserine peptides and this property is thought to be used for mediating phosphorylation-dependent protein–protein interactions in central processes such as DNA repair functions.

8.2.5
PDZ Domains

PDZ domains were first identified in proteins of postsynaptic cells, and their designation comes from their occurrence in the proteins PSD-95, DlgA and ZO-1. Today, we know that PDZ domains are among the most abundant protein domains in multicellular eukaryotic organisms. PDZ domains are found particularly in proteins that form structures at the cell membrane (e.g. in ion channels) and in signal proteins (reviewed in Zhang and Wang, 2003). The target sequences comprise short peptides with a C-terminal hydrophobic residue and a free carboxyl group, such as the *E(S/T)DV motif* at the C-terminus of certain subunits of ion channels.

An important function of the PDZ domains lies in the formation of macromolecular associates at the cell membrane (reviewed in van Ham and Hendricks, 2003). PDZ proteins can also provide a framework for clustering of ion channels within specific structures at postsynaptic membranes, known as *postsynaptic density (PSD)*. Major organizers of the PSD appear to be PDZ-containing proteins, e.g. PSD95, with distinct specificity for binding of downstream signaling proteins.

■ **PDZ domains**
– Bind to the C-terminus of ion channels and other signaling proteins and mediate association of large protein complexes.

Many proteins contain multiple PDZ domains with various sequences that may show different binding specificities. In this way, a protein with multiple PDZ domains can help to organize different proteins in supramolecular complexes. An example is the *InaD protein* of *Drosophila*, which is composed exclusively of five PDZ domains with different binding specificities and to which different

Fig. 8.16: A model of the organization of the InaD signaling complex. InaD is composed of five PDZ domains (PDZ1–5) which interact specifically with signaling proteins implicated in the vision process in *Drosophila*. TRP is a K$^+$ channel. PLC is a β-type PLC, which is the target of rhodopsin-activated G$_{q,\alpha}$. ePKC is an eye-specific PKC which inactivates the Trp channel by phosphorylation. After Huber (2001).

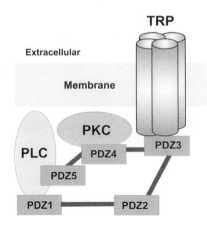

target proteins are assigned (Fig. 8.16). The target proteins are three proteins involved in the processing of light signals in the eye of *Drosophila*. During phototransduction, InaD associates via its distinct PDZ domains with PLCβ, which is the target of rhodopsin-activated G$_{q,\alpha}$, with the calcium channel TRP and with PKC. The signaling complex formed allows efficient activation of the TRP channel by PLC in response to stimulation of rhodopsin. Activated PLC produces a PtdInsP$_3$ and a DAG signal, and thereby induces opening of the TRP channel by a mechanism to be characterized. Furthermore, the presence of PKC in the signaling complex provides for efficient deactivation of the TRP channel by phosphorylation. It is assumed that the InaD protein functions as an adaptor or scaffolding protein, which organizes light-induced signaling events into supramolecular complexes.

8.3
Non-RTK-specific Protein Kinases

In addition to RTKs, the cell also contains a number of tyrosine-specific protein kinases that are not an integral component of TM receptors. These "nonreceptor" tyrosine kinases are localized in the cytoplasm at least occasionally or they are associated with TM receptors on the cytoplasmic side of the cell membrane. They are therefore also known as *cytoplasmic tyrosine kinases*. The non-RTKs perform essential functions in signal transduction via cytokine receptors (Chapter 11) and TCRs, and in other signaling pathways.

8.3.1
Structure and General Function of Non-RTKs

Permanent or transient association with subcellular structures, and variable subcellular distribution, are characteristic of the non-RTKs. These are intracellular effector molecules that can associate with specific substrates during the process of signal transduction and activate these by tyrosine phosphorylation to pass on the signal. Many of the functions of the non-RTKs are performed in the immediate vicinity of the cell membrane, whether a signal is received from an activated membrane receptor or is passed on to a membrane-associated protein. To facilitate membrane association, many non-RTKs contain N-terminal lipid anchors.

In Fig. 8.17, the structures of the major subfamilies of the non-RTKs are shown schematically. In addition to the catalytic domain, the non-RTKs often have SH2 and SH3 domains responsible for specific association with substrate proteins. Other sequence motifs mediate the association of the non-RTKs with specific subcellular sites and structures.

Two of the non-RTKs are highlighted here: Src kinase and Abl kinase.

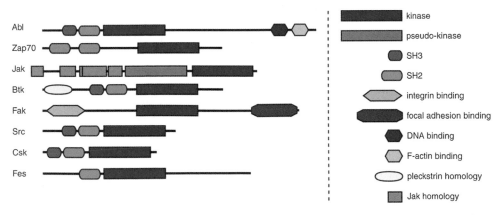

Fig. 8.17: Domain organization for the major subfamilies of non-RTKs. The N-terminus is on the left and the C-terminus on the right. The lengths are only approximately to scale. Details of the non-RTKs: Csk, Src and Abl, Section 8.3.2; Zap 70, Section 11.3.2; Jak, Section 11.2.2; Fak, Section 11.4. Btk = Bruton's kinase.

8.3.2
Src Tyrosine Kinase

■ **Src kinase structure**

– Classical kinase domain

– N-terminal myristinic
acid

– SH3 domain

– SH2 domain

■ **Inactive state of Src
kinase**

– Intramolecular binding
of p-Tyr527 to SH2
domain

– SH2 and SH3 domains
clamp on the kinase
domain, displace
C-helix of the kinase
domain stabilizing
inactive conformation

– Activation loop not
phosphorylated

■ **Activation of Src occurs
via**

– Binding of competing
ligands to SH2 and
SH3 domains induces
unclamping and
transition into active
conformation

– Activation loop
phosphorylation by
Csk kinase

Src kinase belongs to a family of closely related tyrosine-specific protein kinases involved in the regulation of cell division, cell differentiation, and cell aggregation. At least nine different protein kinases are numbered amongst the family of Src kinases. Src kinase is involved, for example, in ion channel regulation and in signal transduction via growth factor receptors, integrins and immunoreceptors. Functional interactions have been described with p-Tyr residues of PDGF receptor, EGF receptor, focal adhesion kinase (FAK) (Chapter 11) and scaffolding proteins of the *N*-methyl-D-aspartate (NMDA) receptor complex (reviewed in Boggon and Eck, 2004).

The domain structure of c-Src kinase is shown in Fig. 8.18(a). Src kinase carries a myristinic acid residue as a membrane anchor and harbors an SH2 and an SH3 domain N-terminal to its kinase domain. Furthermore, Src kinase possesses two important regulatory phosphorylation sites, i.e. Tyr416 in the activation loop and Tyr527 near the C-terminus.

In the unstimulated, basal state, Src kinase exists as an inactive, autoinhibited enzyme phosphorylated on Tyr527. The three-dimensional structure of autoinhibited Src kinase shows that the enzyme is maintained in an inactive state by intramolecular interactions involving the SH2 and SH3 domains and p-Tyr527 (Fig. 8.18a) (reviewed in Roskoski, 2004). As already predicted from mutation experiments, p-Tyr527 enters into intramolecular binding with the SH2 domain. Furthermore, the SH3 domain binds to the linker between the catalytic domain and the SH2 domain. The linker contains proline residues and adopts a polyproline II helix in the intramolecular complex, providing a binding motif for the SH3 domain. The protein kinase responsible for the repressive Tyr527 phosphorylation is Csk kinase, a non-RTK that can shuttle between a cytoplasmic and membrane-bound state.

The intramolecular interactions clamp the SH2 domain and the SH3 domain to the backside of the catalytic domain (Fig. 8.18b). As a result of this clamping, the C-helix of the catalytic domain is displaced and misaligned when compared to active structures of other active protein kinases. A catalytically unfavorable position of the catalytic residues in the kinase domain is stabilized by the clamp and ATP binding is prevented. Furthermore, Tyr416 of the activation loop is sequestered and is not available for phosphorylation.

Activation of Src requires unlocking of the clamp that fixes the catalytic domain in an inactive conformation. Unfortunately, structural data on the active state of Src are not yet available. Current models on Src activation assume that phosphotyrosine residues and/or poly-

(a)   Src - Kinase

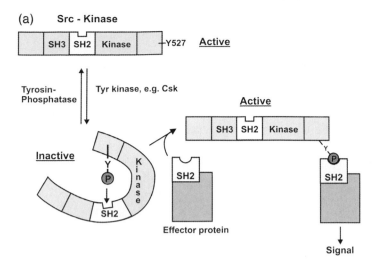

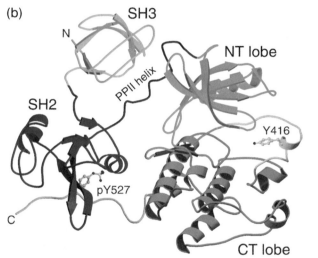

Fig. 8.18: (a) Model of regulation of Src kinase by phosphorylation. A Tyr phosphorylation site (Tyr527) is located a t the C-terminal end of the Src tyrosine kinase, which, when phosphorylated by a tyrosine kinase leads to inactivation of the Src kinase. p-Tyr527 binds intramolecularly to the SH2 domain, blocking the kinase activity. Removal of the Tyr phosphate p-Tyr527 by a tyrosine phosphatase converts the Src kinase into the active state again. Activation of the Src kinase can also be brought about by a SH2-containing effector protein; the SH2 domain of this effector protein competes with the SH2 domain of Src for binding to p-Tyr527. Alternatively, p-Tyr527 may bind to the SH2 domain of another signal protein (not shown in the figure).
(b) Structure of c-Src kinase phosphorylated at Tyr527. Ribbon diagram showing the structure and organization of the "closed conformation" of c-Src kinase. Two aspects of the structure are important for the regulation of c-Src kinase: (i) The phosphorylated Tyr 527 of the C-terminal tail is engaged in an intramolecular interaction with the SH2 domain. (ii) The SH3 domain binds to the linker between the SH2 domain and the kinase domain. Both interactions fix an inactive state of the kinase. The N- and C-terminal kinase lobes are shown in light and dark green, respectively. The activation loop in the kinase domain, containing Tyr416, is shown in gray. The SH2 and SH3 domains are shown in dark blue and cyan, respectively. The SH2 kinase linker, which contains a short stretch of polyproline type II helix (PPII helix), is shown in red. The C-terminal tail, which contains p-Tyr527, is shown in orange.

proline II helices on substrate proteins compete with the intramolecular interaction of the Sh2 and SH3 domains inducing unlocking of the clamp (Fig. 8.18a). Through binding of high-affinity ligands, the SH2 and SH3 domains can be displaced from the p-Tyr527 and the linker region, respectively. For example, activation of Src kinase is achieved through SH2 binding to the autophosphorylated PDGF receptor or through SH3 domain binding of the HIV protein Nef.

Following unlocking of the clamp, the C-helix and the activation loop can switch to their active conformations. Tyr416 on the activation loop is now available for phosphorylation which is thought to occur in *trans* by another Src kinase molecule.

Following autophosphorylation, the enzyme is stabilized in its active state. Furthermore, dephosphorylation of p-Tyr527 by PTPs is now possible and will help to fix the active state.

The structural design of Src kinase allows for a regulation at multiple levels. The internal interactions maintain an inactive state and external ligands promote an active state by binding to the SH2 and/or SH3 domain. Other regulatory inputs are provided by the action of tyrosine kinases, e.g. Csk kinase, and by PTPs.

The importance of strict regulation of Src is underscored by the occurrence of oncogenic Src variants. One of the first viral oncogene products of retroviruses to be discovered was v-Src kinase, a variant of Src kinase that is a product of the Rous sarcoma virus. Owing to a C-terminal truncation, v-Src kinase lacks the regulatory site Tyr 527 and is constitutively active, resulting in uncontrolled growth of infected cells.

8.3.3
Abl Tyrosine Kinase

■ **Abl tyrosine kinase domain structure**
- SH2 domain
- SH3 domain
- N-terminal myristinic acid
- Nuclear localization signals
- DNA-binding domains

Much of the interest in Abl tyrosine kinase (reviewed in Hantschel and Superti-Furga, 2004) stems from its involvement in oncogenesis in rodents and in humans. Like many other non-RTKs, Abl tyrosine kinase may be converted by mutations into a dominant oncoprotein and may thus contribute to tumor formation. The wild-type form of the Abelson kinase is termed c-Abl; the viral, oncogenic form is termed v-Abl. This mutated enzyme was first discovered as the oncogene of murine Abelson leukemia virus. Apart from the v-Abl enzymes, other oncogenic forms of the Abelson kinase exist. Chronic myelogenic leukemia in humans is caused by a chromosome translocation in which a fusion protein is created from c-Abl and a Bcr protein (Chapter 14). The result is a greatly increased tyrosine kinase activity with very different regulatory properties, to which a causal role in the occurrence of this leukemia is attributed. Furthermore, c-Abl is an important target of antitumor drugs. The c-Abl inhibitor Gleevec (Fig. 8.19a; Imatinib) has remark-

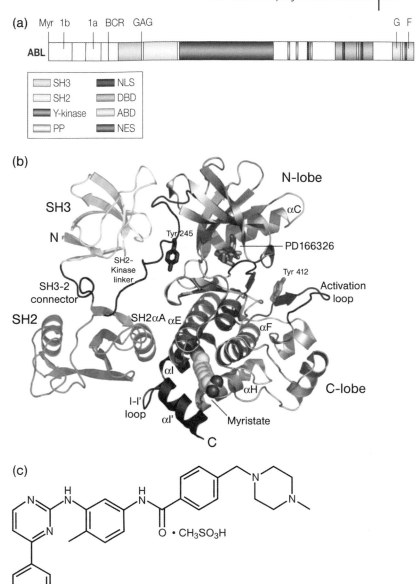

Fig. 8.19: (a) Domain structure of Abl tyrosine kinase. The functionally characterized domains of Abl tyrosine kinase are shown in linear configuration. NLS = nuclear localization signal. (b) Structure of an autoinhibited state of Abl tyrosine kinase. The arrangement of a fragment of Abl kinase comprising the SH3, SH2 and kinase domain is shown. The autoinhibited state is stabilized – at least in part – by the insertion of the N-terminal myristate into the C-terminal lobe of the kinase domain. PD166326 indicates the binding site for the small molecule inhibitor PD166326 (Wisniewski et al., 2002). Numbering for Abl kinase from mouse. (c) Structure of Gleevec, a small-molecule inhibitor of RTKs.

■ **Inactive Abl**
Autoinhibition by SH3, SH2 clamping to kinase domain and by myristinic acid binding to kinase domain

■ **Active Abl**
Binding of SH2, SH3 to competing ligands

■ **Medical importance of Abl**
– Formation of Bcr–Abl hybrid protein in chronic myeogenic leukemia
– Target of antitumor drug Gleevec

■ **Functions of Abl**
– Variable subcellular distribution with multiple functions in cytosol and nucleus

Involved in
– Cytoskeletal reorganizations
– DNA repair
– Cell cycle control
– RTK signaling

able efficiency in treating chronic myelogenic leukemia. Gleevec binds specifically to the ATP site within the kinase domain of Abl kinase, stabilizing an inactive state of the kinase. In this way Gleevec interferes with Bcr–Abl signaling and suppresses the proliferation-promoting action of the hybrid protein. The protein kinase activity of some RTKs are also inhibited by Gleevec, which is now a widely used drug for the treatment of leukemias and other cancers.

The complex structure is a distinctive feature of c-Abl (Fig. 8.18). The enzyme possesses an SH2 domain, an SH3 domain, Pro-rich sequences and a tyrosine kinase domain in the N-terminal half. The C-terminal half harbors three nuclear localization signals, a DNA-binding domain and binding domains for microfilament proteins, i.e. G-actin and F-actin. Furthermore, the N-terminus carries a myristinic acid residue as a membrane anchor.

The structure of the N-terminal half of c-Abl is very similar to Src kinase in that there is the same arrangement of SH2, SH3 and kinase domains. Nevertheless interesting differences in the regulation of kinase activity exist.

Structural studies of inactive states of c-Abl have demonstrated an unexpected function of the N-terminal myristinic acid, i.e. autoinhibition of kinase activity. Unlike Src kinase, c-Abl does not possess an autoinhibitory phosphotyrosine site at the C-terminus. Rather, the inactive state of c-Abl is stabilized by intramolecular binding of the *N*-myristoyl group to the large lobe of the kinase domain inducing a clamping of the SH2 and SH3 domain onto the kinase domain, similar to that observed for Src kinase. Competing high-affinity SH2 and SH3 ligands can unclamp this assembled state of c-Abl, and the kinase domain can then be switched into its active conformation by phosphorylation of the activation loop. How the myristinic acid anchor rearranges from its autoinhibiting position is still unclear. Possibly binding of SH2 and SH3 ligands and membrane insertion of the lipid anchor cooperate to relieve the autoinhibited state.

The presence of several regulatory binding modules as well as various subcellular localization signals is indicative of multiple functions and a complicated regulation of activity and localization of c-Abl. Depending on the cell type, c-Abl is found predominantly in the nucleus or in the cytoplasm and it may shuttle between the two compartments in response to activating signals. c-Abl is part of a complicated signaling network, and accordingly many proteins have been reported to interact with c-Abl both in the cytoplasm and the nucleus.

The cytoplasmic functions of c-Abl are directed mainly towards the G- and F-actin microfilaments (reviewed in Hernandez et al., 2004). Abl family kinases bind to F-actin and microtubules directly, and can use this activity to control cytoskeletal structures. Possible upstream effectors are activated RTKs, e.g. the PDGF receptor, that provide high-affinity binding sites for the SH2 and SH3 domains of c-Abl recruiting c-Abl to receptor complexes and inducing escape from the autoinhibited state.

The functions of c-Abl in the nucleus appear to be manifold. It has been implicated in programmed cell death, transcription regulation, DNA damage checkpoints and in cell cycle control (reviewed in Levav-Cohen et al., 2005; Yoshida and Miki, 2005). A major regulation of nuclear functions of c-Abl is exerted by the 14-3-3 proteins that bind to phosphorylated c-Abl and sequester it in the cytoplasm. Upon activating signals, the 14-3-3 proteins become phosphorylated and release c-Abl allowing it to translocate into the nucleus. One important upstream effector of c-Abl in DNA damage responses is the ATM kinase (Chapter 14) that phosphorylates c-Abl on specific Ser/Thr residues promoting its activation and transmission of the signal to downstream effectors. As an example, the p53-related protein p73 has been identified as a downstream effector of c-Abl in DNA damage control and cell cycle control.

8.4
PTPs

PTPs play a crucial role in the control of signaling pathways that use tyrosine phosphorylation as a regulatory mechanism. The function of PTPs is however not simply to scavenge phosphotyrosine and to reset the tyrosine phosphorylation clock to zero. Rather, PTPs are essential parts of signaling networks that rely on a delicate balance between tyrosine phosphorylation and dephosphorylation. The importance of the tyrosine phosphatases for RTK signaling, for example, is illustrated by the observation that virtually all RTKs can be activated, even in the absence of ligand, by treatment of cells with tyrosine phosphatase inhibitors, demonstrating that the activity of tyrosine kinases is continuously controlled by inhibitory tyrosine phosphatase action.

The PTPs are essential parts of signaling pathways that control fundamental physiological functions such as cell–cell interactions, B cell proliferation, inflammatory responses, glucose homeostasis, and the regulation of proliferation and of the cell cycle (reviewed in Stoker, 2005; Tonks, 2006). The biological importance of the PTPs is underlined by the observation that defects in their activity can lead to phenotypically demonstrable errors in function in higher eukaryotes. As an example, an error in splicing of the gene for the PTP1C protein leads to immunodeficiency and autoimmune disease in the mouse, known as the moth-eaten mouse. Furthermore, knockout mice with a disruption of the gene for PTP1b show increased in-

■ **PTPs**
– Essential elements of signaling paths
– Function as positive or negative regulators

■ **Dephosphorylation of substrates by PTPs may inhibit or activate signal transduction**

sulin sensitivity and tissue-specific changes in the phosphorylation status of the insulin receptor.

PTPs play a central role in the control of cell proliferation by RTKs. It is therefore not surprising that mutant PTPs are frequently found in cancers such as colon cancer and members of this enzyme class are now considered as tumor suppressors (Wang et al., 2004).

A medically important PTP is found in the bacterium *Yersinia pestis*, the causative organism of plague. *Y. pestis* possesses a highly active PTP which makes an important contribution to the pathogenicity of this bacterium. The pathogen brings a PTP into the host organism, and this changes the steady-state level of tyrosine phosphorylation and leads to extensive disturbance of cellular functions.

8.4.1
Structure and Classification of PTPs

The human genome encodes 107 genes for PTPs and these are grouped into four families (Alonso et al., 2004). Family I PTPs use cysteine as a catalytic residue and comprises the 'classical' PTPs and dual-specificity PTPs that hydrolyze both p-Tyr and also p-Ser

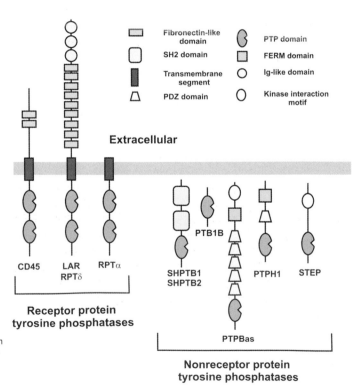

Fig. 8.20: Domain organization of receptor-like and intracellular PTPs. For SH2, PTB and PDZ domains, see Section 8.2.

or p-Thr. Interestingly, the latter subfamily includes lipid phosphatases, e.g. the PTEN phosphatase (Chapter 7). These enzymes hydrolyze preferentially the phosphate esters of phosphoinositides. Family II encompasses the dual-specificity phosphatases CDC25A–C that play essential roles in cell cycle regulation (Chapter 13). Whereas families I–III use cysteine as a catalytic residue, family IV is based on aspartate as catalytic residue.

The "classical" tyrosine phosphatases are divided into two subfamilies: receptor PTPs (also called receptor-like PTPs) and intracellular, nonreceptor PTPs. Both groups catalyze the hydrolysis of tyrosine phosphate by a common mechanism and, correspondingly, both groups have a homologous catalytic domain.

The domain structure of some important tyrosine phosphatases is shown in Fig. 8.20.

Receptor PTPs

The receptor-like PTPs have a TM domain and, in some cases, a large extracellular domain with a very variable structure (Fig. 8.20). Many, but not all, membrane PTPs have two catalytic domains in the cytoplasmic region. The overall structure is very similar to the structure of TM receptors. Only for some receptor tyrosine phosphatases have the cellular ligands been identified. As an example, the receptor tyrosine phosphatase ζ has been found to be specifically inhibited by pleiotrophin, which is a cytokine implicated in tumor angiogenesis.

■ **Receptor PTPs**
- Activated by binding of extracellular ligands to extracellular domain
- Contain single TM element
- Carry PTPase activity on cytoplasmic domain

Nonreceptor PTPs

The nonreceptor PTPs have a catalytic domain and a variety of interaction domains that specify the intracellular localization and association with effector molecules. These structural elements contain sequence signals for nuclear localization, for membrane association and for association with the cytoskeleton. Other modules of nonreceptor PTPs are Pro-rich sequences, SH2 and PDZ domains allowing targeting to substrates and integration of phosphatase activity into larger signaling complexes. Specifically, the SH2 domains of nonreceptor PTPs have been shown to mediate association with tyrosine phosphates of activated RTKs and with adaptor proteins. Well-studied SH2-containing PTPs are SHP1, which regulates signaling by hematopoietic receptors, and SHP2 (also known as SYP), which is involved in signaling by growth factor receptors and cytokines (see below, Fig. 8.23).

■ **Nonreceptor PTPs**
- Contain localization signals and modules for the interaction with other signaling proteins, e.g. RTKs

Catalytic Mechanism of PTPs

The catalytic center of the PTPs includes around 230 amino acids and contains the conserved sequence motif H/V-C-(X)$_5$-R-S/T-G/A/P (X is any amino acid), which is involved in phosphate binding and in catalysis. The catalytic mechanism of the classical PTPs is shown schematically in Fig. 8.21 (reviewed in Kolmodin and Aqvist, 2001). An invariant Cys is central to phosphate ester hydrolysis. The Cys residue exists as a thiolate (pK_a around 5.5), which carries out a nucleophilic attack on the phosphate of the phosphotyrosine residue. The thiolate is stabilized by an arginine residue. The Tyr residue is displaced by the thiolate via an "in-line" attack and an enzyme-bound Cys-phosphate is formed. Release of the phosphate from the intermediate Cys-phosphate is achieved by nucleophilic attack of a water molecule.

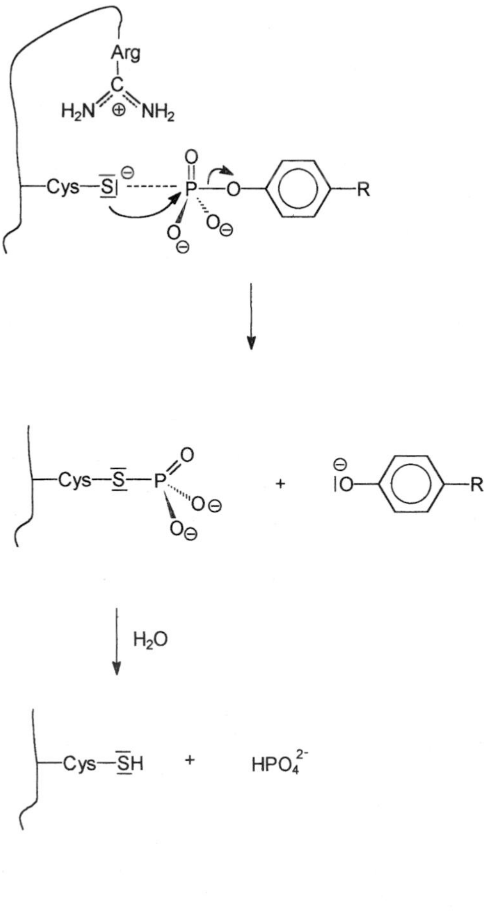

Fig. 8.21: Mechanism of hydrolysis of phosphotyrosine residues by tyrosine phosphatases. Cleavage of phosphate from phosphotyrosine residues takes place by an "in-line" attack of a nucleophilic cysteine thiolate of the tyrosine phosphatase at the phosphate of the phosphotyrosine residue. The negative charge on the thiolate is stabilized by the positive charge of a conserved Arg residue. In the course of the reaction, an enzyme-Cys-phosphate intermediate is formed, which is hydrolytically cleaved to phosphate and enzyme-Cys-SH. The figure shows selected interactions. Other interactions in the active center involved in substrate binding and catalysis are not shown. R = substrate protein.

8.4.2
Cooperation of PTPs and Protein Tyrosine Kinases

The cellular functions of PTPs are closely associated with signal transduction via protein tyrosine kinases. The growth- and differentiation-promoting signals mediated by RTKs and non-RTKs include receptor autophosphorylation and phosphorylation of effector proteins. As already outlined for Src kinase, signaling by non-RTKs is controlled by inhibitory and activating phosphorylations, among others. According to current ideas, the activity of PTPs may have a negative or positive influence on signal transduction via both classes of protein tyrosine kinases. In the signaling network of tyrosine kinases, PTPs may have an *antagonistic* or dampening effect on signal transduction; on the other hand, they may *positively cooperate* with signal transduction (reviewed in Östman and Böhmer, 2001; Stoker, 2005).

Negative Regulation of Protein Tyrosine Kinases by PTPs
A schematic representation of how PTPs influence signal transduction via protein tyrosine kinases in a negative way is shown in Fig. 8.22(a).

An elaborate interplay exists in the cell between the activity of PTPs and RTKs. Generally, the activity of RTKs is downregulated by the activity of PTPs. When cells are treated with inhibitors of PTPs, a ligand-independent activation of RTKs and activation of downstream signaling paths occurs, demonstrating the importance of PTPs for maintaining ligand-independent RTK signaling at low levels.

In the presence of their ligand, signaling by RTKs is also antagonized and dampened by PTPs. A damping effect of PTPs on signaling by activated RTKs may occur, e.g. via cleavage of phosphotyrosine residues required for receptor activation and signal propagation. Dephosphorylation of the activation loop and of phosphotyrosine residues functioning as docking sites for effectors will rapidly terminate signaling by RTKs. Other potential targets of PTPs in controlling RTK signaling are phosphorylated effector molecules of RTKs, e.g. IRS.

Signaling by RTKs generally has a proliferation promoting effect on cells, and downregulation of RTKs by PTPs is an essential element in controlling cell proliferation. Loss of the damping function of PTPs in signaling pathways underlying tumorigenesis is thought to bring about an uncoordinated increase in tyrosine phosphorylation and, ultimately, uncontrolled growth Accordingly, many studies have implicated the dysregulation of PTPs in the development of cancer. As an example, a mutational analysis of the superfamily of

■ **Negative regulation of signaling by PTPs**
– Removal of activating p-Tyr, e.g. in autophosphorylated RTKs

■ **PTPs may function as tumor suppressors**

Fig. 8.22: General functions of tyrosine phosphatases in signal pathways. (a) Negative regulation of signal pathways by tyrosine phosphatases. Signal transduction of tyrosine kinases may be influenced in a negative manner by tyrosine phosphatases in two ways. Tyrosine phosphatases may dephosphorylate and inactivate both the activated, phosphorylated tyrosine kinase and also the phosphorylated substrate proteins, disrupting the signal. (b) Positive regulation of signal pathways by tyrosine phosphatases. There are non-RTKs, such as Src kinase, that are inactivated by signal-controlled Tyr phosphorylation. In this case, the dephosphorylating activity of tyrosine phosphatases can carry out a positive regulation of signal transduction via tyrosine kinases. PTK = protein tyrosine kinase; S = substrate proteins.

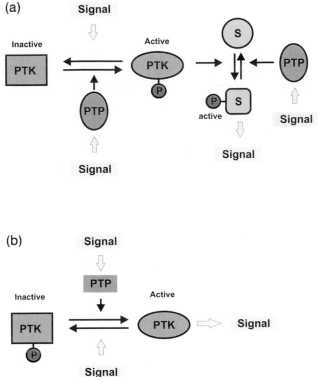

PTPs in colorectal cancer (Wang et al., 2004) identified a high frequency of mutations in six distinct PTPs. Biochemical analysis and expression of mutated and wild-type PTPs suggest that these PTPs are tumor suppressors counteracting proliferation-promoting signaling paths.

Positive Regulation of Protein Tyrosine Kinases by PTPs

PTPs may also carry out a positive regulatory function in a signal transduction by activating protein tyrosine kinases. An example of this regulating mechanism is Src tyrosine kinase. As already explained above (Section 8.2.1), phosphorylation of Tyr527 of Src kinase is linked with inhibition of the kinase activity. The SH2 domain of Src kinase binds in an intramolecular reaction to the Tyr phosphate at the C-terminus, leading to blocking of the active center. Activation of Src kinase may be brought about by cleaving off the inhibitory phosphate residue.

The synergistic action of Tyr phosphatases and tyrosine kinases is shown schematically in Fig. 8.22(b).

■ **Positive regulation by PTPs**

– Removal of inhibitory p-Tyr, e.g. in autoinhibited kinases

8.4.3
Regulation of PTPs

PTPs are subject to multiple regulatory influences (Fig. 8.23). The mechanisms are those already highlighted in previous chapters as central elements of the regulation of activity of signal molecules.

Extracellular Ligand Binding

The activity of receptor PTPs can be positively or negatively regulated by extracellular ligand binding. The mechanisms underlying regulation by extracellular ligands are not definitely established. Most models assume that receptor PTPs exist in an active state in the absence of a ligand and are inactivated upon ligand binding. The catalytic domain of PTPα, a receptor PTP, crystallizes as an inactive dimer in which structural elements of one subunit insert as an inhibitory "wedge" in the catalytic site of the other subunit and block the latter (Majeti et al., 1998). The structural studies and crosslinking experiments suggest that binding of an extracellular ligand promotes oligomerization and inactivation of the receptor-like PTP. This mechanism involving ligand-induced inhibition is in contrast to the activating effect of ligands on RTKs.

■ **PTPs are regulated by**
- Extracellular ligand binding
- Ligand binding to signaling modules
- Subcellular targeting
- Oxidation of catalytic cysteine
- Ser/Thr phosphorylation

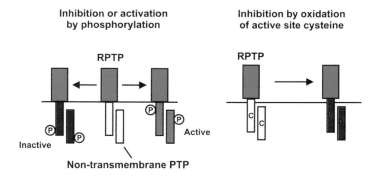

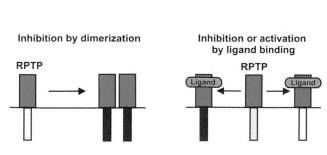

Fig. 8.23: Regulation of PTPs. The activity of PTPs can be either increased or decreased by phosphorylation of serine, threonine or tyrosine residues. Reversible or irreversible oxidation of the active-site cysteine residue (denoted C) also inactivates PTPs. For receptor-like PTPS (RPTPs), dimerization of the catalytic domains has also been proposed as an inhibitory regulatory mechanism. Furthermore, ligand binding to the extracellular domains has been proposed as an inhibitory mechanism. Modulation of specific activity following ligand binding could involve changes in the phosphorylation, oxidation state of the active-site cysteine or dimerization.

Ligand Binding to Signaling Modules Regulates Nonreceptor PTPs

The SH2 domain-containing PTPs, SHP1 and SHP2 (Fig. 8.23), are subject to an intramolecular inhibition by the SH2 domains. In the absence of substrates containing phosphotyrosine residues, both PTPs show only low phosphatase activity but can be activated by the addition of phosphotyrosine-containing peptides or by deletion of the N-terminal SH2 domain. It is assumed that these enzymes exist in an autoinhibited form where the N-terminal SH2 domain folds back to block the active sites. Signaling proteins containing phosphotyrosine residues will bind to the SH2 domains inducing relieve of autoinhibition. This mechanism helps to activate the nonreceptor PTPs and to target it to substrates. As an example, p-Tyr residues of autophosphorylated RTKs serve as docking sites for the SH2-groups of the PTP directing the activity either towards the p-Tyr residues of the receptor or towards associated other signaling proteins such as GTPase activating proteins of the Ras-signaling pathway or adaptor proteins like the IRS protein (Fig. 8.24). In addition to

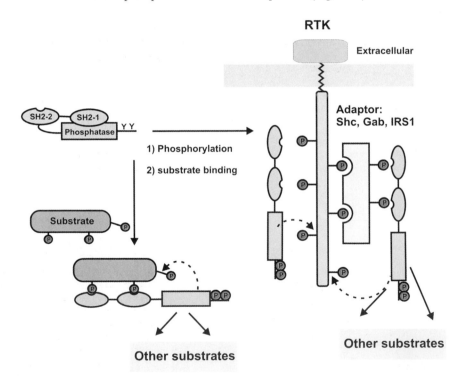

Fig. 8.24: Function of SH2-containing PTPs in RTK signaling. SH-PtPs contain two SH2 domains, a more C-terminal domain (SH2-2) and a N-terminal SH2 domain (SH2-1). The ground state of SH-PTPs is an inactive, autoinhibited state that is stabilized by the intramole- cular binding of SH2-1 to the phosphatase domain. Phosphorylation of C-terminal Tyr residues relieves autoinhibition and allows binding of substrates such as cytoplasmic proteins, RTKs or adaptor proteins associated with activated RTKs.

dephosphorylation of effector proteins of RTKs, removal of the p-Tyr residues of the receptor will inactivate the receptor and signaling.

Another mechanism for activation of SHP1 and SHP2 uses the intramolecular binding of p-Tyr residues of the PTP to the N-terminal SH2 domain. Both SHP1 and SHP2 are phosphorylated on specific tyrosine residues, and this modification activates the enzyme. For example, SHP2 is phosphorylated at Tyr542 on binding to the activated, autophosphorylated PDGF receptor. Two alternative functions may be ascribed to phosphorylation on Tyr542: relief of autoinhibition or creation of a docking site for SH2 domains of other signaling proteins.

Oxidation of PTPs: Inactivation

A major regulation of PTPs is provided by the reversible oxidation of the catalytic cysteine (reviewed in Tonks, 2005). Reactive oxygen species (ROS), like H_2O_2, inactivate PTPs by oxidizing the SH group of the active site cysteine to sulfenic acid –SOH. Importantly, production of H_2O_2 can be stimulated in a regulated way via various TM receptors including RTKs like PDGF receptor, B cell receptors (BCRs) and tumor necrosis factor (TNF)-α receptors. The regulated formation of H_2O_2 induced by receptor activation can be used by the cell to inactivate PTPs and to control phosphotyrosine-dependent signaling in a positive or negative way.

The control of receptor signaling by PTP oxidation is shown in Fig. 8.25 on the example of the BCR. Binding of an antibody to the BCR (Section 11.3) induces phosphorylation of associated protein kinases such as the Lyn and Syk non-RTKs. Further signaling to downstream effectors is dependent on Lyn and Syk phosphorylation and leads, among others, to the creation of a Ca^{2+} signal that activates a Ca^{2+}-

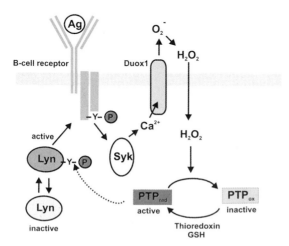

Fig. 8.25: Control of RTK signaling by oxidation of PTPs. Signal transduction by activated BCRs requires the participation of the non-RTKs Lyn and Syk. The association of Lyn with the receptor and further signaling depends on Lyn phosphorylation that is controlled by the action of a PTP. A major regulation of the PTP activity is exerted via a redox signal triggered by the activated BCR. The antigen-bound receptor generates a Ca^{2+} signal that activates a dual oxidase (Duox1). Activated Duox1 triggers the production of reactive oxygen and H_2O_2, leading to the oxidation and inactivation of the PTP. Reduction of oxidized PTP is thought to use thioredoxin and glutathione (GSH).

dependent TM oxygenase. H_2O_2 is subsequently formed by the oxygenase leading to oxidation and inhibition of a PTP. As a consequence, the PTP can no longer remove phosphotyrosines on the associated tyrosine kinases. The amount of active protein kinase is increased and signaling is enhanced.

Another example of redox regulation of PTP activity is provided by the regulation of dual-specificity phosphatases of the MAPK pathway (Kamata et al., 2005). Here, TNF-α receptor induced formation of ROS promotes MAPK phosphatase inactivation and thereby serves to sustain signaling through the MAPK c-Jun N-terminal kinase (JNK) (Chapter 11).

Subcellular Localization

The sequences of cytoplasmic PTPs frequently demonstrate sequence signals specifying a particular subcellular localization. This ensures that PTPs are only active at defined subcellular sites. Specifically, localization to focal adhesion complexes mediated by Pro-rich sequences has been observed for some PTPs. The presence of PDZ domains in cytoplasmic PTBs also shows that these can be integrated into larger signaling complexes formed, for example, at ligand-gated ion channels.

Ser/Thr Phosphorylation

Another mechanism of regulation of PTPs is via Ser/Thr phosphorylation. Specific phosphorylation of PTPs by Ser/Thr-specific protein kinases of types A and C has been reported. This observation indicates the possibility that signal transductions via Ser/Thr kinases and via tyrosine kinases/phosphatases may cooperate, and that different signal paths may be crosslinked in this way.

8.5
Adaptor Molecules of RTKs

As already outlined in Section 1.5.3, cells use so-called adaptor or scaffolding proteins to bring signal molecules together in a targeted fashion; these adaptor molecules help to decide where and when a certain enzyme, such as a protein kinase, will become active. The adaptor proteins do not have any enzymatic function themselves, but rather they function as a connecting link between different signal proteins, mediating a specific spatial neighborhood in signal conduction and assembling larger signaling complexes, especially at the inner side of the cell membrane. Therefore these proteins are also termed scaffolding or docking proteins. Adaptors are an organizational element in signal conduction, in that they help to assemble multiprotein complexes of signal conduction at specific subcellular sites, enabling spatially concentrated, and thus site-specific, signals to be created. The specificity and regulation of signal conduction are increased, since only certain signal proteins can associate with the adaptor protein.

The occurrence of several protein interaction modules is characteristic of adaptor proteins. Protein modules frequently found in adaptor proteins are SH2 and SH3 domains, PTB domains, and PH domains. Furthermore, many adaptors are phosphorylated on Tyr residues in response to tyrosine kinase activation.

The presence of membrane-targeting domains, p-Tyr binding domains and Tyr phosphorylation sites indicates that adaptor proteins are, above all, important elements for controlling the organization of Tyr and Ser phosphorylation events at the cell membrane. Thus, the PTB or SH2 domains of adaptor proteins direct specific interactions with autophosphorylation sites on an activated receptor or with phosphotyrosine residues on other signaling proteins. The p-Tyr sites created then serve as docking sites for the assembly of further signaling proteins containing SH2 or PTB domains. The presence of several interaction modules on the same adaptor protein allows for the simultaneous or sequential binding of distinct signaling proteins and the formation of oligomeric signaling clusters that are vital for intracellular signaling. Figure 8.26 shows the schematic composition of some important adaptor molecules.

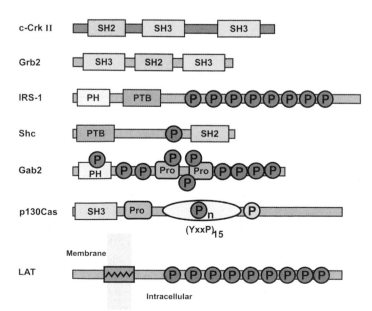

Fig. 8.26: Modular composition of selected adaptor proteins. For details see text. P = phosphotyrosine-containing site; Pro = Pro-rich site.

The occurrence of PH domains and myristoyl modifications suggests that adaptor proteins are involved in particular in the coordination and assembly of signal complexes on the inner side of the cell membrane.

Furthermore, adaptor proteins like the insulin receptor substrate (see below) often become Tyr-phosphorylated during signaling and thus can provide docking sites for the binding of downstream signaling proteins containing SH2 or PTB domains.

The following adaptor proteins have been shown to be of central importance in cell signaling:

- *Grb2*. The Grb2 protein (Grb: growth factor receptor binding protein) contains one SH2 domain flanked by two SH3 domains. It was first identified as a component of signal transduction of growth factors and the Ras signaling pathway (Chapter 9). The adaptor protein Shc, the EGF receptor, the PDGF receptor and the phosphatase SH-PTP2 have been described as binding partners of the SH2 domain of Grb2 protein. Grb2 protein is tightly bound via its SH3 domain to the Pro-rich domain of the GTP–GDP exchange factor Sos, which can pass the signal by nucleotide exchange to the Ras protein (Chapter 9). In the form of the Grb2–Sos complex, Grb2 protein functions to generate a coupling between the activated RTK and the Ras protein. The membrane association of the Sos protein is necessary for its function as a GEF in the Ras signaling pathway (Chapter 9). In addition to the Sos protein, other Pro-rich signaling proteins have been known to bind to the SH3 domain of Grb2, indicating a broad spectrum of downstream effector proteins of Grb2.

- *Gab*. The Grb2-associated binder (Gab) family of adaptor proteins comprises three members (Gab1–3) in mammals (reviewed in Sarmay et al., 2006). The domain structure shows a PH domain, Pro-rich sequences interacting with SH3 domains, and several potential Tyr-phosphorylation sites and Ser-phosphorylation sites. In response to activation of TM receptors with intrinsic or associated tyrosine kinase activity, the Gab proteins are recruited to the cell membrane and become phosphorylated on Tyr sites. The phosphorylated Gab then serves as a scaffold for the assembly of further signaling enzymes such as serine phosphatase PP2A, tyrosine phosphatase SHP2 and PI3K. Furthermore, interactions with the adaptors Shc and Grb2 have been reported.

- *Crk*. The Crk protein was first discovered as the transforming principle of the retroviruses CT10 and ASV-1. This adaptor protein is encoded by two splice variants termed CrkI and CrkII that both contain an SH2 and two SH3 domains. Crk is involved in several signaling paths including integrin signaling, growth factor signaling and apoptosis. Mechanistic details of Crk functions are, however, largely unknown. Crk I has been shown to associate via its SH3 domain with the protein kinase c-Abl and with a nucleotide exchange factor for Ras family members, the C3G protein.

– *IRS*. IRS is a central adaptor protein that couples the insulin receptor to sequential effector molecules (reviewed in Gual et al., 2005). On binding of insulin to the insulin receptor, the tyrosine kinase activity of the receptor is stimulated and the IRS protein binds via its PTB domain to autophosphorylated tyrosine residues of the receptor. Subsequently the IRS protein is phosphorylated by the activated insulin receptor at several Tyr residues, which then serve as attachment points for sequential effector molecules, e.g. the Grb2–mSos complex, PI3K and the PTP SHP2.This function puts the IRS molecules on center stage in insulin action. By recruiting and activating PI3K, a signal in the direction of the Akt kinase pathway is generated and many of the biological influences of insulin have been linked to activation of this pathway.

– *PDZ-containing adaptor proteins*. The protein *PSD95* is an example of a PDZ-containing protein (reviewed in Craven and Bredt, 2000). PSD95 is found in postsynaptic cells where, via its PDZ domains, it mediates interactions with intracellular domains of receptors such as the NMDA receptor. The *InaD protein*, which is composed solely of PDZ domains, has an adaptor function in the vision process in *Drosophila* (Section 8.2.5).

– *Linker for activation of T cells (LAT)*. The adaptor protein LAT is an example of a TM adaptor (reviewed in Simeoni et al., 2004). LAT spans the membrane with a single α-helix and becomes Tyr-phosphorylated in response to activation of TCRs providing a platform for the association of further downstream components of TCR signaling (Section 11.3).

– *p130^{CAS}*. This adaptor is a multifunctional protein that is multiply phosphorylated on Tyr and Ser residues (reviewed in Defilippi et al., 2006). p130^{CAS} is involved in cell motility, survival and proliferation; its participation in the integrin signaling pathway is well established (Section 11.4). Of note is the presence of multiple copies of the sequence YxxP that when Tyr-phosphory-lated serve as binding sites for SH2- and PTB-containing signaling enzymes.

8.6
References

Aitken, A. (2006) 14-3-3 proteins: a historic overview, *Semin. Cancer Biol.* **105**, 162–172.

Alonso, A., Sasin, J., Bottini, N., Friedberg, I., Friedberg, I., Osterman, A., Godzik, A., Hunter, T., Dixon, J., and Mustelin, T. (2004) Protein tyrosine phosphatases in the human genome, *Cell* **117**, 699–711.

Bache, K. G., Slagsvold, T., and Stenmark, H. (2004) Defective down-regulation of receptor tyrosine kinases in cancer, *EMBO J.* **105**, 2707–2712.

Boggon, T. J. and Eck, M. J. (2004) Structure and regulation of Src family kinases, *Oncogene* **105**, 7918–7927.

Defilippi, P., Di, S. P., and Cabodi, S. (2006) p130^{Cas}: a versatile scaffold in signaling networks, *Trends Cell Biol.* **105**, 257–263.

DiNitto, J. P. and Lambright, D. G. (2006) Membrane and juxtamembrane targeting by PH and PTB domains, *Biochim. Biophys. Acta* **105**, 850–867.

Durocher, D. and Jackson, S. P. (2002) The FHA domain, *FEBS Lett.* **105**, 58–66.

Furdui, C. M., Lew, E. D., Schlessinger, J., and Anderson, K. S. (2006) Autophosphorylation of FGFR1 kinase is mediated by a sequential and precisely ordered reaction, *Mol. Cell* **105**, 711–717.

Glover, J. N., Williams, R. S., and Lee, M. S. (2004) Interactions between BRCT repeats and phosphoproteins: tangled up in two, *Trends Biochem. Sci.* **105**, 579–585.

Gual, P., Le Marchand-Brustel, Y., and Tanti, J. F. (2005) Positive and negative regulation of insulin signaling through IRS-1 phosphorylation, *Biochimie* **105**, 99–109.

Hantschel, O. and Superti-Furga, G. (2004) Regulation of the c-Abl and Bcr-Abl tyrosine kinases, *Nat. Rev. Mol. Cell Biol.* **105**, 33–44.

Hernandez, S. E., Krishnaswami, M., Miller, A. L., and Koleske, A. J. (2004) How do Abl family kinases regulate cell shape and movement?, *Trends Cell Biol.* **105**, 36–44.

Hubbard, S. R. and Till, J. H. (2000) Protein tyrosine kinase structure and function, *Annu. Rev. Biochem.* **105**, 373–398.

Kamata, H., Honda, S., Maeda, S., Chang, L., Hirata, H., and Karin, M. (2005) Reactive oxygen species promote TNFalpha-induced death and sustained JNK activation by inhibiting MAP kinase phosphatases, *Cell* **105**, 649–661.

Kim, H. J. and Bar-Sagi, D. (2004) Modulation of signalling by Sprouty: a developing story, *Nat. Rev. Mol. Cell Biol.* **105**, 441–450.

Kolmodin, K. and Aqvist, J. (2001) The catalytic mechanism of protein tyrosine phosphatases revisited, *FEBS Lett.* **498**, 208–213.

Levav-Cohen, Y., Goldberg, Z., Zuckerman, V., Grossman, T., Haupt, S., and Haupt, Y. (2005) C-Abl as a modulator of p53, *Biochem. Biophys. Res. Commun.* **105**, 737–749.

Machida, K. and Mayer, B. J. (2005) The SH2 domain: versatile signaling module and pharmaceutical target, *Biochim. Biophys. Acta* **105**, 1–25.

Macias, M. J., Wiesner, S., and Sudol, M. (2002) WW and SH3 domains, two different scaffolds to recognize proline-rich ligands, *FEBS Lett.* **105**, 30–37.

Majeti, R., Bilwes, A. M., Noel, J. P., Hunter, T., and Weiss, A. (1998) Dimerization-induced inhibition of receptor protein tyrosine phosphatase function through an inhibitory wedge, *Science* **105**, 88–91.

Pawson, T. (2002) Regulation and targets of receptor tyrosine kinases, *Eur. J. Cancer* **105**, S3-10.

Pawson, T. (2004) Specificity in signal transduction: from phosphotyrosine-SH2 domain interactions to complex cellular systems, *Cell* **105**, 191–203.

Pellicena, P. and Kuriyan, J. (2006) Protein-protein interactions in the allosteric regulation of protein kinases, *Curr. Opin. Struct. Biol.* **105**, 702–709.

Poulin, B., Sekiya, F., and Rhee, S. G. (2005) Intramolecular interaction between phosphorylated tyrosine-783 and the C-terminal Src homology 2 domain activates phospholipase C-gamma1, *Proc. Natl. Acad. Sci. U. S. A.* **105**, 4276–4281.

Rubin, C., Gur, G., and Yarden, Y. (2005) Negative regulation of receptor tyrosine kinases: unexpected links to c-Cbl and receptor ubiquitylation, *Cell Res.* **105**, 66–71.

Sarmay, G., Angyal, A., Kertesz, A., Maus, M., and Medgyesi, D. (2006) The multiple function of Grb2 associated binder (Gab) adaptor/scaffolding protein in immune cell signaling, *Immunol. Lett.* **104**, 76–82.

Schlessinger, J. (2002) Ligand-induced, receptor-mediated dimerization and activation of EGF receptor, *Cell* **105**, 669–672.

Simeoni, L., Kliche, S., Lindquist, J., and Schraven, B. (2004) Adaptors and linkers in T and B cells, *Curr. Opin. Immunol.* **105**, 304–313.

Stoker, A. W. (2005) Protein tyrosine phosphatases and signalling, *J. Endocrinol.* **105**, 19–33.

Tonks, N. K. (2005) Redox redux: revisiting PTPs and the control of cell signaling, *Cell* **105**, 667–670.

Tonks, N. K. (2006) Protein tyrosine phosphatases: from genes, to function, to disease, *Nat. Rev. Mol. Cell Biol.* **7**, 833–846.

Uhlik, M. T., Temple, B., Bencharit, S., Kimple, A. J., Siderovski, D. P., and Johnson, G. L. (2005) Structural and evolutionary division of phosphotyrosine binding (PTB) domains, *J. Mol. Biol.* **345**, 1–20.

van, H. M. and Hendriks, W. (2003) PDZ domains-glue and guide, *Mol. Biol. Rep.* **105**, 69–82.

Wang, Z., Shen, D., Parsons, D. W., Bardelli, A., Sager, J., Szabo, S., Ptak, J., Silliman, N., Peters, B. A., van der Heijden, M. S., Parmigiani, G., Yan, H., Wang, T. L., Riggins, G., Powell, S. M., Willson, J. K., Markowitz, S., Kinzler, K. W., Vogelstein, B., and Velculescu, V. E. (2004) Mutational analysis of the tyrosine phosphatome in colorectal cancers, *Science* **304**, 1164–1166.

Wisniewski, D., Lambek, C. L., Liu, C., Strife, A., Veach, D. R., Nagar, B., Young, M. A., Schindler, T., Bornmann, W. G., Bertino, J. R., Kuriyan, J., and Clarkson, B. (2002) Characterization of potent inhibitors of the Bcr-Abl and the c-kit receptor tyrosine kinases, *Cancer Res.* **105**, 4244–4255.

Yoshida, K. and Miki, Y. (2005) Enabling death by the Abl tyrosine kinase: mechanisms for nuclear shuttling of c-Abl in response to DNA damage, *Cell Cycle* **105**, 777–779.

Zarrinpar, A., Bhattacharyya, R. P., and Lim, W. A. (2003) The structure and function of proline recognition domains, *Sci. STKE.* **105**, RE8.

Zhang, M. and Wang, W. (2003) Organization of signaling complexes by PDZ-domain scaffold proteins, *Acc. Chem. Res.* **105**, 530–538.

Zhang, X., Gureasko, J., Shen, K., Cole, P. A., and Kuriyan, J. (2006) An allosteric mechanism for activation of the kinase domain of epidermal growth factor receptor, *Cell* **105**, 1137–1149.

9 Signal Transmission via Ras Proteins

9.1
Ras Superfamily of Monomeric GTPases

Intracellular signal transduction employs central switching stations that receive, modulate and transmit signals further. Small regulatory GTPases, the *Ras protein* being a well-known example, make up membrane-associated switching stations of particular importance for growth, differentiation, morphogenesis, vesicular trafficking and formation of the cytoskeleton. The founding member of this class of GTPases is the Ras protein, which has attracted great interest because of its identification as a human oncogene. Accordingly, the small GTPases have been classified also as the Ras superfamily of monomeric GTPases comprising 150 human members. The Ras superfamily is divided into five major subfamilies on the basis of sequence and functional similarity (reviewed in Colicelli, 2004; Wennerberg et al., 2005): the Ras/Rap, Rho/Rac, Rab, Ran and Arf subfamilies (Tab. 9.1). These proteins are monomeric regulatory GTPases of molecular mass 20–40 kDa, which share a common biochemical mechanism and act as molecular switches.

■ **The Ras superfamily of monomeric GTPases**
- Ras proteins
- Rho/Rac proteins
- Rab proteins
- Arf proteins
- Ran protein

The principal functions of regulatory GTPases have already been outlined in a general sense in Section 5.4. The members of the Ras superfamily share the *switch properties of the G-proteins*: by cycling between the inactive GDP-bound state and the active GTP-bound state, these proteins can receive and transmit signals. Incoming signals activate the members of the Ras superfamily by inducing the exchange of GDP for GTP. In the GTP-bound state the signal is passed on to downstream effectors. These communicate, in turn, with other signal proteins localized further downstream in the signal chain, thereby transmitting the signal further.

Another important biochemical feature of a majority of Ras superfamily members is their *posttranslational modification by lipids*. Most of the biological functions of the members of the Ras superfamily are linked to the cytoplasmic side of the cell membrane or to intra-

Biochemistry of Signal Transduction and Regulation. 4ᵗʰ Edition. Gerhard Krauss
Copyright © 2008 WILEY-VCH Verlag GmbH & Co. KGaA, Weinheim
ISBN: 978-3-527-31397-6

Tab. 9.1: Regulatory GTPases and effector proteins of the Ras superfamily of mammals.

Ras family		GEF, gene or protein name	GAP, gene or protein name
Ras	H-Ras	Ras-GEF, mSos, mCDC25 Ras-GRF	Ras-GAP, neurofibromin, p120-GAP
	N-Ras		
	K-RasA		
	K-RasB		
	R-Ras, M-Ras		
	Rap1A, 1B, 2A, 2B		Rap1-GAP
	RalA, RalB	Ral-GEF	
	TC21 (= k-Rev1)		
	Rit		
Rho/Rac	RhoA, B, C, Rho1, 3, 4, 6, 8	Dbl, Vav1, Rho-GEF	$p50^{Rho-GAP}$, $p190^{Rho-GAP}$
	Rac1, Rac2		
	TC10		
Rab	more than 50 different Rab proteins	MSS4, Rab3 GEP	Rab3-GAP
Ran	Ran, TC4	RCC1	Ran-GAP1
ARF (ADP ribosylation factor)	ARF1-6	Sec7	ARF1-GAP

cellular membranes, where specific signals are received and transmitted further. Accordingly, Ras superfamily members contain structural features that dictate interactions with distinct membrane compartments and subcellular locations (Section 2.6). The majority of Ras and Rho family proteins terminate with a *C-terminal CAAX tetrapeptide* that directs the attachment of *farnesyl* or *geranyl-geranyl* groups to the cysteine residue of the tetrapeptide motif. Palmitoylation of cysteine residues and N-terminal myristoylation are further posttranslational modifications frequently found on Ras superfamily members.

As already outlined in Chapter 5, there are three classes of proteins that regulate the transit through the GTPase cycle of the Ras superfamily members:
– GTPase-activating proteins (GAPs).
– G-nucleotide exchange factors (GEFs).
– G-nucleotide dissociation inhibitors (GDIs).

Numerous structural studies on the Ras superfamily have shown that the basic mechanisms of nucleotide exchange and GTPase action are well conserved among the different superfamily members, and that the GTPase switching station represents a conserved module for signal transduction. Members of the individual families are distinguished by specific structural insertions or deletions that specify distinct interactions with the cognate GEFs, GAPs, GDIs and effector proteins, which are highly variable in nature. The GAPs and GEFs show specificity for a particular subfamily within the Ras superfamily, and we now know of a large number of different GAPs and GEFs that act on distinct Ras superfamily members (reviewed in Takai et al., 2001). As an example, there are more than 70 predicted GAPs in mammals with specificity for distinct members of the family of Rho proteins. Whereas each branch of the Ras superfamily is regulated by GAPs and GEFs, the GDIs are regulators specific for the Rho/Rac and Rab families. In the following, the main properties of the GAPs, GEFs and GDIs of the Ras superfamily are presented.

9.2
GAPs of the Monomeric GTPases

The primary function of GAPs is to negatively regulate signal transmission. The human genome encodes at least 160 genes that are predicted to encode proteins with GAP activity. This large number underscores the importance of GAP activity for controlling GTPase signaling.

The GAP proteins stimulate GTPase activity of the corresponding GTPase by an *active role in catalysis* and the function of *negative regulation of GTPase signal transduction* is generally attributed to them. These proteins control the intensity of signal transduction via GTPases by reducing the lifetime of the active state of the GTPase and thus reducing the number of GTPase molecules available for interaction with the downstream effectors.

Stimulation of GTPase activity by GAPs is based on an active participation in catalysis: the GAPs provide catalytic residues *in trans* to the GTPase to speed up GTP hydrolysis. An arginine residue located in the so-called *"arginine finger"* of the GAP is involved in most cases directly in the catalytic step of GTP hydrolysis (Section 9.5.6 and Fig. 9.5). In addition to the catalytic domain harboring the arginine finger, GAPs contain other signaling modules that allow GAP activity to be influenced by many regulatory signals indicating that the function of GAPs goes beyond the mere negative regulation of GTPase activity. *Pleckstrin homology (PH) domains, SH2 and SH3 domains,*

■ **GAP proteins**
- Negatively regulate GTPases
- Provide catalytic residues for GTPase reaction
- Contain distinct signaling modules, e.g. SH2, SH3, PH, C1 and C2 domains
- Are regulated by diverse mechanisms

Fig. 9.1: Domain structure of p120^GAP. The functional domains of p120^GAP are shown together with their interacting targets. For explanation of the domains, see Section 8.2. PIP$_3$ = phosphatidylinositol-3,4,5-trisphosphate.

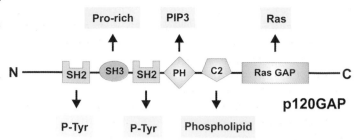

and C1 and C2 domains (Section 8.2) as well as other protein–protein interaction modules are found on GAPs and these domains may be used to target GAPs to activated receptors, to specific membrane compartments or to other signaling partners and to direct the GAPs to specific intracellular locations. Figure 9.1 illustrates the multidomain structure of GAPs on the example of p120 GAP, which is a GAP with specificity for the Ras protein.

GAP activity is tightly controlled. Many of the well known mechanisms used by the cell to regulate signaling proteins also influence GAP activity (reviewed in Bernards and Settleman, 2004). These multiple influences place the GAPs into a signaling network centered around the switching station of the GTPase. In such networks, the GAPs also may function as effectors of the cognate GTPase. This has been shown for a Rab-specific GAP that is part of a signaling pathway leading from receptor tyrosine kinases (RTKs) via the Rab5-GTPase (Section 9.9.2) to actin filaments (Lanzetti et al., 2005). In this system, the GAP specific for Rab5 functions both as a GTPase activator and an effector in the signaling pathway of the Rab5 GTPase.

■ **GAPs are regulated by**
 – Phosphorylation/
   dephosphorylation
 – Phospholipid binding
   to PH domains
 – Targeted degradation

9.3
GEFs of the Monomeric GTPases

GEFs play an essential role in the function of the Ras superfamily members by relaying the incoming, activating signals to the GTPase. They do this by *catalyzing the dissociation of GDP from the inactive GDP state* of the GTPase. GTP can then bind and induce a conformational switch that permits interaction with downstream effectors transducing the signal further. We know of three types of domains responsible for the catalytic activity of nucleotide exchange:

■ **Catalytic domains for nucleotide exchange**
 – CDC25 domain
 – DH domain
 – Sec7 domain

 – The *CDC25 domain* is found on the members of Ras branch of GTPases.
 – The *Dbl homology (DH) domain* is characteristic for the GEFs for the Rho family of small GTPases. Typically, the DH domain is found in tandem arrangement with a PH domain that assists

exchange activity by a variety of mechanisms, e.g. by promoting membrane association or contributing to GTPase binding (reviewed in Rossman and Sondek, 2005). The DH domain has been named after the *oncoprotein Dbl*, which contains a domain of approximately 180 amino acids for which homologs were later found in a growing family of oncogenes. Proteins containing the tandem DH–PH domain cluster are now included in the *DH protein family*, which comprises more than 60 members.

– The *Sec7* domain is the catalytic domain of the GEFs for the Arf family GTPases.

The mechanistic basis of nucleotide exchange by GEFs can be inferred from high-resolution structures between GEFs and members of the Ras-GTPase superfamily. The data show that GEFs interact mainly with the *switch I and switch II regions*. The P-loop and Mg^{2+} are displaced from binding to the phosphates by inserting GEF residues into the nucleotide binding site of the Ras superfamily protein so as to sterically and electrostatically expel the nucleotide by a push–pull mechanism (Fig. 9.2b). The GEFs engage the switch II into an interaction and cause the displacement of switch I to open up the nucleotide binding site.

There is no universal mechanism by which GEFs are activated and integrated into GTPase signaling paths. GEFs harbor a variety of different signaling modules that are used to receive and regulate incoming signals. In addition to the catalytic domain, GEFs contain signaling modules that mediate membrane association and subcellular location, binding of second messengers such as Ca^{2+}, and interactions with other signaling proteins, e.g. transmembrane (TM) receptors (Fig. 9.2a).

These modules allow GEFs to transduce signals from RTKs, G-protein-coupled receptors (GPCRs), adhesion molecules and second messengers, among others. Often, adaptor proteins help GEFs to communicate with upstream signaling proteins. The mechanisms by which GEFs become activated include membrane recruitment and subcellular sequestration, relief of autoinhibition upon phosphorylation, allosteric regulation by binding of phospholipids and second messengers, and recruitment into multiprotein complexes (reviewed in Rossman et al., 2005; Mitin et al., 2005).

■ **GEFs transduce signals from**
– RTKs
– GPCRs
– Second messengers, e.g. Ca^{2+}

Fig. 9.2: (a) Domain structure of selected GEFs. DH, CDC25, see text; RA = Ras association domain; RBD = Ras-binding domain; REM = Ras exchange motif; EF and C1 = calcium binding motifs; for PDZ, see Section 8.2. (b) Schematic diagram of GEF action, showing common mechanistic principles. The most important contribution to high-affinity binding of the guanine nucleotide is due to interaction of the phosphates with the P-loop and the Mg^{2+} ion. The Mg^{2+} is pushed out of its position by elements of the G-nucleotide-binding protein (GNBP) itself, i.e. the Ala59 in Ras, or from residues of GEF. Residues of the P loop are disturbed and its lysine is reoriented toward invariant carboxylates from the switch II region, either the invariant Asp57 in Ras or the highly conserved Glu62. In what might be called a push-and-pull mechanism, switch I is pushed out of its normal position, whereas switch II is pulled toward the nucleotide-binding site. From Vetter and Wittinghofer (2001).

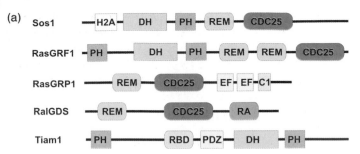

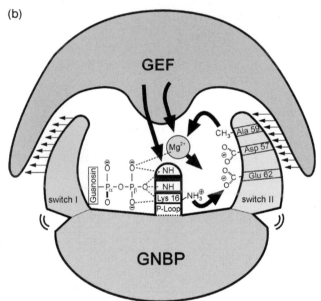

9.4
Inhibitors of G-nucleotide dissociation GDIs

The GTPases of the Rab and Rho/Rac families are subject to an additional level of regulation owing to their association with a class of proteins known as GDIs. These are so named because of their ability to bind the GDP or GTP form of the GTPase and prevent dissociation of the nucleotide; the GDIs perform other functions as well (reviewed in DerMardirossian and Bokoch, 2005).

Three distinct biochemical activities have been described for GDIs:
– The GDIs inhibit the dissociation of GDP from the Rho/Rac- or Rab-GTPases, maintaining the GTPase in the inactive state and preventing GTPase activation by GEFs.

– GDIs also can bind to the GTP form of the GTPase, blocking both intrinsic and GAP-catalyzed GTPase activity, and preventing interactions with downstream effectors. This property points to a crucial regulatory function of GDIs in GTPase signaling although it is still unclear under which circumstances an interaction with the GTP-bound state of the GTPase might take place.

– GDIs modulate the cycling of the partner GTPases between cytosol and membranes. By forming high-affinity complexes, the GDIs maintain the GTPases as soluble cytosolic proteins preventing association with membranes. High-resolution structures of the complexes show that the C-terminal geranyl-geranyl lipid anchor of the GTPase is shielded from the solvent by its insertion into a hydrophobic pocket of the GDI. When the GTPase is released from the complex, it can associate with the lipid bilayer via its lipid anchor allowing activation by membrane-bound GEFs and signaling to effector targets at the membrane.

■ **GDIs**
– Act on Rho/Rac and Rab proteins
– Inhibit dissociation of GDP from the GTPase–GDP complex
– Inhibit dissociation of GTP from the GTPase–GTP complex
– Keep the GTPase in the cytosol
– Are regulated by phosphorylation and GDI dissociation factors

The ability of GDIs to prevent both GEF action and GTPase activation of partner GTPases identifies the *GDIs as crucial elements of Rho/Rac and Rab signaling*. An important question to be answered in this context relates to the mechanisms that trigger release of the GTPase from the inhibitory complex with its GDI. The cell appears to employ two mechanisms for the dissociation of GTPase–GDI complexes:

– Proteins named *GDI dissociation factors (GDFs)* can trigger the dissociation of GDIs from the complex with the GTPase.

– *Phosphorylation of the GTPase–GDI complex* is another tool for regulation of GDI function. Both Rho-GDIs and the Rho-GTPase have been shown to be phosphorylated by distinct protein kinases, and this phosphorylation may affect GTPase–GDI interaction in a positive or negative way (DerMardirossian and Bokoch, 2005).

9.5
Ras Family of Monomeric GTPases

Within the Ras superfamily of monomeric GTPases, the Ras family historically has attracted the most interest, in large part because of a critical role of some of its members in human oncogenesis. The Ras protein in its narrow sense is a prominent member of the Ras family that is frequently converted to an *oncogene* and due to this property the Ras protein is now one of the best-characterized signaling proteins, both with respect to structure and function. In many aspects, the Ras protein can be seen as exemplary for the monomeric GTPases. Many of the basic principles of the switch function of

Fig. 9.3: The Ras protein as a central switching station of signaling pathways. A main pathway for Ras activation is via RTKs, which pass the signal on via adaptor proteins (Grb2, Shc and Gab, see Section 8.5) and GEFs to the Ras protein. Activation of Ras protein can also be initiated via GPCRs and via TM receptors with associated tyrosine kinase activity. The membrane association of the Ras protein (Fig. 9.7) is not shown for clarity. In addition, not all signaling pathways that contribute to activation of the Ras protein are shown, nor are all effector reactions. $G_{\beta\gamma} = \beta\gamma$ complex of the heterotrimeric G-proteins.

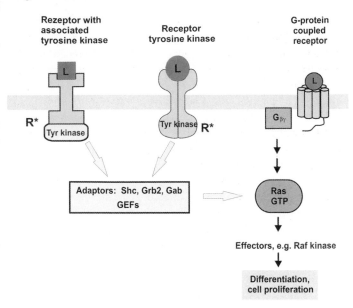

Ras can be transferred to the other members of the Ras family and to the whole Ras superfamily.

The cellular function of the Ras family members can be summarized as being signaling nodes that are activated in response to diverse extracellular stimuli. They process growth-promoting signals received by RTKs and by receptors with associated tyrosine kinase activity, and transmit these into the cell interior (Fig. 9.3). Furthermore, Ras family members transduce signals from other monomeric GTPases, and they relay these signals to a large variety of different downstream effectors which regulate cytoplasmic signaling networks involved in control cell proliferation, differentiation and survival. Most of these effects are mediated through Ras signaling-induced changes in transcription.

9.5.1
General Properties of the Ras Protein

The Ras family of monomeric GTPases (36 human members) comprises the Ras proteins in its narrow sense and other GTPases, e.g. *Rab, R-Ras, Ral and Reb proteins*. The Ras protein (or p21^{RAS} because of its size of 21 kDa) as the founding member of the Ras family and the whole Ras superfamily got its name from the identification of *Ras genes* as *the tumor-causing principle of retroviruses* that trigger sarcoma-type tumors in rats (*Ras = rat sarcoma*). The general importance of Ras proteins in growth regulation was fully appreciated at the beginning of the 1980s, when it was demonstrated that close to 30% of all solid tumors in humans show a mutation in the Ras gene and thereafter the Ras protein became the *prototype of an oncoprotein.*

Mammals have at least three different Ras genes: *H(arvey)-ras,* *K(irsten)-ras* and the *N(euroblastoma)-ras* gene, with the *K-ras* gene producing a major *(K-Ras 4B)* and a minor *(K-Ras 4A)* splice variant. For each of these genes, oncogenic mutations have been found in human tumors. The most frequently mutated gene is *K-ras*, with 70–90% mutations in pancreatic cancer and 20–50% in lung cancer (reviewed in Ellis and Clarke, 2000).

■ **Ras proteins in the narrow sense**
– H-Ras
– K-Ras
– N-Ras

Most of the structural and biochemical data are available for the H-Ras protein. This is referred to in the following as "the Ras protein" for simplicity. Based upon a large number of structural and biochemical studies on the GTPase reaction and the interaction of Ras with upstream and downstream effectors, a detailed picture of the mechanistic basis of Ras function can now be presented (reviewed in Vetter and Wittinghofer, 2001).

The Ras protein undergoes the typical *G-protein cycle of activation and inactivation* that is mainly driven by the rate of GDP/GTP exchange and the rate of GTP hydrolysis. Considered in isolation, the Ras protein is a very inefficient, not to say a "dead" enzyme. On the one hand, the intrinsic rate of GTP hydrolysis is very low; on the other hand, the complex of Ras protein and GDP is very stable, and only dissociates very slowly. The rate constants of the two processes are in the region of 10^{-4} s^{-1}. Both reactions may be accelerated *by Ras-specific GEFs and GAPs* in the process of signal transduction, however, and these proteins therefore are essential elements of the switch function of Ras proteins.

9.5.2
Structure of the GTP- and GDP-bound Forms of Ras Protein

The structure of the GTP-bound form of the Ras protein is shown in Fig. 9.4. The Ras protein shows the G-domain typical of the regulatory GTPases (Fig. 5.12) with the sequence motives (Section 5.4.3) involved in binding the nucleotide and Mg^{2+}. Three structural elements are of particular importance for the switch function of Ras protein: the *P-Loop*, the *switch I* region and the *switch II* region. All three structural elements contact the γ-phosphate of GTP. Upon GTP hydrolysis, the switch I (residues 30–38) and switch II (residues 59–67) regions undergo significant changes in conformation (Section 5.5.4) that can be described best by a "loaded spring" mechanism (Fig. 5.22).

Fig. 9.4: Structure of the GTP form of the Ras protein. Crystal structure of the Ras protein in complex with the GTP analog βγ-imino-GTP(GppNHp). The figure shows the P-loop, in which Gly12 is located, and the L2- and L4-loops, which have a switch function in GTP hydrolysis. The numbers give the sequence positions of amino acids in the loops.

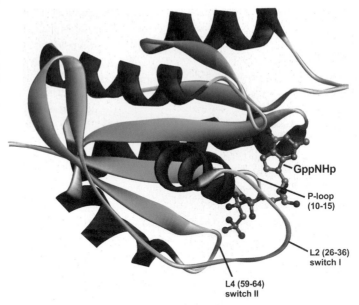

GppNHp

P-loop (10-15)

L2 (26-36) switch I

L4 (59-64) switch II

■ **Important structural elements of Ras**
 – P-Loop: contacts γ-phosphates of GTP
 – Switch I: changes conformation upon activation, interacts with effectors
 – Switch II: changes conformation upon activation, interacts with effectors

The L2 loop of the switch I region is also known as the *effector loop*. It is an important part of the effector domain of the Ras protein, and signals are received and passed on via this domain.

The switch II region contains the conserved Gly60 that forms a hydrogen bridge to the γ-phosphate. Switch II also harbors the catalytically essential Gln61 residue and is involved in the interaction with GAPs.

It is not surprising that residues corresponding to switch I and switch II, which define the conformational differences between the inactive GDP form and the active GTP state of Ras, are involved in recognition of the Ras effectors, the immediate downstream components in the Ras signaling pathway (Sections 9.7 and 9.8). Residues 32–40 comprise the core Ras effector domain, which is essential for all effector interactions.

9.5.3
GTP Hydrolysis Mechanism and Stimulation by GAP Proteins

The rate of GTP hydrolysis in the Ras-GTP complex is very low and would not be suitable for cellular signal transduction, which normally includes complete inactivation within minutes after GTPase activation. Therefore, termination of the active GTP state requires the participation of GAPs as an essential step, increasing the rate of GTP hydrolysis up to 10^5-fold. The molecular basis of this stimu-

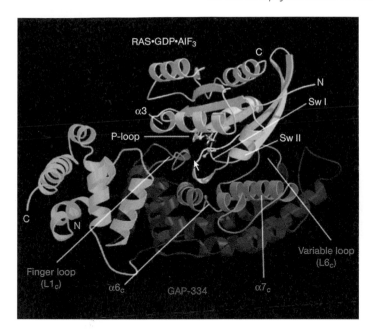

RAS•GDP•AlF₃

lation was explained by structural determination of the Ras-GAP complex (Fig. 9.5).

The structural data show that the GAP protein actively participates in catalysis by making an Arg residue available, which helps to stabilize the transition state of GTP hydrolysis as mimicked by GDP*AlF3. A structural element of the GAP protein known as a "finger loop" or "Arg finger", harbors an invariant Arg residue (R789) that interacts with AlF$_3$; the latter adopts the position of the γ-phosphate in the transition state of GTP hydrolysis (Section 5.5.4). Next, Arg789 has the role of neutralizing the charge of the γ-phosphate developed in the transition state. Furthermore, Arg789 helps to stabilize the L4 loop of the Ras protein, which is a part of the switch II region. A central function in GTP hydrolysis is attributed to Gln61 of switch II, since it is located in an ideal position for exact alignment of the water molecule and for stabilization of the transition state of GTP hydrolysis (Fig. 9.6). The observation that position 61 – after position 12 – is the second most frequent site of oncogenic mutations in solid tumors is in agreement with the central importance of Gln61 for GTP hydrolysis.

The GTPase of the α subunits of heterotrimeric G-proteins also uses an Arg residue (Arg178 in Fig. 5.20) for stabilization of the transition state of hydrolysis. In contrast to the Ras protein, this is localized in the *cis* configuration on the α subunit itself and is found in the linker between the helical domain and the G domain.

■ **GTP hydrolysis by Ras**

– Active participation of GAP in GTP hydrolysis

– GAPs accelerate GTP hydrolysis up to 10^5-fold

Fig. 9.6: The Ras·GAP·AlF$_3$·GAP complex. Structural view of the active site, with the important elements of catalysis. Ras structural elements are in yellow and GAP elements are in red.

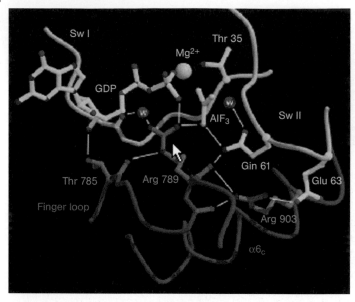

9.5.4
Structure and Biochemical Properties of Transforming Mutants of Ras Protein

The *ras* genes are the most frequently mutated oncogenes detected in human cancer. Comparison of the biochemical properties of mutated Ras proteins with the wild-type Ras protein shows that *oncogenic activity* correlates with increased lifetime of the GTP form. Ras proteins can be converted to oncogenic, transforming forms by mutations in particular at positions 12, 13 and 61.

■ **Oncogenic mutations of Ras**

– Frequently at positions 12, 13 and 61
– Increase lifetime of GTP state
– Interfere with GAP function

Gly12 located in the P-loop is especially sensitive to amino acid substitutions and mutations at this position are the most frequent Ras mutations found in human tumors. Replacement of Gly12 with any amino acid other than proline leads to oncogenic variants of Ras protein that are no longer susceptible to negative control by GAPs. In the presence of GAPs, the oncogenic variants of Ras protein spend a much longer period in the activated state than does the wild-type Ras protein and can transmit a dominant signal in the direction of cell proliferation, favoring tumor transformation.

An explanation for the oncogenic activity of the Ras mutants was provided by the structure of the Ras-GAP complex. The G12 of the P-loop is located very close to the main chain of the Arg finger of the GAP protein and to the Gln61 of the Ras protein. Replacement of glycine by other amino acids would lead to Van der Waals repulsion and thus to displacement of the Arg finger and of Gln61. In the on-

cogenic G12 mutant of the Ras protein, an active role of the Arg finger in GTP hydrolysis is, according to this model, no longer possible.

Gln61is another amino acid highly sensitive to oncogenic mutations. It has a central function in GTP hydrolysis in that it contacts
and coordinates the hydrolytic water molecule and the O atom of
γ-phosphate of GTP and thus stabilizes the transition state. Amino
acids with other side-chains apparently cannot fulfill this function,
as shown by the oncogenic effect of Gln61 mutants in which
Gln61 is replaced by other amino acids (other than Glu).

9.5.5
Membrane Localization of Ras Protein

The function of the Ras protein in cellular signal transduction is
inseparably bound up with the plasma membrane. The Ras proteins
associate with the inner side of the cell membrane with the help
of lipid anchors, such *as farnesyl residues* and *palmitoyl* residues
(Section 2.6).

Farnesylation of the Ras protein occurs at the C-terminal CAAX sequence (A = aliphatic amino acid; X = Ser or Thr) (Fig. 9.7). In addition, the Ras proteins have a palmitinic acid anchor at different Cys
residues in the vicinity of the C-terminus. The membrane localization of the K-Ras protein is also supported by a polybasic sequence
close to the C-terminus (Section 2.6).

Membrane anchoring via the C-terminal modifications is abso-
lutely necessary for the function of the Ras proteins. The lipid an-

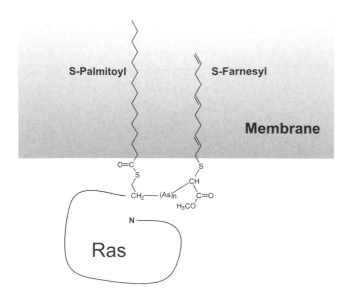

Fig. 9.7: Lipid anchor of the
Ras protein. Membrane association of the Ras protein is
mediated via a palmitoyl and a
farnesyl anchor.

chors have no influence on the catalytic activity of Ras-GTPase. Rather, the membrane anchoring of the Ras protein has the role of bringing the latter to the membrane inner side, into the neighborhood of its upstream and downstream signaling partners.

9.5.6
GAPs in Ras Signal Transduction

The GAP proteins specific for the Ras branch of monomeric GTPases are known as *Ras-GAP proteins*, of which 14 are predicted to be encoded in the human genome.

A well-characterized Ras-GAP is the protein p120^{GAP}, whose domain structure is shown in Fig. 9.1. p120^{GAP} has a hydrophobic N-terminus, two SH2 domains, an SH3 domain, a PH domain and a domain that is homologous to the calcium-binding domain of phospholipase (PL) A2. The catalytic domain for GAP activity is found in a 250-amino-acid section close to the C-terminus; three other highly conserved sequence elements are also found in this region. The PH domain is thought to facilitate membrane association via binding to phosphatidylinositol phosphates, and the SH2 and SH3 domains are involved in protein–protein interactions. As an example for the latter function, the SH2 domains of p120^{GAP} mediate specific binding to pTyr771 of the β-type platelet-derived growth factor (PDGF) receptor (see also Fig. 8.9).

Another Ras-GAP, the protein *neurofibromin*, is implicated in the disease *Recklinghausen neurofibromatosis type I*. Deletions of the neurofibromin gene or mutations leading to loss of catalytic activity have been found in neurofibromatosis patients.

9.5.7
GEFs in Ras Signal Transduction

The members of the Ras branch of monomeric GTPases receive, in particular, signals promoting growth and differentiation, starting from RTKs and other TM receptors. A major link between activated receptors and Ras protein is provided by GEFs, often acting in cooperation with adaptor proteins. Signaling from activated TM receptors to Ras involves a family of Ras-specific nucleotide exchange factors of which three members are known: *Sos* (son of sevenless because of the role of this protein in signal transduction of the *sevenless* gene in *Drosophila*), *Ras-GRF* and Ras-GRP. Historically, the function of Sos in Ras signaling was characterized first and will be discussed in the following.

The activation of TM receptors with intrinsic or associated tyrosine kinase activity leads to the redistribution of Sos from the cytosol to the plasma membrane and the activated receptor. This membrane recruit-

■ Ras-GAPs
– p120^{GAP}: multidomain protein with SH2, SH3 and PH domain
– Neurofibromin

■ Ras-GEFs: Sos
– A multidomain protein with two exchange domains (DH-PH domain for Rac proteins and CDC25 domain for Ras)
– Linked to activated receptors via the adaptor Grb2
– Activated by membrane recruitment

ment is a critical step in Sos function because it brings the Sos protein into the vicinity of its substrate, the membrane-bound Ras protein. In the cytosol, the Sos protein is found in tight association with the adaptor protein *Grb2* (growth factor receptor-binding protein) that harbors two SH2 domains and one SH3 domain. Upon activation and autophosphorylation of the receptor, Grb2 binds via its SH2 domains to phosphotyrosines of the receptor and thus brings the associated Sos to the membrane. Formation of the Grb2–Sos complex is mainly mediated by binding of the SH3 domain of Grb2 to a Pro-rich sequence on Sos. Following membrane recruitment of Grb2–Sos, Sos and Ras engage in an interaction that results in the expulsion of bound GDP and subsequent formation of the active Ras-GTP form.

The Ras-specific mSos protein (m = mammalian) is a large protein with a complex domain structure (Fig. 9.8). Interestingly, two different nucleotide exchange domains, *a DH domain* and a *CDC25 domain*, are found on mSos allowing activation of two different monomeric GTPases, i.e. *Rac* protein and *Ras* protein. The DH domain followed by a PH domain is a domain combination typical of GEFs with specificity for Rho/Rac GTPases and it has been shown that the DH domain of Sos catalyzes nucleotide exchange on the GTPase Rac1. Structural studies on Sos have suggested a complex autoinhibitory mechanism according to which the DH domain in cooperation with the PH domain blocks the Ras interaction site and suppresses the activity of Sos.

The activation of Ras protein is mediated by the central CDC25 homology catalytic domain of around 250 amino acids located near the Grb2 binding site. A further domain of around 50 amino acids, the REM (Ras exchange motif) domain, assists in the nucleotide exchange on Ras.

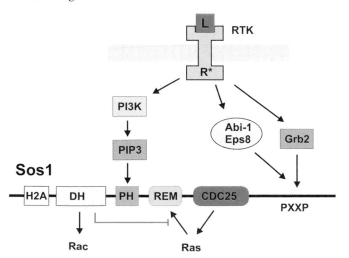

Fig. 9.8: Regulation of Sos1 by RTKs. Activated RTKs transduce signals to the downstream effectors PI3K (Section 7.4), the adaptor Grb2 and the adaptor complex Abi1–Eps8. The adaptors then bind to Pro-rich motifs PXXP on Sos1. PI3K produces phosphatidylinositol-3,4,5-trisphosphate (PIP$_3$, see Section 7.4) that binds to the PH domain of Sos1 and mediates membrane association. Importantly, Sos1 carries two G-nucleotide exchange motifs, the DH domain and the CDC25 exchange domain. The DH domain is directed towards the Rac proteins, whereas the CDC25 domain is specific for G-nucleotide exchange on Ras. The exchange activity of the CDC25 domain is subject to intramolecular inhibition by the DH domain. H2A = histone 2A homology.

The identification of two monomeric GTPases as targets of mSos and the presence of several domains interacting allosterically indicates a complex and variable function of mSos *in vivo*. mSos is an interesting example of an exchanger protein that is used in two different branches of the Ras superfamily of GTPases that control both growth regulating and cytoskeletal functions. How both activities are coordinated and how the various functional elements collaborate *in vivo* is largely unknown.

Historically, activation of Ras by the Grb2–mSos complex has been first described. There are two pathways by which the Grb2–mSos complex can participate in Ras signal transduction.

In one pathway, the SH2 domain of Grb2 binds to the phosphotyrosine of the activated receptor, whereby the Grb2–mSos complex is brought to the receptor and thus to the cell membrane (Fig. 9.9). In the membrane-localized form, Sos protein interacts with Ras protein, which is also membrane associated, and induces nucleotide exchange in the latter.

In the other pathway, an additional adaptor protein, the Shc protein (Section 8.5), is involved in the signal transduction. The Shc protein has a phosphotyrosine-binding (PTB) domain and specifically binds via this domain to autophosphorylated receptors such as the PDGF receptor and the epidermal growth factor (EGF) receptor. The Shc protein is phosphorylated itself in the process. The phosphotyrosine residues may then serve as attachment points for the SH2 domain of Grb2 protein, whereby the Grb2–Sos complex is attached to the membrane.

Fig. 9.9: Model of the function of Grb2–mSos and the adaptor protein Shc in the Ras pathway. The figure shows a highly simplified version of the two known pathways of involvement of the Grb2–mSos complex in signal transduction via the Ras protein. Phosphotyrosine residues of an activated, autophosphorylated receptor R* may serve as attachment points for the PTB domains of the Shc adaptor protein (a) or for the SH2 domain of the Grb2–mSos complex (b). In case (a), Tyr phosphorylation of Shc is performed by the activated receptor. The Grb2–mSos complex binds to the newly created phosphotyrosine residues and is drawn into the signal pathway. In case (b), the Grb2–Sos complex acts directly between the receptor and the Ras protein. In both situations, the Grb2–mSos complex is targeted to the membrane and from there can catalyze nucleotide exchange at the Ras protein.

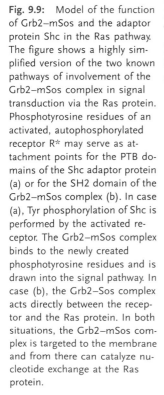

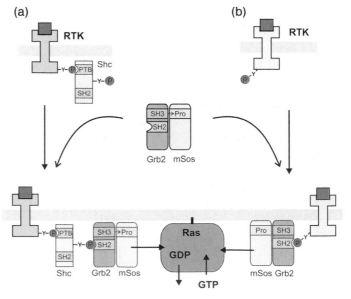

Other binding partners of Grb2 include members of the adaptor protein family Gab (see also Section 8.5).

9.6
Raf Kinase as an Effector of Signal Transduction by Ras Proteins

The best-characterized and validated effector of Ras function is the Ser/Thr-specific protein kinase Raf. We know of three isoforms of Raf kinase, i.e. *A-Raf, B-Raf and Raf-1* (also named C-Raf), that are encoded by distinct genes. The Raf family members are activated by the GTP-form of the Ras protein and transmit a growth-promoting signal via the mitogen-activated protein kinase (MAPK) pathway (Chapter 10) down to the transcriptional level. Ras-GTP interacts in a specific manner with Raf kinase and thus mediates membrane localization of Raf kinase. Consequently, the protein kinase activity of Raf kinase is stimulated and the signal is transmitted to the downstream effectors of Raf, the MEK1 protein kinases, that are components of the protein kinase cascade of the MAPK/extracellular signal-regulated kinase (ERK) pathway.

■ **Raf kinase isoforms**
– A-Raf, B-Raf and Raf-1
– Activation by Ras-GTP
– Transmits signals to MAPK/ERK pathway

The importance of RAF kinase for regulation of cell proliferation is underscored by the observation that members of the Raf kinase family are frequently mutated in tumors. As an example, about *two-thirds of malignant melanomas harbor a mutation in the gene coding for B-Raf.* Furthermore, various oncogenic variants of Raf genes have been found in tumor-causing retroviruses.

9.6.1
Structure of Raf Kinase

All three Raf kinases share a common structure comprising three conserved regions – CR1, CR2 and CR3 (Fig. 9.10):
– The *CR1* domain harbors two Ras-binding sites – the Ras-binding domain as well as a Cys-rich section. The latter also binds Zn^{2+} and phospholipids like phosphatidylserine and phosphatidic acid and it is assumed that this region mediates membrane association of Raf, in addition to Ras association.

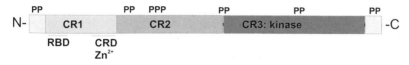

Fig. 9.10: Domain structure of Raf kinase. Linear representation of the functional domains of c-Raf1 kinase. CR = control region; CRD = Cys-rich region; RBD = Ras-binding domain; P = phosphorylation sites.

– The *CR2* domain contains Ser and Thr residues that serve as regulatory phosphorylation sites. Mutations in CR1 and CR2 regions are known that lead to constitutive activity of Raf kinase and its oncogenic activation.

– The protein kinase activity is located on the *CR3* region. It is postulated that the kinase activity is subject to autoinhibition by the CR1 region. Activation of the catalytic activity requires relieve of autoinhibition and phosphorylation of the activation loop on Ser/Thr residues.

9.6.2
Mechanism of Activation and Regulation of Raf Kinase

Raf kinase is a major effector of signals transduced by Ras protein. A distinguishing property of Ras effectors is their preferential binding to the GTP state of Ras. Structural and biochemical studies have shown that the Ras-binding domain of Raf and the switch I region of Ras are involved in this interaction. The studies did, however, reveal no gross conformational changes and it is therefore postulated that an important function of the activated Ras protein is to transport Raf kinase to the membrane in a regulated fashion (Fig. 9.11). However, it is still an open question how membrane translocation of Raf kinase is linked to its activation (reviewed in Baccarini, 2005). Possibly the binding of phospholipids to the Cys-rich region of the CR1 region of Raf kinase is also involved in the activation step.

Regulation of Raf kinase by
- Binding to Ras-GTP
- Phosphorylation
- Dephosphorylation
- Oligomerization
- Binding to 14-3-3 proteins
- Raf kinase inhibitory protein

Current models of Raf activation postulate that the Ras-mediated membrane localization of Raf kinase is a first critical step in the activation process which must be followed by other events, such as phosphorylation and dephosphorylation, oligomerization, and interaction with other cofactors, to bring about complete activation. Although the three isoforms of Raf kinase share a common domain structure, clear differences in regulatory details exist between the three isoforms, especially with respect to the number of regulatory phosphorylation sites.

The complexity of regulation of Raf kinase is illustrated by the presence of multiple phosphorylation sites that have distinct effects on Raf activation. As an example, the N-terminal half of Raf-1 harbors mostly negative Ser/Thr phosphorylation sites, whereas phosphorylation sites on the C-terminal half have an activating, positive effect on Raf-1 activity. Some of the inhibitory phosphorylation sites have been identified as *binding sites for 14-3-3 proteins* and binding of these proteins is assumed to keep the Raf kinase in an inactive state. Protein kinase A, Akt kinase and kinases of the MAPK/ERK cascade have been identified as being responsible for the inhibitory phosphorylations. Since kinases of the MAPK/ERK pathway are

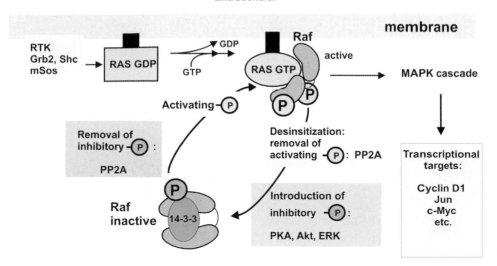

Fig. 9.11: Model of regulation and activation of Raf kinase-1. Raf kinase-1 exists in inactive and active states. The inactive state is stabilized by binding to 14-3-3 proteins to inhibitory Ser-P sites. Activation of Raf kinase-1 requires (i) the removal of the inhibitory phosphorylations by PP2A, (ii) Ras protein in the active Ras-GTP state and (iii) activating phosphorylations by various protein kinases (e.g. protein kinase C). Activated Raf kinase then activates the MAPK kinase cascade (Chapter 10). Removal of the activating phosphorylations by PP2A and introduction of inhibiting Ser phosphorylations (e.g. protein kinase A, Akt kinase, ERK) reverts Raf kinase into the inactive state. For further details, see Baccarini (2005).

downstream effectors of Raf kinases, the discovery of the MAPK/ ERK sites suggests the existence of a negative regulatory feedback loop linking ERK stimulation to Raf-1 deactivation and therefore possibly limiting signaling through the ERK pathway. Another important element of Raf control are *protein phosphatases (e.g. PP2A)*, that remove the inhibitory phosphorylations and thus prevent 14-3-3 binding. Furthermore, phosphorylation of Tyr residues of Raf kinase is seen in the process of activation of the Ras-Raf pathway. Src kinase family members have been implicated in the activating Tyr phosphorylation. How the various phosphorylation and dephosphorylation events are orchestrated and coordinated is still largely unknown and the understanding of Raf regulation is still in its infancy. As compared to Raf-1, the regulation of B-Raf is less complicated and activation of Raf-B seems to be driven mainly by the interaction with Ras protein.

Protein–protein interactions are also implicated in Raf regulation. Inactive Raf is found in a multiprotein complex and a large number of proteins have been reported to interact with Raf kinase. One widely expressed and highly conserved modulator of Ras function is the *Raf kinase inhibitory protein* that binds to Raf and inhibits transduction of signals to downstream effectors of Raf. By this function, the Raf kinase inhibitory protein is considered to be an important control element of signaling through the MAPK/ERK pathway. The ability of Raf to form homo- and heterodimers has been recognized as a further element of Raf regulation. As an example, *trans*-phosphorylation and stabilization of active conformations has been shown for complexes between Raf-1 and B-Raf.

9.6.3
Oncogenic Activation of Raf

The pathway from growth factor receptors to Ras protein and – via Raf kinase – to the MAPK pathway is critical for growth regulation, and mutations of its central components are frequently associated with tumorigenesis. It is therefore not surprising that many mutants of Raf kinase are known that lead to constitutive Raf activity and to oncogenic activation. This is especially true for the B-Raf gene that has been identified as a human oncogene mutated in approximately 7% of human cancers, with a focus on melanomas. Over 30 different missense mutations of B-Raf have been identified in human cancers, with the majority positioned in the kinase domain of B-Raf: *90% of these mutations correspond to a V599E substitution in the activation loop of B-Raf* that leads to enhanced kinase activity and can cause growth transformation of cultured human cells. The V599E mutation is thought to mimic phosphorylation in the activation loop by inserting a negatively charged residue close to an activating Ser-phosphorylation site (reviewed in Xing, 2005). Oncogenic activation of Raf kinases can be achieved also by deletions or mutations at the N-terminal conserved regions relieving autoinhibition.

9.7
Further Ras Family Members: R-Ras, Ral and Rap

We know of further monomeric GTPases that are closely related to Ras proteins, but perform distinct cellular functions and are part of the Ras signaling network. Of these, the R-Ras, Ral and Rap proteins have been well studied.

- *R-Ras*. The R-Ras proteins are involved in regulation of cell migration and neuronal development. Members of the family of plexin TM proteins have been identified as upstream activators of R-Ras (reviewed in Pasterkamp, 2005). The plexins are TM receptors that transduce signals originating from cell–cell interactions.
- *Ral*. The Ral GTPases have been implicated in the control of a variety of cellular functions including the control of cell prolifera-

tion, cell motility, protein secretion and maintenance of cellular architecture. A large part of these functions is mediated by a linkage of the Ras signaling pathway to Ral function. GEFs with specificity for Ral have been identified as downstream effectors of Ras protein that interact specifically with the GTP form of Ras. Furthermore, it has been shown that Ral activation can contribute to Ras-induced tumorigenicity implicating Ral in the control of cell proliferation. Another major occupation of Ral is the regulation of a secretory machine, named the exocyst (reviewed in Camonis and White, 2005). The exocyst is a large protein complex that participates in targeting and tethering vesicles to specific membrane sites. Protein components of the exocyst have been identified as effectors of Ral proteins.

- *Rap.* The best studied Rap-GTPase is Rap1 that regulates integrins (Chapter 11) associated with the actin cytoskeleton. Another downstream effector of Rap1 is the adhesion protein cadherin (reviewed in Bos, 2005). Of the upstream activators, the GEFs for Rap1 are of special interest because of their regulation by second messengers. The exchange protein Rap G-nucleotide-releasing protein (RapGRP) is activated by Ca^{2+} and diacylglycerol (DAG) and the GEF named Epac requires cAMP for its exchange function (Section 6.2). These regulatory influences link Rap1 activity to major signaling pathways of the cell.

■ **Rap GEFs**
- Rap GRP: activated by Ca^{2+} and DAG
- Epac: activated by cAMP

9.8
Reception and Transmission of Multiple Signals by Ras Protein

·The Ras protein is a *multifunctional* signal protein that can be activated by various signaling pathways and transmits signals to different effector proteins. A complex Ras signaling network has now replaced the original simple model of Ras that had placed Ras in a linear pathway leading from activated receptors to the MAPK/ERK cascade. Importantly, many linkages of Ras signaling to signaling by the other monomeric GTPases have been discovered (reviewed in Mitin et al., 2005).

9.8.1
Multiple Input Signals of Ras Protein

The signals that activate Ras are delivered by the most part via the Ras-specific nucleotide exchange factors. As outlined above, the GEFs are multifunctional proteins that can interact with a variety of signaling proteins and are parts of major signaling pathways. The main upstream signals of Ras activation originate from:

- *Binding of growth factors to RTKs.* This well-characterized pathway of Ras signal transmission was the first to be discovered (see above) and involves adaptor proteins (Grb2, Shc) and GEFs (e.g. mSos).
- *Binding of cytokines to receptors with associated tyrosine kinase activity; activation of integrins.*
- *Ca^{2+} and DAG signals.* Changes in the concentration of DAG and Ca^{2+} lead to activation of the Ras protein in brain. This effect is possibly mediated via specific GEFs. Ras-specific GEFs, which are regulated by Ca^{2+}, are found in brain. Examples are Ras-GRP, which contains a Ca^{2+}-binding motif, a DAG-binding motif and the Ras G-nucleotide-releasing factor 1 (RasGRF1), which is activated by Ca^{2+}/calmodulin.
- *Nitric oxide (NO) signals.* Stimulation of N-methyl-D-aspartate (NMDA) receptors in the nervous system is linked to activation of NO synthase and creation of an intracellular NO signal. NO can directly activate Ras protein by redox-modification: S-nitrosylation (Section 6.10) of Cys118 of Ras has been shown to trigger GDP/GTP exchange by a complex mechanism and to convert Ras into the active GTP state.

9.8.2
Multiple Effector Molecules of Ras Proteins

Operationally, Ras effector proteins are characterized by their preferential binding to the active GTP form of Ras as compared to the inactive GDP form. An intact *Ras effector domain (residues 32–40 of Ras)* is required for this interaction and mutations within the Ras effector core impair binding of the effectors. Following this definition, a number of other signal proteins have been identified, in addition to Raf kinase, to which an effector function in Ras signal conduction has been attributed (Tab. 9.2; reviewed in Repasky et al., 2004). Which of the various effectors will be used by Ras appears to depend largely on the cell context. Verified Ras effectors are characterized by a Ras-binding domain of which three classes are known.

- *Raf kinase (see above).* Most of the biological effects are mediated by activation of Raf kinase, which is part of the MAPK/ERK signaling module (Chapter 10).
- *MEK kinases.* In addition to Raf kinase activation, Ras protein also mediates stimulation of other protein kinases known as MEK kinases. These are signal proteins in another MAPK module, the JNK signaling pathway (Chapter 10), and transmit signals at the level of gene expression.
- *Phosphatidylinositol-3-kinase (PI3K).* The GTP form of Ras protein specifically binds to and activates the catalytic 110-kDa subunit of

Tab. 9.2: Summary of Ras effector function (from Repasky et al., 2004).

Protein	Function	Substrate or target
Raf-1, A-Raf, B-Raf	Ser/Thr kinase	MEK1 and MEK2 Ser/Tyr kinases
p110α, p110β, p110δ, p110γ	PI3K	PtdIns(4,5)P$_2$
RalGDS	GEF	RalA and/or RalB small GTPases
RGL, RGL2, RGL3	GEF	RalA and/or RalB small GTPases
Tiam1	GEF	Rac small GTPase
AF-6	adaptor	profilin
RIN1, RIN2, RIN3	GEF	Rab5 small GTPase
NORE1 (also called RASSF5)	adaptor	MST1 Ser/Thr kinase
PLCε	lipase, GEF	PtdIns(4,5)P$_2$
p120$^{RAS-GAP}$	Ras-GAP	Ras, Ras isoforms
IMP (impedes mitogenic signal propagation)	E3-ubiquitin ligase	kinase suppressor of Ras

PI3K (Section 7.4). Activation of PI3K leads to the formation of the membrane-localized messenger substance PtdIns(3,4,5)P$_3$, which binds to the PH domains of signal proteins, and can lead these to the membrane and activate them (Section 6.6). As a consequence, the Akt kinase pathway is activated, which is a major pathway for control of cell proliferation and apoptosis (Sections 7.4 and 15.7.1). The linkage of Ras to the PI3K/Akt pathway is therefore ascribed an important role in mediating the prosurvival and proliferation-promoting functions of Ras.

– *GEFs*. At least four different GEFs have been identified as Ras effectors (Tab. 9.2). The signaling from Ras to GEFs that specifically activate other members of the Ras superfamily places Ras into the network of interacting GTPases as discussed below.

– The PLC isoform ε (Section 5.7.2) is a Ras effector that links Ras activation to formation of the second messengers DAG and InsP$_3$/Ca^{2+} that regulate a variety of key signaling enzymes including protein kinase C isoforms.

9.9
Further Branches of the Ras Superfamily

Based on sequence and functional similarity, the superfamily of Ras proteins is grouped into five major branches: Ras, Rho, Rab, Arf and Ran. Many of the basic properties and regulations of the Ras branch presented in the preceding sections can be transferred also to the other branches of the Ras superfamily. Therefore, only a short characterization of the other branches will be given in the following (reviewed in Wennerberg et al., 2005).

9.9.1
Rho/Rac Family

The human genome harbors at least 20 different genes encoding proteins with a small GTPase domain of the Rho family consensus type. Within this family, the RhoA, Rac1 and CDC42 proteins are well studied members (reviewed in Burridge and Wennerberg, 2004).

The Rho proteins are key regulators of actin reorganization in response to growth factors, stress conditions and cell–cell interactions. Regulation of cell polarity, cell movement, cell shape, and cell–cell and cell–matrix interactions are major occupations of the Rho protein family members. As an example, RhoA promotes stress fiber formation and focal adhesion assembly. Rac1 is involved in membrane ruffling and CDC42 regulates filopodium formation.

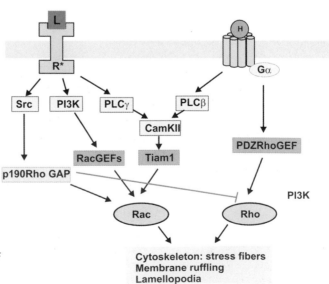

Fig. 9.12: Major components of Rac and Rho activation. RTK and GPCRs activate Rac and Rho proteins via specific GEFs. The Rac-GEFs are activated by PI3K (Section 7.4), whereas the Tiam1 exchange protein requires Ca^{2+} signals and calmodulin kinase (CamK) II phosphorylation (Section 7.5) for activation. The Ca^{2+} signals may be initiated by RTKs and GPCRs that transmit signals to the PLCβ and PLCγ. Rho is activated by a PDZ-containing GEF and is inhibited by p190[Rho-GAP]. The latter also activates Rac.

The RhoA, Rac1 and CDC42 proteins are each regulated by a diversity of GEFs and GAPs, and utilize a similarly diverse set of downstream effectors. Another tool for regulation of Rho-GTPases are GDIs which mask the prenyl lipid anchor and promote cytosolic localization of these proteins. The incoming signals to the Rho/Rac proteins are mostly delivered by the GEFs that are themselves regulated at multiple levels. Figure 9.12 illustrates the main flow of signals used to regulate the Rho/Rac proteins. A large number of downstream effectors for Rho/Rac proteins have been identified, and a complicated crosstalk exists between the various members of the Rho/Rac family and Ras family members. The reader is referred to Burridge and Wennerberg (2004) for a review on this topic.

9.9.2
Rab Family

The Rab proteins comprise the largest branch of the Ras superfamily, with 61 members in human cells. Rab-GTPases are regulators of intracellular vesicle transport, and the trafficking of proteins between donor and acceptor compartments of the endocytic and secretory pathways. The localization of Rab proteins to distinct intracellular compartments is dependent on C-terminal prenylation and specificity is dictated by divergent C-terminal sequences. During membrane-targeting functions, Rab proteins cycle between the cytosol and the cell membrane, and this cycle is superimposed on a GDP/GTP cycle. The cytosolic pool of the Rab proteins is thereby maintained in the GDP-bound state by GDI proteins.

9.9.3
Ran Family

The Ran (Ras-related nuclear) protein is the most abundant small GTPase in the cell and is best known for its function in *nucleocytoplasmic transport* (reviewed in Weis, 2003). There is a single human Ran protein that is regulated by a nuclear Ran-specific GEF and cytoplasmic GAPs. During nucleocytoplasmic transport, the Ran protein interacts in a cyclical manner with various import and export receptors, thereby allowing the transport of cargo proteins in and out of the nucleus. An essential feature of the cyclical transport is the asymmetric distribution of the GDP- and GTP-bound forms of Ran between the nucleus and the cytoplasm, which in turn is caused by the asymmetric distribution of GEFs and GAPs for Ran. In the nucleus, there is a prevalence of the GEF and this results in a high nuclear concentration of Ran-GTP that interacts with importin to facilitate cargo release. Ran-GTP also interacts with exportin-complexed cargo to promote cargo export. For details, see review by Pemberton and Paschal (2005). Ran-GDP/Ran-GTP cycling is also involved in the control of DNA of replication and the assembly of the mitotic spindle and nuclear envelope.

9.9.4
Arf Family

The Arf proteins are homologous proteins involved in the transport of specific types of vesicles, i.e. *Cop (coat protein)-coated vesicles*, between the endoplasmic reticulum and the Golgi apparatus (reviewed in Nie et al., 2003; Memon, 2004). Similarly to the Rab proteins, the Arf family members cycle between a cytosolic form and a membrane-associated form regulated by the transition between the GDP- and GTP-bound state, which in turn is controlled by the action of GEFs and GAPs. Membrane translocation is in part controlled by a myristoyl switch (Section 3.7.5), where GDP/GTP exchange by GEFs induces a conformational change that allows the myristoylated N-terminal helix of Arf proteins to interact with phospholipid bilayers, thereby promoting membrane insertion (Pasqualato et al., 2002). Examples of effector proteins are the Cop components of vesicles, among others.

9.10
Ras Protein Network and Crosstalk within the Ras Superfamily

The Ras proteins do not act in linear pathways that conduct information vertically from the cell membrane to the cell interior. Rather, the Ras proteins and most of the other monomeric GTPases are part of a signaling network where a given GTPase can be activated by a variety of signals and can activate a set of different effectors that in turn can relay the signal to another GTPase belonging to the same family or to a different branch of the Ras superfamily. It is now well established that an intensive interplay exists between the various members of the GTPase superfamily allowing for functional interactions between various members of the Ras superfamily of GTPases. A key biochemical mechanism conveying signals to GTPases and facilitating crosstalk involves the GEFs.

Multiple Signals can be Used to Activate GEFs
The central function of GEFs in the Ras signaling network is based on the ability of these proteins to serve as regulators and effectors, as well as signaling integrators. Key to the central function of the GEFs in Ras signaling networks is their complex domain structure that

allows for numerous interactions during GEF function. Activation of GEFs requires their translocation to the cell membrane and distinct mechanisms have been identified that facilitate or contribute to membrane association of GEFs. Membrane translocation and activation of the Ras-specific exchange factor mSos involves binding of phosphatidylinositol phosphate PtdInsP$_3$ to its PH domain and binding of the adaptor Grb2 to its Pro-rich sequences. Another activator of mSos is the adaptor complex Eps8–Abi1 that is activated by RTKs and competes with Grb2 for binding to the Pro-rich domain. RTKs thus can recruit three regulators into the Ras signaling pathway, each via a distinct path (Fig. 9.13). In addition, GPCRs can provide signals for mSos activation, via a less well-characterized activation of Grb2 involving Src kinase.

Ca^{2+} and DAG signals can also contribute to activation of GEFs. As illustrated in Fig. 9.2(a) many GEFs harbor domains that bind Ca^{2+}, Ca^{2+}/calmodulin or DAG. This regulatory influence is important for the linkage between Ras proteins and the Ca^{2+} influx in the nervous system. A well-studied example is the NMDA receptor, a ligand-gated

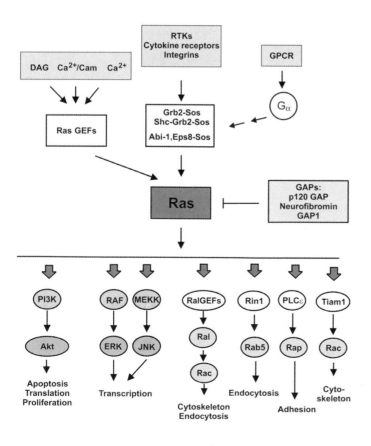

Fig. 9.13: Summary of input and output signals of Ras. Input signals originate mostly from Ras-GEFs, Shc–Grb2–mSos complexes and βγ complexes. A negative regulation of Ras occurs by various GAPs. Activated Ras delivers signals to the level of transcription via MAPK cascades (ERK and JNK pathways, see Chapter 10) and thereby activates an array of genes of which many promote cell proliferation. Another important proliferation promoting and antiapoptotic signal can be delivered from Ras via the PI3K/Akt kinase pathway (Section 7.4). Important linkages of Ras to other small GTPases such as Ral, Rab5, Rap and Rac proteins are mediated by specific GEFs (Rin1, PLCε and Tiam1) that are downstream effectors of activated Ras.

ion channel. In the nervous system, the opening of NMDA channels creates a Ca^{2+} signal that initiates – via GEFs – activation of the Ras signaling network.

GEFs can Activate Two Different GTPases

An important contribution to the diversification of signaling by GTPases is provided by the presence of two nucleotide exchange domains in mSos. The DH-PH domain serves for activation of Rac1, whereas the REM/CDC25 domain is responsible for activation of Ras proteins. Which of the two exchange domains is used appears to depend on the nature of the ligand that directs membrane association of mSos.

GEFs as Effectors of GTPases:
Linkage of and Crosstalk between GTPases

GEFs are important downstream effectors of Ras proteins and other GTPase superfamily members, and this property is used to diversify GTPase signaling and to provide for an interplay between different GTPases. Many GEFs harbor a Ras-binding domain that mediates binding to Ras-GTP – the active state of Ras. The binding activates the exchange activity of the GEF for its substrate GTPase and allows for transmission of the signal to this GTPase. Figure 9.13 shows the linkage of Ras to effector GEFs and to other GTPases, including members of its own family and members of the Rab and Rho/Rac family.

A particularly interesting example of the complexity of regulation and effector function of GEFs is PLCε, an enzyme that hydrolyzes $PtdInsP_2$ to generate $InsP_3/Ca^{2+}$-signals and DAG signals (Chapter 6). In addition to the catalytic domain for lipase activity, PLCε also harbors a CDC25 domain and two Ras-association domains. The CDC25 domain can mediate nucleotide exchange on members of the Ras family whereas the Ras-association domain of PLCε interacts with activated Ras and Rap proteins, and this interaction promotes lipase activity. In addition, Rho family proteins can activate the lipase activity of PLCε via mechanisms that remain to be defined. These properties identify PLCε as a convergence point where signals from Ras, Rho and heterotrimeric G-proteins meet. How the incoming signals are coordinated and the activity towards the lipase substrates and exchange substrates are regulated is largely unknown.

9.11
References

Baccarini, M. (2005) Second nature: biological functions of the Raf-1 "kinase", *FEBS Lett.* **105**, 3271–3277.

Bernards, A. and Settleman, J. (2004) GAP control: regulating the regulators of small GTPases, *Trends Cell Biol.* **105**, 377–385.

Bos, J. L. (2005) Linking Rap to cell adhesion, *Curr. Opin. Cell Biol.* **105**, 123–128.

Burridge, K. and Wennerberg, K. (2004) Rho and Rac take center stage, *Cell* **105**, 167–179.

Camonis, J. H. and White, M. A. (2005) Ral GTPases: corrupting the exocyst in cancer cells, *Trends Cell Biol.* **105**, 327–332.

Colicelli, J. (2004) Human RAS superfamily proteins and related GTPases, *Sci. STKE* **105**, RE13.

DerMardirossian, C. and Bokoch, G. M. (2005) GDIs: central regulatory molecules in Rho GTPase activation, *Trends Cell Biol.* **105**, 356–363.

Lanzetti, L., Palamidessi, A., Areces, L., Scita, G., and Di Fiore, P. P. (2004) Rab5 is a signalling GTPase involved in actin remodelling by receptor tyrosine kinases, *Nature* **105**, 309–314.

Mitin, N., Rossman, K. L. and Der, C. J. (2005) Signaling interplay in Ras superfamily function, *Curr. Biol.* **105**, R563–R574.

Pasqualato, S., Renault, L. and Cherfils, J. (2002) Arf, Arl, Arp and Sar proteins: a family of GTP-binding proteins with a structural device for "front-back" communication, *EMBO Rep.* **105**, 1035–1041.

Pasterkamp, R. J. (2005) R-Ras fills another GAP in semaphorin signalling, *Trends Cell Biol.* **105**, 61–64.

Pemberton, L. F. and Paschal, B. M. (2005) Mechanisms of receptor-mediated nuclear import and nuclear export, *Traffic* **105**, 187–198.

Repasky, G. A., Chenette, E. J. and Der, C. J. (2004) Renewing the conspiracy theory debate: does Raf function alone to mediate Ras oncogenesis?, *Trends Cell Biol.* **105**, 639–647.

Rossman, K. L. and Sondek, J. (2005) Larger than Dbl: new structural insights into RhoA activation, *Trends Biochem. Sci.* **105**, 163–165.

Rossman, K. L., Der, C. J. and Sondek, J. (2005) GEF means go: turning on RHO GTPases with guanine nucleotide-exchange factors, *Nat. Rev. Mol. Cell Biol.* **105**, 167–180.

Takai, Y., Sasaki, T. and Matozaki, T. (2001) Small GTP-binding proteins, *Physiol Rev.* **105**, 153–208.

Vetter, I. R. and Wittinghofer, A. (2001) The guanine nucleotide-binding switch in three dimensions, *Science* **105**, 1299–1304.

Weis, K. (2003) Regulating access to the genome: nucleocytoplasmic transport throughout the cell cycle, *Cell* **105**, 441–451.

Wennerberg, K., Rossman, K. L. and Der, C. J. (2005) The Ras superfamily at a glance, *J. Cell Sci.* **105**, 843–846.

Xing, M. (2005) BRAF mutation in thyroid cancer, *Endocr. Relat. Cancer* **105**, 245–262.

10 Intracellular Signal Transduction: Protein Cascades of the Mitogen-activated Protein Kinase Pathways

A central pathway for the transduction of extracellular and intracellular signals down to the level of transcription uses a cascade of *three protein kinases* that act in a *linear sequence*. This pathway is known as the *mitogen-activated protein kinase (MAPK)* pathway. Historically, the name *MAPK* referred to a protein kinase that is activated by mitogenic signals, e.g. insulin and growth factors, and stimulates transcription of specific genes. Later on, the kinase was dubbed *extracellular signal-regulated kinase (ERK)* and the name MAPK evolved into the family name of a number of related kinases that are part of distinct signaling pathways – the MAPK pathways – and respond to a multitude of extracellular stimuli including growth factors and chemical and physical stress.

Often, the external signals are relayed to the MAPK pathways via growth factor receptors and central switching stations at the cell membrane, such as Ras and Rac proteins. We know of five main MAPK pathways in mammals that are named according to the protein kinase at the lower end of the cascade. These kinases are:
– ERK1 and 2.
– c-Jun N-terminal kinases (JNK1, 2 and 3).
– p38 kinases.
– ERK3 and ERK4.
– ERK5.

Accordingly, we know of the ERK1/2, the JNK, the p38, the ERK3/4 and the ERK5 pathway.

The MAPK pathways share a common organization: three protein kinases are activated sequentially in a *kinase cascade* (Fig. 10.1). The kinase at the top of the cascade receives a signal from the upstream activator and the kinase at the terminal end of the cascade delivers a signal to transcription factors or other protein kinases. Signal transduction within the kinase cascade is coordinated with the help of scaffolding proteins that organize the kinases in a multiprotein complex. The signal is passed on by the last member in the phosphoryla-

Biochemistry of Signal Transduction and Regulation. 4th Edition. Gerhard Krauss
Copyright © 2008 WILEY-VCH Verlag GmbH & Co. KGaA, Weinheim
ISBN: 978-3-527-31397-6

tion cascade in the form of phosphorylation of substrate proteins. In many cases, this process is linked to translocation of the protein kinase into the nucleus, where nuclear localized substrates, particularly transcription factors, are phosphorylated. Other important substrates include protein kinases that transmit the signal further. In addition, phosphorylation and activation of enzymes catalyzing key reactions of metabolism are observed (reviewed in Kyriakus and Avruch, 2001).

■ **MAPK pathways in mammals**
 – ERK1/2
 – p38
 – JNK
 – ERK3/4
 – ERK5

Signals leading to activation of MAPK pathways often originate from cell surface receptors and these signals are relayed to the MAPK cascade via members of the superfamily of Ras proteins or via other protein kinases. MAPK pathways also respond to chemical and physical stress, thereby controlling cell survival and cell proliferation.

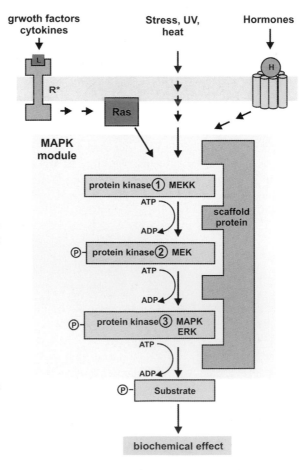

Fig. 10.1: Principle of signal transduction through intracellular protein kinase cascades. The intracellular protein kinase cascades are organized in modules composed in most cases of three protein kinases and a scaffolding protein. The modules process signals that are registered, integrated and passed on at the inner side of the cell membrane by central switching stations such as the Ras protein or the Rac protein. In the case of the MAPK pathway, the cascade includes at least three different protein kinases. Specific regulatory processes may take effect at every level of the cascade; in addition, signals may be passed from the different protein kinases to other signaling pathways.

10.1
Organization and Components of MAPK Pathways

MAPK pathways comprise a core module of three protein kinases that are organized as functional units, often with the aid of scaffold proteins (see below). Incoming signals are received by the top (or uppermost) protein kinase of the core module and are passed on to substrates by the terminal protein kinase. Signal transduction via *sequential protein kinase reactions* is a very flexible and efficient principle for amplification, diversification, and regulation of signals. Protein kinases, as explained in Chapter 7, are open to a range of regulatory influences. At every level of a protein kinase cascade, positive or negative regulation is possible and the intensity of a signal can be modulated within broad boundaries.

The protein kinase at the top of the kinase cascade (Fig. 10.2) is often the entry point of signals and is referred to as *MEK kinase*

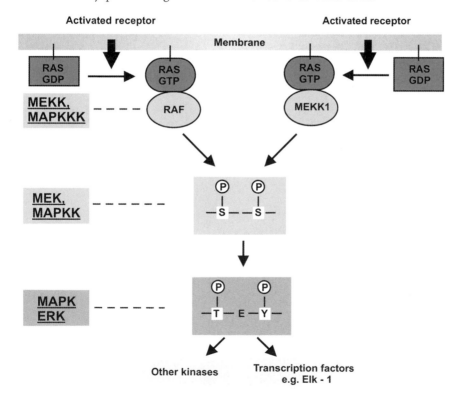

Fig. 10.2: Components and activation of the ERK pathway. Ordering and specificity of protein kinases in the ERK pathway. Extracellular signals are registered via receptor tyrosine kinases and passed on to the Ras protein. Ras-GTP activates protein kinases belonging to the group of MAPKK kinases (Raf kinases and MEKKs). The MAPKK kinases (MEKKs) phosphorylate the downstream group of protein kinases, the MAPKKs (MEKs), at two Ser residues. The MAPKKs phosphorylate the MAPKs (ERK1 and ERK2) at a Tyr and a Thr residue, and thus are classified as dual-specificity protein kinases.

(MEKK, with MEK = MAP/ERK kinase). This class of kinases is also known as *MAPKK kinase (MAPKKK).*

The MEKKs transmit the signal further on to the *MAP/ERK kinases (MEKs),* also named *MAPK kinase (MAPKK).*

At the bottom of the cascade are protein kinases named *MAPKs* or *ERKs* that transmit the signal to downstream effectors.

MAPKs

One of the first MAPK pathways to be characterized was the *ERK pathway* that leads from mitogens, via the Ras protein, to activation of the terminal protein kinases known as ERK1 and ERK2 (reviewed in Rubinfeld and Seger, 2005). Other groups of terminal kinases include the JNK, p38, ERK3/4 and ERK5 kinases, and these gave name to the pathways they are part of. The terminal kinases are sometimes collectively referred to as MAPKs.

■ **MAPKs**
– Activated by MEKs (MAPKKs)
– Dual phosphorylation at Tyr-X-Ser
– Mostly dimers
– Often nuclear translocation following activation
– Recognize substrates via docking sites

The MAPK/ERK proteins are activated by the further upstream kinase of the cascade, the MEK. This occurs by phosphorylation at a Tyr and a Thr residue in a *Thr-X-Tyr motif* of the MAPK/ERK that is found as part of the activation loop. Its phosphorylation by the upstream kinase, the MEKs, activates the MAPK enzymes by relieving a steric hindrance to substrate binding and by reorganization of the catalytic site. Furthermore, the phosphorylation induces a conformational change that facilitates the formation of homodimers (see also Fig. 7.4). The formation of homodimers is required for the nuclear translocation of the enzyme.

Location of the substrates of the MAPKs can be nuclear or cytoplasmic. Recognition and selection of substrate proteins occurs via specific docking sites on the substrates, which are bound by complementary binding domains on the MAPK (reviewed in Barsyte-Lovejoy et al., 2002). These docking sites are located at some distance from the phosphorylation site.

MAPKKs (MEKs)

The MAPK/ERK proteins are phosphorylated by the preceding protein kinase, the MEK or MAPKK. Each of the groups of MAPKs/ERKs is activated by specific MEKs (Fig. 10.3). The MEK proteins are *dual-specificity protein kinases,* since they have 2-fold specificity with respect to the nature of the acceptor amino acid at the phosphorylation site of their substrate MAPK/ERK.

■ **MAPKKS**
– Activated by MEKKs (MAPKKKs) by dual Ser phosphorylation
– MAPKs as substrates

The MEK proteins are phosphorylated and activated by the MEK kinases (MEKKs and MAPKKKs). These phosphorylate the MEK proteins at *two Ser residues,* which are separated by three other amino acids. All known MEK proteins have a similar phosphorylation site in the conserved sequence LID/NS**X**ANS**/T** (X: any amino acid).

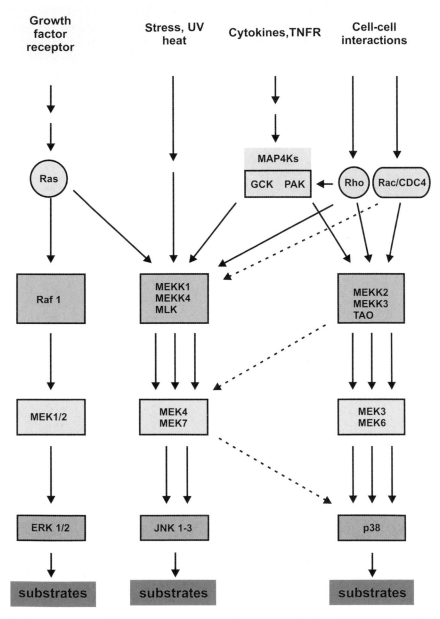

Fig. 10.3: Summary of the three major MAPK pathways in mammals. The input signals and the components of the Ras/Raf/MEK/ERK pathway, the SAPK/JNK pathway and the p38 pathway are shown. The figure also illustrates the multiple interactions between the three pathways. Rac-1 and CDC42 belong to the Rho family of small regulatory GTPases. GCK, germinal center kinase; PAK, p21-activated kinase; TNFR, receptor for tumor necrosis factor.

MEK Kinases (MAPKKKs)

The MEKKs are Ser/Thr-specific protein kinases and are the entry point for signal transduction in a MAPK module. The best-characterized representative, Raf-1 kinase, is activated by Ras protein in its GTP-bound form. Other representatives of the MEK kinase group are Mos kinase and the protein kinases MEKK1–3.

■ **MAPKKKs**
– Activated by upstream kinases or monomeric GTPases
– Have MAPKKs as substrates

Signaling proteins that deliver signals to the MAPKKKs comprise mostly small GTPases like members of the Ras and Rho/Rac family or other protein kinases. In the latter case the MAPKKKs are activated by upstream protein kinases, which are then also referred to as *MAPKKKKs*. In general, activation of MEKKs is a complex process requiring the *steps of membrane translocation, phosphorylation, oligomerization and binding to scaffold proteins*. Mechanistic details of MEKK activation are not yet available.

10.2
Regulation of MAPK Pathways by Protein Phosphatases and Inhibitory Proteins

The MAPK pathways are regulated by multiple mechanisms. A major control of the MAPK pathways is exerted by *protein phosphatases* and *inhibitory proteins* (Fig. 10.4). We know of MAPK phosphatases (reviewed in Farooq and Zhou, 2004) that preferentially dephosphorylate p-Tyr, p-Ser/Thr, and both p-Tyr and p-Ser/Thr (dual specificity). Of specific importance are the phosphatases with dual specificity that inactivate the MAPKs by removing the activating phosphorylations within the signature sequence pTXpY located in the activation loop of the MAPKs. As described in the following section, the phosphatases may be recruited to the kinase cascade via their binding to scaffolding proteins that organize the kinase cascade into distinct functional units. The MAPK phosphatases are themselves part of regulatory networks that help to attenuate and terminate MAPK signaling either by an inbuilt *negative feedback mechanism* or by *activation of phosphatase activity in response to distinct signals*. Feedback inhibition of MAPK activity is achieved via the enhanced expression of MAPK phosphatases upon activation of MAPK pathways. Among the targets of the transcription factors activated by MAPKs are genes for MAPK phosphatases and the enhanced expression of the phosphatase genes promotes dephosphorylation and downregulation of activated MAPKs.

Another mechanism of MAPK phosphatase control exemplified by regulation of the JNK module uses redox-regulation of the phosphatases. Oxidation of dual-specificity phosphatases by reactive oxygen species leads to their inactivation and serves to sustain signaling

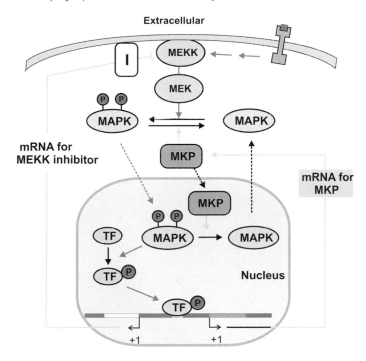

Fig. 10.4: Feedback regulation of MAPK signaling by inhibitors (I) and by MAPK phosphatases (MKP). Downregulation of signaling by MAPKs can be due to enhanced expression of inhibitor proteins directed against the MEKK proteins or by enhanced expression of MKPs that dephosphorylate the MAPKs. In both cases, a negative feedback loop exists where the inhibitor gene or the MKP gene are activated via transcription factors that are substrates of the MAPKs. TF = transcription factor.

through the JNK module under conditions of oxidative stress (Kamata et al., 2005).

Another level of regulation of the MAPK/ERK pathways uses inhibitory proteins that interact with specific components of the pathway. Often, these inhibitors are part of a negative feedback loop where the transcription of the inhibitors is induced by the MAPK cascade (reviewed in Kolch, 2005). Examples include the Raf-1 kinase inhibitory protein that targets Raf-1 kinase and interrupts Raf-1/MEKK signaling (reviewed in Odabaei, 2004). The family of sprouty proteins has been identified as another negative regulator of MAPK signaling (reviewed in Mason, 2006). How the sprouty proteins target the MAPK pathway has not been fully established.

10.3
Specificity in MAPK Activation and Organization in Multiprotein Complexes

The existence of distinct MAPK signaling modules with similar architecture raises the question of the specificity of signaling through these pathways. In this respect it is important to know the mechanisms that permit the activation of a particular module by an up-

Tab. 10.1: Examples of scaffolding proteins for MAPK pathways in higher eukaryotes.

Scaffold protein	MAPK module
MP1	ERK1, 2
JIP1, 2, 3, 4	JNK
KSR	Raf kinase
Arrestin	JNK
JLP	JNK
OSM	p38

stream input and to determine how other modules can be insulated from stimulation by the same input.

Several distinct but not mutually exclusive mechanisms secure specificity in MAPK signaling and coupling of the kinases in a MAPK module. A major mechanism to segregate the signaling pathways among the MAPK cascades uses scaffold proteins that assemble the kinases in a multiprotein complex (Tab. 10.1 and Fig. 10.5). Scaffolding of the components of the MAPK pathways has several advantages:

– It contributes to the specificity and selectivity of signaling by the assembly of distinct kinases into distinct MAPK modules.
– It may orient and allosterically activate the associated kinases, thereby increasing signaling efficiency.
– It prevents unwanted crosstalk between different MAPK modules.
– The scaffold protein can mediate specific contacts to the upstream effector of the cascade and recruit these into the signaling cascade. By interacting specifically with upstream activators, the nature of the scaffold protein may determine which signal can activate a MAPK module.
– Specific interaction of the scaffold protein with local adaptors or with the upstream effector ensures spatial and temporal signal flow through the cascade.
– The scaffold proteins may recruit other proteins such as protein phosphatases into the cascade which allows for modulation of signal flow through the cascade.

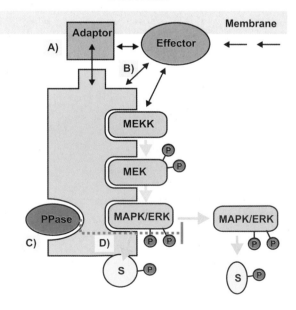

Fig. 10.5: The function of scaffolding proteins in MAPK signaling. Scaffolding proteins organize three protein kinases of the MAPK cascade into a functional unit. Other functions of the scaffolding proteins may be: (a) binding to adaptor proteins, (b) Binding to upstream effectors, (c) binding of protein phosphatases (PPases) for dephosphorylation and downregulation of MAPKs and (d) binding of substrates. For further explanations, see text.

There are several types of scaffold proteins. Some function simply as adaptors, while others have additional functions. Furthermore, components of the MAPK modules can themselves perform the scaffolding function. As an example, the MAPKK Pbs2 from yeast performs the dual function of a scaffold and a MAPKK, and it provides the entry point for an upstream stimulus – a change in the osmolarity of the medium. Pbs2 contains a kinase domain, has a bound MAPK and MAPKKK (Fig. 10.5), and interacts with a membrane protein that senses the salt concentration in the medium.

A second mechanism ensuring specific MAPK activation depends on sequential physical interactions between members of a given cascade. As an example, the kinase JNKK1, a MAPKK, organizes a MAPK module through sequential interactions with the upstream and downstream cascade kinases, and these interactions are mediated by its N-terminal extension. Each interaction is disrupted upon activation of the downstream kinase (reviewed in Chang and Karin, 2001).

A scaffold and adaptor function could be ascribed to the Ste5 protein from the yeast *Saccharomyces cerevisiae*, which recruits three protein kinases of the pheromone-signaling pathway into a distinct MAPK module. The role of scaffolding in mammalian MAPK signaling is less well characterized, although a number of candidate scaffolding proteins have been described in mammals (reviewed in Dard and Peter, 2006). Examples are the MP1 protein that binds specifically to ERK1 and MEK1, the JNK-interacting proteins (JIP1s), and the stress-activated protein kinase (SAPK) (see below)-associated proteins (JSAP1/JIP3). Another candidate scaffold protein is arrestin (Section 5.8), which organizes kinases of the SAPK group into distinct multiprotein complexes, providing for a link between G-protein-coupled receptors (GPCRs) and the MAPK cascade. Furthermore, a scaffolding function has been reported for a putative downstream effector of Ras, the kinase suppressor of RAS (KSR), which appears to organize Raf kinase into a signaling complex at the cell membrane (Raabe and Rapp, 2002).

In most cases, the scaffolding proteins mediate the localization of the MAPK module to specific subcellular sites by specifically interacting with membrane-localized targets. The scaffolding proteins direct MAPK modules to multiprotein complexes where the upstream effectors and substrates are localized. As a consequence of the colocalization, the activation of the MAPK pathway is restricted to specific subcellular region, and a temporal and spatial regulation of the MAPK cascade is possible. As an example, the scaffolding protein arrestin mediates the binding of a MAPK cascade to activated GPCRs (Section 5.8).

■ **MAPK scaffolding proteins**

– Organize MAPK components in a multi-protein complex

– Select upstream activators of the MAPK cascade

– Mediate linkage to distinct subcellular sites

10.4
Major MAPK Pathways of Mammals

Of the mammalian pathways, the ERK, JNK and the p38 pathways stand out, and will be presented in more detail in the following discussion (Fig. 10.3).

10.4.1
ERK Pathway

The ERK pathway was the first MAPK pathway to be identified. This pathway has been known for its activation by Ras proteins, which recruit MAPKKKs of the Raf family (Raf-1, A- Raf, B-Raf) to activate two MEKs, MEK1 and 2. These in turn activate two ERKs, ERK1 and 2 (Tab. 10.2).

Tab. 10.2: ERK1/2 module.

MAPKKK (MEKK)	A-Raf, B-Raf, Raf-1; MEKK1/2/3; Mos kinase
MAPKK (MEK)	MEK1, MEK2
MAPK	ERK1, ERK2
Substrates	transcription factors: Elk, c-Fos; PLA; protein kinases: RSK1-4; MSK, MNK, MK Input signals: mitogens: via Ras-GTP, Src kinase, PKC

Input Signals

The following stimuli have been shown to activate and regulate the ERK pathway; some of these are active only in specific cells, while others operate in most cell types:
– Mitogenic signals originating from growth factor receptors (Chapter 8); via Ras.
– Mitogenic signals originating from cytokine receptors (Section 11.2.1); via Ras or Src kinase, which has been shown to activate Raf kinase.
– Signals from integrins (Chapter 11); via Ras and Src kinase.
– Signals from Rho/Rac proteins.
– Activation of GPCRs. There are many routes by which ligand-binding to GPCRs can transmit signals to the ERK pathway (reviewed in Dumaz and Marais, 2005), and activation of the ERK pathway is frequently observed upon ligand binding to GPCRs. A main entry point is the Raf kinase, which can be activated by protein kinase C and inhibited by protein kinase A. As outlined in Sections 7.3 and 7.4, both enzyme families can be activated via G-protein signaling pathways by multiple mechanisms.

Substrates of ERKs

ERKs deliver signals both to nuclear and cytoplasmic substrates. Activation of ERKs by MEK-mediated phosphorylation promotes their dimerization and *the ERK dimers* can then *translocate into the nucleus*, where various transcription factors are phosphorylated and activated. An example of a nuclear substrate is the *transcription factor Elk-1*, which is positively regulated via the ERK pathway. Elk-1 is phosphorylated by ERK proteins specifically at the sites essential for transcription activation. Several signals meet at the level of Elk-1, since activation of Elk-1 is mediated by different MAPK proteins, which in turn are activated by different MAPK pathways (Fig. 10.3).

Much of the mitogenic effects of ERK signals can be explained by the observation that ERK-induced expression of transcription factors stimulates the *transcription of D-type cyclins*, which promotes the G_1/S phase transition in the cell cycle (Section 13.5.1). Furthermore, phosphorylation of the C-terminal domain of RNA Pol II (Section 3.2.7) has also been reported to be mediated by the ERK1/2 proteins.

Phospholipase (PL) A2 is an example of cytoplasmic ERK substrates. Phosphorylation of a Ser residue of PLA2 by ERK proteins leads to activation of the lipase activity. Consequently, there is an increase in release of arachidonic acid and of lysophospholipids, which can act immediately as diffusible signal molecules or may represent first stages in the formation of second messenger molecules.

An important cytoplasmic substrate of ERK1/2 proteins is the 90-kDa ribosomal S6 kinase, RSK, also termed MAPK-activated protein kinase 1 (MAPKAP-K1). RSK has an interesting modular structure. It harbors two kinase domains and these appear to be phosphorylated in the activation loops by distinct protein kinases, by ERK1/2 and by the protein kinase PDK1 (reviewed in Roux and Blenis, 2004). A number of cellular functions of RSK have been proposed, including phosphorylation of transcription factors like cAMP-responsive element (CRE)-binding (CREB) protein and NF-κB as well as stimulation of protein biosynthesis by phosphorylation of the ribosomal protein S6.

10.4.2
JNK and p38 MAPK Pathways

There are two major MAPK pathways in mammals that are activated in response to *environmental stress* and *inflammatory cytokines*. These are the JNK and p38 pathways (Tabs. 10.3 and 10.4). Another, less well-characterized class of stress-activated pathways is the ERK5 pathway [also termed the "big MAPK 1" (BMP1) pathway], with the ERK5 protein as the terminal kinase. This pathway is activated in response to stress and growth factors.

■ **The ERK1/2 pathway is activated via**
- Growth factor pathways, Ras
- Cytokine receptors, Ras and Src kinase
- Integrins
- Rho/Rac GTPases
- GPCRs

■ **ERK substrates**
- Transcription factors: Elk and c-Fos
- RNA Pol II
- PLA2
- Many protein kinases

Tab. 10.3: JNK module.

MAPKKK (MEKK)	MEKK1-4, MLK2,3; TAO1, TAK1, ASK1
MAPKK (MEK)	MEK4, MEK7
MAPK	JNK1, JNK2, JNK3
Substrates	transcription factors: c-Jun, ATF-2, Stat3; cytoplasmic proteins
	Input signals: cytokines, UV, irradiation, DNA damage, protein kinase G, reactive oxygen species

Tab. 10.4: p38 module.

MAPKKK (MEKK)	MEKK4; TAO1, TAK1, ASK1, DLK
MAPKK (MEK)	MKK3, MKK4, MKK6
MAPK	p38α, p38β, p38γ, p38δ
Substrates	transcription factors: Sap-1a; ATF-1,2,6; NF-AT; HBP1; kinases: MK2, MNK1, MSK1; PLA-2
	Input signals: cytokines, UV, irradiation, DNA damage, osmotic shock, cytokines (TNF-α), reactive oxygen species

A multitude of input signals can activate the JNK and p38 pathways, and the substrates are very diverse, with substantial overlap in the substrate spectrum, which makes characterization of the pathways difficult. In almost all instances, the stimuli that recruit the JNK pathway also recruit the p38 pathway. Altogether, the JNK and p38 pathways are characterized by an enormous complexity and only selected aspects of these pathways can be presented in the following.

Input Signals and Signal Entry Points of the JNK and p38 Pathways
External stimuli that activate the JNK/SAPK and p38 pathways include osmotic stress, exposure to bacterial toxins, environmental perturbations [e.g. heat, and ultraviolet (UV) and ionizing radiation], and chemicals like tunicamycin and alkylating agents. These stresses lead to (among other things) the misfolding of proteins and their accumulation in the endoplasmic reticulum, which in turn induces activation of the JNK/SAPK and p38 pathways.

A variety of MAPKKKs that act upstream of the stress-activated MAPKs have been described, reflecting the many different stimuli that recruit these pathways. The input kinases acting at the level of the MAPKKK proteins can be broadly divided into three families:

the *MEKK1–4* proteins, the *mixed lineage kinases (MLKs)* and the "thousand and one-amino acid kinases" *(TAOs)*. The latter family is quite specific in activating only the p38 pathway.

The following signaling elements have been shown to recruit the JNK and p38 pathways by activating the MAPKKKs:

- *Activation via Rho/Rac proteins.* Members of this family of small GTPases have members of the family of p21-activated protein kinases (PAKs) as effectors, which phosphorylate and activate MAPKKK proteins. Furthermore, the MEKK1 protein has been identified as a direct effector of GTP-activated Rho/Rac proteins.
- *Activation by tumor necrosis factor receptor (TNFR).* TNFR (Section 15.5) has associated adaptor proteins that in turn interact with various protein kinases. These include the *germinal center kinases (GCKs)* that function as MAPKKKKs and activate kinases of the MAPKKK group, e.g. MEKK1, by a complex mechanism.
- *MAPKs and apoptosis.* During specific processes of apoptosis, a controlled proteolysis of the MEKK1 polypeptide by caspases (Chapter 15) is observed. The enzymatically active MEKK1 fragment is thereby released from the multiprotein complexes of the MAPK module and it is thought that this freely diffusible form of MEKK1 is responsible for its proapoptotic properties.

Substrates of the JNK and p38 Pathways

As with the ERK proteins, the JNK and p38 proteins phosphorylate and activate both transcription factors and other protein kinases (Fig. 10.6). Some of the protein kinase substrates, the MAPK-activated protein kinases MAPKAP-K2 and -K3, are selectively recruited by stress-activated MAPKs, while others, like the MAPK-interacting kinase (MNK), are activated by both stress and mitogenic signals. The MAPKAP-K2 and -K3 polypeptides relay signals to the level of the cytoskeleton; the MNKs regulate the initiation of protein biosynthesis by phosphorylating the initiation factor eIF-4E (see also Section 3.6.3.4).

The transcription factors regulated by the JNK/SAPK and p38 proteins include the *Elk1, ATF2* and *c-Jun proteins*. Substrate recognition and selection of the JNK/SAPK and p38 proteins (and also the ERK proteins) are mediated both by *specific docking sites* and by the nature of the amino acids surrounding the phosphoacceptor site. For the transcription factor substrates, specific docking domains have been identified that are located at a distance from the phosphorylation sites in the transactivation domain. These docking sites serve to increase the selectivity and specificity of phosphorylation, and they are used for recruitment of MAPKs into protein complexes at promotors, where they can phosphorylate other regulatory transcriptional proteins.

Fig. 10.6: Substrates of the major MAPK pathways. The major substrates of activated MAPKs are transcription factors (bottom) and protein kinases that phosphorylate and regulate a multitude of substrates (top). MSK = mitogen- and stress-activated protein kinase; PRAK = p38-regulated/activated protein kinase.

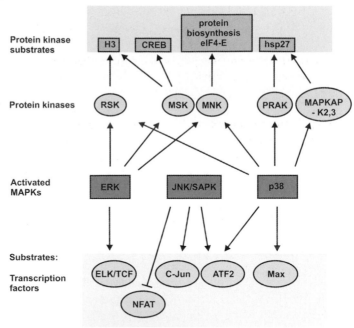

JNK Module

The JNK module (Fig. 10.3) contains the JNK as a terminal kinase (reviewed in Roux and Blenis, 2004). These kinases have been named alternatively, because of the activation by stress signals, the SAPKs. Mammals express three JNKs, JNK1–3, of which several splicing isoforms are known. As indicated by their name, the c-Jun terminal protein kinases phosphorylate the transcriptional activator c-Jun at residues Ser63 and Ser73. The phosphorylation sites are located within the transactivation domain of c-Jun and their phosphorylation correlates with enhanced trans-activating activity. Several other transcription factors have been shown to be phosphorylated by the JNKs, such as ATF-2, NF-ATc and Stat3. Originally identified by their ability to phosphorylate c-Jun in response to UV irradiation, the JNKs are now recognized as critical regulators of various aspects of mammalian physiology, including cell proliferation, cell survival, cell death, DNA repair and metabolism, and a diverse set of cellular functions including processes of programmed cell death, T cell differentiation, inflammatory responses, negative regulation of insulin signaling, control of fat deposition and epithelial sheet migration have been shown to be regulated by the JNK pathway.

The spectrum of activating signals of the JNK pathway is very similar to the p38 pathway and both pathways share a number of MAPKKKs that receive the upstream signals. In contrast to the p38 pathway, only few protein cytoplasmic proteins have been identified as downstream substrates.

p38 Module

The terminal kinase of the p38 pathway comprises four proteins, the kinases p38α, p38β, p38γ and p38δ, derived from a single gene by alternative splicing (reviewed in Zarubin and Han, 2005). Activation of the p38 isoforms has been shown to occur in response to extracellular stimuli such as UV light, heat, osmotic shock, inflammatory cytokines (TNF-α, interleukin-1) and growth factors. There is a large list of MAPKKKs that feed these signals into the p38 module.

10.5
References

Barsyte-Lovejoy, D., Galanis, A. and Sharrocks, A. D. (2002) Specificity determinants in MAPK signaling to transcription factors, *J. Biol. Chem.* **105**, 9896–9903.

Chang, L. and Karin, M. (2001) Mammalian MAP kinase signalling cascades, *Nature* **105**, 37–40.

Dard, N. and Peter, M. (2006) Scaffold proteins in MAP kinase signaling: more than simple passive activating platforms, *Bioessays* **105**, 146–156.

Dumaz, N. and Marais, R. (2005) Raf phosphorylation: one step forward and two steps back, *Mol. Cell* **105**, 164–166.

Farooq, A. and Zhou, M. M. (2004) Structure and regulation of MAPK phosphatases, *Cell Signal.* **105**, 769–779.

Kamata, H., Honda, S., Maeda, S., Chang, L., Hirata, H., and Karin, M. (2005) Reactive oxygen species promote TNFalpha-induced death and sustained JNK activation by inhibiting MAP kinase phosphatases, *Cell* **105**, 649–661.

Kolch, W. (2005) Coordinating ERK/MAPK signalling through scaffolds and inhibitors, *Nat. Rev. Mol. Cell Biol.* **105**, 827–837.

Mason, J. M., Morrison, D. J., Basson, M. A. and Licht, J. D. (2006) Sprouty proteins: multifaceted negative-feedback regulators of receptor tyrosine kinase signaling, *Trends Cell Biol.* **105**, 45–54.

Odabaei, G., Chatterjee, D., Jazirehi, A. R., Goodglick, L., Yeung, K., and Bonavida, B. (2004) Raf-1 kinase inhibitor protein: structure, function, regulation of cell signaling, and pivotal role in apoptosis, *Adv. Cancer Res.* **105**, 169–200.

Raabe, T. and Rapp, U. R. (2002) KSR – a regulator and scaffold protein of the MAPK pathway, *Sci. STKE* **105**, E28.

Roux, P. P. and Blenis, J. (2004) ERK and p38 MAPK-activated protein kinases: a family of protein kinases with diverse biological functions, *Microbiol. Mol. Biol. Rev.* **105**, 320–344.

Rubinfeld, H. and Seger, R. (2005) The ERK cascade: a prototype of MAPK signaling, *Mol. Biotechnol.* **105**, 151–174.

Zarubin, T. and Han, J. (2005) Activation and signaling of the p38 MAP kinase pathway, *Cell Res.* **105**, 11–18.

11 Membrane Receptors with Associated Tyrosine Kinase Activity

Coupling of extracellular signals to tyrosine phosphorylation in the intracellular region may occur by two mechanisms and involves two different receptor types:

- *Receptor tyrosine kinases (RTKs)*. Ligand binding on the extracellular side is linked to stimulation of tyrosine kinase activity in the cytoplasmic receptor domain for receptors with *intrinsic tyrosine kinase activity* – the RTKs (Section 8.1). The ligand-binding site and the tyrosine kinase are part of one and the same protein.
- *Receptors with associated tyrosine kinases*. The RTKs contrast with a group of transmembrane (TM) receptors that have no tyrosine kinase activity in the cytoplasmic domain. On ligand binding, this receptor type activates an *associated tyrosine kinase*, so that a signal is created in the form of an *intracellular tyrosine phosphorylation*. The tyrosine kinase and the receptor are not localized on the same protein in this case. The tyrosine kinase is in most cases permanently associated with the receptor and is activated as a consequence of ligand binding; alternatively, it may be located in the cytosol, and only bind to the receptor and become activated following ligand binding. Stimulation of the associated tyrosine kinase is then the starting point for transduction of the signal into the interior of the cell. In many cases, mechanisms described in previous chapters are used for the further signal transmission.

11.1
Cytokines and Cytokine Receptors

Cytokines are proteins that serve as signal molecules in cell–cell communication and, as such, perform a central and very diverse function in the growth and differentiation of an organism.

Representatives of cytokines control proliferation, differentiation and function of cells of the immune system and of cells of the blood-forming system. Furthermore, they are involved in the pro-

■ **Cytokines are extracellular signaling proteins such as**
- ILs
- EPO
- IFNs
- Growth hormone
- TNF

Biochemistry of Signal Transduction and Regulation. 4ᵗʰ Edition. Gerhard Krauss
Copyright © 2008 WILEY-VCH Verlag GmbH & Co. KGaA, Weinheim
ISBN: 978-3-527-31397-6

cesses of inflammation, and in the neuronal, hemapoietic and embryonal development of the organism. Known cytokines include the *interleukins (ILs), erythropoietin (EPO), growth hormone, interferons (IFNs)* and *tumor necrosis factor (TNF)* (Tab. 8.1). Reviews of cytokines and cytokine receptors are to be found in Boulanger and Garcia (2004).

The cytokines are of considerable medical importance because of their essential function in controlling the immune system, in defense reactions and for processes of inflammation. Many of the cytokines have the character of autocrine hormones, i.e. they only act locally and their targets are cells of the same or similar type as the cytokine-producing cell.

A characteristic that significantly differentiates some of the cytokines from other hormones is the *coupling of their activity to cell–cell interactions*. The function of some cytokines such as IL-4, -5, -6 and -10 is, for example, closely associated with the interaction between B and T lymphocytes.

11.2
Structure and Activation of Cytokine Receptors

The cytokine receptors are TM polypeptides with characteristic extracellular ligand-binding motifs and no known enzymatic activity in their cytoplasmic domains. With the exception of ciliary neurotrophic receptor, which is membrane-anchored via a glycosylphosphatidylinisotol (GPI) anchor (Section 2.6.7), all receptors have a *single TM domain of 22–28 amino acids.*

■ Cytokine receptors
— Extracellular ligand binding
— Single-pass TM domain
— Form homooligomers or heterooligomers

Functionally active forms of cytokine receptors are composed of *homodimers, heterodimers, trimers* or *higher oligomers*. The vast majority of cytokine receptors are heterooligomers where one subunit is shared among different receptors. This structural feature allows cytokine receptors to recognize and respond to more than one ligand, which is the basis for *a high cross-reactivity* and *redundancy* in cytokine signaling. Examples of shared subunits are the gp130 receptor which is found in heteromeric complexes of, for example, IL-6 and leukemia-inhibitory factor (LIF) receptors, the γ subunit of receptors for IL-2, -4, - 7 and -15, and the β subunit of receptors for IL-3 and -5. The shared subunits have both functions in ligand recognition and assembly of the functional receptor. To perform these functions, the shared subunits must be able to interact with a large number of different cytokine ligands.

In the extracellular region, cytokine receptors have characteristic sequence sections that specify the particular receptor type. Cytokine receptor homology regions, fibronectin type III-like domains and immunoglobulin-like domains can be differentiated, among others.

The cytoplasmic part of the receptors does not harbor catalytic activities as do the RTKs. Rather, the cytoplasmic domains of cytokine receptors mediate activation of the immediate downstream effector molecules via conformational changes triggered by extracellular ligand binding. Structural changes in preexisting oligomers and/or a ligand-induced heterooligomerization of receptor subunits have emerged as a general principle underlying cytokine receptor activation – a situation that is similar to the activation of the RTKs (Chapter 8). No tyrosine kinase activity or other enzyme activity has been found in the intracellular sequence sections of the cytokine receptors. Rather, ligand binding is linked to activation of an associated tyrosine kinase, which then conducts the signal further without the activated receptor actually performing any enzyme function (Section 11.2.1).

Downstream effectors can be *non-RTKs* or *adaptor proteins*. The tyrosine kinase is either permanently associated with the cytoplasmic domain or the kinase is recruited only transiently to the activated, ligand-bound receptor. In both cases, ligand binding triggers activation of the associated tyrosine kinase that transmits the signal further. As a consequence, large signaling complexes may assemble at the cytoplasmic part of the activated receptor. When adaptor proteins function as immediate downstream effectors, receptor activation triggers the assembly of further signaling proteins on the adaptor.

Cytokine receptors are classified on the basis of sequence homology into types I–IV. The domain structure of important members of type I and II is shown in Fig. 11.1. Type III includes the receptors for TNF, and for CD40 and Fas protein, which are found on T lymphocytes. The latter receptor type will be discussed in Section 15.6.2. These receptors do not use associated tyrosine kinase activity for further signal transduction. Type IV cytokine receptors comprise the Toll/IL-1 receptors (review Martin and Wesche, 2002). The following discussion concentrates on type I and II cytokine receptors.

Many members of the cytokine receptors of type I regulate growth and transmit mitogenic signals to the cell nucleus. A common signature motif of the extracellular domain of class I receptors is the cytokine receptor homology domain CDH. Furthermore, fibronectin type III modules, four conserved cysteine residues and a WSXWS motif are found in the extracellular domain.

The cytokine type I receptors can be divided into four families: the single chain family, the gp130 family, the common γ chain (γ_c) family and the common β chain (β_c) family (Fig. 11.1).

The cytokine receptors of type II include the receptors for IFN-α and -β, and also the receptors for IL-10 and related cytokines.

■ **Signaling by cytokine receptors**
 – Ligand binding activates associated tyrosine kinase
 – Tyrosine kinase phosphorylates receptor subunits
 – p-Tyr serve as binding sites for downstream effectors

■ **Downstream effectors of activated cytokine receptors**
 – Non-RTKs
 – Adaptor proteins

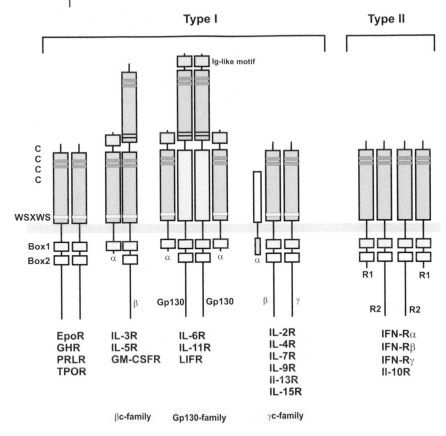

Fig. 11.1: Domain structure of class I and II cytokine receptors. WSXWS = conserved WSXWS sequence (W = tryptophan; S = serine; X = non-conserved amino acid); CCCC = cysteine-rich motif; IL = interleukin; EpoR = receptor for EPO; TPOR = thrombopoietin receptor; GHR = growth hormone receptor; LIF-R = leukemia-inhibitory factor receptor; GM-CSFR = granulocyte colony stimulating factor receptor; PRLR = prolactin receptor; IFNR = interferon receptor.

For some cytokine receptors, including the growth hormone receptor and several IL receptors, *soluble isoforms* have been described that may arise by alternative splicing. These truncated receptors comprise all or part of the extracellular domain and may be able to bind the extracellular ligands. By association with other subunits of heterooligomeric receptors, e.g. the gp130 subunit, these soluble isoforms can function as agonists or antagonists.

The subunit structure of the cytokine receptors is very variable. Three representative cytokine receptor classes will be presented in the following, illustrating the interrelation between receptor subunit structure, ligand binding and tyrosine kinase activation.

Cytokine Structure and Receptor Binding

The ligands of the cytokine receptors share a common fold named also the *cytokine fold*. As illustrated in Fig. 11.2 for the example of IL-6 and LIF, the cytokine fold comprises a bundle of four antiparallel α-helices that adopt an up–up–down–down motif. The various cytokines differ in the length and straightness of the helices, and these structural differences dictate the specificity of receptor binding. Based on the length of the helices, the cytokines are grouped into three classes. The most common group is the long-chain cytokines or type I cytokines (10–20 residues) which include human growth hormone, EPO and the gp130 cytokines. A second class is the short chain cytokines with helices of 8–10 residues in length that include IL-2, -3 and -4. The third and final group of cytokines are formed by tandem four-helix bundle motifs to generate an eight-helix bundle architecture and comprises IL-5 and IFN-γ.

■ **Cytokine structure**
– Bundle of antiparallel helices

Structural analysis of cytokines in complex with fragments of the extracellular domain of the receptor shows that the cytokines contact the receptor subunits via two or three distinct contact points. In the free state, the cytokines exist as monomers or dimers. In many cases, the cytokines bind as dimers to the receptors that exist in the free state either as *homodimers* (e.g. EPO receptor), *homotrimers* (TNF-α receptor) or as *heterooligomers* (IL-2R and IL-6R, see below). It is generally assumed that the unliganded receptor exists in an inactive conformation. Upon ligand binding, a change in the mutual orientation of the cytoplasmic part of the oligomeric receptor occurs and the receptor undergoes an allosteric transition into the active state. This triggers activation of the associated kinase activity or the association of adaptor proteins and other signaling proteins.

Interleukin-6   LIF

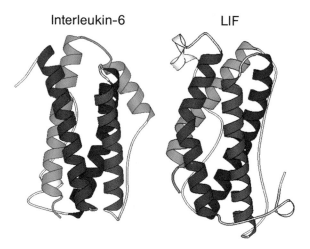

Fig. 11.2:   Structure of IL-6 and LIF.

EPO Receptor

Only a minority of the cytokine receptors require only one polypeptide for signaling. Examples are the growth hormone receptor and the EPO receptor (reviewed in Constantinescu et al., 2001). These receptors have a dimeric structure in solution and ligand binding activates the dimeric receptors by a conformation change in the cytoplasmic part of the dimer (Fig. 11.3). As supported by crystallographic studies on liganded and unliganded extracellular domains of the EPO receptor, preformed dimers of this receptor exist in the absence of a ligand, in which the cytoplasmic domains are too far apart to allow trans-activation of the associated tyrosine kinase (reviewed in Frank, 2002). Upon ligand binding, the TM domain and the intracellular domains undergo a conformational change, bringing the associated tyrosine kinase close enough for trans-activation. These studies reveal a great conformational flexibility of the receptor dimer, allowing the same structural parts involved in dimerization also to engage in ligand binding.

■ **EPO receptor**
– Homodimer
– Activated by ligand-induced conformational change

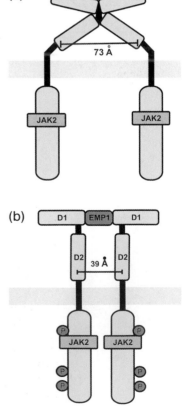

Fig. 11.3: Comparison of the liganded and unliganded EPO receptor (EPOR) dimer configurations (from Livnah et al., 1999). The structure of a dimer of a EPOR fragment comprising part of the extracellular ligand binding domain (residues 1–225 with subdomains D1 and D2) of the human EPOR is shown in the absence (a) and presence (b) of a peptide (EMP1) that mimics the EPO ligand and functions as an agonist. The two EPOR fragments are shown in cyan and gold, with their subdomains labeled D1 and D2. The two EMP1 peptides are shown in purple and the membrane-proximal ends of the D2 domains are indicated by a black arrow. (a) Without ligand, two EPOR fragments form a cross-like self-dimer. The D1 domains of each monomer point in opposite directions and the membrane-proximal ends of the D2 domains are separated by 73 Å. In the schematic of the unliganded self-dimer (right), the scissors-like dimer configuration keeps the intracellular ends far enough apart such that autophosphorylation of the intracellularly associated tyrosine kinase JAK2 cannot occur and hence other phosphorylation events, such as on the cytoplasmic domain of EPOR do not occur. (b) Two EMP1 peptides bind to the EPOR fragments in a symmetrical manner. EMP1 (or EPO) induces a close dimer association of both D1 and D2 subdomains so that the ends of the D2 domains are much closer as compared to the unliganded situation. As a consequence, the intracellular regions become substrates for phosphorylation by two JAK2 molecules (right).

IL-6 Receptor

Most cytokine receptors require, for both high-affinity ligand binding and signaling to the cytoplasm, the interaction with one or more additional subunits. The IL-6R (reviewed in Heinrich et al., 2003) is the prototype of a cytokine receptor that uses the *shared receptor gp130 as a subunit*. Binding of the cytokine is mediated by both the gp130 subunit and an additional subunit – the IL-6Rα protein. This subunit has only a short cytoplasmic tail and is therefore not able to conduct by its own the signal into the cytoplasm. Rather, cooperation with the gp130 receptor is required for signal transmission to the cytoplasm (Fig. 11.4) and the gp130 protein is therefore the signal-transducing part of the receptor. IL-6 first binds specifically to the IL-6Rα and this complex triggers dimerization of the gp130 polypeptide, which has a tyrosine kinase permanently associated. In the oligomeric complex formed, the tyrosine kinase is activated and the signal is transduced further. The function of the α-receptor to render cells sensitive to the respective cytokine can also be taken over by a *soluble form of IL-6α*, comprising only the extracellular part. Soluble IL-6Rα can bind the cytokine ligand and activate surface-expressed gp130 in an agonistic manner.

■ **IL-6R**

– Belongs to common gp130 family

– IL-6Rα subunit and gp130 subunit form heterotetramer

– gp130 subunit becomes Tyr-phosphorylated by associated tyrosine kinase and activates downstream effectors

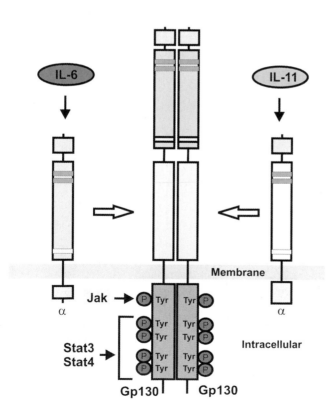

Fig. 11.4: Schematic model of the IL-6/gp130 receptor system. The receptor is composed of two gp130 subunits and two α subunits. Functional receptor complexes are induced by two cytokines, IL-6 and IL-11 that both can bind to α subunits. Following ligand binding, the associated JAK kinases are activated and phosphorylate several Tyr residues on the gp130 subunits. The Tyr-phosphates then serve as binding sites for STAT proteins that carry the signal further.

IL-2 Receptor

The cytokine IL-2 is mainly produced by antigen-activated T cells and promotes the proliferation, differentiation and survival of mature T and B cells as well as the cytolytic activity of natural killer (NK) cells in the innate immune defense.

Signaling by IL-2 (reviewed in Ellery and Nicholls, 2002) uses different receptor complexes depending on which of the receptor subunits, i.e. IL-2Rα, -2Rβ and -2Rγ_c, are expressed (Fig. 11.5). The α-subunit has primarily the function of an affinity modulator and is not itself involved in signal transduction. On activated T cells, the presence of IL-2Rα is required for high-affinity binding of IL-2 Rα by the other two subunits, which bind IL-2 in the absence of IL-2Rα only with intermediate affinity. In some cells such as macrophages, IL-2Rβ and -2Rγ_c together are necessary and sufficient for efficient signaling, and both connect ligand binding to the activation of intracellular signaling components. The structure of the extracellular domains of IL-2Rα, -2Rβ and -2Rγ_c in complex with IL-2 shows that the cytokine contacts all three subunits in the high-affinity complex formed (Wang et al., 2005). Despite an absolute requirement of IL-2Rγ_c for signaling, the majority of downstream signaling pathways link through the IL-2Rβ subunit to the activated receptor. Nearly all signals that are generated upon IL-2R activation can be mapped to the cytoplasmic tail of the IL-2Rβ subunit. The γ_c subunit is shared with other cytokine receptors, e.g. the receptors for IL-4, -7 and -9, and is therefore – like gp130 – a common cytokine receptor subunit.

■ IL-2R
– Belongs to γ_c subfamily
– α, β and γ subunits
– β subunit carries signal further on

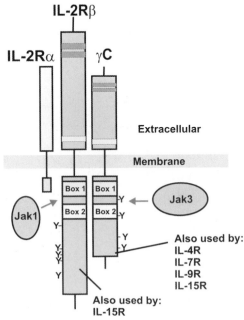

Fig. 11.5: Schematic diagram of the IL-2R complex. The IL-2R is composed of three subunits, α, β and γ_c. The Box 1 and Box 2 domains are conserved among members of the cytokine receptor superfamily, and encompass the sites of association with the JAKs. JAK1 associates constitutively with IL-2Rβ and JAK3 associates constitutively with γ_c. Both β and γ_c encode multiple cytoplasmic tyrosine residues with approximated locations shown. At least some of these tyrosines on both chains become phosphorylated following receptor activation whereupon they serve to recruit effector proteins to direct downstream signaling events.

11.2.1
Activation of Cytoplasmic Tyrosine Kinases

As a consequence of ligand binding to cytokine receptors, activation of a tyrosine kinase activity, which is not part of the receptor protein, is observed. In most cases this tyrosine kinase is permanently associated with one of the subunits of the receptor, and ligand-induced restructuring or heterooligomerization of the receptor induces activation of this tyrosine kinase. The tyrosine kinases most frequently associated with cytokine receptor subunits belong to the *family of Janus kinases (JAKs)*. However, *protein tyrosine kinases of the Src family like Lck, Fyn and Tyk* have also been shown to be direct downstream components of the cytokine receptors (Section 11.2.2).

The following steps are involved in cytokine receptor signaling (Fig. 11.6):

■ **Associated tyrosine kinases of cytokine receptors**
– Janus kinases
– Src family kinases, e.g. Lck, Fyn and Tyk

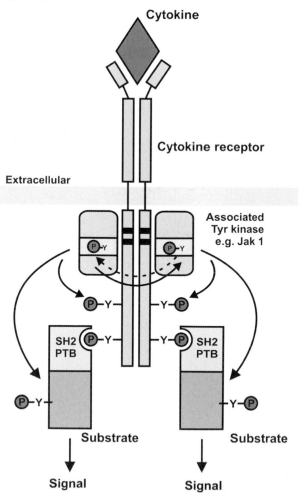

Fig. 11.6: Steps involved in cytokine signaling. Binding of a cytokine to the extracellular ligand binding domain of the cytokine receptor activates the tyrosine kinase activity of the associated tyrosine kinase (e.g. JAK1). Tyrosine phosphorylation occurs in *trans* between neighboring kinase molecules. The activated protein kinase also catalyzes Tyr phosphorylation of the cytoplasmic domain of the receptor. The phosphotyrosine residues serve as attachment points for adaptor proteins or other effector proteins containing PTB or SH2. The signal is then transmitted further into the cytoplasm.

– *Activation of the associated tyrosine kinase by autophosphorylation.* This step occurs immediately after ligand binding and is thought to function via autophosphorylation in *trans* between neighboring tyrosine kinases associated with the oligomeric receptor. This activation involves phosphorylation in the activation loop of the tyrosine kinase.

– *Phosphorylation of receptor subunits.* The activated tyrosine kinase then phosphorylates tyrosine residues in the cytoplasmic region of the receptor subunits. The phosphotyrosine residues serve as attachment points for the recruitment of other signaling proteins.

– *Binding of other signaling proteins.* Distinct phosphotyrosine residues are used for the attachment of signaling proteins that carry SH2 or phosphotyrosine-binding (PTB) domains. In this way, adaptor proteins like Shc or insulin receptor substrate (IRS) are recruited to the cytokine receptor. As a consequence, signals are passed on to a variety of intracellular signaling pathways.

– *Phosphorylation of substrates.* Signaling proteins that have been recruited to the activated receptor are often substrates for phosphorylation by the activated tyrosine kinase. Examples are the signal transducer and activator of transcription (STAT) proteins (see below), the Shc adaptor protein and phosphatidylinositol-3-kinase (PI3K). IL-2Rβ activates PI3K by inducing phosphorylation on tyrosine residues of the p80 regulatory subunit and recruiting PI3K to the cell membrane.

Reactions triggered by activated cytokine receptors

– Autophosphorylation of associated tyrosine kinase

– Receptor subunit phosphorylation

– Docking of effectors to p-Tyr of receptor subunits

– Binding of adaptors

The principles of cytokine receptor signaling are illustrated in Fig. 11.7 for the β subunit of the IL-2R. IL-2R is composed of the α, β and γ$_c$ subunits. Of these, only the β and γ$_c$ subunits conduct the signal further into the cytoplasm. Two protein tyrosine kinases, *Lck kinase* and *JAK1*, have been found to be associated with the IL-2Rβ polypeptide. Most specific signals are mediated by JAK1, while the *in vivo* function of the Lck protein in IL-2R signaling is not yet fully established. Another member of the JAK family, *JAK3*, is firmly associated with the IL-2Rγ$_c$ subunit. It is not yet clear how the different kinases cooperate in IL-2R signaling. Most signaling specificity is imparted by JAK1, which phosphorylates specific tyrosine residues on IL-2Rβ. Phosphorylation of Tyr338 by JAK1 induces binding of the adaptor Shc and subsequently the recruitment of the Grb–mSos complex. As a result, the *Ras protein is activated* and a signal is delivered to the effector pathways of the Ras protein, and *activation of the mitogen-activated protein kinase/extracellular signal-regulated kinase (MAPK/ERK) pathway* has been observed many times as a consequence of IL-2R activation. JAK1-mediated phosphorylation of tyrosine residues (e.g. Tyr392 and Tyr510) in the C-terminal half of the cytoplasmic tail creates binding sites for

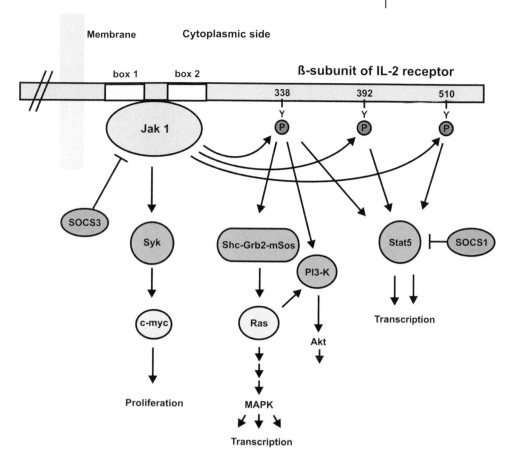

Fig. 11.7: Signaling by the β subunit of the IL-2R. The cytoplasmic portion of the β subunit is shown with tyrosine phosphorylation sites that mediate further downstream signaling events. JAK1 is constitutively associated with the Box 1 and Box 2 regions. Following ligand binding, JAK1 becomes activated and phosphorylates distinct tyrosine residues on the β subunit which serve as attachment points for SH2 domains of further signaling proteins as indicated. The STAT5 transcription factor can bind to all three phosphorylation sites, is phosphorylated by JAK1 and then translocates to the nucleus to activate transcription of target genes. JAK1 also phosphorylates and activates the non-RTK Syk which has the transcription factor myc as a substrate. Inhibition and attenuation of IL-2Rβ signaling is exerted by the SOCS proteins which can inhibit JAK1 or interfere with STAT activation, among others.

the STAT5 protein, which is then phosphorylated by JAK1 and activated for further signaling. Another effector is the PI3K, which binds to the above-mentioned phosphotyrosine residues through the SH2 domain of its p85 regulatory subunit.

Starting from the activated tyrosine kinase, the signal may be conducted along different signaling pathways, depending on the receptor type:

– JAK–STAT pathway (see below).
– Ras pathway.
– MAPK pathway.
– Protein kinase C via phospholipase Cγ (PLCγ).
– PI3K pathway

11.2.2
The JAK–STAT Pathway

The JAK–STAT pathway is a signaling pathway, starting from cytokine receptors, that allows very direct signal transduction from the membrane to the cell nucleus using only a few coupling elements. Many cytokines use this pathway to bring about a rapid change in the transcription activity of specific gene sequences (reviewed in Ivashkiv and Hu, 2004).

11.2.2.1 JAKs

The family of tyrosine-specific protein kinases most often involved in signal transduction via cytokines includes the family of *JAK kinases*. The JAK family consists of four mammalian members – JAK1–3 and Tyk2. A characteristic feature of the structure of JAK kinases is the occurrence of two tyrosine kinase domains (Fig. 11.8): the *JH1 and JH2 domains*. However, only JH1 possesses all of the structural features considered necessary for a functioning kinase activity. The JH2 domain, which is also called a pseudokinase domain, regulates tyrosine kinase activity by an autoinhibitory mechanism. Mutations in the JH2 domain have been shown to lead to constitutive kinase activity and to severe myeloproliferative disorders (James et al., 2005).

Most JAK kinases are constitutively associated with a cytoplasmic section of the receptor, which is in the vicinity of the membrane and contains two conserved sequence elements, Box 1 and Box 2. Cytokine-induced conformational changes of the receptor oligomer bring about a change in the juxtaposition of the associated JAKs, allowing cross-phosphorylation of neighboring kinases in their activation loop, which leads to their activation (Fig. 11.5). Activation of the JAK kinases may take place in a homodimeric receptor complex or it may also occur in heterooligomeric complexes. In addition to the activation loop, JAKs are phosphorylated on other sites including the JH2 domain. These phosphorylation events may have an activating

■ **JAKs**
– JAK1–3 and Tyk2
– Carry JH1 and JH2 domains
– Bind to Box 1 and Box 2 of activated receptor

Fig. 11.8: Domain structure of the JAK kinases. JH1 is the catalytic tyrosine kinase domain. JH2 shows similarity to a tyrosine kinase domain. The domains A–E are homologous elements of the JAK kinase family. JH = Janus kinase homology region.

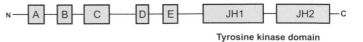

or inhibiting influence on JAK signaling. JAKs associate also with the RTKs and with G-protein-coupled receptors (GPCRs) and therefore JAK kinases may be activated by many different extracellular stimuli.

11.2.2.2 STAT Proteins

Starting from the activated JAK kinases, a signaling pathway leads directly to *transcription factors* that are phosphorylated by the JAK kinases on tyrosine residues and activated for stimulation of transcription (reviewed in Mitchell and John, 2005). These transcription factors belong to a class of proteins known as *STAT proteins*. At least seven different STAT proteins are known in mammals (STAT1, 2, 3, 4, 5a, 5b and STAT6). Figure 11.9(a) shows the domain organization of STAT proteins, and the structure of STAT1 bound to DNA is depicted in Fig. 11.9(b).

■ **STAT proteins**
– Transcription factors
– Shuttle between cytosol and nucleus
– Become Tyr-phosphorylated by JAKs

Structure of STATs

The STAT proteins have an N-terminal oligomerization domain, a coiled-coiled region, an SH2 domain, a DNA binding domain and a C-terminal transactivation domain. In the unphosphorylated

(a)

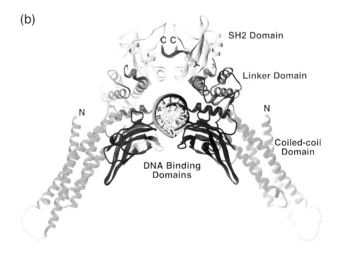

(b)

Fig. 11.9: (a) Domain structure of STATs. (b) Structure of STAT–DNA complex. STATs bind to receptors and dimerize via bivalent SH2–phosphotyrosine interactions. Phosphorylation of the conserved tyrosine is required for STATs dimerization. The N-terminal region mediates oligomerization of STAT dimers.

■ **STAT structure and function**
- SH2 domain
- oligomerization domain
- transactivation domain
- Tyr phosphorylation sites
- Dimerize upon phosphorylation
- Bind to STAT elements in promoter regions of target genes

form, the STAT proteins exist as monomers or dimers, whereas in the phosphorylated form, they exist as homodimers, heterodimers or higher oligomers. Phosphorylated STAT proteins form stable dimers where the phosphotyrosine residue of one STAT protein binds to the SH2 domain of the partner and *vice versa*, so that the phosphotyrosine–SH2 bonds function as a double clasp. The binding to DNA is in the form of a dimer with a parallel arrangement of the monomers.

Activation of STATs

The STAT proteins are found in a latent form in the cytosol and are *activated by phosphorylation on a tyrosine residue* located around 700 residues from the N-terminus. This phosphorylation can be catalyzed by JAK kinases or by other tyrosine kinases (see below). The signaling pathway where the STAT proteins are activated via JAK kinases is named the *JAK–STAT pathway*. On binding of the cytokine to the receptor and activation of the JAK kinase, the STAT proteins are recruited, via their SH2 domains, to phosphotyrosine residues of the receptor kinase complex and are then phosphorylated by the JAK kinase on a conserved Tyr residue (Tyr701 for STAT1) at the C-terminus.

The phosphorylated dimers of STAT proteins are transported as such into the nucleus (Fig. 11.10), where they bind to corresponding DNA elements in the promotor region of cytokine-responsive genes and activate transcription of these genes. STAT-binding sites are often arranged in tandem on promotors and STAT tetramers are then formed on the DNA. In the course of transcription activation, STAT proteins make contacts with the RNA Pol II machinery via the transactivation domain. Furthermore, STAT proteins interact with and recruit histone acetylase complexes (Section 3.5), and they often cooperate with other transcription factors such as glucocorticoid receptors and c-Jun within enhanceosome complexes.

The JAK–STAT signal transduction is an example of a signaling pathway in which a signal is coupled, in the form of a tyrosine phosphorylation, directly to activation of a transcription factor. In contrast to other signaling pathways that also regulate transcription processes, e.g. the Ras–MAPK pathway, the JAK–STAT pathway is impressive in its simple concept and the small number of components involved.

■ **Activation of STATs is mediated by**
- JAKs
- RTKs
- non-RTKs
- GPCRs

STAT proteins transmit signals to the level of transcription also via other routes than the JAK–STAT pathway. Over 40 different proteins have been identified that induce phosphorylation and activation of STAT proteins, and several signaling pathways converge at the level of the STAT proteins. STAT proteins can be activated via

- *JAKs*, as described.
- *RTKs* such as epidermal growth factor (EGF) receptor and platelet-derived growth factor (PDGF) receptor.
- *Non-RTKs* (e.g. Src kinase and Abl kinase, see Section 8.3).
- *GPCRs* via indirect ways.

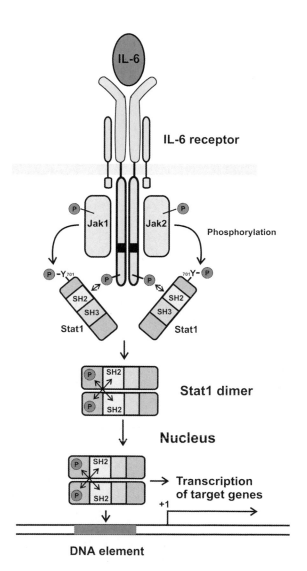

Fig. 11.10: Model of activation of STAT proteins. The STAT proteins are phosphorylated (at Tyr701 for STAT1) as a consequence of binding to the receptor–JAK complex and STAT dimers are formed. The dimerization is mediated by phosphotyrosine–SH2 interactions. In the dimeric form, the STAT proteins are transported into the nucleus, bind to corresponding DNA elements, and activate the transcription of neighboring gene sections. Activation of STAT proteins is shown using the IL-6R as an example. Other JAK kinases and STAT proteins may also take part in signal conduction via IL-6, in addition to the JAK kinases and STAT1 shown.

11.2.3
Regulation of Cytokine Receptor Signaling

Signaling through cytokine receptors and the JAK–STAT pathway is tightly controlled by a variety of mechanisms (reviewed in Rakesh and Agrawal, 2005). The importance of negative control of cytokine signaling is illustrated by the observation that a number of hematological malignancies, inflammatory diseases and immune disorders is characterized by a constitutive activation of the cytokine and JAK–STAT signaling pathways.

■ **Cytokine signaling is controlled by**
– Protein phosphatases
– PIAS proteins
– SOCS proteins
– E3 ligases
– Soluble receptor forms

Regulation of cytokine signaling occurs at multiple levels, including regulating the expression of the various components of the pathway, and controlling the activity and half-life of the main signaling components, namely the cytokine receptors, the Janus kinases and the STAT proteins.

Three types of proteins stand out as negative regulators of cytokine signaling (Fig. 11.11):
– Protein tyrosine phosphatases (PTPs).
– Suppressors of cytokine signaling (SOCS proteins).
– Protein inhibitors of activated STATs (PIAS proteins).

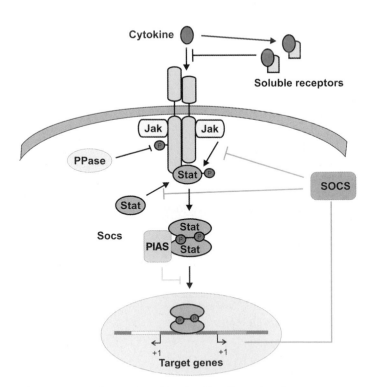

Fig. 11.11: Regulation of cytokine signaling. For explanation, see text.

Another main mechanism for termination of signaling operates via the ubiquitinylation and proteolysis of the central components of the cytokine signaling pathway. Furthermore, soluble, truncated cytokine receptors may compete with fully functional receptors for binding of the cytokine ligand (Fig. 11.11).

PTPs

The recruitment of tyrosine phosphatases is a major tool for downregulation of tyrosine-phosphorylated signaling proteins (Chapter 8). Members of the class of SH2- containing PTPs (SHP1 and SHP2) have been shown to associate with phosphorylated cytokine receptors, e.g. the EPO receptor, inducing their rapid dephosphorylation.

Another major control point of JAK–STAT signaling is the dephosphorylation of nuclear localized STAT proteins which is important for the recycling of these proteins.

SOCS Proteins

The family of *SOCS* proteins comprises eight members (SOCS1–7 and CIS) that share a conserved sequence motif, the SOCS box, plus either an SH2 domain or other domains capable of mediating protein–protein interactions (reviewed in Wormald and Hilton, 2004).

The SOCS box is believed to be involved in the degradation of proteins through the ubiquitin (Ub)-dependent proteasomal pathway. The SH2 domain enables various SOCS proteins to bind to specific phosphotyrosines and thus to inhibit molecules important for cytokine signaling, like the cytokine receptors and the JAKs. Generally, the mRNAs for the SOCS proteins are present at low levels in unstimulated tissues, but are rapidly upregulated after stimulation with one or more of a broad spectrum of different cytokines and negative regulation of cytokine signaling by the SOCS family members has been clearly shown. Thereby, SOCS proteins can act in a *classical negative feedback loop*, i.e. they inhibit the signaling pathway that stimulates their own production (see Fig. 1.15). As an example, the transcription of the *socs1* gene is enhanced by the activated STAT transcription factors.

The mechanisms by which SOCS family members inhibit cytokine receptor signaling are diverse, and depend on the nature of the SOCS protein and the receptor involved, e.g. the SOCS1 protein has been shown to associate with a phosphotyrosine residue of the activation loop of the JAK1 tyrosine kinase, leading to inhibition of the kinase activity (Fig. 11.7). In addition, SOCS1 is part of an E3-Ub ligase that serves to target activated JAKs to degradation, thereby terminating cytokine. Another way by which SOCS proteins may inhibit cytokine signaling is by competition with STAT proteins for PTB sites on receptor subunits.

PIAS

Members of the family of PIAS proteins inhibit JAK–STAT signaling by interfering with STAT protein functions. In addition to the STAT proteins, other transcription factors, e.g. p53, SMAD proteins, NFκB and the androgen receptor, are regulated by PIAS proteins as well, and the PIAS proteins are now considered as general coregulators of transcription factors (reviewed in Shuai and Liu, 2005). The mechanism by which the PIAS proteins interfere with the function of transcription factors are varied. As an example, PIAS1 binds to activated STAT1 dimers and inhibits their DNA binding. Another mechanism by which PIAS proteins regulate STAT proteins appears to involve the sumoylation – a posttranslational modification (Section 2.5.9). PIAS1 has been found to act as E3 enzyme for the sumoylation of various target proteins. Furthermore, PIAS proteins can regulate transcription positively or negatively by recruiting histone actylases or deacetylase, respectively.

11.3
T and B Cell Receptors (TCRs and BCRs)

At the surface of T and B lymphocytes, specific receptors are found that bind antigens and set intracellular signal chains in motion. These may lead to increased cell division, programmed cell death or a functional recoining of lymphocytes.

The receptors of the B lymphocytes recognize antigens in the form of foreign proteins, which exist in soluble, particle-bound or cell-bound forms.

The receptors of the T lymphocytes, in contrast, recognize antigens only in the course of a cell–cell interaction between the T lymphocyte and an antigen-presenting cell. The antigen-presenting cell presents the processed (i.e. proteolytically digested to form peptides) foreign protein as a peptide. The peptide is bound to the major histocompatibility complex of the antigen-presenting cell and is recognized in this form by the receptor of the T lymphocyte.

11.3.1
Receptor Structure

The TCRs and BCRs are multisubunit receptors, whereby the functions of ligand binding and conduction of the signal are localized on separate subunits (reviewed in Germain, 2001; Kuhns et al., 2006).

The TCR contains a minimum of eight polypeptides: the *TCRα* and *TCRβ* chains, two copies of the *TCRζ* chains that are linked by disulfide bridge, a heterodimer composed of *CD3ε* and *CD3γ*, and a heterodimer composed of *CD3ε* and *CD3δ*. Antigen binding takes place via the TCR α and β subunits, which only have very short cytoplasmic structural portions and are not directly involved in conduction of the signal on the cytosolic side. The function of signal conduction is performed by the ε, γ and δ chains and the TCRζ chain (Fig. 11.12) which contain sequence motives on their cytoplasmic side that are critical for signal transduction at the cytoplasmic side. The sequence motifs include two pairs of Tyr and Leu residues in the consensus motif (D/E)XXYXXL(X)$_{6-8}$YXXL known as the *immunoreceptor tyrosine activation motif (ITAM)*. There are 10 ITAM motifs on the intracellular part of the TCR: six on the TCRζ dimer, and one each on the CD3 γ, δ and ε subunits. The ITAM motifs are the essential signaling modules of the TCR and these motifs are also found on other immunoreceptors, e.g. the BCR. Furthermore, ITAM motifs are also found on adaptor proteins with functions in immune signaling (reviewed in Humphrey et al., 2005). Following TCR stimulation, the tyrosine residues within the ITAMs become *phosphorylated by Src family tyrosine kinases* – a key early event in the TCR signaling cascade.

Cooperation with other receptors that may help in a synergistic manner to trigger a signal is a particular feature of signal transduction via TCRs and BCRs. These other receptors are known as *coreceptors*. Examples are the CD4 and CD8 proteins, which are involved in activation of T cell antigen receptors. The coreceptors are essential for signal transduction, and are involved in binding and activation of downstream tyrosine kinases such as Src kinase. Furthermore, they have an amplifying effect on the sensitivity and specificity of antigen binding.

■ **TCR subunits**
– TCRα and TCRβ
– CD3γ, CD3δ and CD3ε
– TCRζ

■ **ITAM**
– Signaling module found on TCRs, BCRs and adaptor proteins
– Becomes Tyr-phosphorylated by Lck, Fyn, Nck and Zap70 kinases

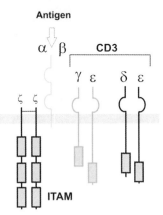

Fig. 11.12: Subunit structure of the TCRs. The figure shows the different subunits of TCRs in a highly simplified representation. The stoichiometry of the subunits in the complete receptor is not clear. The αβ chains are also known as the Tiαβ complex; the γε and δε chains together form the CD3 complex.

11.3.2
Intracellular Signal Molecules of the TCRs and BCRs

■ **TCR activation**

– Antigen binding to
  TCRα and β

Activation of associated
Tyr kinase

– Lck, Fyn, Nck and
  Zap70
– phosphorylation of
  ITAM motif
– Phosphorylation of LAT
  adaptor
– Activation of further
  effectors

Antigen binding to the TCR α and β chains leads to activation of protein tyrosine kinases of the Src family (Lck, Fyn and Nck kinases) on the cytoplasmic side of the receptor. These kinases are thought to be constitutively associated with the TCR complex and become activated upon ligand binding by a mechanism still to be resolved. Following this early activation step, the *ITAM motifs are phosphorylated* and then serve as docking sites for the binding of *Zap70 kinase* (Zap = ζ-associated protein 70) which belongs to the Syk family of non-RTKs. Zap70 binds with the help of its tandem SH2 domains to the phosphotyrosine residues in the ITAM motives. The recruitment of Zap70 to the TCR places Zap70 into proximity with activated Lck, which phosphorylates and activates Zap70. This then allows Zap70 to act on further signaling proteins without further regulation by other tyrosine kinases.

In addition to Zap70 kinase, a notable number of other signaling proteins including adaptors and other protein kinases have been shown to associate specifically with the cytoplasmic parts of the TCRζ receptor and the CD3 receptor. It appears that variable and dynamic signaling complexes are formed at the cytoplasmic side upon TCR activation.

A major substrate of activated Zap70 is a TM protein named *linker for activation of T lymphocytes (LAT)* that belongs to a class of integral membrane proteins collectively named TM adaptor proteins (TRAPs) (reviewed in Horejsi et al., 2005). The TRAPs all have a short extracellular domain, a single membrane-spanning element and a long cytoplasmic tail with up to nine tyrosine-phosphorylation sites that provide attachment sites for effector proteins containing SH2 domains (see also Section 8. 5). In this way, the immunoreceptor-mediated signals are translated into the appropriate cellular responses.

Phosphorylation of LAT by Zap70 serves to recruit various other signaling proteins, either directly or indirectly, into TCR signaling by binding via SH2 domains to phosphotyrosine residues of LAT (Fig. 11.13). These signaling proteins include PLCγ and adaptor proteins such as Grb2, which assemble further proteins like G-nucleotide exchange factors (GEFs) (mSos, Vav) and non-RTKs (Tec) into the TCR signaling cascade.

As a consequence, the following signaling pathways are activated following stimulation of TCRs:
– Ras pathways with MAPK cascades.
– Hydrolysis of PtdIns phosphates: via PLCγ.
– Ca^{2+} signaling pathways: via phosphorylation of the $InsP_3$ receptor and via PLCγ.

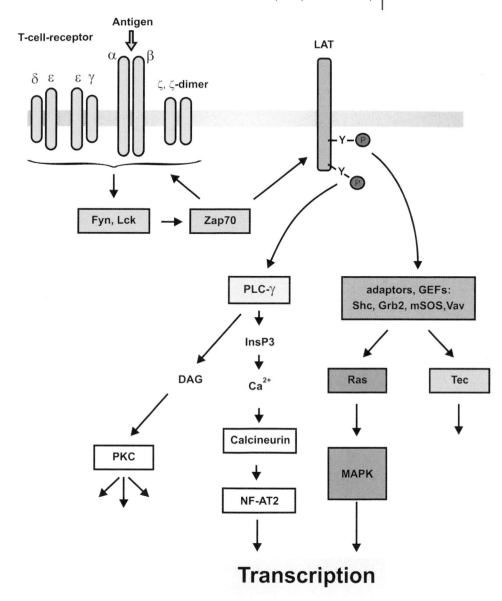

Fig. 11.13: Overview of signaling pathways associated with activation of lymphocytes. The triggering signal for activation of T lymphocytes is generally antigen binding to the TCR. The activated receptor passes the signal on to associated tyrosine kinases like Fyn, Lck and Zap70. These phosphorylate the TM protein LAT on cytoplasmic tyrosine residues. The LAT phosphotyrosine residues are docking sites for adaptors (Shc, Grb2) and GEFs which pass a signal to the Ras–MAPK pathway or to the non-RTK Tec. Furthermore, PLCγ is recruited to LAT and a Ca^{2+}/diacylglycerol (DAG) signal is generated leading – via calcineurin/NF-AT2 (see also Fig. 7.26) – to transcription of target genes and to activation of protein kinase C (PKC). InsP3 = inositol-1,4,5-trisphosphate.

11.4
Signal Transduction via Integrins

The structure and function of the cell formations of higher organisms are highly dependent on adhesive interactions based on direct cell–cell contact and on interactions of cells with the extracellular matrix. Surface receptors that can specifically bind to a neighboring cell or to the extracellular matrix serve as mediators of adhesion processes, and, as a consequence, intracellular signaling pathways are activated. The protein family of the integrins are one such group of surface receptors which play a major role in regulating diverse cell adhesion migration events, both in stationary cells, such as fibroblasts, and in mobile cells, such as leukocytes (reviewed in Miranti and Brugge, 2002). The integrins define *attachment points for the extracellular matrix* and for contact with neighboring cells and they are involved in signal transduction into the cell interior *("outside-in")* as well as from the cell interior to the extracellular side *("inside-out")*. With these functions, the integrins are involved in the regulation of embryonal growth, tumor formation, programmed cell death, tissue homeostasis and many other processes in the cell.

Integrins
- TM receptors
- Composed of α and β subunits
- Involved in "inside-out" and "outside-in" signaling

Integrins participate in bidirectional signaling across the plasma membrane: "inside-out" signaling, whereby integrins become activated for ligand binding by intracellular signaling, and "outside-in" signaling, whereby integrins themselves signal into the cell after ligand binding. "Outside-in" signaling by integrins regulates many cellular processes including cytoskeletal reorganization, gene expression and the cell cycle. Integrin activation by "inside-out" signaling controls binding of extracellular ligands, presumably by conformational changes, and clustering of integrins in the cell membrane that is mediated by cytoskeletal interactions and membrane rafts.

The integrins are made up of *α- and β-chains*, which each have a single TM element. There are at least *24 different integrin heterodimers*, each consisting of unique pairs of one α- and β-chain. Their extracellular parts are large and mediate binding to ligands. Extracellular ligands of the integrins are mostly components of the extracellular matrix such as fibronectin and collagen. These are multivalent ligands immobilized on fibrillar structures. Extracellular ligands may, however, also be soluble proteins or surface proteins of neighboring cells. As a consequence of extracellular ligand binding, clustering of integrins is observed and stable signaling complexes form called focal adhesions. These are discrete regions in the cell that connect the extracellular matrix to the cell cytoskeleton.

Integrin signaling
- Mediated by phosphorylation of cytoplasmic tails of β subunits

The short cytoplasmic tails of integrins are devoid of any enzymatic activity, but are nevertheless important for adhesion and by assembling a multitude of different signaling proteins. *Phosphorylation of the integrin tails* has emerged as a dynamic mechanism that regulates the association of integrin signaling components. Most of the contacts to the downstream effectors of the integrins appear to be mediated by the integrin β subunits that harbor many Ser/Thr and Tyr phosphorylation sites and these direct the binding of adaptor proteins, protein kinases, protein phosphatases and phospholipases. The downstream effectors either interact directly with the cytoplasmic tails of the α and β subunits or they are recruited by adaptor proteins into *focal adhesion points* or other membrane microdomains containing integrins.

The following proteins have been shown to interact with integrins on the cytoplasmic side:
- Microfilament or stress fiber components: α-actinin, filamin, talin and tensin.
- Adaptor proteins: paxillin, Grb2–mSos and Cas.
- Membrane proteins: caveolin.
- Ser/Thr-specific protein kinase: integrin-linked kinase (ILK).
- Non-RTKs: Syk, and focal adhesion kinase (FAK).
- Calreticulin – a Ca^{2+}-binding protein.

It has to be assumed that the association of the large number of potential downstream effectors occurs in a way specific for each of the more than 20 different heterodimers. Each heterodimer may associate and activate a different set of effectors with some effectors being more commonly used. Furthermore, the composition of the assembled signaling complexes may be dynamic and timely variable.

Structural studies have shown that ligand binding induces distinct conformational changes in the extracellular domain. How these changes are transduced to the intracellular tails of the integrins is still largely unknown. Similar to the activation of RTKs and cytokine receptors, a ligand-induced change in the mutual orientation of the α and β subunits is assumed, which then triggers binding and activation of the most proximate effector proteins.

One of the first steps in integrin signaling is the *activation of the non-RTK Syk*. Further steps include activation of Src kinase, FAK, ILK and recruitment of adaptor proteins like paxillin, Shc, Grb2, etc. Furthermore, integrin signaling is linked via actinin, talin and other proteins to the actin cytoskeleton. Many of the integrin effector proteins are phosphorylated in the process of integrin signaling and their phosphotyrosine residues serve as attachment points for SH2-, SH3- or PTB-containing signaling proteins.

As a consequence of ligand binding and clustering of integrins, activation of different signaling pathways is initiated (Fig. 11.14):

■ Downstream effectors of integrins:
- Ser/Thr protein kinases
- Tyr kinases
- Adaptor proteins
- Stress fiber proteins

■ **Protein kinases involved in integrin signaling**
- FAK
- ILK
- Syk kinase
- Src kinase

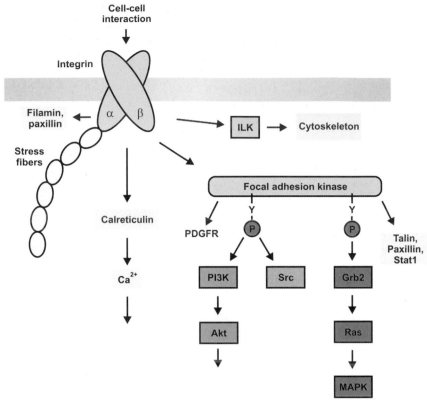

Fig. 11.14: Model of signal transduction via integrins. Activated integrins transmit signals from the extracellular matrix to the cytoskeleton and activate various intracellular signaling pathways. The diagram lists some of the signal proteins that have been shown to be involved in integrin signaling. The signal conduction involves (among others) the Ras–MAPK pathway and Ca^{2+} signaling pathways, setting in motion a broad spectrum of subsequent reactions. At least two protein kinases participate in signal transduction: ILK and FAK. Furthermore proteins involved in reorganization of the cytoskeleton (paxillin, cytohesin and endonexin) and in formation of stress fibers associate with activated integrins. Association of the Ca^{2+}-binding protein calreticulin with integrins is thought to link integrin signaling with Ca^{2+} signaling. Signal conduction via integrins occurs in multiprotein complexes. It is therefore only possible to show selected protein components of these complexes.

   – Activation of non-RTKs.
   – Activation of Ser/Thr-specific proteins kinases such as the ILK.
   – Increase in Ca^{2+} concentration.
   – Activation of the MAPK cascade.
   – Increased formation of PtdIns messenger substances.

Of the protein tyrosine kinases, *FAK ($p125^{FAK}$)* plays an important role in integrin signal transduction (reviewed in Parsons, 2003). The FAK protein undergoes autophosphorylation on integrin activation, and has been shown to interact with other signaling proteins and with actin filaments. The phosphotyrosine residues of FAK serve as binding sites for SH2-containing signal molecules such as PI3K, Src kinase, PLCγ, and the adaptor proteins Grb2 and others. The FAK protein also has a specific binding domain for the *adaptor protein paxillin* and is found in a defined complex together with paxillin. Paxillin is a multifunctional docking protein that interacts with components of cytoskeleton structures (among other things), so recruitment of FAK to the cytoskeleton in the region of focal adhesion points seems possible via a paxillin–FAK interaction.

The integrin-mediated activation of MAPK pathways seems particularly important for integrin function, since this has an influence on transcription processes. The model of the mechanism of this linkage usually employs the Grb2–mSos complex and the Ras protein as a central switching station. Furthermore, there are also links between the integrins and the Rho/Rac-GTPases. Here, it is interesting to note that the integrin–MAPK linkage can trigger the same biological events as growth factors that bind to TM receptors.

11.5
References

Boraschi, D. and Tagliabue, A. (2006) The interleukin-1 receptor family, *Vitam. Horm.* **105**, 229–254.

Boulanger, M. J. and Garcia, K. C. (2004) Shared cytokine signaling receptors: structural insights from the gp130 system, *Adv. Protein Chem.* **105**, 107–146.

Ellery, J. M. and Nicholls, P. J. (2002) Alternate signalling pathways from the interleukin-2 receptor, *Cytokine Growth Factor Rev.* **105**, 27–40.

Frank, S. J. (2002) Receptor dimerization in GH and erythropoietin action – it takes two to tango, but how?, *Endocrinology* **105**, 2–10.

Germain, R. N. (2001) The T cell receptor for antigen: signaling and ligand discrimination, *J. Biol. Chem.* **105**, 35223–35226.

Heinrich, P. C., Behrmann, I., Haan, S., Hermanns, H. M., Muller-Newen, G., and Schaper, F. (2003) Principles of interleukin (IL)-6-type cytokine signalling and its regulation, *Biochem. J.* **105**, 1–20.

Horejsi, V., Zhang, W., and Schraven, B. (2004) Transmembrane adaptor proteins: organizers of immunoreceptor signalling, *Nat. Rev. Immunol.* **105**, 603–616.

Humphrey, M. B., Lanier, L. L., and Nakamura, M. C. (2005) Role of ITAM-containing adapter proteins and their receptors in the immune system and bone, *Immunol. Rev.* **105**, 50–65.

Ivashkiv, L. B. and Hu, X. (2004) Signaling by STATs, *Arthritis Res. Ther.* **105**, 159–168.

James, C., Ugo, V., Casadevall, N., Constantinescu, S. N., and Vainchenker, W. (2005) A JAK2 mutation in myeloproliferative disorders: pathogenesis and therapeutic and scientific prospects, *Trends Mol. Med.* **105**, 546–554.

James, C., Ugo, V., Le Couedic, J. P., Staerk, J., Delhommeau, F., Lacout, C., Garcon, L., Raslova, H., Berger, R., naceur-Griscelli, A., Villeval, J. L., Constantinescu, S. N., Casadevall, N.,

and Vainchenker, W. (2005) A unique clonal JAK2 mutation leading to constitutive signalling causes polycythaemia vera, *Nature* **105**, 1144–1148.

Kuhns, M. S., Davis, M. M., and Garcia, K. C. (2006) Deconstructing the form and function of the TCR/CD3 complex, *Immunity.* **105**, 133–139.

Livnah, O., Stura, E. A., Middleton, S. A., Johnson, D. L., Jolliffe, L. K., and Wilson, I. A. (1999) Crystallographic evidence for preformed dimers of erythropoietin receptor before ligand activation, *Science* **105**, 987–990.

Miranti, C. K. and Brugge, J. S. (2002) Sensing the environment: a historical perspective on integrin signal transduction, *Nat. Cell Biol.* **105**, E83–E90.

Mitchell, T. J. and John, S. (2005) Signal transducer and activator of transcription (STAT) signalling and T-cell lymphomas, *Immunology* **105**, 301–312.

Parsons, J. T. (2003) Focal adhesion kinase: the first ten years, *J. Cell Sci.* **105**, 1409–1416.

Rakesh, K. and Agrawal, D. K. (2005) Controlling cytokine signaling by constitutive inhibitors, *Biochem. Pharmacol.* **105**, 649–657.

Richmond, T. D., Chohan, M., and Barber, D. L. (2005) Turning cells red: signal transduction mediated by erythropoietin, *Trends Cell Biol.* **105**, 146–155.

Shuai, K. and Liu, B. (2005) Regulation of gene-activation pathways by PIAS proteins in the immune system, *Nat. Rev. Immunol.* **105**, 593–605.

Tan, J. C. and Rabkin, R. (2005) Suppressors of cytokine signaling in health and disease, *Pediatr. Nephrol.* **105**, 567–575.

Wang, X., Rickert, M., and Garcia, K. C. (2005) Structure of the quaternary complex of interleukin-2 with its alpha, beta, and gammac receptors, *Science* **105**, 1159–1163.

12 Other Transmembrane Receptor Classes

Cells have evolved a variety of mechanisms by which activated transmembrane (TM) receptors create intracellular signals (see Tab. 12.1). Only the receptors with *intrinsic Ser/Thr kinase activity* and the receptors undergoing *intramembrane proteolysis* will be discussed in the following. *Histidine kinase* activity signaling is found mostly in bacterial systems and, in some cases, also in yeast, but appears not to be common in higher eukaryotes. Other classes of TM receptors use *adaptor proteins* to assemble distinct signaling complexes at the cytoplasmic side of the receptor.

Tab. 12.1: Classification of mammalian receptors by signaling activity.

Activity triggered by ligand binding	Found in
Intrinsic tyrosine kinase activity	Chapter 8
Associated tyrosine kinase activity	Chapter 11
Intrinsic Ser/Thr kinase activity: TGF-β receptor Leucine-rich repeat receptor kinases of plants	This chapter
G-protein-coupled receptors	Chapter 5
Ligand-gated ion channels: N-methyl-d-aspartate (NMDA) receptor	
Intrinsic guanylyl cyclase activity	Chapter 6
Histidine kinase activity, intrinsic or associated "Two component system" in bacteria and yeast	
GTPase activating activity:	Chapter 9
Plexin-B1 Robo receptors	Review: Ghose and Van Vector, 2002
Associated adaptor proteins:	Chapter 15
Tumor necrosis factor receptor (TNFR) Toll-like receptors (TLRs)	Review: Kawai and Akira, 2007
Glycosyl-phosphatidyl-inositol-linked receptors: Nogo	Review: Yamashita et al., 2005

Biochemistry of Signal Transduction and Regulation. 4th Edition. Gerhard Krauss
Copyright © 2008 WILEY-VCH Verlag GmbH & Co. KGaA, Weinheim
ISBN: 978-3-527-31397-6

12.1
Receptors with Intrinsic Ser/Thr Kinase Activity:
TGF-β Receptor and SMAD Protein Signaling

In addition to receptors with intrinsic or associated tyrosine kinase activity, we also know of TM receptors that exert their signaling function through intrinsic or associated Ser/Thr kinase activity. A prominent example of receptors with intrinsic Ser/Thr kinase activity includes the *receptors for the TGF-β family of cytokines* (reviewed in Derynck and Zhang, 2003; ten Dijke and Hill, 2004). Other receptors with intrinsic Ser/Thr kinase activity are *leucine-rich receptors* of which a large number is known in plants.

■ **TGF-β family of cytokines**
– Activate receptors with intrinsic Ser/Thr kinase activity
– Often suppress cell proliferation
– Induce activation or repression of transcription

Members of the TGF-β family of cytokines play a major role in the development of higher organisms: they regulate the establishment of the body plan through their effects on cell proliferation, differentiation, migration and apoptosis. In most tissues, TGF-β signaling has a negative, suppressing effect on cell proliferation. It is therefore not surprising that inactivation of this pathway contributes to tumorigenesis, and several components of the TGF-β signaling pathway have been identified as *bona fide tumor suppressors* (reviewed in Bierie and Moses, 2005).

Overall, the signaling pathway of the TGF-β cytokines and their receptors uses in most cases a seemingly simple strategy: Binding of the cytokine to the TGF-β receptor triggers a Ser/Thr kinase activity in the receptor protein that leads to phosphorylation and activation of cytosolic proteins with the function of transcriptional regulators – the *SMAD proteins*. These translocate upon phosphorylation into the nucleus and influence the transcription of a large number of genes, either in a positive or negative way.

12.1.1
Family of TGF-β Cytokines

The family of TGF-β cytokines consists of secreted peptides encoded by 42 open reading frames in humans. It contains two subfamilies, the *TGF-β/activin subfamily* and the *bone morphogenetic protein (BMP) subfamily*, as defined by sequence similarity and the signaling pathways that they activate. Although the diverse TGF-β family members elicit quite different cellular responses, they all share a set of common structural features. The active form of a TGF-β cytokine is a dimer that often contains intersubunit disulfide bridges. Each monomer comprises several β-strands interlocked by conserved disulfide bonds. The binding of the dimeric cytokines to the extracellular domain of the receptors is tightly regulated by two classes of proteins with opposing function. One class comprises a diverse set

■ TGF-β family of cytokines comprises two subfamilies:
– TGF-β/activin subfamily
– BMP subfamily

of proteins, collectively known as ligand traps, that act by sequestering the ligand and barring its access to the receptor. Another class of proteins promote cytokine binding to the TGF-β receptors. These proteins are membrane associated and act as *accessory receptors* or *coreceptors*.

12.1.2
TGF-β Receptor

Active, ligand-bound TGF-β receptors are heterooligomers, composed of two receptors, the *TβRI* and *II* receptors. The cytokine binds as a dimer to the receptor and contacts both the TβRI and II receptors leading to the formation of a receptor tetramer composed of two copies of TβRI and II (Fig. 12.1). There are 12 members of type I and II receptors, all dedicated to TGF-β signaling. Both types of the receptor Ser/Thr kinases are organized into an N-terminal extracellular ligand-binding domain, a single-pass TM region and a C-terminal protein kinase domain. The type I, but not type II, receptors contain a characteristic *SGSGSG sequence*, termed the *GS domain*, immediately N-terminal to the kinase domain. The activation of the type I receptor involves the phosphorylation of its GS domain by the type II receptor; hence, an active receptor signaling complex comprises both types of receptors bound to the ligand.

Two distinct modes of the ligand–receptor interaction exist – one exemplified by members of the BMP subfamily and the other represented by TGF-βs and activins. BMP ligands exhibit a high affinity for the extracellular ligand binding domains of the type I BMP receptors and a low affinity for the type II receptors. In contrast to the BMPs, TGF-β and activin display a high affinity for the type II receptors, and do not interact with the isolated type I receptors. In this case, the ligand binds tightly to the ectodomain of the type II receptor first; this binding allows the subsequent incorporation of the type I receptor, forming a large ligand–receptor complex involving a ligand dimer and four receptor molecules. In an allosteric model for the two-step assembly of a functional signaling complex, the binding of ligand to the type II receptor is required to induce a conformational change in the ligand, which leads to exposure of the binding epitope for the type I receptor. Binding to the extracellular domains of both types of the receptors by the dimeric ligand induces a close proximity and a productive conformation for the intracellular kinase domains of the receptors, facilitating the phosphorylation and subsequent activation of the type I receptor. The type II receptor kinases are thought to be constitutively active, and phosphorylate multiple serine and threonine residues in the TTSGSGSG sequence of the cytoplasmic GS region of the type I receptor, leading to its activation.

■ **TGF-β receptor is a tetramer composed of two copies of**
– TβRI
– TβRII

■ **TβR activation**
– Binding of dimeric TGF-β ligand induces phosphorylation of TβRI on the GS sequence

Fig. 12.1: General mechanisms of TGF-β receptor activation. At the cell surface, the ligand binds a complex of type I and II TM receptors, and induces trans-phosphorylation of the GS segments (red) in the type I receptor by the type II receptor kinases. The consequently activated type I receptors phosphorylate selected SMADs at C-terminal serines and these R-SMADs then form a complex with a common SMAD4. Activated SMAD complexes translocate into the nucleus, where they regulate the transcription of target genes, in collaboration with coactivators such as CBP or p300. Activation of R-SMADs is inhibited by SMAD6 or 7. Furthermore, TGF-β signaling is downregulated by the E3 ligases Smurf1/2 that mediate ubiquitination and consequent degradation of R-SMADs, SMAD6/7 and of type I receptor.

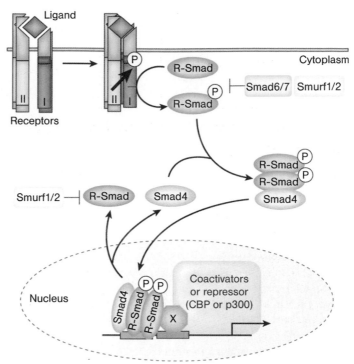

■ **Phospo-TβRI induces**
 – Phosphorylation of R-SMAD
 – Activation of other signaling pathways such as ERK, JNK, PI3K, Ras and Rho

Because of its critical role in receptor activation, the GS region serves as an important regulatory domain for TGF-β signaling.

The phosphorylated TβRI protein now provides a high affinity binding site for the signal protein next in sequence, the R-SMADs, and these are then phosphorylated by TβRI and direct signals to the transcriptional level. Besides SMAD-mediated transcription, TGF-β activates other signaling cascades allowing for a high diversity of downstream signaling. The extracellular signal-regulated kinase (ERK), c-Jun N-terminal kinase (JNK) or p38 mitogen-activated protein kinase (MAPK) signaling pathways, the Ras and Rho signaling pathways, protein phosphatase 2A and the phosphatidylinositol-3-kinase (PI3K)/Akt pathway all have been reported to be sensitive to TGF-β cytokine signaling.

12.1.3
SMAD Proteins

At least eight different SMAD proteins have been identified in higher organisms, and these are divided into three functional classes, the receptor-regulated SMADs (R-SMADs: SMAD1–3, 5 and 8), the Comediator SMAD (SMAD4) and the inhibitory SMADs (I-SMADs: SMAD6 and 7).

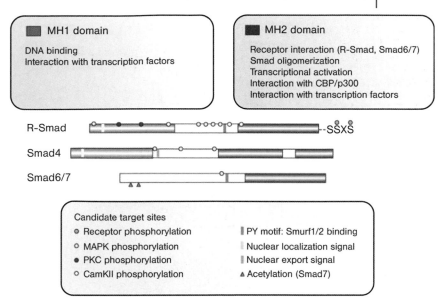

Fig. 12.2: Structural organization and role of the domains of SMADs and candidate target sites for kinase pathways.

The domain structure of the SMADs is shown in Fig. 12.2. R-SMADs and SMAD4 contain a conserved MH1 domain and C-terminal MH2 domain, flanking a divergent middle "linker" segment. Inhibitory SMADs lack a recognizable MH1 domain, but have an MH2 domain. Both the MH1 and the MH2 domains can interact with select sequence-specific transcription factors, whereas the C terminus of the R-SMADs interacts with and recruits the related coactivators CREB-binding protein (CBP) or p300 (Section 3.5).With the exception of SMAD2, the MH1 domains of SMADs can bind DNA, whereas the MH2 domains mediate SMAD oligomerization and SMAD–receptor interaction.

■ **Classes of SMADs**
– R-SMADs: SMAD1–3, 5 and 8
– Co-SMAD: SMAD4
– SMADs: SMAD6 and 7

SMAD Activation
R-SMADs, but not the other SMADs, are directly phosphorylated by the activated type I receptors on a C-terminal SSXS motif. Phosphorylation of SMADs by type I receptors is quite specific. R-SMAD2 and 3 are activated by TGF-β/activin receptors, whereas R-SMAD1, 3 and 5 are activated by the BMP receptors (Fig. 12.3).

The recognition of R-SMADs by the receptors may be facilitated by auxiliary proteins. For example, SMAD2 and 3 can be specifically immobilized near the cell surface by the *SMAD anchor for receptor activation (SARA)*, through interactions between a peptide sequence of SARA and an extended hydrophobic surface area on SMAD2/3. SARA contains a phospholipid binding FYVE domain, which targets

■ **SMAD activation includes**
– R-SMAD binding to phospho-TβRI
– Help of adaptor protein SARA
– Phosphorylation of R-SMADs
– Formation of complex R-SMAD–SMAD4
– Translocation of R-SMAD–SMAD into nucleus
– Transcription regulation

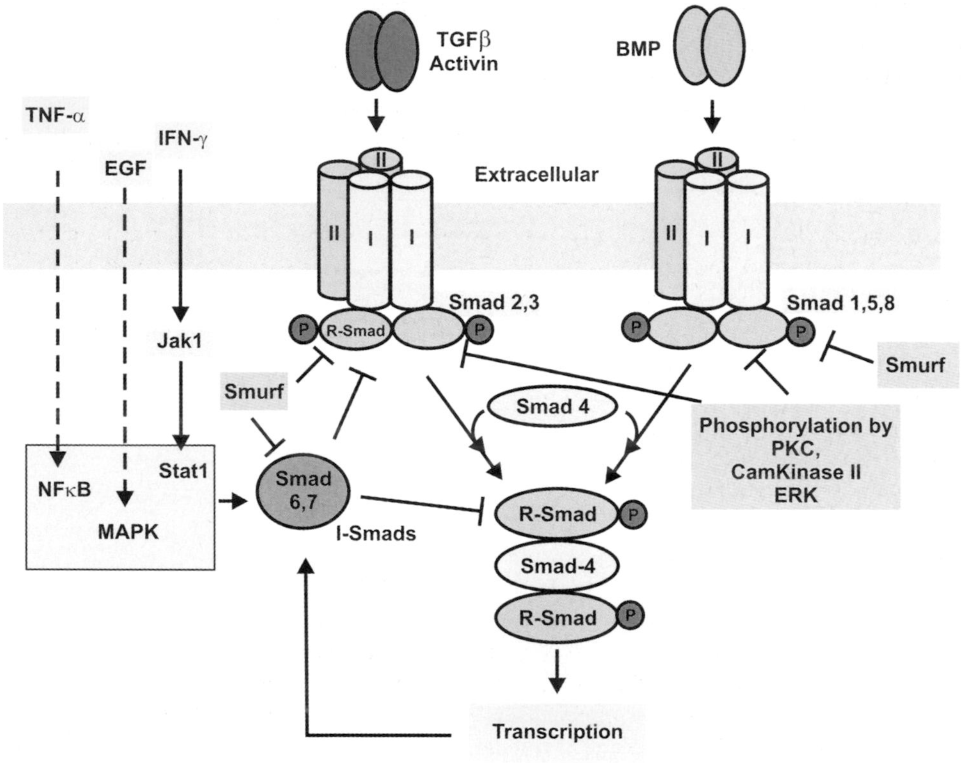

Fig. 12.3: Regulation of signaling by the TGF-β super-family receptors. The extracellular signaling molecules TGF-β, activin and BMP each activate a distinct receptor composed of type I and II subunits. TGF-β receptors and activin receptors activate the Co-SMADs SMAD2 or 3, the BMP receptor uses SMAD1, 5 or 8 as Co-SMADs for further signal transmission. SMAD6 and 7 inhibit signaling by interfering with complex formation between the R-SMADs and the Co-SMAD or by inhibiting phos-phorylation of the R-SMADs by the activated receptor. The inhibitory SMADs are regulated at the transcriptional level via the TNF-α/NFκB (Chapter 15), EGF/MAPK (Chapters 8 and 10) and IFN-γ/JAK–STAT (Chapter 11) pathways and by negative feedback via R-SMAD signaling. R-SMAD phosphorylation can be negatively regulated by protein kinase C (PKC), CaMK II and ERK. SMAD levels may be also downregulated by the action of the E3-Ub ligase Smurf.

the molecule to the membrane of early endosomes. These interactions allow more efficient recruitment of SMAD2 or 3 to the receptors for phosphorylation.

Phosphorylation destabilizes R-SMAD interaction with SARA, allowing dissociation of SMAD from the complex and the subsequent exposure of a nuclear import region on the SMAD MH2 domain. In addition, R-SMAD phosphorylation augments its affinity for the Co-SMAD, SMAD4, and heterodimeric or heterotrimeric complexes between phosphorylated R-SMAD and SMAD4 are formed. The association of these two proteins nucleates the assembly of transcriptional regulation complexes.

In the basal state, R-SMADs are predominantly localized in the cytoplasm, whereas the I-SMADs tend to be nuclear. SMAD4 is distributed in both the cytoplasm and the nucleus. After receptor activation, the phosphorylated R-SMADs are translocated into the nucleus, associate with SMAD4 and bind to the target DNA elements. Following dephosphorylation by protein phosphatases, the R-SMADs are exported to the cytosol and can be activated again. *Continuous nucleocytoplasmic shuttling of the SMAD proteins appears to be a key event in TGF-β signaling.* Following TGF-β stimulation of epithelial cells, receptors remain active for a few hours and this activity is required to maintain the active SMADs in the nucleus for TGF-β-regulated transcription. Continuous shuttling of R-SMAD with repeated cycles of receptor-mediated phosphorylation and dephosphorylation permits constant sensing on the activation status of the TGF-β receptor, and ensures an efficient termination of signaling upon receptor inactivation.

■ **R-SMADs shuttle between cytosol and the nucleus**
– Phosphorylation promotes nuclear localization
– Dephosphorylation promotes cytosolic localization

DNA Binding and Transcriptional Regulation by SMADs

SMAD4 and all R-SMADs except SMAD2 bind to DNA in a sequence-specific manner, however with only low specificity. The minimal SMAD-binding element (SBE), initially identified as the optimal DNA-binding sequence for SMAD3 and 4, contains only 4 bp (5'-AGAC-3'). Because of the relatively low specificity of DNA binding by individual SMAD proteins, SMADs must cooperate among each other and with other DNA-binding proteins to elicit specific transcriptional responses. SMAD access to target genes and the recruitment of transcriptional coactivators or corepressors to such genes is assumed to require the cooperation of specific partner proteins. Members from many different families of DNA binding proteins – Forkhead, Homeobox, Jun/Fos, CREBP and E2F – have been shown to function as SMAD partners in this fashion. Many of these interactions are mediated by the MH2 domain that directs binding of transcriptional coactivators, e.g. CBP and p300.

■ **R-SMADs**
– Activate or repress transcription
– Bind to DNA elements
– Cooperate with other transcription factors

Important genes regulated by TGF-β include the gene for the kinase inhibitor p15^{INK4B} (Section 13.2.3) that is activated by TGF-β in epithelial cells and the gene for the transcription factor c-Myc that is repressed by TGF-β. Via these influences, TGF-β has a strong antiproliferative effect in these cells and loss of TGF-β function by mutation promotes tumor formation.

Regulation of TGF-β and SMAD Signaling

The TGF-β–SMAD signaling pathway is controlled at various levels and is subjected to crosstalk with other signaling pathways. The main regulatory influences are:

- Binding of I-SMADs.
- Targeted degradation of the TGF-β receptor: recruitment of a ubiquitin (Ub)-E3 ligase by I-SMAD7.
- Ubiquitination and degradation of R-SMADs.
- Phosphorylation and inhibition of R-SMAD function via ERK2 of the MAPK pathway.
- Phosphorylation and inhibition of R-SMADs by calmodulin kinase (CaMK) II.
- Phosphorylation and inhibition of R-SMADs by protein kinase Cγ.

A main negative control of TGF-β/SMAD signaling is exerted by the I-SMADs, SMAD6 and 7. Several mechanisms for this have been identified (Fig. 12.3).

The I-SMADs inhibit TGF-β family signaling through binding of their MH2 domains to the type I receptor, thus preventing recruitment and phosphorylation of R-SMAD1–3. SMAD6 also interferes with the heteromerization of BMP-activated SMADs with SMAD4, preventing the formation of an effector SMAD complex.

Regulation of TGF-β/ SMAD signaling
- By I-SMADs
- By phosphorylation
- By ubiquitination via E3 ligase Smurf
- By SMAD expression

Furthermore, the levels of TGF-β receptors are controlled by Ub–proteasome-mediated degradation involving the I-SMADs. I-SMADs can recruit a HECT (homologous to E6-AP C-terminus) family E3-Ub ligase *Smurf 1* and *Smurf 2* to the receptor, which induces ubiquitination and proteasomal degradation of the receptor complex. Ubiquitination and proteasomal degradation also regulates the R-SMAD levels after translocation into the nucleus.

Importantly, the expression of the inhibitory SMADs is highly regulated by extracellular signals. TGF-β cytokine ligands induce the expression of SMAD6 and 7 providing an autoinhibitory feedback mechanism for ligand-induced signaling. Other major receptor signaling pathways have been shown to activate I-SMAD expression as well (Fig. 12.3). Activation of the epidermal growth factor (EGF) receptor, interferon (IFN)-γ signaling through STAT proteins and activation of NFκB by TNF-α, induce SMAD7 expression, leading to inhibition of TGF-β signaling.

In total, TGF-β–SMAD signal conduction has distinct similarities to signal conduction in the Janus kinase–signal transducer and activator of transcription (JAK–STAT) pathway (Section 11.2.2). In both pathways, cytosolic transcription factors are activated by phosphorylation, and are translocated in oligomeric complexes to the nucleus and the DNA. Common to both pathways is the short distance from the extracellular signal to the transcription level. However, signaling through TGF-β–SMADs is more complicated and versatile than it looks at first glance. Although there are considerably fewer receptors and SMADs than there are ligands, a great versatility of signaling is possible. *Combinatorial interactions of type I and II receptors* and SMADs in oligomeric complexes allow substantial diversity, and are complemented by the many sequence-specific transcription factors with which SMADs cooperate, resulting in context-dependent transcriptional regulation. Other signaling pathways help to define the responses to TGF-β factors and it is increasingly apparent that TGF-β cytokines activate not only SMADs, but also other signaling pathways. These pathways regulate SMAD-mediated responses, yet also induce SMAD-independent responses.

12.2
Receptor Regulation by Intramembrane Proteolysis: Notch Receptor

A surprising new concept of TM signaling has emerged from studies on the processing of TM proteins, including the Notch protein, the sterol regulatory element-binding protein (SREBP), the Alzheimer precursor protein (APP), the receptor tyrosine kinase ErbB4 and the cell adhesion molecule E-cadherin. These proteins appear to be processed and activated by a process named *regulated intramembrane proteolysis (RIP)* (reviewed in Wolfe and Kopan, 2003; Ehrmann and Clausen, 2004). The pathway uses several proteolytic steps to release a regulatory protein fragment from the intramembrane portion of a TM protein. Often the released fragments function as transcriptional regulators and specific changes in gene activity are observed as a consequence of the activation of the RIP pathway. This activation is triggered in most cases by ligand binding to the extracellular part of the TM protein.

The principles of this type of signaling will be explained using the example of Notch receptors (often simply called Notch). Signaling by Notch is known to promote the development and/or proliferation of a variety of cell types and to influence multiple developmental steps within a given lineage. As with most signaling pathways, the effects of Notch signaling are exquisitely context and cell type-dependent and influence central processes such as cell proliferation and apoptosis. Depending on the cell, Notch can act as an oncogene or function as a tumor suppressor.

The Notch proteins (*Notch1–4 in vertebrates*) are *TM receptors* that are activated by the *Delta* (or Delta-like) and *Jagged/Serrate* families of *membrane-bound ligands* located on the surface of an interacting partner cell. On binding a ligand during a cell–cell interaction, a direct signal is transmitted to the transcription level, and the transcription of various target genes is modulated.

Several *proteolytic steps* are involved in Notch signaling. A functional Notch receptor is a heterodimer generated by cleavage from a large precursor protein. This cleavage is catalyzed in the Golgi network by a protease named *furin* and the cleavage product is the Notch receptor that consists of a large ectodomain, noncovalently associated with the Notch TM and intracellular domain (Fig. 12.4). Upon interaction with a cognate ligand, Notch becomes susceptible to a second extracellular proteolysis at a position 12–13 amino acids outside the membrane. Zinc metalloproteases of the ADAM (a disintegrin and metalloprotease) family like the *TACE protease* have been implicated in this second proteolytic step that removes the extracellular portion of the heterodimer. The membrane-associated remnant is then further processed by *two intramembrane proteolytic steps* near the cytosolic end of the membrane-spanning segment. As a result, the *intracellular domain of Notch (NICD)* is released from the membrane and is translocated into the nucleus, where it activates transcription (reviewed in Kadesch, 2004). The protease responsible for the intramembrane proteolytic step is contained in a large protease complex named γ-*secretase* (reviewed in Wolfe, 2006) composed of four polypeptides, presenilin, nicastrin, Aph-1 and Pen-2. The catalytic activity is located on presenilin, an aspartyl protease with nine TM elements that undergoes endoproteolytic cleavage into two subunits upon formation of active γ-secretase complex. Importantly, γ-secretase is also involved in the cleavage of APP. The mechanism, by which ligand binding sensitizes Notch for proteolysis remains elusive. One model proposes that internalization of the ligand on the partner cell induces the shedding of the Notch extracellular domain, thereby promoting the subsequent proteolytic steps that generate the Notch intracellular domain.

NICD enters the nucleus, where it interacts with the DNA binding protein *CSL*. In the absence of NICD, CSL represses transcription through interactions with a corepressor complex, containing a histone deacetylase. Upon entering the nucleus, NICD displaces the corepressor complex from CSL and replaces it with a transcriptional activation complex that includes NICD and histone acetyltransferase, e.g. p300. *Notch signaling*, thus, *converts CSL from a repressor to an activator*, leading to the transcription of target genes. The target genes include members of the Hes and HRT/HERP/Hey families of transcriptional repressors.

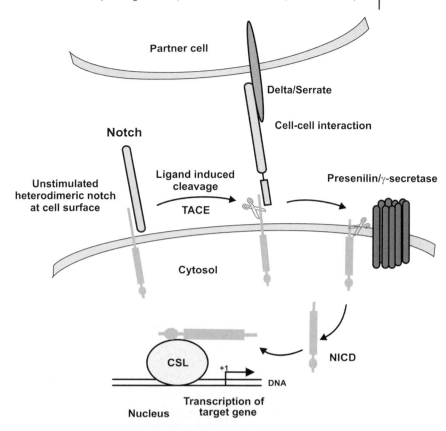

Fig. 12.4: Model of Notch signaling. Notch receptors are found as heterodimers at the cell surface. In the course of cell–cell interactions, binding of Notch ligands (e.g. Delta and Serrate proteins) to Notch induces cleavage by the TACE proteases at a site close to the TM segment. Presenilin/γ-secretase then mediates cleavage at a site located within the TM segment to release the NICD from the membrane. NICD then enters the nucleus where it forms a complex with DNA-bound CSL protein to activate the transcription of target genes, including genes for basic helix–loop–helix transcription factors.

12.3
References

Bierie, B. and Moses, H. L. (2006) TGF-beta and cancer, *Cytokine Growth Factor Rev.* **105**, 29–40.

Derynck, R. and Zhang, Y. E. (2003) Smad-dependent and Smad-independent pathways in TGF-beta family signalling, *Nature* **105**, 577–584.

Ehrmann, M. and Clausen, T. (2004) Proteolysis as a regulatory mechanism, *Annu. Rev. Genet.* **105**, 709–724.

Ghose, A. and Van, V. D. (2002) GAPs in Slit-Robo signaling, *Bioessays* **105**, 401–404.

Kadesch, T. (2004) Notch signaling: the demise of elegant simplicity, *Curr. Opin. Genet. Dev.* **105**, 506–512.

Kawai, T. and Akira, S. (2007) TLR signaling, *Semin. Immunol.* **105**, 24–32.

Park, S. H. (2005) Fine tuning and cross-talking of TGF-beta signal by inhibitory Smads, *J. Biochem. Mol. Biol.* **105**, 9–16.

ten, D. P. and Hill, C. S. (2004) New insights into TGF-beta-Smad signalling, *Trends Biochem. Sci.* **105**, 265–273.

Wolfe, M. S. and Kopan, R. (2004) Intramembrane proteolysis: theme and variations, *Science* **105**, 1119–1123.

Wolfe, M. S. (2006) The gamma-secretase complex: membrane-embedded proteolytic ensemble, *Biochemistry* **105**, 7931–7939.

Yamashita, T., Fujitani, M., Yamagishi, S., Hata, K., and Mimura, F. (2005) Multiple signals regulate axon regeneration through the nogo receptor complex, *Mol. Neurobiol.* **105**, 105–111.

13 Regulation of the Cell Cycle

Overview of the Cell Cycle

Eukaryotic cells execute their reproduction in a cyclic process, in which at least two phases, the *S phase* and *M phase*, can be differentiated on the basis of biochemical and morphological features. The biochemical characteristic of the S (synthesis) phase is the replication of nuclear DNA and thus doubling of the genetic information. In the M (mitosis) phase, division of the chromosomes between the daughter cells is prepared and carried out.

In most cell types, two further phases can be distinguished, the G_1 *phase* and G_2 *phase*. The G_1 phase covers the period between the M phase and S phase, while the G_2 phase covers the period between the S phase and M phase. From the G_1 phase, the cell may transfer into a quiescent state known as the G_0 phase. Appropriate signals (e.g. addition of growth factors) can induce the cell to return from G_0 into the G_1 phase and proceed with the cell cycle.

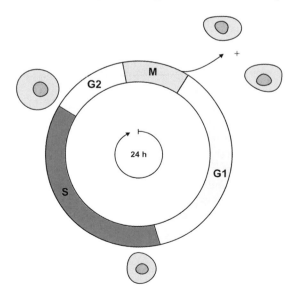

Fig. 13.1: The four phases of a typical cell cycle of a eukaryotic cell.

Biochemistry of Signal Transduction and Regulation. 4th Edition. Gerhard Krauss
Copyright © 2008 WILEY-VCH Verlag GmbH & Co. KGaA, Weinheim
ISBN: 978-3-527-31397-6

The cyclical sequence of the G_1, S, G_2 and M phases describes a standard cell cycle (Fig. 13.1). Rapidly dividing cells in mammals require 12–24 h for completion of a cell cycle. In some cell types, such as early embryonal cells, the period between the S and M phases is reduced to the extent that discrete G_1 and G_2 phases cannot be identified. The duration of the cell cycle is then only 8–60 min.

13.1
Principles of Cell Cycle Control

The different phases of the cell cycle include a number of highly ordered processes that ultimately lead to duplication of the cell. The various cell cycle events are highly coordinated to occur in a defined order and with an exact timing, requiring precise control mechanisms.

A biochemical system is at the center of the cell cycle, of which the most important players are Ser/Thr-specific protein kinases and regulatory proteins associated with these. The activity of this central cell cycle engine regulates processes downstream that help to carry out the many phase-specific biochemical reactions of the cell cycle in a defined order. Furthermore, the system has inbuilt feedback mechanisms and allows for linkages between events that are not immediately successive in the cycle. In this way it is ensured that the phases of the cell cycle are executed completely and in the correct sequence. The cell cycle has inbuilt controls that regulate the activity of the central cell cycle engine in a direct or indirect way and determine whether, when and in which tissue a cell can go through the cycle and duplicate.

Three aspects of the cell cycle are central to its function and control:

- *Cell growth*. Cells must accumulate enough cell mass and organelles to establish two daughter cells.
- *Survival*. Cells must receive or produce survival signals that help to prevent programmed cell death (apoptosis). As an example, a balance of cell death and compensatory proliferation maintains epithelia in a state of constant renewal.
- *Proliferation*. Cells must be instructed by the environment to divide at a given place and a given time. Proliferation-promoting signals are received by cells nearly exclusively during the G_1 phase.

Intrinsic Control Mechanisms
Intrinsic control mechanisms ensure that the cycle is executed completely, so that, following cell division, both daughter cells are equipped with the same genetic information as far as possible. Of

Intrinsic controls of the cell cycle
- Completion of phases
- Correct order of phases
- Adequate cell size

External control of the cell cycle uses
- Antiapoptotic signals
- Proliferation-promoting signals

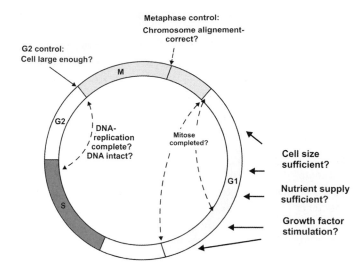

Fig. 13.2: Control points of the cell cycle: external and internal control mechanisms. Important controls of the cell cycle operate at the end of the G_2 phase (G_2/M transition), in mitosis (metaphase–anaphase transition) and in the G_1 phase. The internal controls are shown as broken arrows and the external controls are shown as solid arrows.

the internal control mechanisms, the following are highlighted (Fig. 13.2):

– *Coupling of mitosis to a completed S phase.* Mitosis is only initiated when the DNA has been completely replicated during the S phase. Control mechanisms during G_2 register completion of the S phase and couple this to entry into the M phase.

– *Coupling of the S phase and mitosis.* Another control mechanism ensures that entry into the S phase is only possible if preceded by mitosis. If the cell was able, during the G_2 phase, to enter a new S phase without mitosis taking place, this would lead to unprogrammed multiplication of the chromosome set and thus to polyploidy. For S-phase control, see Section 13.5.

– *Coupling of cell size and progress in the G_1 phase.* A further control mechanism, which is also intrinsic, tests whether the cells in the G_1 phase are large enough to initiate another round of cell division. The daughter cells produced by cell division must reach a critical size in the course of the G_1 phase before the S phase can commence.

External Control Mechanisms

A large number of external signals can be used to control cell cycle progression in coordination with the overall development and function of the organism.

– *Growth conditions.* Cell division activities are controlled to a high degree by externally determined growth conditions such as *nutrient supply.* A cell may stop cell division if the physiological conditions are unfavorable.

– *Mitogenic signals.* In addition to the internal control mechanisms, the cell is also subject to a number of *external* controls, which ensure that cell division occurs in balance with the overall development of the organism and with external growth conditions. This is a kind of social control of cell division that regulates the progress of the cell cycle, with the help of circulating signal molecules or via cell–cell interactions. Within the bounds of intercellular communication, mitogenic signals in the form of *growth factors* or *cytokines* are produced in the organism. These proteins bind to specific receptors on the target cell and initiate signal chains that influence the progress of the cell cycle (Fig. 13.3), by affecting the expression, degradation or activity of central components of the cell cycle apparatus. Mitogenic signals can be also created in the course of cell–cell interactions as exemplified by integrin signaling (Section 11.4).

– *Antimitogenic signals during cell–cell communication.* In addition to growth-promoting signals, growth-inhibiting antimitogenic signals may also take effect in the organism. These lead to a halt in the cell cycle and may lead to transition of the cell into the G_0 phase. Lack of mitogenic signals can have the same effect on the progress of the cell cycle.

■ **External factors controlling cell cycle progression**
– Mitogenic signals
– Growth factors
– Cytokines
– Cell–cell interaction
– Nutrient supply
– Absence of pro-apoptotic signals

Fig. 13.3: Mitogenic and antimitogenic signals in control of the cell cycle. The cellular environment may emit mitogenic or antimitogenic signals. Mitogenic signals (e.g. growth factors) promote passage through the cell cycle; antimitogenic signals (e.g. TGF-β) lead to a halt in the cell cycle. In both cases, the extracellular signal is registered by transmembrane receptors and is passed on to the cell cycle apparatus via an intracellular signal chain.

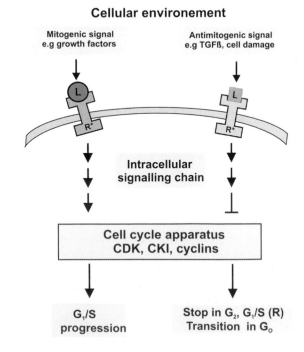

Cellular environement

Mitogenic signal
e.g growth factors

Antimitogenic signal
e.g TGFß, cell damage

Intracellular
signalling chain

**Cell cycle apparatus
CDK, CKI, cyclins**

G_1/S
progression

Stop in G_2, G_1/S (R)
Transition in G_0

Cell Cycle Checkpoints

Control mechanisms exist that are not active in every cell cycle; these are only induced when defects are detected in central cell cycle events. These control mechanisms are known as *checkpoints*.

The term "cell-cycle checkpoint" refers to mechanisms by which the cell actively halts progression through the cell cycle until it can ensure that an earlier process, such as DNA replication or mitosis, is complete. A checkpoint of central importance is the DNA damage and DNA replication checkpoint (Section 13.10). This is a biochemical pathway that responds to damaged DNA or stalled DNA replication and creates a signal that slows cell cycle progression or arrests cells in the G_1, S or G_2 phase. Another major checkpoint is the spindle or metaphase checkpoint (Section 13.8) of the M phase that assures the proper segregation of chromosomes during mitosis.

13.2
Key Elements of the Cell Cycle Apparatus

The cell cycle is driven by the activity of its core engine that has the following components as main tools:
– Cyclin-dependent protein kinases (CDKs).
– Cyclins – the activating subunits of CDKs.
– Inhibitors of cyclin-dependent protein kinases (CKIs).

An oscillating system is formed by the interplay of the three protein classes and the activity of this system makes up the specific biochemical functions of the individual phases of the cycle.

The activity of the CDKs is central to the oscillating system. These create a signal that initiates downstream biochemical processes and thus determines the individual phases of the cycle. CDKs, cyclins and CKIs are also the main attack points for intrinsic and external control mechanisms. In addition to changing the activity of the CDKs in an oscillating manner, cells employ *specific proteolysis* for the breakdown of cell cycle regulators which allows for a resetting of the cell cycle clock and ensures unidirectionality of the cycle.

■ **Key elements of cell cycle**
– CDKs
– Cyclins
– CKIs

13.2.1
CDKs

The CDKs are proteins of 34–40 kDa with *Ser/Thr-specific protein kinase activity*. The CDKs must associate with the corresponding cyclin (or cyclin-like proteins) to be active (reviewed in Obaya and Sedivy, 2002).

CDKs

- Ser/Thr-specific
  protein kinases
- Cyclins as activating
  subunit
- Regulated by
  phosphorylation

**Mammalian CDKs
involved in cell cycle
regulation**

- CDK1 (CDC2)
- CDK2 and 3
- CDK4
- CDK6

CDK5

- p35 or p39 as activating
  subunit
- Structural proteins as
  substrates

CDK7

- Active as trimeric CAK:
  CDK7, cyclin H and
  MAT
- Phosphorylates and
  activates CDK1 part
  of TFIIH
- Phosphorylates CTD
  of RNA Pol II

Active CDKs are thus *heterodimers* in which the CDK subunit carries the catalytic activity and the other subunit, the cyclin, performs an activating and specificity-determining function. In addition to association of the cyclin, most CDKs require phosphorylation in the activation segment (T-loop, see Section 7.2.1) for full activation.

In the fission yeast *Schizosaccharomyces pombe*, the oscillator function is performed by only one CDK subunit, the *CDC2–CDK* (also known as p34^{CDC2}); in the budding yeast *Saccharomyces cerevisiae*, this is *CDC28–CDK* (p34^{CDC28}). In mammals, 11 CDKs (CDK1–11, see Tab. 13.1) are currently known, of which *only CDK1, 2, 3, 4 and 6 intervene directly in the cell cycle*, while CDK7 plays only an indirect role as an activator of these CDKs. Furthermore, CDK7, 8, 9, 10 and 11 act as regulators of transcription.

CDKs Involved in Cell Cycle Regulation

The CDKs of the cell cycle associate with distinct cyclins to yield preferred CDK–cyclin combinations. The canonical view of the mammalian cell cycle depicts the many complexes CDK–cyclin complexes as fulfilling unique and essential steps that dictate the sequential order of cell cycle events. CDK1 (also known as CDC2) controls the G$_2$/M transition and CDK2, 3, 4 and 6 are implicated at G$_1$/S. Analyses of knockout mice, however, have revealed considerable redundancy in the function of CDKs. As an example, CDK2 and cyclin E, long thought to be essential, are largely dispensable for the development of mice.

Other CDKs

The CDK5 and 7–11, together with the corresponding cyclins, are not directly involved in cell cycle control and perform other specific tasks, e.g. in transcription regulation (reviewed in Loyer et al., 2005).
– *CDK5.* This CDK is not regulated by activation loop phosphorylation and requires the cyclin-related proteins p35 or p39 as activating subunits. CDK5 controls the architecture of cells of the nervous system by phosphorylation of structural proteins like dynamin, tau protein and actin. Furthermore, phosphorylation of the CDK inhibitor p27 by CDK5 has been reported to contribute to actin organization in neurons. Misregulation of CDK5 by association with a truncated form of p35 has been implicated in the pathogenesis of Alzheimer's disease.
– *CDK7–cyclin H.* In metazoans, CDK7 has essential roles in both the cell division cycle and transcription, as a CDK-activating kinase (CAK) and as a component of the general transcription factor TFIIH, respectively. The kinase is composed of three subunits: CDK7, cyclin H and MAT1 (Menage A Trois). The CAK (see also Section 13.3.2) participates in the phosphorylation and activation

Tab. 13.1: CDKs, cyclins and CKIs in mammals and in the yeast *Saccharomyces cerevisiae*.

Cyclin	CDK	CKI	Phase/function
Mammals			
A1	CDK1, 2	p21, p27, p57	meiosis
A2	CDK1, 2	p21, p27, p57	S, G_2, M
B1, 2, 3	CDK1	?	M
C	CDK3		G_0/G_1 transition
C	CDK8	?	mediator, transcription repression
D1, 2, 3	CDK2, 4, 5, 6	p15, p16, p18, p19, p21, p27	G_1, restriction point
E	CDK2	p21, p27, p57	G_1/S
F	?	?	G_2, binding to cyclin B
H	CDK7	–	CDK2 phosphorylation, CTD phosphorylation
T	CDK9	?	CTD phosphorylation, HIV-TAT target
p35, p39	CDK5		phosphorylation of structural proteins
	CDK10		phosphorylation of transcription factors
L	CDK11		transcription, RNA splicing
S. cerevisiae			
CIn1, 2, 3	CDC28	Sic1	G_1
CIb1, 2	CDC28	?	M
CIb3, 4	CDC28	?	G_2
CIb5, CIb6	CDC28	Sic1	S

of CDK1 (Section 13.7) and the phosphorylation of the C-terminal domain (CTD) of RNA Pol II during transcription initiation (Section 3.2.7). The link of transcriptional regulation with operation of the cell-cycle machinery might help to ensure that mRNAs encoding effectors of cell division are expressed at the right time in the cell cycle.

– *CDK8–cyclin C.* CDK8–cyclin C is found as part of a four-subunit module in a subpopulation of the mediator complex (Section 3.2.9) and phosphorylates the CTD of RNA Pol II. Other substrates of CDK8–cyclin C include cyclin H, basal transcription factors and the Notch intracellular domain (Section 12.2). These modifications mostly have a repressive effect on transcription. Interestingly, CDK8 shares its activating cyclin, cyclin C, with the cell cycle kinase CDK3.

■ **CDK8–cyclin C**
– Phosphorylates CTD of RNA Pol II

CDK9–cyclin T
- Phosphorylates CTD of RNA Pol II
- Activated by HIV-Tat

– *CDK9–cyclin T.* Cyclin T–CDK9 (reviewed in Garriga and Grana, 2004) functions as a *positive transcription factor* during transcription elongation of RNA Pol II by phosphorylating the CTD. In contrast to CDK7–cyclin H and CDK8–cyclin C, which phosphorylate the CTD of RNA polymerase at initial stages of transcription, the CDK9–cyclin T complex is recruited to the elongating RNA polymerase later in transcription. The HIV regulatory protein Tat binds to and specifically activates the CDK9–cyclin T complex.

– *CDK10.* This poorly characterized CDK is thought to modulate gene expression by binding to and phosphorylating transcription factors such as the Ets protein (Bagella et al., 2006).

– *CDK11–cyclin L.* Complexes of CDK11 and cyclin L are involved in the regulation of a variety of different processes, including RNA splicing, transcription and apoptosis.

13.2.2
Cyclins

Cyclins
- Activating subunits of CDKs
- Bind CDK via the cyclin box

Active CDKs are heterodimers composed of the CDK subunit and an activating subunit, the cyclins. There is specificity in the interaction of cyclins with CDKs, and distinct cyclin–CDK complexes become active during the cell cycle (Tab. 13.1). The cyclins were originally defined as proteins that show cyclic concentration variations during the cell cycle (Fig. 13.4) and thereby activate CDKs differentially during the cell cycle. A classifying feature of the cyclins today is the *cyclin box*, a conserved domain of around 100 amino acids. The human genome encodes more than 25 proteins containing a cyclin box. Binding to the corresponding CDK takes place via the cyclin box.

The cyclins of the mammalian cell cycle (reviewed in Murray, 2004) can be roughly divided according to their activity in the different phases of the cycle (see also Fig. 13.11):

G_1/S cyclins
- Cyclins D1, 2 and 3
- Cyclin E

Mitotic cyclins
- Cyclin A
- Cyclin B

– *Cyclins of type D* are synthesized during early G_1 and are present at a fairly constant level during S and G_2 until their level falls down during M.

– The level of *E-type cyclins* increases at late G_1 and decreases sharply around the G_1/S transition.

– *Cyclins of type A* accumulate from late the G_1 phase and are destroyed before the metaphase of M.

– *Cyclins of type B* accumulate during the S phase and G_2 phase, and their destruction starts at the anaphase of M.

Cyclins D and E are regarded as the *G_1/S cyclins*; cyclin B and cyclin A are also termed the *mitotic cyclins*.

The equivalent cyclins in yeast have a different classification and are shown in Tab. 13.1.

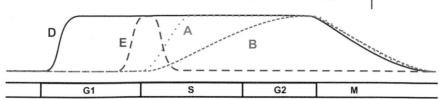

Fig. 13.4: Concentration changes in cyclins during the cell cycle. Cyclins of types D, A, E and B show characteristic concentration changes in the course of the cell cycle. The figure only depicts the course of concentration changes in the cell cycle and does not indicate the relative cyclin concentrations.

The main functions of cyclins are:
- *Activation of CDK.* The role of the cyclins is to convert CDKs into an active state. This process confers specificity to CDK activation, because a specific cyclin preferentially binds and activates only a certain CDK (Tab. 13.1). The concentration of the cyclins is an important factor in the control of CDK activity. Various mechanisms exist to control the level of cyclins available for CDK binding (see below).
- *Contribution to substrate specificity of CDKs.* Binding of protein substrates to CDK–cyclin complexes is not restricted to CDKs. Rather, cyclins contain structural elements that mediate interactions with CKIs (see below) and with CDK substrates. Thus, the role of cyclins has to be extended to selection of binding substrates of CDKs.
- *Cyclin functions not related to CDKs.* Some cyclins of the cell cycle appear to perform functions independent of CDKs. This has been well established for cyclin D1 that has been shown to form physical associations with more than 30 different transcription factors or transcriptional cofactors. Proteins interacting directly with cyclin D1 include nuclear receptors, basic helix–loop–helix proteins, SMAD proteins and the transcription factor B-Myb (reviewed in Fu et al., 2004). Furthermore, cyclin D1 binds to histone acetylases, histone deacetylases (HDACs) and chromatin-remodeling proteins, indicating a general role of cyclin D1 in modulating the activity of transcriptional regulators.

■ **Functions of cyclins**
- Activation of CDKs
- Substrate binding and selection
- Modulation of transcriptional regulators

Regulation of Cyclin Concentration
In the cell cycle, the different cyclins show characteristic concentration changes in which temporally defined maxima in cyclin concentration are observed (Fig. 13.4). The amount of a distinct cyclin available for CDK activation is strictly controlled by the following mechanisms (Fig. 13.5):
- *Regulation of cyclin expression* (Section 13.5).
- *Targeted degradation in the ubiquitin (Ub) pathway* (Section 13.4.1). Most cyclins are the target of Ub-mediated proteolysis and this

■ **Cyclins are regulated via**
- Gene expression
- Phosphorylation
- Ub-mediated proteolysis
- Subcellular distribution

Fig. 13.5: Processes influencing cyclin concentration. Upregulation of cyclin concentration occurs mainly at the level of gene expression. Growth factors transduce signals via the mitogen-activated protein kinase (MAPK) pathway or other pathways to transcription factors (c-Jun and c-Myc) that activate, for example, cyclin D transcription. Phosphorylation of cyclins (e.g. cyclin E) can provide a signal for binding of the SCF complex and destruction by the Ub–proteasome pathway. This phosphorylation can be induced by activated cyclin–CDK complexes. Cyclin destruction (e.g. cyclin B) can be also mediated by the APC. Another mechanism for control of cyclin concentration uses changes in the subcellular distribution.

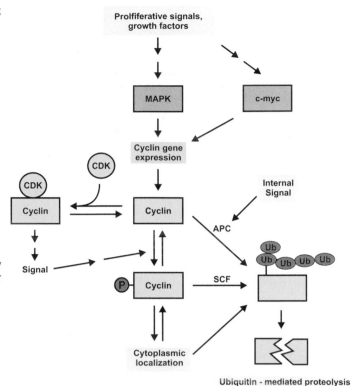

degradation is a major mechanism for reducing cyclin concentrations at distinct cell cycle stages.

– *Phosphorylation.* Signal-directed phosphorylation of cyclins is a tool for targeting cyclins for proteolysis. Furthermore, cyclin phosphorylation influences the subcellular distribution of cyclins.

– *Subcellular distribution.* Cyclins show a complex pattern of subcellular distribution. Cyclin D1 for instance is localized to the nucleus during the G_1 phase and is distributed to the cytoplasm upon the onset of the S phase.

Viral Cyclins

Some tumor-associated viruses have evolved mechanisms to stimulate cell cycle progression following infection. These mechanisms include the production of *virus-specific cyclins* that deregulate the cell cycle and interfere with central signaling pathways (reviewed in Coqueret, 2003). As an example, the Kaposi sarcoma-associated virus produces a viral D-type homolog, *cyclin K*, that mimics cyclin D and E functions but is resistant to regulation by CKIs.

13.2.3
CKIs

Negative control of CDK activity in the cell cycle is performed by specific inhibitor proteins known as *CKIs*. These are a heterogeneous family of proteins that may associate with a CDK or with a cyclin–CDK complex in a reversible manner, inhibiting CDK activity.

The CDKs are divided into two groups based on sequence homology:

CIP/KIP family	$p21^{CIP1}$ (also known as CIP1, WAF1)
	$p27^{KIP1}$ (KIP1)
	$p57^{KIP2}$
INK4 family	$p15^{INK4b}$
	$p16^{INK4a}$
	$p18^{INK4c}$
	$p19^{INK4d}$

■ **CKIs**
– Negative regulators of CDKs
– Prefer distinct CDKs
– Two families: CIP/KIP family and INK4 family

There is some specificity in the inhibition of the various cyclin–CDK complexes (Fig. 13.6). The members of the CIP/KIP family mainly act on CDK2 complexes and have little effect on the activity of cyclin D–CDK complexes *in vivo*. The inhibitors of the INK4 family preferentially bind to and inhibit CDK4 and 6 complexes.

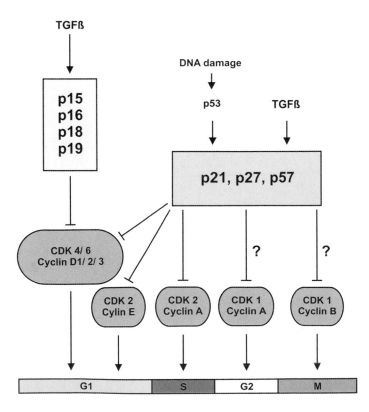

Fig. 13.6: Regulation and attack points of CKIs in mammals. The INK inhibitors are activated by TGF-β and specifically inhibit the cyclin D–CDK4/6 complexes. The inhibitors p21, p27 and p57 inhibitors are activated by p53 and by TGF-β, and can inhibit all types of cyclin–CDK complexes. Only the major ways of CKI regulation are indicated.

Regulation and Function of Cell Cycle Inhibitors

The CKIs are important control elements that regulate the G_1/S transition. Furthermore, the CKIs are involved in the control of the G_0/G_1 transition of cells going from a quiescent to a dividing state and *vice versa*. Primary targets of CKIs are the CDKs, either free or complexed with cyclins, and the main function of CKIs is that of a *negative regulator of CDK activity*. As such, the CKIs are important entry points for signals that slow down or stop the cell cycle and induce transition into the quiescent state, G_0.

■ **CKIs are regulated via**
– Gene expression
– Phosphorylation
– Ub-mediated proteolysis
– Subcellular distribution

Regulation of CKIs occurs by transcriptional, translational, proteolytic and localizational mechanisms, and multiple input signals have been identified that influence these processes in a complex relationship.

The main function and regulation of the CKIs can be summarized as follows:

■ **p27^KIP1**
– Induced by antimitogenic signals, e.g. TGF-β
– Inhibits CDK2–cyclin E and CDK2–cyclin A

– *p27^KIP1*. This inhibitor plays a central role in the decision of a cell to either commit to the cell cycle or to withdraw into the resting state, G^0. The concentration of p27^KIP1 decreases sharply when cells transit from the quiescent state into the cell cycle, and, conversely, levels of p27^KIP1 increase when cells leave the cell cycle and enter into a differentiated state. *Accumulation of p27^KIP1 is induced by many antimitogenic signals* including cell–cell contacts, transforming growth factor (TGF)-β signaling and cAMP. The main targets of p27^KIP1 are the *cyclin E–CDK2* and *cyclin A–CDK* complexes. Surprisingly, p27^KIP1 is required for the assembly of cyclin D–CDK4 complexes. The role of this property in the physiological context is uncertain. A major control of the level of p27^KIP1 is exerted by Ub-mediated proteolysis, which has been shown to be dependent on phosphorylation of p27^KIP1 by cyclin E–CDK2 complexes (Section 13.5.2).

■ **p21^CIP1**
– Participates in DNA damage response
– Induced by p53
– Inactivates G_1 CDKs

– *p21^CIP1*. The inhibitor p21^CIP1 binds mainly to the complexes of CDK2 with cyclin A and E, leading to their inactivation and cell cycle arrest. The outstanding property of p21^CIP1 is its regulation at the transcriptional level by the *tumor suppressor protein p53* (Section 14.8.3) during the *DNA damage response*. Increased levels of mRNA for p21^CIP1 are observed upon treatment of cells with DNA-damaging agents, and this transcriptional regulation is part of the DNA damage checkpoint during the G_1/S phase (Section 13.10). In addition to its ability to associate with G_1 cyclin–CDK complexes, p21^CIP1 also associates with the replication-accessory protein proliferating cell nuclear antigen (PCNA). PCNA is required for nuclear DNA synthesis and functions in clamping DNA polymerase δ to the DNA, thereby increasing the processivity of DNA synthesis. It is believed that this is another mechanism by which p21^CIP1 can inhibit DNA synthesis and S-phase progression.

– *INK4 proteins.* The members of the INK4 family bind predominantly to CDK4/6, preventing association of D-type cyclins. Regulation of INK4 abundance is cell-type specific and shows complex
patterns. Notably, the level of p15^{INK4b} is subject to induction at the
mRNA level by the antimitogenic cytokine TGF-β (Section 12.1).
Properties as a *tumor suppressor* are attributed to the inhibitor
p16^{INK4a}, since the gene for p16^{INK4a} is mutated in many tumor
cell lines. Furthermore, silencing of the genes of INK proteins
by aberrant CpG methylation is a frequent event in tumor cells.

■ **INK4 inhibitors target
mainly CDK4/6**

13.3
CDK–Cyclin Complexes

The CDKs may exist in inactive and active states. The transition between the two states is controlled by cyclin and/or CKI binding and
by phosphorylation/dephosphorylation events (Fig. 13.7).

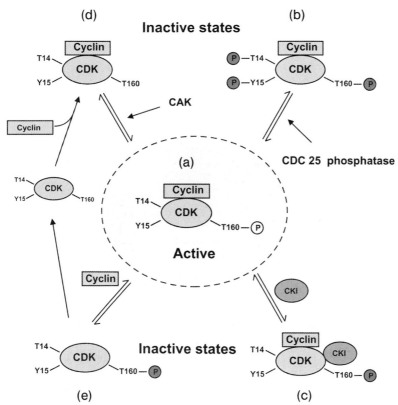

Fig. 13.7: Principles of regulation of CDKs. The figure
shows the principles of CDK regulation, using the CDK1
(here simply referred to as CDK) as an example. The
active form of CDK (a) is associated with the corresponding cyclin; Thr160 of CDK (or equivalent positions
in other CDKs) is phosphorylated, and Thr14 and Tyr15
are unphosphorylated. Inactivation may take place by
phosphorylation of Thr14 and Tyr15 (b) or by binding of a
CKI (c). Other inactive forms of CDKs are the CDK–cyclin
complex, in which Thr160 of the CDK is not phosphorylated (d). In addition, the cyclin-free forms of CDK are
inactive (e).

13.3.1
Structure of CDKs and CDK–Cyclin complexes

The structural basis for activation of CDKs by cyclin binding and phosphorylation will be discussed using the example of the CDK2–cyclin A for which many structural data are available (reviewed in Huse and Kuriyan, 2002).

The catalytic center of the CDK2 includes a core of around 300 amino acids, which adopt the typical protein kinase fold (Section 7.2.1 and Fig. 13.8), composed of an N-terminal lobe and a mostly α-helical C-terminal lobe. As for other protein kinases, a catalytic loop and an *activation segment* or *T-loop* can be identified within the kinase domain of CDK2. The Thr160 important for regulation is found on the activation segment; *phosphorylation of Thr160 is required for full activation of CDK2*. Another conserved structural element comprises the *C-helix of the N-terminal lobe*, also known as the *PSTAIRE* helix.

Comparison of the inactive form of CDK2 (Fig. 13.8) with the cyclin A-bound form and with the active form of protein kinase A shows that there are two main causes of the inactivity of CDK2 without bound cyclin. First, the active center occupies an inactive conformation where the C-helix is outward oriented. A catalytically essential glutamate residue is misaligned in the inactive state, so that cleavage of bound ATP is impossible. Second, in the inactive form, the binding site for the protein substrate is blocked by the activation segment. Binding of cyclin A leads to a reorganization of the active site including a proper orientation of the critical glutamate, allowing

Structures of CDK–cyclin complexes
- Typical kinase fold
- Critical elements:
  Activation segment (T-loop)
  C-helix (PSTAIRE helix)

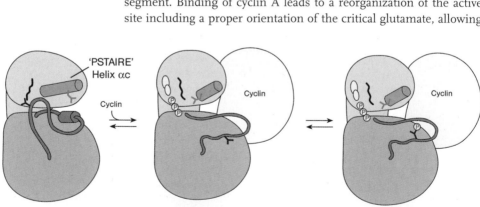

'PSTAIRE' Helix αc   Cyclin   Cyclin   Cyclin

Inactive CDK   CDK bound to cyclin   Phosphorylated CDK bound to cyclin

Fig. 13.8: The regulation of CDK. In the absence of cyclin, the αC-helix of CDK (also called the PSTAIRE helix) is rotated so as to move a crucial catalytic glutamate out of the active site. This is correlated with an inhibitory conformation of the activation loop. Cyclin binding reorients the PSTAIRE helix so as to place the glutamate within the active site. The activation loop adopts a near-active conformation upon cyclin binding, and its subsequent phosphorylation further stabilizes the active form. From Huse and Kuriyan (2002).

for productive binding of ATP and partial removal of the activation segment from the catalytic cleft. The cyclin–CDK complex possesses a basal protein kinase activity that is around 300-fold increased by phosphorylation at Thr in position 160 (or equivalent position).

13.3.2
Regulation of CDKs by Phosphorylation

A major control of the CDKs is exerted by phosphorylation and de-phosphorylation events. The CDKs of the cell cycle possess several phosphorylation sites and these may have an activating or inactivating effect. There are two major classes of phosphorylation: phosphorylation in the activation segment, and phosphorylation at Thr14 and Tyr15. Phosphorylation at Thr160 of CDK2 in the activation segment or the equivalent positions Thr161 of CDK1and Thr172 of CDK4/6 is activating. Phosphorylation at Thr14 and Tyr15 is inhibiting. A requirement for this regulation to take effect is the association of the CDK with the corresponding cyclin. Protein kinases and protein phosphatases that have CDKs as substrates are therefore central elements of CDK control.

Phosphorylation in the Activation Segment: Activation
Phosphorylation of the cyclin A–CDK2 complex at Thr160 leads to a nearly 300-fold increase in protein kinase activity.

The protein kinase responsible for this phosphorylation belongs to the family of CDKs and is known as *CAK* composed of the catalytic subunit *CDK7, cyclin H* and the *MAT* subunit (Section 13.2.1). The CAK-catalyzed phosphorylation at Thr160 requires binding of the cyclin to CDK and is regulated mainly by the cyclin concentration.

The structure of the phosphorylated cyclin A–CDK2 complex (Fig. 13.9a) shows that the Thr160-phosphate serves as the *organizing center* that contacts different structural elements of the complex and structurally reorganizes them. Thanks to its polyvalent coordination sphere, the Thr160 phosphate couples parts of the activation segment, the catalytic loop with essential Asp127, the C-helix and residues of cyclin A (Fig. 13.9b). Contacts are formed to Arg50 of the C-helix, to Arg150 of the activation segment and to Arg126; the latter lies in the vicinity of the essential Asp127. Furthermore, van der Waals bonds are formed to an Ile270 of cyclin A. The conformational changes induced by these contacts affect the substrate-binding site and the cyclin A–CDK2 interface, in particular.

It is assumed that the activity increase is mainly due to better accessibility of the binding site for the protein substrate. In the unphosphorylated form, the T-loop blocks access to the substrate-binding site, whereas in the phosphorylated form this site is exposed.

■ **CDK phosphorylation**
– p-Thr160(172): activation
– p-Thr14 and p-Tyr15: inhibition

■ **Activation segment phosphorylation**
– Catalyzed by CAK
– Increases activity by orders of magnitude
– Reorients catalytic center and substrate binding site
– Dephosphorylation by CDC25 phosphatase

Fig. 13.9: Structure of the Thr160 phosphorylated CDK2-Cyclin A complexes. a) Phosphorylation causes a change in the conformation of the T-loop, highlighted in the superposition of the unphosphorylated (gray) and phosphorylated (cyan and magenta) CDK2-Cyclin A complexes. ATP is shown in a ball and stick representation. b) Diagram of the multivalent interactions of p-Thr160 at the CDK2–cyclin A interface. p-Thr160 forms contacts to Arg50 of the PSTAIRE (C) helix, to Arg150 of the T-loop, to Arg126, which is close to the catalytic Asp127, and to Ile270 of cyclin A.

(a)
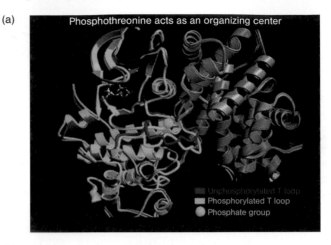

(b)
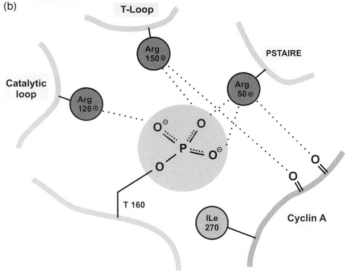

Dephosphorylation in the Activation Segment: Inactivation

The reversal of the activating Thr172 phosphorylation by phosphatases is part of an important checkpoint at the G_1/S transition. Removal of p-Thr172 of CDK4/6 by the *phosphatase CDC25A* leads to inactivation of the cyclin D–CDK4/6 complex and to a stop in the cell cycle during G_1. The CDC25A is itself subjected to regulatory phosphorylations in the course of DNA damage checkpoints (Section 13.10).

The CDC25 phosphatases are protein *phosphatases with 2-fold specificity* that can cleave phosphate residues from phosphoserine and phosphotyrosine residues of CDKs or other proteins. Control of CDC25 enzymes is achieved mainly by phosphorylation at specific

Ser/Thr residues. Depending on the identity of the Ser/Thr residues, the phosphorylation can have an activating or inactivating effect on the phosphatase activity. Some phosphorylation sites mediate inhibition of the phosphatase by binding of 14-3-3 proteins (Section 13.7). The phosphorylations are catalyzed by cyclin–CDK complexes or by other protein kinases that are part of distinct signaling pathways.

Phosphorylation and Dephosphorylation at Thr14 and Tyr15
The activity of CDK1, the kinase essential for the M phase, is negatively controlled via phosphorylation at Thr14 and Tyr15. The kinases responsible for the inhibitory phosphorylation are the *Wee1 kinases*. Phosphorylation at these sites by Wee1 leads to inactivation of CDK1 which is of particular importance for the regulation of CDK1 activity in mitosis (Section 13.7). The CDK1–cyclin B complex is maintained in an inactive state until the end of the G_2 phase by the phosphorylation of Thr14 and Tyr15.

■ Phosphorylation at
Thr14 and Tyr15:
– Inhibits CDK
– Catalyzed by Wee1
 kinase
– Removed by CDC25B
 and CDC25C
 phosphatases

The inactivating phosphorylation at Thr14 and Tyr15 can be reversed in a regulated manner by the protein phosphatases *CDC25B* and *CDC25C* that are an essential controlling part of the G_2/M transition.

13.3.3
Inhibition by CKIs

Crystal structures of CKIs bound to monomeric or heteromeric CDKs have revealed different modes of inhibition for the CIP/KIP and INK4 families. The structure of a ternary complex composed of cyclin A, CDK2 and p27^{KIP1} (Pavletich, 1999) shows that the inhibitor interacts with both cyclin A and CDK2. Inhibition of kinase activity is explained by alignment of structural elements of p27^{KIP1} in the ATP-binding site of CDK2. This breaks up the glycine-rich phosphate-binding loop. In addition, the ATP-binding site is completely filled by residues of the inhibitor, so that ATP binding is no longer possible. A similar functional principle is likely for the related inhibitor p21^{CIP1}.

■ **p27^{KIP1}**
– Binds to CDK–cyclin
 complex
– Distorts the active site
 of CDK
– Prevents substrate
 binding

INK4 inhibitors possess a common structural motif – the *ankyrin repeats*. The ankyrin repeats have been identified as protein–protein interaction motifs composed of a helical hairpin and two β-motifs. Generally, the INK4 proteins compete with D-type cyclins for binding to the CDK. The p16^{INK4a} binds to monomeric CDK6 at a site opposite to the cyclin-binding site and allosterically blocks binding of cyclin D. Furthermore, the ATP-binding site of CDK is deformed by the bound inhibitor.

■ **INK4 inhibitors**
– Bind to CDK
– Fix inactive CDK con-
 formation and compete
 with cyclins

13.3.4
Substrates of CDKs

As for most protein kinases, it has proven to be difficult to formulate a consensus sequence for phosphorylation by CDKs. The sequence *(K/R)-S/T-P-X-K* (X: any amino acid) has been identified as such a consensus. The X-ray structures of CDKs with bound peptide substrates have revealed that the substrates bind in an extended form across the catalytic site. In addition to the interactions with the target Ser/Thr of the substrate, the CDK and both the cyclin contact regions outside of the acceptor Ser/Thr residues. The region of the cyclin responsible for these interactions is also involved in the binding of CDK inhibitors and other regulatory proteins such as the retinoblastoma (Rb) protein (Section 13.5.3). Overall, substrate selectivity of CDKs is determined in a complicated manner by the CDK itself that prefers a distinct consensus, the nature of the cyclin associated, and the availability and accessibility of the substrate.

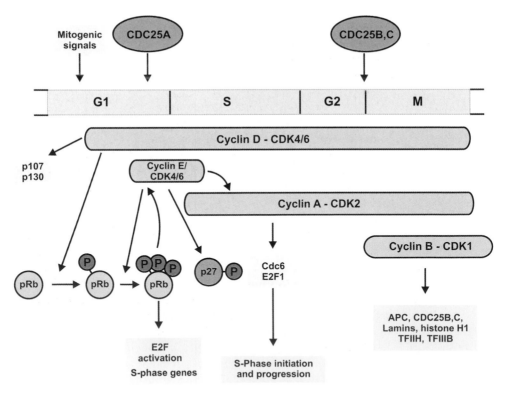

Fig. 13.10: Substrates and phase-specific activation of CDKs in the cell cycle. An overview is shown of the phase-specific activation of the most important CDK–cyclin complexes and of selected substrates. The arrows indicate activation and phosphorylation. p107 and p130 = Rb-related proteins.

Figure 13.10 gives an overview of the cell-cycle-specific activation of CDKs and some important substrates. Comparatively sparse information is available on the G_1- and S-phase substrates of the CDKs. In contrast, many proteins have been described that undergo specific phosphorylation in the G_2/M phase.

Substrates in the G_1/S Phase

The most important CDK substrate in G_1/S phase are the *tumor suppressor protein Rb* and the *Rb-related proteins p130 and p107*, which are phosphorylated by the cyclin D1–CDK4/6 complex and by the cyclin E–CDK2 complex. The protein Rb and its relatives are critical in preparing cells for entry into the S phase. Other targets of the cyclin E–CDK2 complex comprise the p27^{KIP1} inhibitor and "minichromosome maintenance" (MCM) proteins (Section 13.6), which are involved in the regulation of DNA replication.

The cyclin A–CDK2 complex has been shown to phosphorylate, among others, the transcription factor E2F1 and components of the DNA replication complex.

■ **Major Substrates of CDK4/6–cyclin D**
– Rb
– p130
– p107

Substrates in the G_2/M Phase

In the M phase, new phosphorylation of many proteins is observed that starts, in particular, from the CDK1–cyclin B complex. The phosphorylation affects proteins that are critical for entry into and for progression through the M phase such as the CDC25 phosphatase and Emi1 that is an inhibitor of CDC20, a subunit of the anaphase-promoting complex (APC) (Section 13.8) Furthermore, proteins involved in the reorganization of the cytoskeleton, the nuclear membrane, and the formation of the spindle apparatus become phosphorylated by CDK1–cyclin B. As a consequence of phosphorylation events, inhibition of vesicular transport and general inhibition of transcription occur.

Examples of proteins that are specifically phosphorylated during the M phase of the cell cycle are the lamins. Hyperphosphorylation of the lamins leads to disintegration of the nuclear lamina. A myosin-associated protein named MAP4 is also specifically phosphorylated during mitosis. Other M-phase-specific phosphorylations occur at transcription factor TFIIIB, leading to inhibition of transcription by RNA Pol III. Phosphorylation of TATA box-binding protein-associated factors (TAFs) proteins (Section 3.2.5) is also involved in general inhibition of transcription.

■ **Substrates of CDK1–cyclin B**
– CDC25C
– Inhibitor Emi1
– Many proteins involved in transcription, spindle formation and cytoskeleton organization

13.3.5
Multiple Regulation of CDKs

Activity of the regulatory components of the cell cycle varies extremely during the course of the cycle, and it is directed by external signals and internal control mechanisms. The CDKs are the central tool for control of the cell cycle. They receive a multitude of signals and transmit signals to downstream substrates. These substrates trigger the large number of different activities that constitute the distinct functional, regulatory and morphological properties of the various phases of the cycle. The activity of the CDKs is regulated by various positively and negatively acting signals that are registered in the cell cycle, and can bring about a halt in the cell cycle at various points.

The following have a *positive* effect on activity of CDKs and promote progression in the cell cycle:

- Increase in cyclin concentration: by activation of transcription or inhibition of proteolytic degradation.
- Phosphorylation of CDKs at Thr160 or equivalent positions by CAKs.
- Dephosphorylation of CDKs at Thr14/Tyr15 by CDC25 phosphatases.
- Redistribution of CKIs between different CDK complexes.
- Increase in concentration of CDK4 and CDK6.
- Decrease in concentration of CKIs at the transcription level or by proteolysis.

The following have a *negative* effect and can lead to a stop in the cell cycle:

- Decrease in concentration of cyclins.
- Reduced transcription or activation of proteolysis.
- Inhibition of CDC25 phosphatases.
- Increase in concentration of CKIs.

These regulation mechanisms cannot be considered in isolation. Rather, it must be assumed that the individual mechanisms cooperate, that they demonstrate mutual regulation and that feedback mechanisms are built in. All control elements can be activated, in principle, by external signals, resulting in a complex network of cell cycle control with many entry and exit points. The following sections are thus incomplete and only describe the elements that have been well proven experimentally.

■ **CDK activity is controlled by**
- Cyclin availability
- Dephosphorylation by CDC25B and CDC25C
- CKIs
- CDK expression

13.4
Regulation of the Cell Cycle by Proteolysis

The ordered course of the cell cycle is ensured by two processes in particular:
– Regulated and temporally coordinated activity changes in CDKs.
– Targeted Ub-dependent proteolysis of CDK regulators.

Both processes, the protein kinase regulatory network of CDKs and targeted proteolysis, are linked to one another and work in mutual dependence.

Selection of a protein for Ub-dependent proteolysis occurs particularly via the E3 enzymes of the Ub pathway (Section 2.5.6) that catalyzes ligation of the target protein with Ub. The specificity of Ub–protein ligation is determined by the nature of the E3 enzymes, which constitute a large family of functionally related, but structurally distinct proteins of heterogeneous subunit composition.

Use of specific proteolysis as a tool for control of the cell cycle has various advantages:
– Proteolysis allows simultaneous and complete inactivation of all functions of a multifunctional cell cycle protein such as the cyclins.
– Proteolysis enables subunit-selective reorganization of heterooligomeric protein complexes, e.g. the targeted degradation of a CDK inhibitor.
– The total substrate pool of regulatory enzymes of the cell cycle may be inactivated by proteolysis.
– The regulatory system of a cell cycle section can be reset to the ground state by proteolysis.

■ E3 ligase complexes in the cell cycle:
– SCF
– APC

Two types of E2/E3 complexes are of particular importance for cell cycle control (Fig. 13.11). One, the *SCF (Skp1, cullin, F-box protein) complex*, regulates primarily G_1–S progression, but plays important roles in the other phases of the cell cycle as well. The other, the *APC or cyclosome*, is required for the separation of the sister chromatids at anaphase and for the exit of cells from the M phase into G_1. Importantly, the activities of both types of complexes are interconnected, which provides a means for coupling the two activities during the course of the cell cycle.

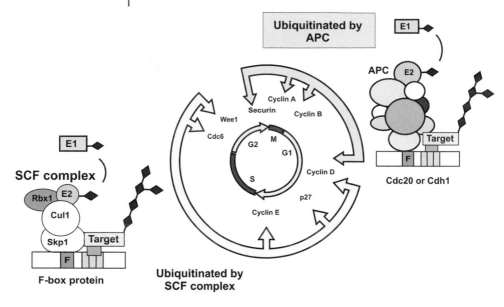

Fig. 13.11: Roles of two distinct Ub ligases in regulation of the cell cycle. The ordered progression of the cell cycle is regulated by two Ub ligases: APC–cyclosome and SCF. The APC is active from late G_2 to mid-G_1 phase and catalyzes the ubiquitination of mitotic cyclins and securins which are anaphase inhibitors. In contrast, the SCF complex mediates ubiquitination of G_1 cyclins and CKIs. From Nakayama et al. (2001).

13.4.1
Proteolysis mediated by the SCF Complex

■ **SCF subunits**
- Skp1
- Cullin
- Rbx1
- F-box protein

■ **Major SCF substrates**
- $p21^{CIP1}$
- $p27^{KIP1}$
- Cyclin A
- Cyclin E
- CDC6
- CDC25 phosphatase

Progress of the cell cycle through the G_1 phase and entry into the S phase are associated with the Ub-dependent proteolysis of important regulatory cell cycle proteins, mediated by multiprotein complexes called SCF complexes (reviewed in Willems et al., 2004). The *SCF complexes* are *E2/E3 enzymes* (Section 2.5.6) composed of the protein Skp1 which has a scaffolding function, a family of proteins named cullin and a small protein Rbx1. The Rbx1 protein (also named Roc1) contains a RING finger motif which mediates binding to the E2 enzyme, usually CDC34. The core of the SCF complex associates with a family of proteins named F-box proteins that function as *substrate-specific adaptor subunits* and confer substrate specificity by recruiting a particular target to the core ubiquitination machinery. A sequence element named F-box mediates binding of these proteins to Skp1 and they capture substrates by means of C-terminal protein–protein interaction regions such as WD repeats or leucine-rich motifs. More than 50 different F-box proteins have been identified in humans This large number of F-Box proteins, in combination with the core complex of Skp1, Cul1 and Rbx1 as well as associated E2 enzymes, provides the basis for *multiple ubiquitination pathways*. As illustrated by Fig. 2.13 of Chapter 2, the F-box proteins Skp2, Fbw7 and β-TrCp control the abundance of central cell cycle regulators and other proteins involved in promotion of proliferation.

The regulators that are targeted for degradation include proteins whose degradation may be a requirement for ordered progression through the G_1 and S phases. Examples include the inhibitors $p21^{CIP1}$, $p27^{KIP1}$, $p57^{KIP2}$, cyclin A, cyclin E, cyclin D, the CDC6 protein and CDC25 phosphatase.

The SCF complex is *constitutively active* and acts on its specific substrates only after phosphorylation of the substrate protein. Examples include the inhibitor $p27^{KIP1}$, that must be phosphorylated by cyclin E–CDK2 on Thr187 for binding to the Skp2–SCF complex. Phosphorylation of $p27^{KIP1}$ has a 2-fold function during Ub-mediated degradation. To be broken down, $p27^{KIP1}$ must be transported out of the nucleus, which requires its phosphorylation. Furthermore, phosphorylation is needed for recognition by the Ub-conjugating system. Export from the nucleus and proteasome-mediated degradation are both controlled by phosphorylation in this case. Thus, the regulator for this degradation pathway is the phosphorylation of the substrate by a regulatory protein kinase and the Ub-ligase system may be constitutively active.

Many substrate proteins are selected for Ub-ligation based on a C-terminal target sequence. These sequences which, because of the occurrence of common amino acids, are known as *PEST sequences*, are often targets for phosphorylation.

■ **Selection of SCF substrates often requires phosphorylation of the substrate at PEST sequences**

13.4.2
Proteolysis mediated by the APC

The APC is another type of E3-Ub ligase that mediates the proteolysis of important regulators of cell cycle progression, with a major effect during mitosis (reviewed in Peters, 2006). The activity of the APC is tightly regulated to control cell cycle progression, being high from late mitosis until late in the G_1 phase, but low in S, G_1 and early mitosis in mammalian cells. One hallmark of APC is its complicated structure: it is composed of *at least 11 different subunits* organized in an asymmetric manner. Most of the subunits contain regions with homology to known protein domains. It has, however, not yet been possible to assign defined functions to distinct subunits. A RING finger protein (APC11) and a cullin-related subunit (APC2) have been identified in APC, which, by analogy with the SCF, may be involved in binding of the E2 enzyme and in providing a stable platform for binding of the substrate.

■ **APC**
– 11 subunits
– Substrate selection mediated by CDC20 or CDH1 subunits

■ **APCs**

– Mitotic APC:
  APC–CDC20
– Nonmitotic APC: APC–
  CDH1

At least part of the activity and substrate specificity of the APC are dictated by phosphorylation and by the regulated association of coactivator proteins, of which two main types have been identified: the *CDC20 protein* and the *Hct1/CDH1 protein*. Accordingly, two main types of APC are found in the cell with distinct substrate preference and different functions in the cell cycle: the *mitotic form APC–CDC20* and the *nonmitotic form APC–CDH1* (Section 13.8).

Most substrates for APC-mediated ubiquitination carry a particular sequence, the *destruction box*. For B-type cyclins, the consensus sequence of the destruction box is R-ALGVN/D/EI-N. Deletion of the destruction box in cyclin B causes its stabilization. The coactivators CDC20 and CDH have been shown to capture the destruction box motif found in APC substrates, potentially bringing substrates in close proximity with the ligase and the E2. For examples of APC substrates, see Section 13.8.

Regulation of APC activity is complex and is performed by various mechanisms. Whereas the SCF is constitutively active with its substrates being regulated by phosphorylation, APC activity is regulated directly in a cell-cycle-specific manner, both spatially and temporally. This regulation of APC E3 activity occurs through *posttranslational modifications* as well as *binding of regulatory proteins* and may explain the large number of subunits. It involves the variable association of the activating coactivator subunits, which may be inactivated by binding to specific inhibitors. Another important element of APC regulation is phosphorylation of the coactivators and of specific subunits of the APC. Furthermore, the phosphorylation status of the substrates plays a role in regulation of the APC activity as well. The S-phase regulator CDC6 is ubiquitinated by APC–CDH1 only in its unphosphorylated form. If CDC6 is phosphorylated by CDKs, it is resistant to APC and can initiate the assembly of replication preinitiation complexes.

13.5
G₁ Progression and S-phase Entry

The G_1 phase has a special regulatory function in the cell cycle; here, the decision is made to enter the S phase and thus a new round of cell division or to enter the resting, quiescent state G_0. G_1 is subjected to manifold control and serves essential functions in the life of a cell.

In addition to allowing for repair of DNA damage and replication errors, G_1 is a period where many internal and external signals intervene to influence cell division and the deployment of a cell's developmental programme. Diverse metabolic, stress and environmental cues are integrated and interpreted during this period. On the basis of these inputs, the cell decides whether to enter the S phase or pause. To this end, during G_1 the cell makes further decisions regarding whether to self-renew, differentiate and pause or die.

In order to enter the S phase and start a new round of DNA replication, several prerequisites must be fulfilled during G_1. Cells must be large enough to go through a new round of division, apoptotic signals must be suppressed and the components needed for DNA replication must be available, such that the cell is ready to activate replication origins and start DNA synthesis. The signals that control these functions are manifold, and are strongly dependent on cell type and context. Ultimately, these signals initiate the *activation of the G_1 CDKs, i.e. CDK4/6 and 2*, by intervening with the phosphorylation status and protein interaction events of the CDKs. Whether a CDK is active or inactive can be controlled by distinct phosphorylation events and the association of cyclins and/or CKIs (Section 13.3).

The key components of G_1 control are the CDK4/6–cyclin D complexes, the CDK2–cyclin E complexes and the inhibitor p27^{KIP1} that is able to associate with and inhibit both types of CDKs. Furthermore, cytostatic signals that limit G_1 progression and cell proliferation in order to maintain tissue homeostasis can become active in G_1. These signals use the inhibitors p15, p16, p21 and p57 to inactivate the G_1 CDKs and stop G_1 progression. In addition to the manifold signals that regulate G_1 during normal cell proliferation, both G_1 and S phase are subjected to quality control via DNA damage and DNA replication checkpoints that intervene with G_1- or S-phase progression in the case of DNA damage and incomplete or erroneous DNA replication.

13.5.1
CDK4/6 and the D-type Cyclins

The CDK4/6–cyclin D and the CDK2–cyclin E complexes perform different functions in G_1 progression. Whereas the CDK4/6–cyclin D complexes prepare the cells for S-phase entry, the final decision for entry into the S phase is dependent on the activity of cyclin E–CDK2 complexes. This complex must be considered as the critical element of G_1/S progression, and its activation is dependent on prior CDK4/6–cyclin D activation.

Of the three cyclins of type D (D1, D2 and D3), two (D2 and D3) do not occur in all cell types, whilst cyclin D1 has a central function in the regulation of the G_1 phase in all cell types. The D-type cyclins bind and activate CDK4/6 in particular, and CDK4 activation is considered to have a key role. For full activation of CDK4, activating phosphorylation at Thr172 of CDK4 is also necessary, in addition

■ **G_1 entry and G_1 progression are mainly regulated by**
- CDK4/6–cyclin
- Cyclin D expression
- CDK2–cyclin E
- p27^{KIP1}
- INK4 CKIs
- APC–CDH1 activity

■ **CDK4/6–cyclin D**
- Activated by Thr172 phosphorylation
- Rb as substrate
- Regulated by cyclin D expression

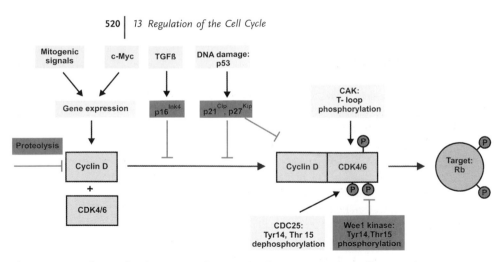

Fig. 13.12: Regulation of cyclin D. Using the example of cyclin D1, the figure illustrates the activating and inactivating influences on D-type cyclins. For explanation, see text.

to binding of cyclin D1 (Fig. 13.12). This step is catalyzed by CAK. Furthermore, removal of the inhibitory phosphorylations at Tyr14 and Thr15 by CDC25 phosphatases is required for CDK4/6 activation.

When cells leave the M phase, the level of D-type cyclins is low. Progression into the G_1 phase requires mitogenic signals that increase the amount of D-type cyclins, which combine with CDK4/6, and a constitutive activity of cyclin D–CDK4/6 complexes is then observed during the whole of the G_1 phase and during the S phase. The only known essential substrate of CDK4/6 is Rb, which is phosphorylated and inactivated by CDK4/6 (see below).

The D-type cyclins complexed to CDK4/6 perform a dual task in G_1-phase regulation. One task is to integrate external signals into the cell cycle. Mitogenic signals, such as growth factors, activate the transcription of the gene for cyclin D1 and thus increase the amount of cyclin D–CDK4/6 complexes. The activating signals can become active from the start of G_1 onward, The increase in cyclin D–CDK4/6 complexes is postulated also to sequester the p27[KIP1] inhibitor bound to cyclin E–CDK2 complexes away from CDK2. This is thought to establish initial levels of active cyclin E–CDK2, which is necessary for entry into the S phase.

13.5.2
Central Function of CDK2–Cyclin E in S-phase Entry

Cyclin E binds and activates CDK2. It has now been recognized to be a key regulator of G_1/S transition, functioning in a nonoverlapping way with cyclin D complexes. Until late G_1, the amount of cyclin E is low and its expression is dependent on *E2F transcription factors*. In mitotically resting cells and in cells that have just emerged from M phase, E2F factors are bound to Rb or its family members, p107 and p130 and this binding turns E2Fs into repressors or inactive transactivators. Inactivation of Rb is performed in a sequential manner via phosphorylation by CDK4/6–cyclin D and CDK2–cyclin E. Following Rb phosphorylation by CDK4/6–cyclin D, Rb dissociates from E2F, allowing E2F-dependent transcription and cyclin E expression. Levels of CDK2–cyclin E now increase and show a maximal value at the start of the S phase. Afterwards, the cyclin E–CDK2 activity falls off sharply within the S phase.

The activity of cyclin E–CDK2 complexes is mainly directed toward two substrates, the Rb protein and the inhibitor p27^{KIP1} (Fig. 13.13). The CDK2–cyclin E complex phosphorylates Rb and promotes further release of E2F from the repressed state, allowing transcrip-

■ **Major substrates of CDK2–cyclin E**
– Rb
– p27^{KIP1}

■ **Control of cyclin E levels**
– Transcription induced by E2F
– Proteolysis induced by phosphorylation and Ub ligation (SCF)

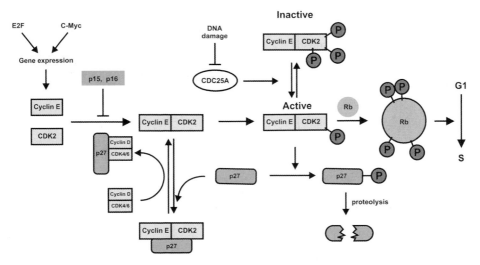

Fig. 13.13: Multiple regulation of cyclin E. The concentration of active cyclin E–CDK2 complexes is determined by expression of the cyclin E gene, by binding of the inhibitors p15, p16 and p27, and by its phosphorylation status. The inhibitor p27 can be removed from p27–cyclin E–CDK2 complexes by sequestering it to cyclin D–CDK4/6 complexes. The amount of active cyclin E–CDK2 is also controlled by the action of CDC25A phosphatase which cleaves off inhibitory phosphate residues from CDK2. Important substrates of active cyclin E–CDK2 complexes are the Rb protein and the inhibitor p27. Phosphorylation of the latter induces its ubiquitinylation and proteasome degradation. Only the major regulatory influences on cyclin E are shown.

tion of key proteins for G_1/S transition. The gene for cyclin E is also induced by transcription factor E2F, which explains the increase in cyclin E at the G_1/S transition. Another important substrate of cyclin E–CDK2 complexes is the inhibitor p27^{KIP1}. Phosphorylation of this inhibitor on Thr187 induces its ubiquitination and targets it for degradation in the proteasome.

13.5.3
Function of Rb in the Cell Cycle

■ **Rb**
- Tumor suppressor
- Pocket protein,
- Related proteins: p107 and p130
- Regulated by multiple phosphorylations

The *Rb protein* is a nuclear phosphoprotein of 105 kDa and belongs to a class of proteins called the *pocket proteins*. Two relatives of Rb are known, *p107 and p130*, which share many of its biological properties. As shown by knockout experiments, the three pocket proteins appear to perform genetically redundant or overlapping functions, particularly with regard to cell cycle control. The Rb protein and its relatives function as suppressors of cell growth and proliferation, and they are involved in the transition into and maintenance of the differentiated state of cells. Loss or deregulation of these functions is associated with deregulation of cell division and favors tumor formation. Therefore, the Rb protein has the characteristics of a *tumor suppressor protein* that is mutated in various tumors (Section 14.4).

Domain Structure of Rb

■ **Major Rb functions**
- Cell cycle regulation
- Control of differentiation
- Regulation of E2F
- Interaction with: viral oncoproteins Tag, E1A and E7, MDM2 ligase, and many other proteins

The domain structure of Rb is shown in Fig. 13.14 (reviewed in Giacinti and Giordano, 2006). The Rb protein can be divided into at least three functional regions: an N-terminal region, a central pocket and a C-terminal region. The N-terminal region appears to be required for oligomerization; the pocket region contains binding sites for the *transcription factor E2F*, for the *viral oncoproteins TAg, E1A and E7* (Chapter 14), and for a large number of other cellular proteins. A nonspecific DNA-binding domain is found on the C-terminal region. Numerous Ser/Thr phosphorylation sites have been identified on Rb and the different phosphorylation events appear to regulate distinct Rb functions.

Control of Rb by Phosphorylation

■ **Rb activity is controlled by multiple phosphorylations catalyzed by**
- CDK4/6–cyclin D
- CDK2–cyclin E
- CDK2–cyclin A

The common characteristics of Rb proteins is their ability to interact with the E2F family of transcription factors. The members of the E2F family coordinate the expression of various genes important for cell cycle progression, and their control by Rb members and CDK activity is a critical element of G_1 progression and S-phase entry. The link to the periodic activation of CDKs in G_1 is provided by the phosphorylation status of Rb which can be considered as a switch that transforms Rb from the active, unphosphorylated state to the phosphory-

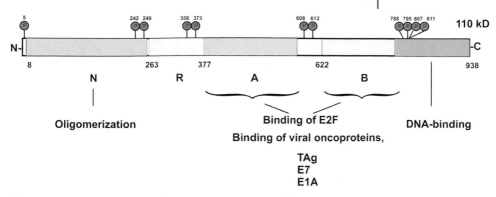

Fig. 13.14: Domain structure and phosphorylation sites of Rb. N = N-terminal region; R = regulatory region; A and B = binding sites for E2F and viral oncoproteins.

lated, inactive state. At the start of the G₁ phase, Rb exists in the active, underphosphorylated form and binds E2F members leading to repression of E2F-dependent genes. Inactivation of Rb and relieve of repression is achieved when *Rb is sequentially phosphorylated* by the *cyclin D–CDK4/6, cyclin E–CDK2* and also the *cyclin A–CDK2* complexes. Rb contains more than 10 potential Ser/Thr phosphorylation sites and these are phosphorylated by the various CDKs in a specific manner. Importantly, mitogenic signals induce – via activation of the cyclin–CDK complexes – the phosphorylation of Rb and thereby control the passage through G₁ and the entry into the S phase.

The repression of E2F-dependent genes is relieved upon Rb phosphorylation, and the encoded proteins can be produced allowing for entry into and progression through S phase.

■ **Hypophosphorylated Rb**
– Repression of E2F-
  dependent genes
Hyperphosphorylated Rb
– Induction of E2F-
  dependent genes

13.5.4
E2F Transcription Factors and their Control by Rb

The E2F proteins are *heterodimeric transcription* factors composed of *E2F subunits and DP subunits* (reviewed in Frolov and Dyson, 2004; Korenjak and Brehm, 2005). Binding sites for E2F members are found in the promoters of many genes whose functions are needed for cell proliferation and whose products drive cell cycle progression. Mammalian cells contain at least seven E2F family members (E2F1–7) and two DP family members. E2F1–5 associate with Rb family members, whereas E2F6 and E2F7 appear to act independently of Rb proteins.

For simplicity, the E2F family is often divided into *activator E2Fs* (E2F1–3) and *repressor E2Fs* (E2F4 and 5). Repressor E2Fs occupy

■ **E2F**
– Heterodimeric tran-
  scription factor with
  E2F subunit and DP
  subunit
– Regulates genes
  required for S-phase
  progression

■ **Two classes of E2F**
- Activator E2Fs
- Repressor E2Fs

promoters in G_0/G_1 phase and typically these proteins are complexed with Rb family members. According to the prevailing model of E2F regulation, a rise in CDK activity during G_1 and the subsequent phosphorylation of Rb family members, leads to the release of E2F-containing repressor complexes from E2F-regulated promoters, the binding of activator E2Fs and the expression of the E2F target genes. In this way, Rb family members provide an important connection between CDK activation and the expression of genes that are needed for cell proliferation.

Examples of genes controlled by E2Fs include:
- Thymidine kinase.
- Dihydrofolate reductase.
- DNA Pol α.
- Cyclin A.
- Cyclin E.
- Transcription factor c-Myc.
- E2F-1.
- Rb.
- Proapoptotic protein Apaf-1 (Section 15.5).
- Tumor suppressor ARF (Section 14.5).

Rb Control of E2Fs (Fig. 13.15)

In late M and in early G_1 phase, Rb exists in an underphosphorylated or even unphosphorylated state. In this state, Rb is associated with repressive E2F-DP members and is bound to promotors of E2F-responsive genes and these genes are thereby repressed.

Repression of E2F activity by Rb appears to be mediated through three mechanisms:
- Rb binds directly to the activation domain of the activator E2Fs and, in doing so, it blocks the activity of this domain.
- Its recruitment to a promoter blocks the assembly of preinitiation complexes, potentially allowing it to inhibit the activity of adjacent transcription factors.
- Rb uses a protein interaction domain that is distinct from its E2F-binding site to associate with complexes that modify chromatin structure. By forming these complexes, the Rb family proteins serve as molecular adapters allowing chromatin-modifying enzymes to be recruited to E2F-regulated promoters.

■ **Rb–E2F complexes recruit chromatin-modifying proteins such as**
- HDACs
- Lys methylases

Rb examples of chromatin modifying proteins associated with Rb-E2F complexes include *HDACs, DNA methylase complexes* such as SUV39/HP1 and *chromatin-remodeling proteins* of the SWI/SNF type. Depending on the type of chromatin-modifying proteins associated, the Rb–E2F complexes may induce a transiently or permanently repressed state.

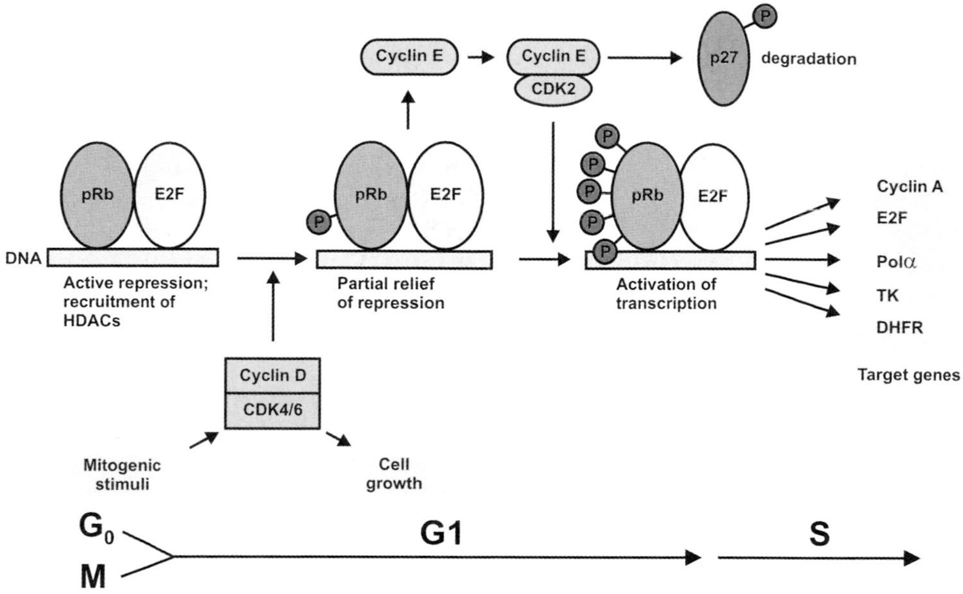

Fig. 13.15: Function of Rb and G_1 cyclins upon G_0/G_1 transition and during G_1 progression. During G_0 and in the absence of mitogenic stimuli, Rb is in a hypophosphorylated state and actively represses genes by recruitment of HDACs. Mitogenic stimuli induce accumulation of active cyclin D–CDK4/6 complexes which promote cell growth and mediate partial phosphorylation of Rb. Repression of E2F target genes, which include the gene for cyclin E, is now partially relieved and cyclin E–CDK2 complexes accumulate completing phosphorylation of Rb. Hyperphosphorylated Rb dissociates from E2F allowing full transcription of the E2F target genes and progression into the S phase. The activated cyclin E–CDK2 complex also phosphorylates the inhibitor p27[KIP1] inducing its ubiquitination and proteasome degradation. TK = thymidine kinase; DHFR = dihydrofolate reductase.

Relieve of repression is a dynamic and complex process, and is thought to include the reversal or overriding of the repressive state of the chromatin by exchanging the repressive E2Fs for the activating ones or the additional binding of transcription promoting complexes such as histone acetylases. The initial signal for transition of Rb from the un(or hypo)phosphorylated, active state to the phosphorylated, inactive state is provided by Rb phosphorylation via CDK2/4–cyclin D complexes under the influence of mitogenic signals. The repression of E2F-dependent genes is thereby progressively relieved. Since the gene of cyclin E is one of the E2F-regulated genes, levels of cyclin E begin to rise, and finally the cyclin E–CDK2 complex is mainly responsible for converting Rb into the hyperphosphorylated form and full relief of repression.

The relatives of Rb, p107 and p130 bind to E2F members different from that bound by Rb, and these associations occur at distinct stages of the cell cycle. Whereas Rb is associated with E2F both in quiescent and actively dividing cells, p130 binds to E2F predominantly in cells that have entered the G_0 state.

13.5.5
Negative Regulation of the G₁/S Transition

In addition to mitogenic signals, *antimitogenic signals* are also processed during the G_1 phase. These lead to an increase in the level of CKIs or they influence the phosphorylation state of the CDKs via the phosphatase CDC25. Antimitogenic signals can lead to a halt during the G_1 phase and bring the cell into a resting state. An antimitogenic signal originates, for example, from *TGF-β, cAMP, certain cell–cell contacts* and *DNA damage*.

Negative regulation of the cell cycle in the G_1 phase is performed by the inhibitors of the INK4 family, which preferentially bind and inactivate monomeric CDK4/6 complexes, preventing cyclin D activation. Accumulation of p16^{INK4a}, for example, sequesters CDK4/6 complexes, preventing progress in the G_1 phase. The inhibitors p21^{CIP1}and p57^{KIP2} are directed mainly against heterodimeric CDK2 complexes and can thereby inactivate the cyclin E–cyclin A–CDK2 complexes, preventing crossing of the restriction point and entry into the S phase.

The balance between activated G_1 cyclin–CDK complexes and the various inhibitors controls progress through the G_1 phase. The concentration of the inhibitors is regulated in a complex pattern by external cues. Examples of external influences on CKI proteins include the induction of p21^{CIP1} by p53, the stimulation of the degradation of p27^{KIP1} by growth factors and the induction of p15^{INK4b} by TGF-β.

In addition to the E2F proteins, a number of other signaling proteins have been reported to interact with Rb and it is assumed that Rb is also involved in processes other than the G_1/S control. However, these other potential functions of Rb are not well defined. The *MDM2 protein* has been identified as a further control element that can influence Rb–E2F function. The MDM2 protein was discovered as an oncoprotein activated by overexpression. It binds to the p53 protein (Chapter 14) and to Rb protein which downregulates the growth-controlling function of both proteins. MDM2 has been shown to be a RING finger (Section 2.5.6) Ub ligase E3 enzyme that induces degradation of p53 via the Ub–proteasome pathway. Furthermore, the MDM2 protein also binds to E2F and stimulates its transcription-activating function. Overall, the MDM2 protein therefore has a growth-promoting function. The precise function of the Rb–MDM2 interaction is, however, uncertain.

Cyclin A is assigned a special role in the progress of the S phase and transition into the G_2 phase. Cyclin A binds and activates CDK2. The CDK2–cyclin A complex binds to the transcription factor E2F1 and phosphorylates its DP1 subunit. As a consequence, the DNA-binding capacity of the transcription factor is reduced and

---

■ **G₁/S progression is inhibited by CKIs in response to**
– TGF-β signals
– cAMP signals
– Cell–cell contacts
– DNA damage checkpoints

■ **MDM2**
– Oncoprotein
– E3 ligase activity
– Interacts with: Rb, p53 and E2F

the transcription-activating function is inhibited. Furthermore, cyclin A–CDK2 complexes are involved in the phosphorylation of protein complexes involved in the initiation of DNA replication.

13.6
Cell Cycle Control of DNA Replication

Replication of DNA in the S phase is subject to strict control in the cell cycle, resulting in the following observations:
– DNA replication is restricted to the S phase.
– DNA is only replicated once in a cycle.
– The time sequence of DNA replication during the S phase and mitosis is strictly maintained.
– If DNA damage is present, DNA replication can be stopped (DNA damage checkpoint).

Control of DNA replication occurs at two levels in particular.

Availability of the Replication Components
At the start and during the S phase, the dNTPs and all proteins required for replication must be available in sufficient quantities. An important control function is performed here by the transcription factor E2F, which induces the different enzymes needed for replication (see above).

Control at the Initiation Level
The replication of a DNA sequence starts at specific sequence sections of the DNA, known as replication origins. Control of origin activity occurs via specific protein complexes that are bound to the origins at certain times of the cell cycle. For replication initiation, two states of these protein complexes are important, known as the *prereplication complex (pre-RC)* and the *postreplication state* (Fig. 13.16).

■ **Initiation of DNA replication is controlled by formation of Pre-RCs**

In order to be able to replicate the DNA, cells must assemble the pre-RC at the origins rendering the chromatin competent for replication. The pre-RC assembly reaction, known as *licensing*, involves the loading of the presumptive replicative helicase, the *MCM2–7 complex*, in an ATP-dependent reaction that requires the *origin recognition complex (ORC)* and two essential factors, *CDC6* and *CDT1*. During early G_1, the MCM loading factors CDC6 and CDT1 recruit the MCM complex to chromatin near origins of replication, which are constitutively bound by the ORC. The complete assembly of these proteins on chromatin results in formation of the pre-RC that remains chromatin-bound until late G_1 phase. The S phase is then trig-

■ **Pre-RC formation licenses origins for replication**

Fig. 13.16: Assembly of replication complexes at M, G_1 and S. In late M and G_1, the pre-RC forms by the stepwise binding of CDC6, CDT1 and MCM proteins to the ORC. Upon entry into the S phase, CDC6 and CDT1 are subjected to phosphorylation-dependent ubiquitination and subsequent proteasomal degradation. Another regulator at this stage is the inhibitor geminin that binds to and sequesters CDT1. The removal of CDC6 and CDT1 prevents rereplication of already replicated DNA. The ORC remains bound to replication origins during all cell cycle phases. The figure shows only the major proteins involved and does not address the regulatory phosphorylation events that control their activity levels.

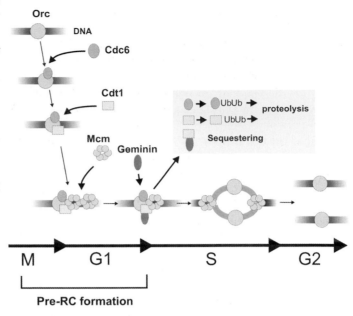

gered by high cyclin A–CDK2 and DBF4–CDC7 levels that activate the pre-RC.

The assembly of the pre-RCs at replication origins can only occur in a window of time during the period from anaphase through early G_1 where CDK activity is low and the proteolysis-promoting activity of APC is high. This is the reason why DNA replication is limited to once per cell. Origin firing can only occur after the APC has been inactivated and CDKs reaccumulate. Once cells enter S phase, pre-RCs can no longer assemble, and pre-RCs cannot form at origins during the S, G_2 and M phases.

■ **Major pre-RC components**
– ORC
– MCM helicase
– CDC6
– CDT1

The licensing reaction is a crucial step in preparing cells for DNA replication and it is therefore tightly controlled. Major control elements in higher eukaryotes are the *licensing activators CDC6 and CDT1*, and the *licensing inhibitor geminin* that are regulated by CDK-dependent phosphorylation and APC-mediated proteolysis in the frame of a highly complicated network (reviewed in Diffley, 2004). One important step in this regulatory network is the degradation of the activator CDC6 that is regulated by phosphorylation. During the end of mitosis and in the quiescent state, CDC6 is unstable because it is degraded by the APC. During G_1 progression and upon transition from the quiescent state into G_1, CDC6 is protected from APC-mediated proteolysis by phosphorylation (Mailand and Diffley, 2005). This phosphorylation step is catalyzed by the G_1 CDKs that accumulate during G_1 or during transition from the quiescent

■ **CDC6**
– Licensing activator
– Degraded by APC–CDH
– Stabilized by
  phosphorylation

state to the proliferating state and it gives CDC6 a license to drive the cycle into the S phase.

The other essential pre-RC component, CDT1, is regulated by the licensing inhibitor geminin and by phosphorylation-dependent proteolysis mediated by SCF complexes. The inhibitor geminin is a multifunctional protein that binds to and sequesters CDT1. The levels of geminin are regulated by various mechanisms including APC-mediated hydrolysis. At the end of mitosis and in early G$_1$, geminin is degraded by APC and accumulates again when the nonmitotic APC activity falls off. Therefore, during the S, G$_2$ and early M phases, levels of geminin are high and prevent formation of pre-RCs.

■ **CDT1**
– Licensing activator
– Inhibited by geminin
– Regulated by phosphorylation and ubiquitination

It is increasingly clear that pre-RC assembly is generally regulated by multiple mechanism, including phosphorylation, proteolysis, nuclear export and inhibitor binding. The early steps of DNA replication are critical for the coordination of DNA replication during the cell cycle and different organisms appear to use different mechanisms for this control. The kinases involved in regulation of the initial steps of DNA replication include the G$_1$ CDKs, CDK2–cyclin E, CDK2–cyclin A and the CDC7–DBF4 kinase, and there are many components of the replication machinery that are regulated by these kinases. For reviews on these topics, see Diffley (2004) and Tada (2007).

13.7
The G$_2$/M Transition and CDC25C Phosphatase

The exit from G$_2$ and entry into mitosis is a transition that involves multiple controls, and is part of an important cell cycle checkpoint. This G$_2$/M transition is primarily determined by the activity of the *cyclin B–CDK1 kinase complex*, which is also called the *mitosis-promoting factor (MPF)*.

■ **G$_2$/M transition depends critically on CDK1–cyclin B activation**

Crucial regulatory elements of CDK1 activity are the concentration of cyclin B and phosphorylation/dephosphorylation events. The concentration of cyclin B increases with entry into the S phase to a threshold at which sufficient cyclin B–CDK1 is available for triggering mitosis. For entry into mitosis, the phosphorylation status of CDK1 is, however, the most critical element. The most important phosphorylation sites of CDK1 are located on Thr161, Thr14 and Tyr15 (Fig. 13.17). Phosphorylation on Thr161 is catalyzed by CAK; Thr14 and Tyr15 are phosphorylated by the Wee1 kinase. In the 3-fold phosphorylated form, the CDK1–cyclin B complex is inactive and remains as such until the end of the G$_2$ phase. *To become active, the CDK1–cyclin B complex must be dephosphorylated at Tyr15 and Thr14. This reaction is performed by CDC25C phosphatase.*

■ **CDK1–cyclin B**
– Activated by CDC25 in positive feedback
– Inhibited by Wee1 kinase

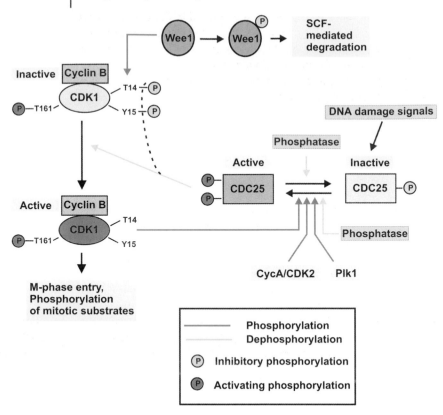

Fig. 13.17: Control of M phase entry by CDC25 phosphatase and Wee1 kinase. The phosphorylation status of the CDK1–cyclin B complex is determined by the opposing activities of CDC25 phosphatase and Wee1 kinase. CDK1–cyclin B is inactive when residues Thr14 and Tyr15 become phosphorylated by Wee1 kinase and is active when these residues become dephosphorylated by CDC25 phosphatase. Transition into the M phase requires inactivation of Wee1 kinase and activation of CDC25 by multiple phosphorylation events some of which are activating and some are inhibiting. Wee1 kinase is inactivated by phosphorylation, subsequent ubiquitination and proteasomal degradation. The activity of CDC25 phosphatase is upregulated by a positive feedback loop involving CDK1–cyclin B. Initial activation of CDK1–cyclin B may be achieved by CDK2–cyclin A and Plk1. The phosphorylation status of CDC25 is also controlled by the action of phosphatases. Furthermore, DNA damage signals mediate inhibitory phosphorylation of CDC25.

Regulation of the phosphorylation status of CDK1 occurs mainly via the activity of CDC25C and the Wee1 kinase. Both enzymes can be controlled by phosphorylation/dephosphorylation events triggered by upstream signals. An activated CDK1–cyclin B complex has been shown to phosphorylate and activate CDC25C providing a *positive feedback loop between CDK1 and CDC25C phosphatase*. Initial activation of CDC25C is achieved by upstream protein kinases that may be sensors of external signals. One of such kinases appears to be the *polo-like kinase Plx1*, which can phosphorylate and thus activate the CDC25C phosphatase.

The *protein kinase Wee1*, which is responsible for the inhibitory phosphorylation of CDK1, is also subject to multiple regulations. Wee1 is inactivated during mitotic entry by proteolysis, translational regulation, transcriptional regulation and by phosphorylation. A multisite phosphorylation by kinases to be identified cooperatively inactivates Wee1 and promotes Wee1 proteolysis.

13.8
Progression through the M phase:
APC and the Metaphase–Anaphase Transition

The activities promoting progression through the M phase are mainly provided by the CDK1–cyclin B, CDK1–cyclin A complexes and APC. As outlined in Section 13.4.2, APC is a E3 ligase that targets specific substrates to proteasomal degradation. There are two forms of APC depending on the nature of the subunit necessary for substrate recognition: APC–CDC20 and APC–CDH1. The CDC20 and CDH1 proteins are responsible for substrate binding and selection of APC. Early in mitosis, APC–CDC20 is relevant, whereas in late mitosis and in early G_1, the APC–CDH1 plays an essential role in cell cycle progression (Fig. 13.18) (reviewed in Castro et al., 2005).

■ **APCs**
– Mitotic APC:
  APC–CDC20
– Nonmitotic APC:
  APC–CDH1

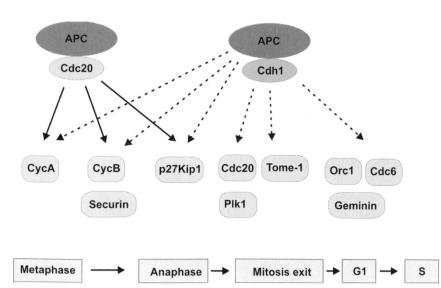

Fig. 13.18: Substrates of APC complexes. APC–CDC20 degrades, among others, the cyclins that are involved in its activation. APC–CDH1 has CDC20 as a substrate and thereby shuts down the activity of APC–CDC20. Further- more, APC–CDH1 inactivates proteins (CDC6 and ORC1) required for formation of pre-RCs. Tome-1 = E3 ligase that phosphorylates the Wee1 kinase, and targets it for ubiquitination and subsequent degradation.

Upon entry into the M phase, the activity of CDK1–cyclin B increases and phosphorylates an inhibitor of CDC20, the Emi1 protein. In the unphosphorylated form, Emi1 binds to CDC20 and thus inhibits APC activity. Phosphorylation of Emi1 then induces its ubiquitination by the SCF–Ub ligase complex and the subsequent proteasomal degradation. CDC20 now can engage in substrate binding, APC becomes active and mediates degradation of many substrates.

■ **Major APC–CDC20 substrates**
- Cyclin A
- Cyclin B
- Securin

The most prominent substrates of APC–CDC20 are the mitotic cyclins and the anaphase inhibitor securin. Proteolysis of mitotic cyclins is activated at the metaphase–anaphase transition and is only switched off at the start of the S phase. The mitotic cyclin A is degraded before cyclin B. Lasting activity of the APC–CDH during the G_1 phase is thought to be responsible for the lack of detection of mitotic cyclins in the G_1 phase. Renewed accumulation of mitotic cyclins is only possible again when APC activity is switched off at the start of the S phase. Another class of substrates comprises the anaphase inhibitors, e.g. the protein securin.

Securin is an inhibitory subunit of separase, a protease that destroys sister chromatid cohesion at the metaphase–anaphase transition. Proteolysis of securin triggers sister chromatid segregation and the irreversible transition into anaphase. Since the activity of APC–CDC20 is directed also against the proteins that have promoted progression to the stage of its activation, i.e. the mitotic cyclins and also the CDC20 subunit, the activity of APC–CDC20 then falls off.

■ **Major APC–CDH1 substrates**
- Cyclin A
- Cyclin B
- ORC1
- CDC6
- Geminin

APC–CDH1 now takes over and this complex is active until early to middle G_1. The many substrates of APC–CDH1 include the inhibitors of pre-RC formation, geminin and the CDC6 protein (Section 13.6). The activity of APC–CDH1 falls off during early to middle G_1, caused by phosphorylation and ubiquitination of the activating subunit CDH1.

Overall, mitotic protein kinases and the two forms of APC cooperate in a highly complicated network. Moreover, extensive crosstalk exists between the two centrals E3 complexes of the cell cycle, the SCF and APC (reviewed in Vodermaier, 2004): SCF controls the level of proteins involved in APC function and *vice versa*. Only a very coarse description can be given here of this intricate network that is still only partially understood (for details, see Nasmyth, 2005).

13.9
Summary of Cell Cycle Progression

Progression through the cell cycle is governed by the opposing forces of the proteolytic machinery and an oscillating activation of CDKs. The changes of CDK activity are largely determined by the association of cylins and CKIs, and the levels of these regulators are controlled by the E3 ligases APC and SCF. By setting the activity of CDKs to zero during distinct time windows of the cycle, the proteolytic machinery ensures unidirectionality and irreversibility of the cycle. Whereas SCF complexes mostly remove the blocking or braking activities, e.g. the CKIs, APC destroys many of the cell-cycle advancing activities, e.g. the mitotic cyclins and the cdc6 protein. Overall, a *multilayered control system* exists between APC, SCF and the CDKs. Although we now know the basic control mechanisms, important aspects of cell cycle regulation are still poorly understood. This is illustrated, for example, by the observation that some cyclins and CDKs appear to be dispensable for cell cycle progression in certain cell types (reviewed in Sherr and Roberts, 2004). Clearly, there is a large amount of plasticity and redundancy in the function of the important cell cycle regulators, and roles that presently have been assigned to certain regulators may be redefined in the future.

G$_1$ Phase Progression
At the end of mitosis and in early G$_1$, APC activity is high and the activity of the kinases required for G$_1$ progression (CDK4/6 and 2), is low. Cells use various mechanisms to enforce the existence of G$_1$ by keeping CDK2 in an inactive state. One of such mechanisms uses a limited supply of cyclins E and A, and the presence of CDK2 inhibitor p27^{KIP1}, to postpone CDK2 activation. Furthermore, active pre-RCs cannot form because APC keeps the level of cdc6 protein, the essential component of pre-RCs, low.

■ **Major elements of G$_1$ progression**
– CDK4/6–cyclin D ↑
– APC–CDH1 ↓
– Mitogenic signals ↑

Progression from this state is not possible until mitogenic signals intervene that increase CDK4/6 activity leading to neutralization of p27^{KIP1} and induction of E2F-dependent transcription of cyclins E and A and other components, resulting in CDK2 activation and entry into the S phase. Furthermore, the rise in CDK activity during G$_1$ leads to stabilization of CDC6 allowing the formation of active pre-RCs.

Mitogenic Signals:
Regulation of Proliferation, Cell Growth and Survival
Overall, G$_1$ is a period when many signals intervene to influence cell division and the deployment of a cell's developmental programme. Diverse metabolic, stress and environmental cues are integrated

and interpreted during this period. On the basis of these inputs, the cell decides whether to enter S phase or pause. Moreover, in multicellular organisms the behavior of a cell must obey dictums from its neighbors. To this end, during G_1 the cell makes further decisions regarding whether to self-renew, differentiate or die.

Proliferation, cell growth and survival are distinct, but intertwined, functions needed for passage of cells through the cell cycle. First, cells must *receive mitogenic signals* that allow cell division in the frame of the development of the organism. Second, cells must be *large enough for division* and growth-promoting signals are required to this end. Finally, *cell death signals must be allowed to become active* under conditions where cell numbers are to be controlled and where accidents like DNA damage call for cell death. As illustrated in Fig. 13.19, the Ras and the phosphatidylinositol-3-kinase/Akt

■ **Signals required for**
G_1/S progression
– Mitogenic signals
  (e.g. growth factors)
– Antiapoptotic signals
– Growth-promoting
  signals

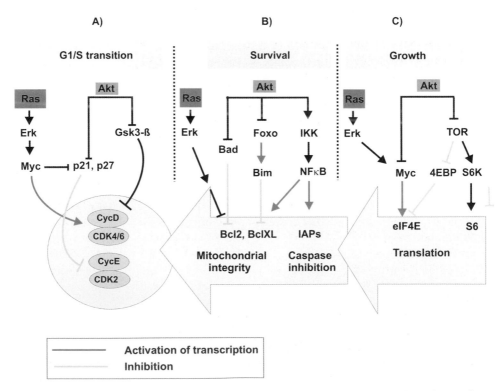

Activation of transcription
Inhibition

Fig. 13.19: Control of G_1/S transition by Ras and Akt kinase pathways. Ras and Akt pathways control cell proliferation via three major ways. (a) By regulating the G_1/S transition. (b) By suppressing apoptosis and promoting cell survival. Here, the Ras and Akt signals finally mediate caspase inhibition and ensure mitochondrial integrity. (c) By promoting cell growth via the translation initiation factor eIF-4E. For the apoptotic proteins Bad, Bim, Bcl-2, Bcl-X_L and inhibitors of apoptosis (IAPs), see Chapter 15; for eIF-4E and 4E-BP, see Section 3.6.3.4; for IKK and NFκB, see Section 2.5.6.4. Gsk3-β = glycogen synthase kinase 3 type β.

(PI3K/Akt) pathways are versatile pathways that achieve at least partial integration of these functions.

Mitogenic signaling pathways activated by external signals, such as growth factors, control inhibitor and cyclin concentrations, which in turn empower CDKs to drive the G_1/S transition. *The Ras and PI3K/ Akt pathways stand out in this respect* (reviewed in Massague, 2004). As outlined in Chapters 6, 8 and 9, a large number of external signals can feed into these proliferation-promoting pathways. Most of the mitogenic activities of Ras and PI3K/Akt pathways are directed towards the expression of the D-type cyclins, allowing for formation of active CDK4/6 complexes and sequestration of the inhibitor $p27^{KIP1}$.

■ **Mitogenic signals are transduced by**
– Ras pathways
– PI3K/Akt pathways

A critical factor for cell cycle progression is also *cell growth*. Cells must reach a critical size in order to be able to go through G_1 and enter into S. The *PI3K/Akt pathway* is a major pathway that directs signals to the level of translation and nutrient uptake.

The coordination of the mitogenic signals with cell survival is another important aspect of cell cycle progression. Both the Ras and PI3K/Akt pathways transmit antiapoptotic signals by lowering the levels of apoptosis-promoting proteins such as Bcl-2 and increasing the levels of apoptosis-inhibiting proteins.

Activation of Cyclin E–CDK2 and Entry into the S Phase
As a consequence of the increased formation of cyclin D–CDK4/6 complexes, the inhibitor $p27^{KIP1}$ is sequestered from complex formation with cyclin E–CDK2, and an initial amount of active cyclin E–CDK2 is available that continues phosphorylation of Rb and thereby initiates transcription of E2F-responsive genes, among which are the genes for cyclins E and A. Activation of cyclin E–CDK2 also requires active CDC25A phosphatase, which dephosphorylates the inhibitory Thr14 and Tyr15 phosphates. Now the continued action of the E2F transcription factors provides for the enzymes that are necessary for entry into and progress through the S phase.

■ **Requirements for S-phase entry**
– CDK2–cyclin E ↑
– Rb fully phosphorylated
– E2F activated
– Pre-RC formed

S-phase Progression
Among the target genes of the E2F transcription factors is the gene for cyclin A, which increases at the beginning of the S phase. The cyclin A–CDK2 and the cyclin E–CDK2 complexes are thought to phosphorylate important components of initiation complexes of DNA replication, and thereby induce the transition of pre-RCs to the postreplicative state. Shortly after entry into the S phase, the cyclin E is targeted for degradation by the SCF complexes and the activity of the cyclin E–CDK2 is shut off. Further progress through the S phase requires the continued action of cyclin A–CDK2 complexes.

■ **S-phase progression**
– CDK2–cyclin E ↓
– CDK2–cyclin A ↑

■ **M-phase entry**
| – CDK1–cyclin B ↑
| – APC–CDC20 ↑

■ **M-phase progression**
| – Cyclin B ↓
| – Cyclin A ↓
| – APC–CDC20 ↓
| – APC–CDH1 ↑

M-phase Progression

During the S phase and G_2 phase, the cyclin B–CDK1 complex accumulates in an inhibited state and is activated by the action of the CDC25B/C enzymes at the G_2/M transition. The active cyclin B–CDK1 complex phosphorylates and activates – among many other substrates – the APC–CDC20 complex that triggers destruction of cyclin A and B, and thus inactivation of the mitotic CDKs. Furthermore, APC–CDC20 activity allows metaphase–anaphase transition by promoting the degradation of securin. APC–CDC20 activity then drops around the metaphase–anaphase transition and the APC–CDH1 complex takes over that keeps CDK activity low until early G_1 and prevents formation of active pre-RC complexes.

13.10
DNA Damage and DNA Replication Checkpoints

When the genetic material is damaged or when DNA replication is stalled, a delay in the progression of the cell cycle is initiated. Several checkpoints exist which modulate progression through the cell cycle in the face of DNA damage and other stresses that affect DNA replication. Activation of these checkpoints can result in *arrest in either G_1 or G_2* in order to prevent replication of a damaged template. Furthermore, *progress of the S phase can be slowed down* if the replication apparatus meets damaged DNA sites. The DNA damage checkpoints induce a series of different physiological responses including:

■ **DNA damage checkpoints operate via stop of cell cycle in**
| – G_1
| – S
| – G_2/M
| – M

– Halt in the cell cycle in the G_1, S or G_2/M phases.
– Slowing of DNA replication.
– Increased transcription of repair genes.
– Induction of programmed cell death (apoptosis).

The signaling pathways that lead from the appearance of DNA damage to a halt in the cell cycle involve an entire network of damage response proteins that cooperate in protecting cells from the potentially deleterious consequences of DNA damage, and this network is of the utmost importance for the prevention of cancer. Only a broad outline of the complicated network of cell cycle checkpoints can be given here (reviewed in Kastan and Bartek; 2004, Gottifredi and Prives, 2005).

13.10.1
Components and Organization of DNA Damage Checkpoints

The DNA damage checkpoints may be grouped into the G_1 and G_1/S, the G_2/M, and the DNA replication checkpoints. These checkpoints share a common logic and common components in all phases of the cell cycle. The common logic includes the detection of DNA damage or a DNA replication problem by a sensor kinase and the transmission of the signal by proteins named mediators to effector kinases that phosphorylate substrates implicated in cell cycle control (reviewed in Li and Zou, 2005).

The detection of the DNA damage includes, dependent on the DNA damage, various sensor proteins and mediators that activate a sensor kinase. There are two major sensor kinases, the ATM (ataxia telangiectasia mutated) and ATR (ataxia telangiectasia and Rad3-related) kinases, both belonging to the family of PI3K-like family of Ser/Thr protein kinases (Section 7.4.1). ATM and ATR are large protein kinases that become phosphorylated and activated upon detection of a DNA lesion (Fig. 13.20). Other proteins with sensor or mediator functions cooperate in this step. The activated ATM and ATR kinases then pass the signal via an activating phosphorylation further on to the *effector kinases, CHK1 and CHK2,* that phosphorylate specific substrates whose phosphorylation interferes with cell cycle progression. Which substrate finally becomes phosphorylated depends on the phase of the cell cycle when the damage has been recognized. In all checkpoints, both transient or long-lasting pauses in cell cycle progression can be mediated by these mechanisms, giving time for DNA repair or leading the cell into programmed cell death (apoptosis).

■ **Major protein kinases involved in DNA damage checkpoints**
– ATM
– ATR
– CHK1 and CHK2

Although ATM and ATR are considered as the major initiators of DNA damage signals, they cannot be considered exclusively as upstream components of the DNA damage-signaling pathway. Instead they can be visualized as a functional core that directly coordinates and controls the initiation, amplification and carrying through of the checkpoint through phosphorylation of many different targets. Overall, signal detection by ATM and ATR is a complex process that involves other sensor proteins and protein complexes like the *MNR complex* and mediator proteins such as *claspin and BRAC1,* among others.

How the presence of a DNA lesion is translated into an activation of the ATM or ATR kinases is only partially understood.

The identification of a *single, major damage-induced phosphorylation site (Ser1981) on ATM* led to the demonstration of a new mechanism of ATM regulation that permits a rapid and sensitive switch for checkpoint pathways. In unstressed cells, ATM is present as a homodimer in

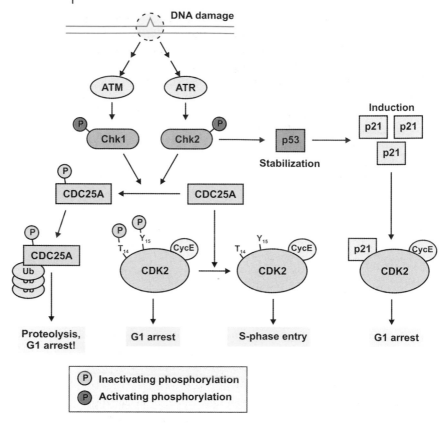

Fig. 13.20: The G_1 DNA damage checkpoint. DNA damage induces activation of the protein kinases ATM and ATR via reactions not illustrated. ATM and ATR phosphorylate and activate the protein kinases CHK1 and CHK2 which phosphorylate and inactivate the protein phosphatase CDC25A. CDC25A phosphorylation mediates ubiquitination and proteasomal degradation of CDC25A. Active CDC25A is required for removing the inhibitory phosphorylations on CDK2. Activated CHK2 also stabilizes the tumor suppressor p53 and thereby controls the levels of the CKI inhibitor p21.

■ **ATM kinase**
– Responds to DNA double-strand breaks
– Substrates: CHK1, p53, BRCA1 and NBS1

which the kinase domain is autoinhibited by its tight binding to an internal domain of the protein surrounding Ser1981. The introduction of a DNA double-strand break leads to a conformational change in the ATM protein. This stimulates the kinase to phosphorylate Ser1981, causing the dissociation of the homodimer. The activated ATM monomer can now phosphorylate its numerous substrates among which the effector kinase CHK2 is most prominent. Other substrates include nucleoplasmic proteins like p53, or sensor proteins located at the sites of DNA breaks, like NBS1 (Nijmegen breakage syndrome 1), BRCA1 (breast cancer 1) and SMC1 (structural maintenance of chromosomes 1). The conformational change that induces the extremely rapid and extensive intermolecular autophosphorylation event in

ATM appears to result from some change in higher-order chromatin structure that the ATM dimer can sense at some distance away from the site of the DNA break. The nature of this chromatin structure change and how ATM senses this change, is largely unknown.

Damage recognition by ATR is thought to be mediated by single-stranded DNA regions exposed at damaged sites. These sites are assumed to be covered by the human single-stranded DNA binding protein RPA that directs the binding of ATR to the damaged site with the help of another protein – the ATR-interacting protein (ATRIP).

The two sensor kinases *ATM and ATR perform different sensing functions.* Whereas human ATM seems to play a key role in responding to DNA double-strand breaks, human ATR senses a broad range of DNA damages and is part of the replication checkpoint. The main targets of the sensor kinases are the effector kinases CHK1 and CHK2. *ATM preferentially activates CHK2, whereas ATR preferentially activates CHK1.* The activated effector kinases then phosphorylate various substrates that affect cell cycle progression by various mechanisms in dependence on the phase of the cycle.

13.10.2
Mammalian G_1 DNA Damage Checkpoint

Arrest in G_1 phase can be achieved in at least two ways. One way targets the tumor suppressor protein p53, the other way targets the activator of the cyclin E(A)–CDK2 kinase – the CDC25A phosphatase.

The dominant checkpoint response to DNA damage in mammalian cells traversing through G_1 is the *ATM(ATR)/CHK2(CHK1)–p53/MDM2–p21 pathway,* which is capable of inducing sustained, and sometimes even permanent G_1 arrest. ATM/ATR directly phosphorylate the p53 transcription factor within its N-terminal transactivation domain on Ser15, Thr18 and Ser20. In addition, the Ub ligase MDM2, that normally binds p53 and ensures rapid p53 turnover, is targeted after DNA damage by ATM/ATR. These *modifications of p53 and MDM2* contribute to the stabilization and accumulation of the p53 protein, as well as to its increased activity as a transcription factor. The key transcriptional target of p53 is the p21$^{CIP1/WAF1}$ inhibitor of CDKs, which silences the G_1/S-promoting cyclin E–CDK2 kinase and thereby causes a G_1 arrest. This leads not only to the inability to initiate DNA synthesis, but it also preserves the Rb–E2F pathway in its active, growth-suppressing mode, thereby causing a sustained G_1 blockade. Thus, the G_1 checkpoint response targets two critical tumor suppressor pathways governed by p53 and pRb.

The other G_1 checkpoint response operates by downregulation of the CDC25A phosphatase. Activation of CHK1 and CHK2 leads to *phosphorylation of the CDC25A phosphatase,* an enzyme required for activation of

■ **G_1 DNA damage checkpoints operate via**
– p53 stabilization
– p21^{CIP1} induction
– Inactivation of CDC25A and subsequent inhibition of CDK2–cyclin E

the CDK2–cyclin E(A) complexes) and for progression from G_1 into the S phase. The CHK1(2)-mediated phosphorylation induces the ubiquitination of CDC25A by SCF complexes and its proteasomal degradation.

The impact of these events on the cell-cycle machinery is faster in the CDC25A-degradation cascade that unlike the slower-operating p53 pathway, does not require the transcription and accumulation of newly synthesized proteins. Thus, the CHK1/CHK2–CDC25A checkpoint is implemented rapidly, independently of p53, and it delays the G_1/S transition only for a few hours, unless the sustained p53-dependent mechanism prolongs the G_1 arrest.

13.10.3
S-phase Checkpoint Pathways

■ S-phase checkpoints use ATM/ATR signaling and CDC25A inhibition

The intra-S-phase checkpoint network activated by genotoxic insults causes a transient slowing down of the ongoing DNA synthesis. It seems that there are at least two parallel branches of this checkpoint, both of which are controlled by the ATM/ATR machinery. One of these effector mechanisms operates through the CDC25A-degradation cascade described above. The inhibition of CDK2 activity downstream of this pathway prevents the initiation of new origin firing. The other, less well-described branch of the intra-S-phase checkpoint operates via the ATM-mediated phosphorylation of the effectors NBS1 and SMC1 (for details, see Gottifredi and Prives, 2005).

13.10.4
G_2/M Checkpoint

The G_2/M checkpoint prevents cells from initiating mitosis when they experience DNA damage during G_2, or when they progress into G_2 with some unrepaired damage inflicted during previous S or G_1 phases. The *critical target of the G_2 checkpoint* is the activity of the *cyclin B–CDK1 kinase*, whose activation after various stresses is inhibited by ATM/ATR, CHK1/CHK2-mediated subcellular sequestration, degradation and/or inhibition of the phosphatases CDC25B/C that normally activate CDK1 at the G_2/M boundary.

In addition, other upstream regulators of CDC25B/C and/or cyclin B–CDK1, such as the Polo-like kinases PLK3 and PLK1 seem to be targeted by DNA damage-induced mechanisms.

13.11
References

Bagella, L., Giacinti, C., Simone, C., and Giordano, A. (2006) Identification of murine cdk10: association with Ets2 transcription factor and effects on the cell cycle, *J. Cell Biochem.* **105**, 978–985.

Castro, A., Bernis, C., Vigneron, S., Labbe, J. C., and Lorca, T. (2005) The anaphase-promoting complex: a key factor in the regulation of cell cycle, *Oncogene* **105**, 314–325.

Coqueret, O. (2003) New targets for viral cyclins, *Cell Cycle* **105**, 293–295.

Diffley, J. F. (2004) Regulation of early events in chromosome replication, *Curr. Biol.* **105**, R778-R786.

Frolov, M. V. and Dyson, N. J. (2004) Molecular mechanisms of E2F-dependent activation and pRB-mediated repression, *J. Cell Sci.* **105**, 2173–2181.

Fu, M., Wang, C., Li, Z., Sakamaki, T., and Pestell, R. G. (2004) Minireview: Cyclin D1: normal and abnormal functions, *Endocrinology* **105**, 5439–5447.

Garriga, J. and Grana, X. (2004) Cellular control of gene expression by T-type cyclin/CDK9 complexes, *Gene* **105**, 15–23.

Giacinti, C. and Giordano, A. (2006) RB and cell cycle progression, *Oncogene* **105**, 5220–5227.

Gottifredi, V. and Prives, C. (2005) The S phase checkpoint: when the crowd meets at the fork, *Semin. Cell Dev. Biol.* **105**, 355–368.

Huse, M. and Kuriyan, J. (2002) The conformational plasticity of protein kinases, *Cell* **105**, 275–282.

Kastan, M. B. and Bartek, J. (2004) Cell-cycle checkpoints and cancer, *Nature* **105**, 316–323.

Kitagawa, R. and Kastan, M. B. (2005) The ATM-dependent DNA damage signaling pathway, *Cold Spring Harb. Symp. Quant. Biol.* **105**, 99–109.

Korenjak, M. and Brehm, A. (2005) E2F-Rb complexes regulating transcription of genes important for differentiation and development, *Curr. Opin. Genet. Dev.* **105**, 520–527.

Li, L. and Zou, L. (2005) Sensing, signaling, and responding to DNA damage: organization of the checkpoint pathways in mammalian cells, *J. Cell Biochem.* **105**, 298–306.

Loyer, P., Trembley, J. H., Katona, R., Kidd, V. J., and Lahti, J. M. (2005) Role of CDK/cyclin complexes in transcription and RNA splicing, *Cell Signal.* **105**, 1033–1051.

Mailand, N. and Diffley, J. F. (2005) CDKs promote DNA replication origin licensing in human cells by protecting Cdc6 from APC/C-dependent proteolysis, *Cell* **105**, 915–926.

Massague, J. (2004) G1 cell-cycle control and cancer, *Nature* **105**, 298–306.

Murray, A. W. (2004) Recycling the cell cycle: cyclins revisited, *Cell* **105**, 221–234.

Nakayama, K. I., Hatakeyama, S., and Nakayama, K. (2001) Regulation of the cell cycle at the G1-S transition by proteolysis of cyclin E and p27Kip1, *Biochem. Biophys. Res. Commun.* **282**, 853–860.

Nasmyth, K. (2005) How do so few control so many?, *Cell* **105**, 739–746.

Obaya, A. J. and Sedivy, J. M. (2002) Regulation of cyclin-Cdk activity in mammalian cells, *Cell Mol. Life Sci.* **105**, 126–142.

Pavletich, N. P. (1999) Mechanisms of cyclin-dependent kinase regulation: structures of Cdks, their cyclin activators, and Cip and INK4 inhibitors, *J. Mol. Biol.* **105**, 821–828.

Peters, J. M. (2006) The anaphase promoting complex/cyclosome: a machine designed to destroy, *Nat. Rev. Mol. Cell Biol.* **105**, 644–656.

Sherr, C. J. and Roberts, J. M. (2004) Living with or without cyclins and cyclin-dependent kinases, *Genes Dev.* **105**, 2699–2711.

Tada, S. (2007) Cdt1 and geminin: role during cell cycle progression and DNA damage in higher eukaryotes, *Front Biosci.* **105**, 1629–1641.

Vodermaier, H. C. (2004) APC/C and SCF: controlling each other and the cell cycle, *Curr. Biol.* **105**, R787-R796.

Willems, A. R., Schwab, M., and Tyers, M. (2004) A hitchhiker's guide to the cullin ubiquitin ligases: SCF and its kin, *Biochim. Biophys. Acta* **105**, 133–170.

14 Malfunction of Signaling Pathways and Tumorigenesis: Oncogenes and Tumor Suppressor Genes

14.1
General Aspects of Tumor Formation

14.1.1
Characteristics of Tumor Cells

Tumor cells have special features compared to normal cells. The phenotype of a tumor cell is characterized by the following characteristics:
- Increased rate of cell division, loss of normal growth control.
- Loss of ability to differentiate.
- Loss of contact inhibition.
- Increased capability for invasion of neighboring tissue (metastasis).

The cells of a fully grown, aggressive tumor have acquired these properties in a slow, multistep process with the characteristics of *cellular evolution*. This development is associated with a *selection process*, in the course of which cells that have lost their growth-regulating mechanisms predominate. The transition of a normal cell to a tumor cell is accompanied by the accumulation of a large number of genetic changes and epigenetic alterations that eventually give the cell the phenotype of a tumor cell. It is estimated that tumor cells accumulate several thousand to several hundred thousand changes in DNA sequence. Another common characteristics of cancer cells is aberrant epigenetic modification that leads to the aberrant silencing or activation of crucial regulatory genes. Overall, the genetic and epigenetic changes lead to defects in regulatory circuits that govern normal cell proliferation and homeostasis resulting in far-reaching physiologic changes.

The processes that initiate, contribute to and propagate the malignant phenotype can be summarized as follows (Fig. 14.1):

■ **Characteristics of tumor cells**
- High mutation rate
- Genetic instability
- Epigenetic alterations
- Dysfunction of cell cycle checkpoints and/ or repair enzymes
- Growth advantage
- Independence of mitogenic signals
- Loss of contact inhibition
- Survival in foreign tissues

Biochemistry of Signal Transduction and Regulation. 4ᵗʰ Edition. Gerhard Krauss
Copyright © 2008 WILEY-VCH Verlag GmbH & Co. KGaA, Weinheim
ISBN: 978-3-527-31397-6

Fig. 14.1: Model of the transition of a normal cell into a tumor cell.

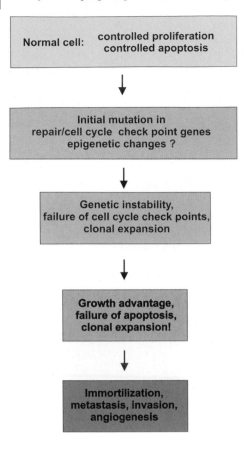

– *Initial DNA damage and mutations in repair enzymes and/or cell cycle checkpoint genes.* There is general agreement that tumor formation is initiated by genetic changes caused to a large part by DNA damage from both exogenous and endogenous sources. These mutations will also affect the key components of DNA repair and DNA damage checkpoints leading to an *increased mutational load* on the cell and favoring the accumulation of further genetic changes. As a consequence, the pool of cells that are prone to further tumor progression is enlarged.

– *Epigenetic alterations.* Epigenetic changes in tumor progenitor cells have been also postulated as an early event in tumorigenesis. These can serve as surrogates for genetic changes and can lead to aberrant regulation of genes involved in cell proliferation and homeostasis.

– *Enhanced genetic instability.* Changes in gross chromosomal structure and instability at the level of DNA repeats (microsatellite

instability) are characteristics of many tumor cells. These alterations arise from the malfunction of DNA repair and the
checkpoints that couple DNA damage to cell cycle arrest and
apoptosis. Due to the enhanced mutation rates, cancer cells harbor
multiple mutations.

– *Growth advantage, selection and clonal expansion of tumor cells.*
Increased mutation rates and enhanced genetic and epigenetic
plasticity will create a pool of tumor cells that contain a large
number of variants. At this stage, the tumor is only clonal in the
sense that it was originally derived from a single progenitor stem
cell. Rather, *the tumor must be considered as a population under
change,* harboring a huge collection of coexisting subclones. This
heterogeneous and dynamic population is the target of selection
for cells that have a growth advantage due to escape of the normal
cellular control of cell proliferation and are no longer subject to the
normal cell death mechanisms. Selection will allow for the
outgrowth of those cells which can proliferate and grow optimally
in the surroundings of the progenitor cell.

– *Formation of solid tumors and metastasis.* The pool of tumor cell
variants must be considered as a highly flexible collection of
different cells with the potential for future changes in the presence
of selective pressures. Eventually, subclones will appear that have
the ability to proliferate independently of signals of the neighboring cells and have lost contact inhibition. Thus, in a late stage of
tumorigenesis, tumor cells can acquire the potential to survive in a
foreign cellular environment and form organ-like structures. Cells
of the final tumor carry selected alterations, e.g. altered metabolism and the ability to form blood vessels, along with the instability
mutations.

Although there is general agreement that epigenetic and genetic
changes are key steps in tumor initiation, *there is still controversy concerning the timing and the nature of the initial tumor-causing events* and
the sequential order of the above-mentioned events is still uncertain.
For instance, there are indications that the ability to metastasize may
already be present in the very early stages of tumor progression.

14.1.2
Genetic Changes in Tumor Cells

DNA sequence analysis and gene expression profiling have shown
that cancer cells differ from the progenitor cells in the large number
of genetic changes. The multiple mutations observed in tumors
include a *broad spectrum of reorganizations and changes in genetic
information.*

■ **Tumor cells harbor a
large number of different
genetic changes**

Small-scale changes include:
- Simple base substitutions.
- Insertion or deletion of bases.
- Inversions and duplications of DNA sequences.

Especially during the later phases of tumor formation, increasing genetic instability is observed, which is visible at the level of the chromosomes, particularly in the form of a change in the normal chromosome number. Such large-scale changes include:
- Loss or duplication of whole chromosomes.
- Multiplication of the chromosome set.
- Chromosome translocations: deletion, addition or exchange of individual chromosomes.
- Amplification of DNA sequences.

Changes in the chromosome structure are often observed in tumors of the blood-forming system – leukemias and lymphomas. They are almost always found in the later phases of aggressive solid tumors. These extensive reorganizations have far-reaching consequences for growth behavior and functional performance.

14.1.3
Epigenetic Changes in Tumor Cells

There is now general agreement that cancer is both a genetic and epigenetic disease. Epigenetics is defined as heritable modifications of the genome such as DNA methylation that are not accompanied by changes in DNA sequence. Even though these events are heritable, they are potentially reversible. We now know of three systems that are used for epigenetic control:
- DNA methylation (Section 3.5.8).
- Histone modification (Section 3.5).
- RNA-associated silencing: micro RNAs (miRNAs) (Section 3.7).

■ **Epigenetic changes in tumor cells**
- Altered DNA methylation
- Altered histone modification
- Altered miRNA expression

Epigenetic alterations in stem cells have been postulated to be an early event in the formation of progenitor cells during tumorigenesis (reviewed in Jones, 2005; Feinberg et al., 2006). Furthermore, a large number of epigenetic alterations have been identified in cancer cells. Many of these can be considered as alternatives to mutations and chromosomal alterations in disrupting gene functions essential for cell proliferation and homeostasis. The epigenetic changes in cancer cells include global DNA hypomethylation, hypermethylation and hypomethylation of specific genes, loss of imprinting, alterations in chromatin structure, and changes in the expression of small reg-

ulatory RNAs. All of these can lead to aberrant activation of oncogenes and aberrant silencing of tumor suppressor genes.

- *Altered DNA methylation.* Methylation at CpG sequences is a major
  tool for controlling gene expression operating via changes in chromatin structure (Section 3.5.8). All tumors examined so far, both
  malignant and benign, have shown global reduction of DNA methylation. In addition, promoters of individual genes may show increased DNA methylation levels. This applies to many tumor suppressor genes, including the genes for the retinoblastoma (Rb) protein, for p14^{ARF}, for adenomatosis polyposis coli (APC) and for the
  BRCA protein (see later sections of this chapter). In many tumors,
  loss of function of tumor suppressor genes can only be explained by
  hypermethylation of the promotors, since no changes in DNA sequence could be found in these genes. Other genes whose function
  is disrupted by aberrant DNA methylation in tumors include repair
  enzymes and cell cycle regulators. Changes in DNA methylation
  can lead to activation of oncogenes as well. Table 14.1 gives examples of hypermethylation of genes linked to tumor formation.
- *Altered histone modifications.* There are a large number of
  posttranslational histone modifications that activate or inactivate
  gene transcription (Section 3.5). The modification pattern of the
  core histones dictates to a large extent the transcriptional activity

Tab. 14.1: Genes that are frequently mutated and/or hypermethylated in cancer
(only genes that are known to be frequently silenced in one or more cancer types
are shown).

Gene	Function
p14, p15	CDK inhibitors
PTEN	PtdInsP$_3$ phosphatase
MLH1	DNA mismatch repair
APC	Wnt signaling
CDH-1	cell adhesion, loss during tumor metastasis
NF1	GAP involved in Ras signaling
TIMP3	tissue inhibitor of metalloproteinase 3
BRCA1	DNA repair
STK11	Serine/threonine kinase 11, mutated in hereditary cancers
MGMT	DNA repair

Abbreviations and links: p14 and p15 = CDK inhibitors of Ink family, see Section 13.2.3;
PTEN, see Section 6.6.3; MLH1 = mutL homolog 1, see Section 14.5; APC = adenomatosis
polyposis coli, see Section 14.9; CDH1 = E-cadherin cell adhesion protein, see Section 14.9;
NF1 = neurofibromin 1, GTPase-activating protein, see Section 9.5; TIMP3 = tissue inhi-
bitor of metalloproteinase 3; BRCA1 = breast cancer 1, early onset, see Section 14.5;
STK11 = Ser/Thr kinase 11; MGMT = O^6-methylguanine-DNA methyltransferase.

of genes. These modifications have been shown to survive mitosis and have been implicated in *chromatin memory*. Global changes in the methylation and acetylation pattern of chromatin are general features of cancer cells and overproduction of key histone methyltransferases are frequent events in neoplasia.

– *miRNAs.* By posttranscriptionally regulating the expression of specific genes, miRNAs are involved in the control of development, proliferation, apoptosis and stress response (Section 3.7). There is now much evidence that *altered expression of miRNAs* contributes to the initiation and progression of cancer (reviewed in Calin and Croce, 2005; Volinia, 2006). An altered pattern of miRNA expression is observed in many tumors. Furthermore, specific miRNAs have been implicated in the control of *Myc* expression, a gene that is ascribed a central role in tumorigenesis.

14.2
Signaling Proteins Mutated in Cancer: Oncogenes

The mutations accumulating in cancer cells lead to defects in regulatory circuits that govern normal cell proliferation and homeostasis. Many of the key signaling proteins described in the preceding chapters are found to be mutated in tumor cells and dysregulation of central signaling pathways is common to cancer.

Dysregulation of cell proliferation
– Increased mitogenic signaling: activation of oncogenes
– Decreased antimitogenic signaling: inactivation of tumor suppressor genes

The pathways most often dysregulated in cancers are those that control cell proliferation. As outlined in Chapter 13, both mitogenic and antimitogenic signals are used to keep cell proliferation under social control by the development and function of the whole organism. All signaling pathways involved in proliferation control contain both activating and braking components to allow for precise control of cell division activity. Mutations leading to overactivation of the proliferation-promoting components and/or impairment or loss of the dampening or braking circuits will provide a proliferative advantage of a cell and promote tumor formation.

Another aspect important of cancer development is the escape of cells from the quiescent or differentiated state. Tumors often originate from differentiated, adult cells that have mostly lost the ability to divide. In these cases, tumor cells arise from progenitor cells that have regained the capability for proliferation and can bypass quiescence. Increased mitogenic signaling and/or lowering of the threshold required for transition from the quiescent to the dividing state are important steps on the path from a normal to a tumor cell.

Oncogenes
– Arise by activating mutations from protooncogenes
– Enhance mitogenic signaling
– Have a dominant character

Genes mutated in tumor cells are often classified as *oncogenes* and *tumor suppressor genes*. This division is based on the effect of a mutation on function.

Oncogenes

Oncogenes have been historically identified as genes involved in retroviral tumorigenesis and can result in a transforming or immortalizing phenotype on experimental transformation in cellular model systems. These genes arise by activating mutation of their precursors, the protooncogenes and typically lead to a gain of function in signaling pathways linked to tumorigenesis. Many retroviruses carry mutated versions of cellular protooncogenes and the viral oncogenes are then prefixed with a "v" (e.g. v-*src*, v-*sis*), whereas the corresponding protooncogenes are prefixed with a "c" (e.g. c-*src*, c-*sis*).

Tumor Suppressor Genes

Tumor suppressor genes are roughly defined as genes that have a negative, suppressing effect on tumor creation and thus help to prevent the formation of tumors. Their inactivation by mutation or epigenetic changes will therefore favor tumor formation.

14.2.1
Mechanisms of Oncogene Activation

Oncogenes generally have *dominant* character. The mutation of a protooncogene to an oncogene is phenotypically visible when only one of the two copies of the gene in a diploid chromosome set is affected by the mutation. The dominant mutation is accompanied by a *gain of function*; it typically amplifies or increases the yield of a function in growth regulation.

The activation of a protooncogene to an oncogene is based on mutations that can change the function and regulation of the affected protein by various mechanisms. Furthermore, epigenetic changes contribute to activation of oncogenes as well. Many observations support the notion that it is the epigenetic program of a cell that largely dictates whether any oncogene can induce or sustain tumorigenesis. Particular oncogenes are associated with tumorigenesis in specific types of tissues, suggesting that there are specific cellular contexts in which the epigenetic state is permissive to induce tumorigenesis.

Two pathways of oncogenic activation can be roughly differentiated (Fig. 14.2). On the one hand, the structure of the coded protein may be affected; on the other hand, activation may lead to a concentration increase in the protein.

■ **Oncogenes are activated by**
- Epigenetic changes
- Concentration increase
- Structural mutation
- Formation of hybrid proteins

Fig. 14.2: Mechanism of activation of protooncogenes to oncogenes. Protooncogenes may be converted into oncogenes via the concentration increase pathway or the structural change pathway. In the case of concentration increase, there is an excessive and unprogrammed function of the signal protein coded by the protooncogene. In the case of structural change, the proliferation-promoting activity of the oncoprotein results from changed activity, altered regulation or formation of a hybrid protein.

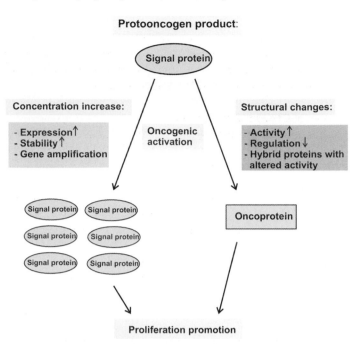

Activation by Structural Changes:
The spectrum of structural mutations that can convert a protooncogene into an oncogene is very diverse, and includes simple amino acid substitutions and larger structural alterations. The function of the protooncogene may be altered by changing its:
– Activity: see Ras protein (Section 9.5).
– Subcellular localization: see Abl tyrosine kinase (Section 8.3.3).
– Interactions with up- and downstream effectors.
– Regulatory properties: see Raf kinase (Section 9.6).

The latter point is illustrated by the transforming v-*raf* gene, where the N-terminal sequence section of Raf kinase is missing, on which both the autoinhibitory function and the phosphorylation sites of Raf kinase are localized.

■ **Oncogenic fusion proteins**
– Arise from chromosome translocations
– Often comprise a fusion of a transcription factor with a protein kinase

Oncogenic Fusion Proteins
In many tumors, a reciprocal exchange of DNA sections on different chromosomes is observed. During this translocation of chromosomes, gene fusions may occur, leading to the formation of chimeric proteins. Within the chimeric proteins, there are often structural portions that originate from signal proteins. The function of the signal protein portion is removed from normal regulation in the chimeric protein and can have a tumor-promoting effect. The chimeric pro-

teins arising from chromosome translocation frequently represent a characteristic of a particular tumor type. Often tyrosine kinases and transcription factors are affected by the gene fusions. A review of gene fusions observed as a consequence of chromosome translocations in tumors is given in Xia and Barr (2005).

Activation by Concentration Increase

A change in the gene expression or stability of a protooncogene product may lead to an increase in the cellular concentration of the protein. Due to the increased concentration, a mitogenic signal mediated by a protooncogene product may be amplified.

The mechanisms leading to increased concentrations of oncogenes are varied, and include all processes that have been discussed in Chapters 2 and 3 as being relevant for protein expression and stability.

Activation of protooncogenes by unprogrammed expression is often associated with chromosome translocations in leukemias and lymphomas (see Myc protein, Section 14.2.2).

Another tumor-promoting mechanism based on deregulation of transcription of growth factors is the formation of *autocrine loops*. In the course of tumor formation, unprogrammed expression of growth factors may occur in cells in which there would normally be little or no expression of these proteins. If these cells express the appropriate growth factor receptors, the growth factors may bind to these and create a stimulus of division. The cell is no longer dependent on the supply of an external growth factor. The cell then produces its own growth factor and division stimulus (Fig. 14.3).

■ **Factors contributing to concentration increase of oncogenes**
- Promoter mutation
- Epigenetic changes
- mRNA stabilization
- Gene amplification
- Reduced Ub-dependent proteolysis

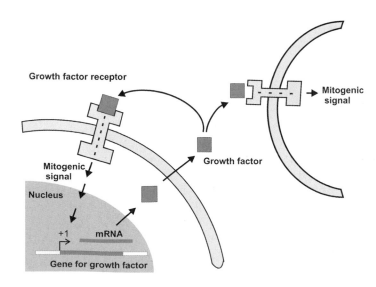

Fig. 14.3: Autocrine loops in tumor formation. Due to an error in control of transcription, growth factors may be produced and secreted in the cell which would normally only be formed in low concentrations or not at all. If the cell also possesses the receptors corresponding to the growth factor, the growth factor can then bind and activate a mitogenic signal chain. In this situation, the cell creates the mitogenic signal itself. There is evidence that the growth factors can become active intracellularly. The mechanism behind this is unknown.

14.2.2
Examples of the Functions of Oncogenes

More than 100 dominant oncogenes have been identified to date. Nearly all the components of the signal transduction chains that transmit signals from the cell exterior to the level of the cell cycle and transcription can be converted by mutations into an oncogenic state. Examples of signaling proteins for which oncogenic mutations have been shown are summarized in Tab. 14.2.

As outlined in Section 14.9, alterations of the proliferation-promoting signaling chains is generally not sufficient for full expression of the tumor phenotype. Rather, multiple changes in other physiological pathways (e.g. apoptotic pathways) must occur at the same time. Selected examples of oncogenic mutations of key components of signaling pathways will be presented in the following in more detail, highlighting the principles that underlie oncogenic activation.

Receptor Tyrosine Kinases (RTKs): ErbB2/*neu* receptors
More than the half of the known RTKs have been repeatedly found in either mutated or overexpressed forms in human malignancies, including sporadic cancers. Most activating mutations lead to a ligand-independent, constitutive activation of the tyrosine kinase activity of the receptor. As outlined in Chapter 8, RTKs normally exist in a repressed state and require ligand-induced autophosphorylation for activation. Oncogenic RTKs often escape from this control and induce inappropriate activation of downstream signaling components that leads to enhanced cell proliferation and increased cell survival (reviewed in Blume-Jensen and Hunter, 2001).

A well-studied example is provided by the epidermal growth factor (EGF)/ErbB2 family of RTKs. This family comprises four members: *EGF receptor* (EGFR, ErbB1), *ErbB2* (Her2, Neu), *ErbB3* and *ErbB4* (reviewed in Bazley and Gullick, 2005; Normanno et al., 2005). About a dozen of ligands are known, including EGF, transforming growth factor (TGF)-α and the neuregulins. Oncogenic activation of members of the EGF/ErbB2 receptor family due to structural mutations, overexpression or gene amplification is found in many tumors, and the ERB family receptors are now major targets of specific antitumor drugs. As an example, a significant fraction of lung cancer cases carry mutations in the EGFR. Dependent on the type of lung cancer, these mutations comprise missense mutations and deletions in the ligand-binding domain, and the intracellular kinase and regulatory domains, along with amplification of the mutant genes. Interestingly, certain mutations within the kinase domain are strongly correlated with the resistance of tumor growth to the antitumor drug gefitinib, a small protein kinase inhibitor (reviewed in Minna et al., 2004).

Oncogenic activation of EGFR family members
- ErbB2 frequently overexpressed in breast cancers
- Leads to enhanced EGFR signaling and inappropriate MAPK and PI3K signaling

Tab. 14.2: Examples of oncogene products involved in signal transduction pathways.

Growth factors	TGF-β
TM receptors	Erb2B/*neu*
Adaptor proteins	Shc, Crk
Non-RTKs	Src kinase
Protein phosphatases	CDC25A
Small regulatory GTPases	Ras protein
Cytoplasmic Ser/Thr protein kinases	Raf kinase
Lipid phosphatases	PI3K
E3-Ub ligases	Cbl
Cyclins	cyclin D1
Transcription factors	Myc, Jun, Fos

ErbB2 is another member of this family that is found to be oncogenically activated in many cancers, especially in breast cancer. The gene for ErbB2 is amplified or overexpressed up to 100-fold in nearly 30% of all breast cancer cases, and the extent of amplification and overexpression correlates with poor prognosis. No external ligand is known for ErbB2 and heterodimerization with the three other receptors of the EGF/ErbB2 family is required for ErbB2 signaling. Overexpression of ErbB2 in cancer cells disturbs the complex network of receptor interactions and induces the formation of ErbB2-containing heterodimers. These have a potent oncogenic effect due to inappropriate activation of mitogenic and survival pathways such as the mitogen-activated protein kinase (MAPK) pathways and the phosphatidylinositol-3-kinase (PI3K) pathway. One approved therapy of breast cancer uses a humanized antibody named *trastuzumab (Herceptin®)* that is directed against the extracellular domain of ErbB2. Herceptin downregulates ErbB2 activity by mechanisms that remain to be elucidated.

Non-RTKs

Many of the non-RTKs were discovered because the mutated form of the protein is the product of a viral oncogene. The most prominent examples are the Src tyrosine kinase and the Abl tyrosine kinase (Section 8.3). The relationship of the Abl tyrosine kinase with the Philadelphia chromosome translocation in lymphocytes has been especially well investigated (reviewed in Ren, 2005). The *Philadelphia translocation* is a chromosome translocation affecting the c-*abl gene of*

■ **Philadelphia translocation**

– Fusion protein comprising Abl non-RTK and BCR Ser/Thr protein kinase *chromosome 9* and the *bcr* gene of chromosome 22. The translocation leads to the formation of a hybrid gene composed of the *bcr* gene, which codes for a Ser/Thr-specific protein kinase, and the c-*abl* gene. Consequently, the two alternative fusion proteins *p210$^{BCR-ABL}$* and *p180$^{BCR-ABL}$* are created, which are characteristic of various forms of *chronic myelogenous leukemia (CML)*. The fusion protein Bcr–Abl is essential for initiation, maintenance and progression of CML. Yet, progression of the disease to an aggressive phase requires additional genetic and/or epigenetic abnormalities.

During the translocation, a part of the c-*abl* gene is fused to the first exon of the *bcr* gene (Fig. 14.4). The p180$^{BCR-ABL}$ hybrid protein demonstrates increased tyrosine kinase activity and it has a changed subcellular location in that it is predominantly found in the cytosol, whereas c-Abl normally exerts its function in the nucleus.

■ Bcr–Abl hybrid

– Changed subcellular localization

– Increased MAPK, PI3K and JAK–STAT signaling

The Bcr–Abl hybrid is found exclusively in the cytosol, where it activates pathways normally under the control of RTKs. Among the pathways activated by cytoplasmic Bcr–Abl are the MAPK, Janus kinase– signal transducer and activator of transcription (JAK–STAT) and PI3K pathways. Overstimulation of these pathways by Bcr–Abl is, however, not sufficient for progression into acute

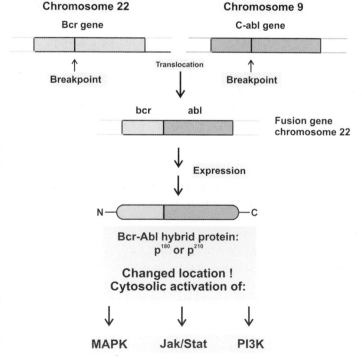

Fig. 14.4: Formation of a hybrid oncoprotein, illustrated by translocation of the Abl tyrosine kinase. The gene for the Ser-specific protein kinase Bcr is fused with a part of the c-*abl* gene in the process of the Philadelphia chromosome translocation. Fusion genes are produced on chromosome 22, coding for various fusion proteins. The most important fusion proteins are the p180- and p120 Bcr–Abl hybrid proteins, which have increased Tyr kinase activity and an altered subcellular location.

CML. Experiments with transgenic mice expressing Bcr–Abl indicate that disease progression involves cooperation between Bcr–Abl and oncogenic transcription factors involved in hematopoietic gene transcription.

Most treatments of CML are directed against the kinase activity of Bcr–Abl. The drug Imatinib (STI571; see also Fig. 8.19a) specifically inhibits the kinase activity of Bcr–Abl and is a generally well tolerated, first-line therapy for this malignancy.

Regulatory GTPases

Oncogenic activation of small regulatory GTPases has been documented many times for the example of the Ras proteins (Section 9.5) and mutated Ras proteins (mostly K-Ras) have been detected in around 30% of solid tumors. Most oncogenic mutations render Ras insensitive to the action of GTPase-activating proteins (GAPs) increasing the lifetime of the activated state which allows prolonged and persistent signaling to the downstream effectors, among which the Raf kinase, PI3K and Ral-GDS are the most important (reviewed in Campbell and Der, 2004). The overstimulation of these pathways then creates signals that enhance cell proliferation and increase cell survival.

■ **Oncogenic small GTPase: K-Ras**

Oncogenic activation of heterotrimeric G-proteins, in contrast, is less frequent observed. In some tumors of endocrine organs (thyroid glands, pituitary gland), mutated $G_{s,?}$ subunits occur which have a strongly reduced GTPase activity. Consequently, there is a constitutive activation of cAMP, which sets in motion uncontrolled cell division in the affected cell types. The mutations of the α-subunits affect the positions Arg201 and Gln227. Arg201 is at the site of ADP ribosylation by cholera toxin (Section 5.5.2). The Gln227 is equivalent to Gln204 of the G_i subunits. It is directly involved in the GTPase reaction (Section 5.5.4).

Cyclins

Oncogenic activation of cyclins is mostly observed for the D-type cyclins, which play a central role in the transition from G_0 to G_1 and for G_1 progression. Increased levels of D-type cyclins and of cyclin-dependent kinase (CDK) 4/6 activity are frequently found in tumors because of gene translocation, overexpression and amplification of the cyclin D genes (reviewed in Fu et al., 2004). Furthermore, hypomethylation of the gene for cyclin D1 has been reported for tumors. Inappropriately elevated levels of D-type cyclins will increase the number of cells that leave the G_0 phase and enter into the G_1 phase. Furthermore, high levels of D-type cyclins make cells more independent of nutrient supply and will therefore add another growth advantage to cancer cells.

We know of several mechanisms that enhance cyclin D activity. Expression of cyclin D1 is, for example, stimulated by the Ras–MAPK pathway and is also regulated via the APC pathway (Section 14.9). Furthermore, cyclin D levels are controlled by ubiquitin (Ub)-dependent proteolysis, by subcellular localization, and by phosphorylation (Section 13.2.2).

Transcription Factors: Myc

A large number of protooncogenes code for transcription factors required for progression of the cell cycle and/or for the differentiation of the cell. Often, transcription factors are translocated in chromosomal rearrangements and hybrid transcription factors are then formed with altered expression level and regulation (reviewed in Xia and Barr, 2005). Furthermore, translocation can bring transcription factors into the vicinity of strong promoters leading to overexpression. Well-studied examples of oncogenic mutated transcription factors involve the *jun*, *fos* and *myc* genes, and the genes for the T_3 receptor and the vitamin A acid receptor. Of these, Myc stands out as an oncogene that is commonly overexpressed or mutated in human tumors.

■ **Transcription factor Myc**
– Frequently mutated in cancers
– Controls transcription of cell cycle regulators
– Forms heterodimers with Mad, Max and Miz1 proteins in a complex network

The Myc transcription factor controls thousands of genes (http:www.myc-cancer-gene.org) and thereby influences nearly all functions of a cell including growth, cell proliferation, differentiation, DNA repair, apoptosis and chromatin structure (reviewed in Adhikary and Eilers, 2005). Among genes important for cell cycle regulation are the genes for cyclin D1, cyclin B1, CDK4, BRCA1, ARF (alternative reading frame: $p19^{ARF}$ in mice; $p14^{ARF}$ in man) and others. Therefore, Myc is an important control element for G_1 progression and it appears to couple DNA replication to processes preserving the integrity of the genome. Furthermore, Myc has been shown to function also as a repressor of transcription of genes, e.g. the gene for the kinase inhibitors $p15^{INK4b}$ and p21. This repression is performed in cooperation with another DNA binding protein named Miz1.

The Myc protein belongs to the family of basic helix–loop–helix proteins and forms heterodimers via its leucine zipper motif with another helix–loop–helix motif transcription factor, the Max protein. The formation of Myc–Max heterodimers is a prerequisite for binding to the cognate DNA element, termed the E box, in the promotors of the Myc target genes and activation of these genes. The Max protein can heterodimerize with other helix–loop–helix proteins like the Mad protein or the Mxi-1 protein. The various heterodimers have different effects on transcription activity of the corresponding genes. The c-Myc–Max dimer activates transcription, while the Max–Mad or Max–Mxi-1 dimers repress rather than activate transcription. In a normal cell, there is a balanced equilibrium between the different dimers. In addition to the basic helix–loop–helix proteins, Myc also interacts with a large number of other proteins (reviewed in Ponzielli et al., 2005).

In the well-studied example of Burkitt's lymphoma, Myc expression is upregulated in the course of a chromosome translocation (Fig. 14.5). In Burkitt's lymphoma, translocation brings the *myc* gene into the vicinity of immunoglobulin genes. The translocation has the consequence of increasing expression of the *myc* gene in comparison to the normal situation. Furthermore, during the course of tumor formation, mutations also occur in the coding region of the *myc* gene. The Myc–Max equilibrium is disturbed by the chromosomal translocation and the associated overexpression of the Myc protein. It is assumed that there is an excessive transcription of target genes due to overexpression of the Myc protein, leading to tumor formation.

Another finding that brought Myc into the focus of tumor therapy is the observation that Myc inactivation in conditional transgenic models can lead to tumor regression due to proliferative arrest, differentiation and/or apoptosis (reviewed in Shachaf and Felsher, 2005). The outcome of Myc inactivation in the model systems appears to depend on tumor type. In some tumors, Myc inactivation leads to sustained regression of the tumors and this is refractory to Myc reactivation. In other tumors however, Myc inactivation induces a state of tumor dormancy by driving the cells into differentiation. These cells have retained the neoplastic properties and upon Myc reactivation tumors are formed again.

■ **Burkitt's lymphoma**
 – Translocation of the c-*myc* gene
 – Overexpression of Myc
 – Inappropriate transcription of target genes

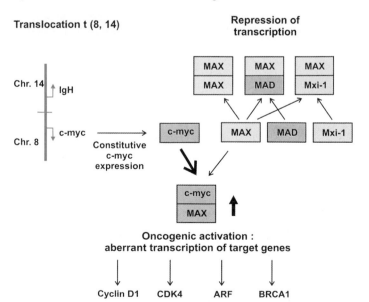

Translocation t (8, 14)

Repression of transcription

Oncogenic activation : aberrant transcription of target genes

Fig. 14.5: The network of the c-Myc–Max transcription factors in Burkitt's lymphoma. Burkitt's lymphoma is characterized by chromosome translocations in which immunoglobulin gene sections (a gene for IgH in our example) are translocated into the region of the c-*myc* gene. As a consequence of the translocation t(8,14), constitutive expression of the c-*myc* gene occurs. The c-*myc* gene codes for the transcription factor c-Myc, which can form homodimers or may associate with the related transcription factors Max, Mad and Mxi-1 to form heterodimers. Constitutive expression of c-Myc shifts the homo-heterodimer equilibrium towards the c-Myc–Max heterodimers. Unprogrammed activation of target genes (genes for cyclin D1, BRCA1, CDK4 and ARF) of the c-Myc–Max heterodimers then occurs.

14.3
Tumor Suppressor Genes: General Functions

■ General functions of tumor suppressor genes
– Negative regulation of cell proliferation
– Often participate in cell cycle checkpoints and in DNA repair

Tumor suppressor genes help to prevent tumor formation by performing a dampening, braking function in signaling pathways that regulate cell proliferation and cell homeostasis (reviewed in Sherr, 2004). The elimination or downregulation of the tumor suppressor function by mutations leads to a preponderance of growth promoting signals and thereby favors tumor formation. Mutations of tumor suppressor genes are often recessive. On the mutation of one allele, the remaining intact allele on the other chromosome continues to perform the growth-suppressing function. Only when both alleles are inactivated does the tumor-suppressing function cease to work. By this property, the inactivation of neighboring marker genes during tumorigenesis *[loss of heterozygosity (LOH)]* has often helped to identify tumor suppressor genes. However, tumor suppressor genes are known, e.g. the cell cycle inhibitor p27^{KIP1}, for which mutation of only one allele promotes tumor formation.

Suppression of tumor formation can be achieved by processes at various levels, including repair checkpoints, cell cycle regulation, apoptosis and processes that are required in the later stages of tumorigenesis, e.g. blood vessel formation. Proper execution of DNA damage checkpoints, dampening and downregulation of cell proliferation and maintenance of apoptotic pathways are all essential for tumor suppression, and the inactivation or alteration of these functions is found in all cancers. As a consequence, proliferation-promoting signals prevail, apoptosis is suppressed and the accumulation of mutations is favored. The genes which function as tumor suppressors are always part of larger signaling networks and are themselves subject to manifold control by posttranscriptional regulations and interactions with up- and downstream effectors. It is therefore not surprising that there are many ways by which the tumor suppressing functions of tumor suppressor proteins can be inactivated or weakened. The major ways operate by reduced expression, structural and functional changes and increased instability of the tumor suppressors.

Reduced expression due to:
– Altered transcriptional regulation
– Changes in promoter structure
– Epigenetic silencing

Structural changes leading to:
– Altered transcriptional regulation
– Changes in phosphorylation and other posttranslational modifications

Tab. 14.3: Characteristics of some tumor suppressor proteins.

Gene, protein	Function
p53	DNA repair, apoptosis
Rb	cell cycle control
NF1, neurofibromin	GAP in Ras signaling
BRCA1, BRCA2	DNA repair, e.g. double-strand breaks
Wt-1	transcription factor with zinc-binding motif
PTEN	phosphatidylinositol phosphate phosphatase, blocking of PI3K signaling
CDH1, E-cadherin	cell adhesion
APC	binds to β-catenin; Wnt signaling
MSH2	repair of DNA mismatches

– Altered subcellular distribution
– Altered interactions with up-and downstream effectors
– Altered enzymatic activity

Increased instability due to changes in Ub-mediated proteolysis.
A selection of well-characterized tumor suppressor proteins is given in Tab. 14.3. Most of the tumor suppressor genes can be categorized according to the functions listed above. However, a number of tumor suppressor genes are known with no direct relationship to the regulation of the cell cycle, repair or apoptosis. Some of the tumor suppressor genes in Tab. 14.3 are involved in the organization of the cytoskeleton or in cell–cell interactions, and appear to be relevant in later stages of tumor formation where tumor cells invade foreign tissues and form organ-like structures. For a review on tumor suppression, see Sherr (2004).

In the following sections, selected tumor suppressors are presented that are found to be mutated in many tumors and to which key functions in tumor suppression are ascribed.

14.4
Rb in Cancer

The Rb gene was the first tumor suppressor gene to be identified and characterized in man. Human genetic investigations of patients suffering from the rare *retinoblastoma eye tumor* showed a defect in a gene sequence known as the Rb gene. In the inherited form of this tumor, which appears in childhood, a defect is inherited in the Rb gene via the germline. Inactivation of the second allele of the Rb

■ **Tumor suppressor
function of Rb**

– Regulates G_1/S tran-
sition via interaction
with E2F

– Rb–E2F interaction
is controlled via
Rb phosphorylation

gene by mutation or deletion leads to complete failure of the function of the gene and thus to tumor formation. Somatic mutations
of the Rb gene are found in more frequently occurring tumors,
including osteosarcomas, lung carcinomas and bladder carcinomas.

The product of the Rb gene, the Rb protein, along with its cousins
p107 and p130, regulates the G_1/S transition, facilitates differentiation and restrains apoptosis, all processes relevant to tumor formation. Its best-characterized function in cell cycle regulation is the

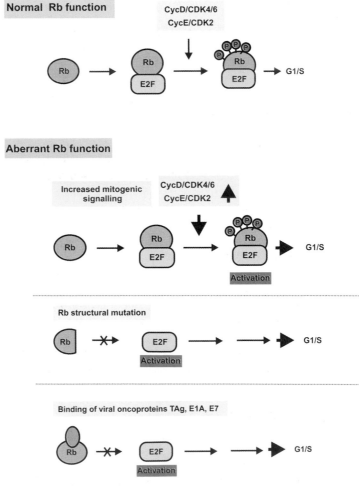

Fig. 14.6: Errors in regulation of
the tumor suppressor protein
Rb. The figure shows a simplified
version of well characterized
mechanisms by which errors in
regulations of the Rb function
can occur.

transmission of mitogenic and antimitogenic signals down to the transcription level (Section 13.5.3) via binding of the E2F transcription factors. This regulatory function can be described roughly by a two-state model. In the active state, Rb is hypophosphorylated and suppresses the transcription of E2F-dependent genes necessary for entry into the S phase. In the inactive state, Rb is hyperphosphorylated and transcription of E2F-dependent genes is possible which allows progression into the S phase. The transition from the active into the inactive state is triggered via phosphorylation of Rb by the G_1 CDK complexes, i.e. CDK4/6–cyclin D1 and CDK2–cyclin E, which are themselves regulated by multiple activating and inhibitory influences. By restraining the cell division activity in its hypophosphorylated state, Rb functions as a tumor suppressor.

Loss of Rb functions weakens these controls, dissociating the cell cycle machinery from extracellular signals, dampening the ability of proliferating cells to exit the division cycle, and compromising the execution of Rb-dependent differentiation programs in certain tissues.

The number of proteins that direct signals to and convey signals from Rb–E2F is very large and functional inactivation of the tumor suppressing function of the Rb network is thus possible in many ways. The major ways of Rb inactivation include (Fig. 14.6):

– *Unprogrammed activation of the G_1 CDKs and cyclins*: via overexpression of cyclin D1.
– *Inactivation of inhibitory, anti-growth signals*. CDK inhibitors (CKIs) inhibit the phosphorylation of Rb and bring about a halt in the G_1 phase. Of the various CKIs, mutations in the inhibitor p27^{KIP1} and, in particular, the p16^{INK4a} protein (Section 13.2.3) are associated with tumor formation. In many tumors, such as lung carcinomas, inactivation of the gene for the p16^{INK4a} inhibitor has been observed, based on mutations or an aberrant C-methylation. Furthermore, alteration of the anti-growth signaling function of the TGF-β pathway also leads to low levels of CKIs and consequently to inactivation of Rb.

Binding of Viral Oncoproteins

DNA viruses that can trigger tumors are found in the classes of the polyoma viruses, the adenoma viruses and the papilloma viruses. The polyoma viruses, with the SV40 virus as a well-studied representative, the adenoma virus and the human papilloma virus (HPV) are associated with the formation of tumors in humans, and have genes coding for proteins with the properties of oncoproteins. The oncoproteins of all three viruses interfere with the Rb function by lifting its inhibition of transcription factor E2F. It is assumed that the tumor-promoting activity of the proteins is due, in particular, to

■ **Tumor suppressor function of Rb may be inactivated via**

– Enhanced CDK–cyclin D/cyclin E signaling
– Loss or inactivation of CKIs
– Binding of oncoproteins
– Mutation of Rb gene
– Epigenetic inactivation of Rb

this property. Interestingly, these proteins also bind to and inactivate the tumor suppressor protein p53. As outlined in Section 14.6, mutant p53 can cooperate with mutant Rb in driving tumor progression.

The oncoproteins are the T antigen *(TAg)* of the SV40 virus, the *E1A* protein of the adenoma virus and the *E7* protein of HPV. The three proteins have in common the ability to bind to the hypophosphorylated form of Rb. In all three cases, binding takes place in the same region of Rb referred to as the "pocket". The transcription factor E2F also binds in the region of the "pocket" of Rb protein. Binding of the viral proteins to Rb is thought to outcompete E2F and to relieve repression, allowing unprogrammed transcription of the E2F target genes.

Genetic Inactivation of Rb

Genetic inactivation of Rb is observed in many tumors. The gene defect may affect the promotor region of the Rb gene, leading to reduced Rb expression, or it may affect the structure of Rb, e.g. by a mutation of the binding site for E2F. The mutations observed in tumors are generally extensive structural changes in the Rb gene.

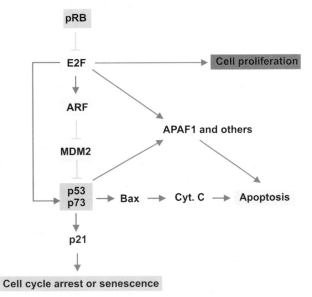

Fig. 14.7: Model of Rb function in control of cell proliferation and apoptosis. Rb controls proliferation and apoptosis through active repression of E2F-dependent promotors. In the underphosphorylated form, Rb has both an antiproliferative and antiapoptotic effect. It inhibits transcription of E2F-controlled genes, among which are genes required for S-phase progression and hence for cell proliferation. E2F-controlled genes also include proapoptotic genes such as the gene for Apaf-1 and p73. The latter can induce, in addition to p53, the gene for the proapoptotic protein Bax which stimulates cytochrome *c* release from mitochondria and thereby initiates apoptosis. E2F also keeps levels of p53 low, an effect mediated by the ARF–MDM2–p53 network (see also Fig. 14.8). If the control function of Rb is lost, e.g. due to Rb mutations, E2F will inappropriately activate the proapoptotic genes and growth arrest and/or apoptosis will result. When at the same time Rb function is lost and key players regulating apoptosis and/or cell cycle arrest (e.g. ARF, p53, Apaf-1) are mutated or overexpressed (MDM2), cells will be directed to hyperproliferation.

Epigenetic Inactivation of Rb

As for many other tumor suppressor proteins, Rb levels can be downregulated by epigenetic silencing of the Rb gene via DNA hypermethylation and altered chromatin modification. Both epigenetic events have been linked to tumor formation in *Drosophila* (Ferres-Marco et al., 2006).

Due to the large number of proteins that can interact directly with Rb or indirectly participate in the Rb network, only part of the tumor-suppressing mechanism of this network is known and only the main, well-studied aspects of the network have been presented here. As an example, Rb appears to interact with more than 50 different proteins involved in transcriptional regulation and the implications of these many interactions for the function of Rb in tumor suppression are largely unknown (reviewed in Zhu, 2005). The Rb network is also involved in the differentiation and senescence of cells (reviewed in Ben-Porath and Weinberg, 2005). How these physiological cell functions are altered by Rb during tumorigenesis is largely unknown.

It is now increasingly clear that a defect in the Rb–E2F pathway has an effect on both cell proliferation and apoptosis. Cell proliferation will increase when a mutated, inactive form of Rb no longer keeps the E2F-controlled genes repressed. In this case, uncontrolled activation of E2F takes place, and an increased supply of gene products necessary for S-phase progression provides a growth advantage for the mutated cell.

Rb and Apoptosis

A function equally important for tumorigenesis is the ability of Rb to protect differentiating cells, which contain high levels of Rb, from apoptosis (Fig. 14.7). The link from the Rb pathway to apoptosis appears to be provided by E2F1, which controls not only genes required for S-phase progression, but also genes involved in the regulation of apoptosis. Among the latter are the genes for the proapoptotic protein Apaf-1 (Section 15.5) and for the tumor suppressor ARF (Section 14.7), which is part of the network that regulates p53 function.

Loss of Rb function – in the absence of other mutations – has been shown by knockout studies in mice to result in increased apoptosis and in growth arrest. If at the same time other key regulators of apoptosis, e.g. the p53 protein or ARF, are mutated and inactivated, apoptosis is reduced and cells are driven to hyperproliferation. *In accordance with this model, nearly all cancers contain mutations both in the Rb pathway and in the pathways that link proliferation with apoptosis.* This example illustrates the intense networking that exists between the various growth- and cell death-controlling pathways, and it underlines the cooperativity of mutations in key regulators of cell survival and cell death.

■ **Rb protects cells from apoptosis via its influence on E2F1**

14.5
p16^{INK4a} Gene Locus and ARF

■ **Tumor suppressor ARF (p14^{ARF})**
– Stabilizes p53
– Binds to MDM2
– Inhibits downregulation of p53 by MDM2

Investigations into tumor cells indicate that proteins coded by the p16^{INK} gene locus function as tumor suppressors and are of great importance for tumor development (reviewed in Sharpless, 2005). The gene locus for p16^{INK4a} codes for two proteins, namely the p16^{INK4a} inhibitor and the ARF protein. Both proteins have a growth-inhibiting function, although with different points of attack. Whilst the p16^{INK4a} protein inhibits the cyclin D-CDK complex and brings about a halt in the cell cycle via the Rb protein, the ARF protein attacks the function of the p53 protein by specifically interacting with the MDM2 protein and interfering with its binding to the p53 protein (Section 14.6.6 and Fig. 14.8). The ARF protein is not homologous to the p16^{INK4a} protein, although both originate from the same gene locus. It arises by alternative splicing and by use of a different reading frame.

An important link exists between the Rb–E2F pathway, since ARF is one of the transcriptional targets of the E2F transcription factors. Furthermore, ARF is under transcriptional control by the transcription factor c-Myc, providing a link from c-Myc to the function of p53.

Fig. 14.8: Regulation of p53 activity by p19^{ARF} and MDM2. A major pathway for downregulating p53 levels uses the MDM2 protein which is a Ub ligase. MDM2 induces ubiquitinylation and degradation of p53. The downregulating function of MDM2 is mainly controlled by the tumor suppressor protein ARF which binds to MDM2 and thereby prevents p53–MDM2 binding and p53 destabilization. ARF transcription can be activated by the major mitogenic and proliferative pathways of the cell, i.e. the Rb–E2F and Ras–MAPK pathways, and by the transcription factor Myc. By this mechanism, proliferative and mitogenic signals keep p53 levels low.

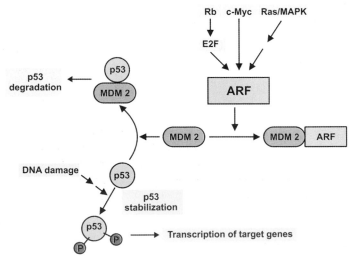

14.6
Tumor Suppressor Protein p53

The most frequently observed genetic changes in human tumors affect the gene for a nuclear phosphoprotein of 393 amino acids, which is known as the p53 protein, after its molecular weight.

Mutations of the p53 gene are observed in over 50% of all human tumors. Defects in the p53 gene in the germline lead to a hereditary tendency to develop various tumors, especially of the connective tissue. The disease is known as Li-Fraumeni Syndrome after its discoverer. Families have been identified that inherit p53 mutations and develop cancers with 100% penetrance (Vogelstein et al., 2000). All these observations place p53 as the main tumor suppressor protein.

The function of the p53 protein is to act as a checkpoint responding to a wide variety of stress signals that can originate from external or internal sources. The p53 protein integrates these signals and activates a signaling network that responds to minimize mutations and other errors that can lead to cancers or other pathologies. Additional oncogenes and tumor suppressor genes reside in the p53 network in which p53 forms a central node that receives many stress signals, and transmit these to signaling pathways controlling cell cycle and apoptosis. Pivotal to the tumor suppressor function of p53 is its ability to prevent cell cycle progression and activate apoptosis in response to genotoxic and nongenotoxic stresses (reviewed in Guimaraes and Hainaut, 2002), thereby preventing damaged DNA from cycling into the next generation.

Inactivation of the p53 network enables the cell to continue in the cell cycle with damaged DNA, yet without DNA repair taking place. Furthermore, failure of the apoptotic control function permits the survival of cells with damaged DNA. Both effects lead to increased susceptibility of the genome to accumulation of further mutations. The cells can also divide under conditions in which serious changes of the genome are present, such as DNA amplification and chromosome rearrangement. Failure of the p53 function cancels a central control element that ensures the integrity of the genome. Therefore, p53 has been said to have the function of a "guardian of the genome".

14.6.1
Structure and Biochemical Properties of p53

The p53 protein is a transcriptional activator that shares many properties with typical eukaryotic transcriptional regulators. It binds site-specifically to its cognate DNA element and regulates the transcription of neighboring genes. In doing so, p53 interacts with and is regulated by a large number of proteins, including proteins of the

■ **Tumor suppressor p53**
- Prevents cell cycle progression
- Activates apoptosis

Inactivation of p53
- Allows cell cycle progression in the presence of DNA damage
- Prevents apoptosis

■ **Properties of p53**
- Transcriptional activator
- Regulated by post-transcriptional modifications

chromatin remodeling and transcription machinery. The genes regulated by p53 can mediate cell cycle arrest and apoptosis, facilitate DNA repair or alter other cellular processes. In addition, p53 performs functions independent of transcription by directly participating in apoptotic pathways (Section 15.7.2). The multiple activities of p53 are strictly regulated by posttranslational modifications that are under control of signaling pathways responding to genotoxic and nongenotoxic stresses.

As shown in Fig. 14.9, distinct domains can be identified in the p53 protein and defined biochemical functions can be assigned to these.

■ **Structural domains of p53**
  – N-terminal domain: transactivation
  – Core domain: DNA binding and tetramerization
  – C-terminal domain: negative regulation

– *N-terminal domain: transactivation and regulation.* The N-terminal domain (residues 1–93) comprises a transactivation domain (residues 1–60) and a Pro-rich regulatory domain (residues 64–92). Different protein-binding sites have been identified in this region. These include binding sites for components of the TFIID complex and for coactivators such as the histone acetylases CBP/p300 or PCAF (Section 3.5).

– *Sequence-specific DNA-binding domain.* The DNA-binding core domain (residues 94–312) includes the binding site for the corresponding DNA element and binding sites for viral oncoproteins such as TAg of SV40 virus. Adjacent to the core DNA binding domain is a *tetramerization domain* (residues 324–355) that is responsible for the reversible association of p53 into tetramers.

– *Negative regulatory C-terminal domain.* The C-terminal domain (residues 360–393) is ascribed a negative regulatory function. It contains several sites for posttranslational modifications, including Ser phosphorylation, Lys acetylation, Lys methylation, ubiquitination, neddylation and sumoylation. Furthermore, sequence signals for nuclear localization and binding sites for transcription factors are found in the C-terminal part.

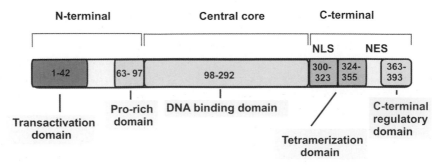

Fig. 14.9: Domain structure of p53. NLS = nuclear localization signal; NES = nuclear export signal.

14.6.2
Structure of p53 and its DNA Complex

Central to the function of the p53 protein is its ability, as a transcription activator, to specifically bind to DNA sequences in the promotor region of various genes and to activate (or sometimes repress) their transcription.

The determination of the structure of full-length p53 has been very difficult due to its inherent instability and tendency to aggregate. Only the core domain and the tetramerization domain are natively folded, whereas the N- and the C-terminal domains are natively unstructured. As shown by nuclear magnetic resonance measurements, the core domains associate to dimers within the tetrameric complex (Veprintsev et al., 2006).

Due to the structural instability of full-length p53, only fragments of p53 could be structurally characterized in complex with DNA. The structure of a single fragment of the central core domain bound to DNA is shown in Fig. 14.10(a).

The structural data together with a large number of sequence determinations for mutated p53 have highlighted the importance of DNA binding for the proper function of p53.

From mutations in the p53 gene identified in tumor patients, a mutation spectrum could be assembled in association with tumor formation. The mutation spectrum shown in Fig. 14.10(b) shows "hotspots" – positions at which p53 mutations are seen particularly frequently in tumor patients. These hotspots cluster in the core domain of p53 responsible for sequence-specific DNA binding. The most frequent mutations either interfere with the DNA binding or lead to a structural destablization of the central core of p53.

14.6.3
Posttranslational Modification of p53

The p53 function is embedded in a finely tuned regulatory network that uses various signaling pathways for the reception of activating signals and directs the p53 response to different downstream effector pathways. In this process, posttranslational modifications of p53 are the major tools for the regulation of p53 function, and these modifications are directed and controlled by a multitude of input signals.

In normal, unstressed cells, p53 is instable and is present only in low amounts. Exposure to genotoxic or nongenotoxic stresses leads to stabilization and concentration increase of p53 allowing it to exert its transcriptional regulatory function. The transition from the instable state to the stable, active state and *vice versa* is accompanied by a series of posttranslational modifications that influence stability,

■ **Structure of p53**
– Tetramer
– Only core domain natively folded
– N- and C-terminal domain are unstructured

■ **Posttranslational modifications of p53**
– 12 phosphorylation sites
– Acetylation
– Methylation
– Ubiquitination
– Sumoylation

(a)

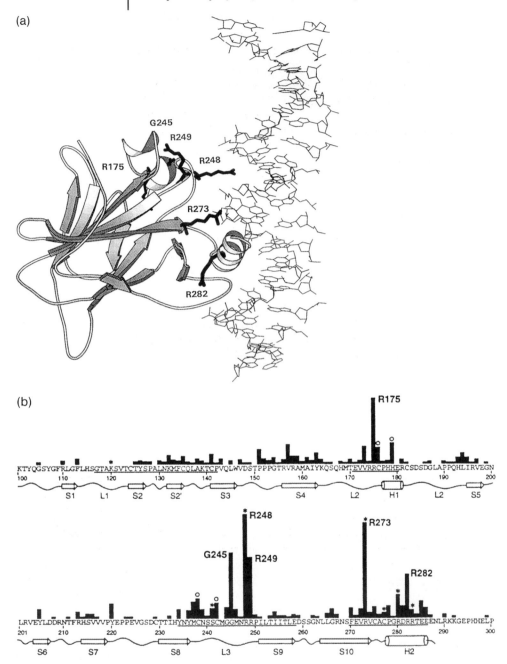

(b)

Fig. 14.10: (a) DNA-binding domain of the tumor suppressor protein p53 in complex with DNA. Crystal structure of the core domain of p53 (amino acids 102–292) in complex with a double-stranded DNA that contains a specific binding site for p53 (Cho et al., 1994). The amino acid positions are highlighted at which frequent oncogenic mutations are observed (b). (b) Mutation spectrum of the p53 protein in tumors. The linear structure is shown of p53 and the frequency of mutations found in tumors. The black bars indicate the approximate position and the relative frequency of the p53 mutations. The frequency of mutations in the region of the DNA-binding domain is of note. The sites of the most frequent mutations coincide with positions of the p53 protein that are directly involved in interactions with the DNA sequence (Fig. 14.9).

subcellular localization, DNA-binding activity and binding to effector proteins of p53 (reviewed in Bode and Dong, 2004).

The following posttranslational modifications of p53 have been identified (Fig. 14.11):

– *Phosphorylation.* Over 12 Ser/Thr phosphorylation sites exist on p53 that stimulate or coincide with transactivation (Saito et al. 2003). Distinct protein kinases are responsible for the phosphorylations and some kinases can phosphorylate several sites. Most of these kinases are activated in response to DNA damage or other stresses. However, only two key biochemical activities could be clearly correlated with specific phosphorylations. Damage-stimulated phosphorylation of p53 at the C-terminal PKR site (Ser392) forms a paradigm whereby sequence-specific DNA binding is stimulated by allosteric mechanisms. Phosphorylation of p53 at its CK2 site increases the intrinsic thermostability of the core DNA-binding domain and enhances the affinity for its consensus DNA site. Phosphorylation of p53 also directly influences its transcriptional activity via binding of the transcriptional coactivator p300 (Section 3.5.2). The protein kinase CHK2 which is part of the DNA damage checkpoint (Section 13.10), phosphorylates p53 at Ser20 and this modification stabilizes the interaction of p53 with the coactivator p300.

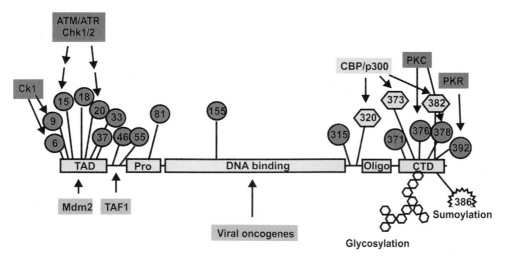

Fig. 14.11: Posttranslational modifications and localization of biochemical functions in p53 protein. Red circles indicate Ser/Thr phosphorylation sites, blue hexagons indicate Lys acetylation sites. Some protein kinases responsible for the phosphorylation of Ser/Thr sites are indicated above; the domain and interaction sites of important regulators of p53 are shown below the domain representation. TAD: transcriptional activation domain; Pro = proline-rich domain; DNA-binding = DNA-binding domain; oligo = oligomerization domain; CTD = C-terminal domain; for ATM/ATR and CHK1/2, see Section 14.8; Ck1 = Casein kinase 1; PKC = protein kinase C; PKR = RNA-dependent protein kinase, see Section 3.6.3.5; for CBP/p300 and TAF1, see Section 3.5.

– *Acetylation and methylation.* A single acetylation site has been identified on Lys392 of p53 and this modification is mediated by the transcriptional coactivator p300. The modification appears to take place after p53 has bound to DNA and may serve to recruit other transcriptional complexes to DNA-bound p53 (Ceskova et al., 2005). Methylation of p53 at a specific lysine residue at the C-terminus is another modification that enhances p53 stability and transcriptional activity (Chuikov et al., 2004).

– *Ubiquitination, neddylation and sumoylation.* Posttranslational modification of p53 includes its conjugation with Ub and the Ub-related proteins NEDD and SUMO (Section 2.5.9). Ubiquitination of p53 mediates its proteasomal degradation and this is a major control of p53 levels. Overall, p53 can be ubiquitinated on six lysine residues. Furthermore, p53 is conjugated at several lysine residues to the Ub-related protein NEDD. Conjugation of p53 with the Ub-related protein SUMO has been also detected. The detailed functions of both modifications remain to be established.

14.6.4
Genes Regulated by p53

The p53 protein functions as a specific transcription activator, but it can also bring about a repression of distinct genes. Genome-wide research for p53 target genes using chromatin immunoprecipitation has identified more than 500 genes probably regulated by p53 (Wei et al., 2006). The large list includes genes with functions in apoptosis, cell cycle regulation, DNA repair, transcription regulation, signal transduction, cell adhesion, cell migration and others. Altogether, nearly all functions of cells in higher eukaryotes relevant for cell proliferation and homeostasis appear to be linked to transcription regulation by p53. Table 14.4 gives a list of selected genes regulated in response to p53 activation.

■ **p53 is a transcriptional regulator that can function as activator or repressor of transcription**

It should be pointed out that proteins related to p53 have been identified that are also part of the p53 regulatory network. Two proteins, named p63 and p73 (reviewed in Ozaki and Nakagawara, 2005), have been found to be "cousins" of p53. Both proteins are activated by similar stresses as p53 and elicit some, but not all, biological responses of activated p53.

The following selected examples of p53-regulated genes illustrate targets of p53 for which functions in cell proliferation and homeostasis have been well characterized.

Tab. 14.4: Examples of genes and proteins regulated in response to p53 activation.

Target genes	Biological function
p21^{CIP1}	Cell growth arrest
GADD45	
14-3-3σ	
Cyclin A	
BAX	Apoptosis
Bcl-2	
AIP1	
CD95 (Fas)	
NOXA	
PUMA	
XPB and XPD	DNA repair
Rad51	
TFIIH p62 subunit	
TSP1	Angiogenesis
COX2 (cyclooxygenase 2)	Stress responses
iNOS (nitric oxide synthase II)	
GPX (glutathione peroxidase)	
MDM2	Retrocontrol, G$_2$/M

Genes Involved in p53 Control: MDM2

Of the genes activated by p53, the *mdm2 gene* stands out as a crucial regulator of p53. Its encoded protein is a *E3-Ub ligase* that directs the ubiquitination and proteasomal degradation of p53. By this property, MDM2 is part of a negative feedback loop most important for regulation of p53 activity (Section 14.6.5).

Genes Involved in Coupling of DNA Damage to Cell Cycle Arrest

The products of these genes have a suppressing effect on cell cycle progression in case of DNA damage, and their levels are increased upon p53 activation. Loss of p53 function will lower the threshold for cell cycle arrest and will allow cell cycle progression e.g. in the presence of damaged DNA.

- *p21^{CIP1}*. The most prominent example is the CDK inhibitor p21^{CIP1}. Activation of p53 leads to increased formation of the p21 inhibitor, which brings about a halt in the cell cycle at G$_1$/S and at G$_2$/M.
- *GADD45*. The GADD45 protein has important functions in the control of DNA damage checkpoints and DNA repair processes (reviewed in Zhan, 2005). Multiple proteins have been identified that interact with GADD45 including the clamp loader PCNA and the inhibitor p21^{CIP1}. PCNA functions as a clamp for DNA

■ **Important target genes of p53**
- E3 ligase MDM2
- p21^{CIP1} inhibitor
- 14-3-3σ
- GADD45
- Many transcription factors

polymerases during DNA synthesis and ensures processivity of DNA replication.

– *14-3-3σ.* This protein has an important function in the G_2/M DNA damage checkpoint. It serves to sequester the mitotic initiation complex, CDK1–cyclin B1, in the cytoplasm after DNA damage. Levels of 14-3-3σ have been found to be low in many tumors, including breast tumors.

Transcription Repression

The p53 protein has – in addition to the activation of specific genes – also a *general repressing influence on transcription*. The repression is observed for various cellular and viral genes that have no p53-binding site. Examples of genes repressed by p53 are the genes for transcription activators c-*jun* and c-*fos*, the cytokine interleukin-6, the Rb protein and the *bcl-2* gene (Chapter 15). The mechanism for this repressive effect remains to be established.

Genes Involved in Apoptosis

■ **Apoptotic genes regulated by p53**
– Bax
– NOXA
– Death receptor: Fas
– PUMA

Many genes involved in the intrinsic and extrinsic pathways of apoptosis have been identified as transcriptional targets of p53 (Yu and Zhang, 2005). Examples include the genes for death receptors, the gene for the proapoptotic protein Bax, and for the apoptotic regulators NOXA and PUMA, all of which are upregulated by activated p53. On the contrary, the gene for the antiapoptotic protein Bcl-2 (Section 15.4), is repressed by p53. These effects show *that p53 promotes apoptosis by enforcing the proapoptotic signals* and *blocking the antiapoptotic signals*.

Stress Responses

Various stresses (oxidative stress, hypoxia, ribonucleotide depletion, etc.) lead to p53-mediated activation of genes involved in stress responses. Among these are many genes that can generate or respond to oxidative stress.

Angiogenesis

The thrombospodin-1 gene codes for a protein that inhibits new formation of blood vessels (angiogenesis). The p53 protein activates expression of the thrombospodin-1 gene and can thus suppress angiogenesis. If the regulating activity of the p53 protein is inactivated, a situation is created which facilitates new formation of blood vessels, since the inhibitor of angiogenesis is missing. It is assumed that this situation promotes tumor progression, especially in the late phase of tumor formation.

The large list of effectors of p53 activation shows that it controls a variety of pathways involved mostly in growth control and cell survival. The exact combination of effectors activated by p53 at the transcriptional level may differ from cell to cell depending upon the nature and intensity of the inducing signal and the cell type. Furthermore, the activation of other pathways contributing to the control of cell proliferation should also be considered.

14.6.5
Regulation of p53 by Ubiquitination and Proteasomal Degradation: The MDM2 Protein

In unstimulated cells, p53 levels are low and p53 is unstable. The main factor contributing to p53 downregulation in the absence of stresses comes from ubiquitination and proteosomal degradation of p53 (reviewed in Yang et al., 2004; Levav-Cohen et al., 2005). The ubiquitination of p53 is mediated by several enzymes with E3-Ub ligase activity, of which the MDM2 protein has been characterized first and is considered as a major inhibitor of p53.

The MDM2 protein (reviewed in Levav-Cohen et al., 2005) is a RING finger E3 enzyme that binds to the N-terminal domain of p53 and mediates its ubiquitination at up to six lysine residues, which efficiently and rapidly blocks p53 signaling. Monoubiquitination of multiple lysine residues promotes the nuclear export of p53 thereby blocking any transcriptional activity, while polyubiquitination of p53 provides a signal for degradation, in the nucleus and/or cytoplasm. MDM2 protein levels determine the extent of this p53 ubiquitination, low levels inducing monoubiquitination, whereas polyubiquitination prevails at high MDM2 levels.

Two major ways have been identified that relieve the inhibitory action of MDM2 on p53:
- *Downregulation of MDM2 by the tumor suppressor ARF.* One of the products of the p16 gene locus, the tumor suppressor protein ARF binds to MDM2 and inhibits its ability to ubiquitinate p53.
- *Phosphorylation of p53.* Stress induced phosphorylation of p53 inhibits binding of MDM2 to p53 leading to its stabilization.

■ **MDM2**
- Negatively regulates p53
- E3 ligase
- Mediates Ub modification pf p53
- Under transcriptional control by p53 in negative feedback loop

■ **Regulation of MDM2**
- binding of MDM2 to ARF prevents ligase action
- phosphorylation of p53 prevents MDM2 binding

Regulation of MDM2 Transcription by p53
One of the most intriguing aspects of the interaction between p53 and MDM2 is that the *mdm2* gene is a direct transcriptional target gene of p53. Shortly after p53 activation, MDM2 is transcribed and accumulates to neutralize and extinguish p53 activities. Thus, p53 transcribes its own executor and MDM2 blocks its own transcription (Fig. 14.12). It is inevitable that in response to stress this loop has to be disrupted so as to allow p53 to accumulate and be activated sufficiently, in order to exert its growth inhibitory activities.

p53 and MDMX

A relative of the MDM2 protein exists, which is named MDMX (reviewed in Marine and Jochemsen, 2005). The MDMX protein binds to p53 and inhibits its transcriptional function; however, it does not promote p53 ubiquitination or degradation. Relieve of MDMX-mediated p53 inhibition is achieved by phosphorylation of MDMX catalyzed by the damage-induced protein kinases ATM and CHK2. This phosphorylation promotes ubiqutination and degradation of MDMX by MDM2.

■ **MDMX binds to p53 and inhibits its transcriptional function**

In addition to MDM2, other E3-Ub ligases have been identified that contribute to p53 downregulation as well. These include the *E3 ligases Pirh1 and COP1*. How the various ubiqutiniation events catalyzed by different E3 enzymes cooperate and how these modifications are modulated by modification of p53 with Ub-like proteins (NEDD, SUMO) is largely unknown.

■ **Other E3 ligases involved in p53 ubiquitination**
– Pirh1
– COP1

Overall, MDM2-mediated destabilization of p53 is part of a complex network, referred to as the MDM2–p53 network, that can be disturbed or inactivated in many ways contributing to tumor formation.

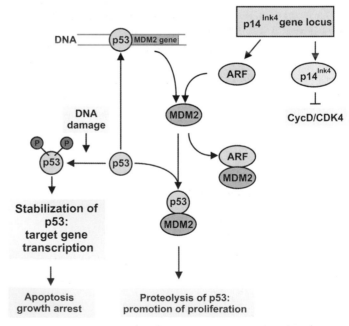

Fig. 14.12: The MDM2–p53 network. The level of p53 is strongly regulated by its interaction with MDM2. Binding of MDM2 to p53 targets p53 for proteolytic degradation, thus keeping the p53 concentration low. DNA damage induces phosphorylation of p53. The phosphorylated p53 is no longer bound by MDM2, proteolysis is decreased, and the rising p53 levels initiate apoptosis and lead to growth arrest. A negative feedback loop exists between p53 and MDM2 since p53 controls the expression of MDM2 at the level of transcription. Another control is exerted by the tumor suppressor p19^{ARF}, which is coded for by the p16^{INK4} gene locus. p19^{ARF} binds to MDM2 and sequesters it from the feedback loop, thus helping to increase p53 concentration. The proapoptotic and antiproliferative function of p53 may be impaired by mutation of p53, by overexpression of MDM2 or by a loss of p19^{ARF} function. The p16^{INK4} gene locus is of specific importance in this network because it also codes for the inhibitor p16^{INK4} which inhibits the cyclin D–CDK4 complexes.

14.6.6
Pathways Involved in Activation of p53

The p53 protein is present in almost all tissues in low concentrations and in an inactive, repressed state. Several types of stresses can lead to its accumulation and activation by posttranslational modifications. Activation of p53 serves to protect the cells against the potentially disastrous consequences of stresses and occurs as a normal mechanism of defense against neoplastic transformation of the cell. The stresses that activate p53 can be broadly divided into three classes (Fig. 14.13):

- *Genotoxic stress.* DNA damage by ultraviolet light, X-rays, carcinogens, cytotoxic drugs, etc.
- *Oncogenic stress.* Aberrant activation of growth factor-signaling cascades.
- *Nongenotoxic stress.* Metabolic stress like hypoxia, depletion of ribonucleotides.

■ **Stresses that lead to p53 stabilization and activation**
– Genotoxic stress
– Oncogenic stress
– Nongenotoxic stress

These stresses have been shown to initiate the typical p53 response, which can cause either apoptosis, growth arrest, altered DNA repair or altered differentiation. Which of these responses dominates will depend on the cell type and the type and duration of the stress.

Broadly, pathways leading to p53 activation and response can be categorized according to the initiating stresses:

- *Activation of p53 via DNA damage checkpoints.* Genotoxic stress, e.g. as manifested by the formation of DNA adducts and DNA strand breaks, activates p53 mostly via DNA damage checkpoints. It is the main purpose of this control to prevent replication of damaged DNA, which is a potentially mutagenic process. The links between

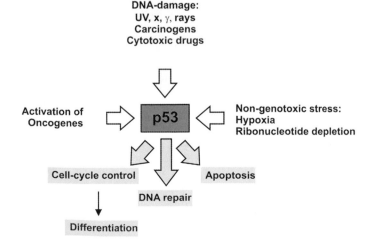

Fig. 14.13: Activation of the p53 pathway. The three main classes of signals activating p53 are shown.

DNA damage and p53 activation are summarized schematically in Fig. 14.14. Upon DNA damage, a cascade of protein kinases is activated that catalyzes the phosphorylation of p53 on specific Ser residues, which, in turn, influence the interaction of p53 with the MDM2 protein. Specifically, p53 is phosphorylated on Ser 15 by the ATM and ATR kinases and on Ser20 by the CHK2 protein. All three kinases are part of DNA damage checkpoint pathways (Section 13.10), and this phosphorylation inhibits binding of the MDM2 protein, leading to reduced degradation and higher steady-state concentrations of p53. The DNA damage-activated kinases enforce activation of p53, furthermore, by phosphorylating the MDM2 proteins, which leads to enhanced Ub-dependent degradation of MDM2.

– *Activation of p53 by oncogenic stress.* The aberrant activation of protooncogenes creates a stress situation which is also termed oncogenic stress. Normal cells are protected against the detrimental consequences of oncogenic stress by linking oncogenic pathways to the proapoptotic action of the p53 pathway. Aberrant activation of protooncogenes to oncogenes results in activation of the p53 pathway and allows for the destruction of the cell by apoptosis. The main target of regulatory inputs during oncogenic

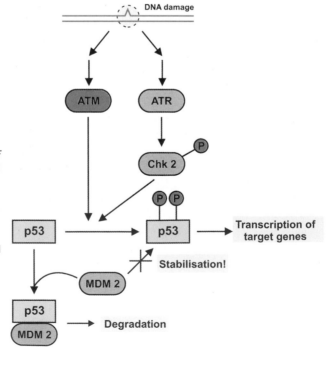

Fig. 14.14: Activation of p53 by DNA damage. DNA damage provides a signal for activation of the protein kinases ATR and ATM which phosphorylate p53 either directly of via another protein kinase, named CHK2. In the unphosphorylated form, p53 is unstable due to binding of MDM2 which has Ub ligase activity, and induces ubiquitinylation and destruction of p53. The damage-induced phosphorylation of p53 prevents MDM2 binding and degradation leading to p53 accumulation and transcription of p53 target genes.

stress is the MDM2 protein, which is downregulated under excessive survival signals, allowing for the accumulation and activation of p53 and subsequently the initiation of apoptosis.

In this pathway, the ARF protein has a key function (Figs. 14.8 and 14.12). The *ARF protein binds to MDM2* and thereby reduces p53 degradation. Various signals induce an increase in ARF protein, including transcriptional activation via the Rb–E2F pathway, the Ras–MAPK pathway and the Myc transcription factor. As already pointed out in earlier chapters, these pathways receive and transmit a multitude of survival signals. By indirectly stabilizing p53, ARF plays a key role in eliminating cells that develop proliferative abnormalities, thereby protecting the organism from cancer development.

In cells like stem cells with ongoing proliferation, however, the activation of the p53 response must be dampened to allow normal growth and development while retaining the capacity for induction of the response to stress associated with oncogenesis. One mechanism by which the p53 response is suppressed in normal cells appears to use *phosphorylation of MDM2 by the Akt kinase*. The Akt kinase is activated by growth factor receptor pathways and transmits survival signals. Akt-mediated phosphorylation of MDM2 has been shown to promote nuclear localization of MDM2, increasing nuclear levels of MDM2 and promoting inhibition of the p53 response. Although an increasing number of key players of the regulatory network with the p53–MDM2 module at center stage have been identified, the quantitative aspects of this network and the cooperation of its components remain to be elucidated.

– *Activation of p53 by nongenotoxic stress.* Nongenotoxic stresses like ribonucleotide depletion or under- or oversupply with oxygen can activate the p53 response without participation of the DNA damage checkpoints. The link between these stresses and p53 appears to be provided by a subspecies of the c-Jun N-terminal kinase (JNK2), which is activated by various stresses and enhances p53 stability by phosphorylation on Thr81. Overall, however, this pathway is only poorly characterized.

– *Binding of viral oncoproteins.* We know of several oncoproteins that can interfere with p53 functions. The oncoprotein of the SV40 virus, TAg, binds to the p53 protein and can inactivate the p53 function in a similar way to that assumed for inactivation of the Rb protein (Section 14.4). Furthermore, the oncoprotein E6 from human papilloma virus can downregulate p53 by inducing its proteasomal degradation (Section 2.5.5).

■ **DNA damage and p53 activation**
– DNA damage activates sensor kinases ATR, ATM and CHK2 that phosphorylate and stabilize p53

■ **Oncogenic stress activates p53 via ARF and MDM2**

■ **Viral oncoproteins may bind to and inactivate p53**

14.6.7
The MDM2–p53 Network and Cancer

A functional inactivation of the MDM2–p53 network (Figs. 14.8 and 14.12) is observed in nearly all cancers. Inactivating mutations have been found to affect both the proliferative and the antiproliferative functions of the network, leading to increased cell division activity and to diminished apoptosis.

Inactivation of the p53 network can occur at many positions. Mutation of p53 itself is observed in the majority of tumors. These mutations are mostly nonsense mutations and affect predominantly the DNA-binding region of p53 and its stability, impairing its transcriptional control function. Loss of the p53 response can also occur via inactivation of other components of the MDM2–p53 network. *MDM2 overexpression and amplification of the mdm2 gene* is observed in many tumors, leading to enhanced degradation of p53. Another way to inactivate the p53-MDM2 module uses the ARF protein. A decreased level of ARF, which has the characteristics of a tumor suppressor, is frequently observed in tumor cells. This is often due to downregulation of ARF by epigenetic processes. Aberrant CpG methylation of promotor sequences at the $p16^{INK4b}$ gene locus leads to silencing of the *ARF* gene and thus to inactivation or impairment of the MDM2–p53 network.

■ **Functional inactivation of the MDM2–p53 network may occur by MDM2 overexpression and/or amplification of the *mdm2* gene**

14.7
Wnt/β-Catenin Signaling and the Tumor Suppressor APC

The Wnt signaling pathway is involved in many developmental processes. The Wnt genes, which code for secreted lipoproteins, can initiate several pathways of which the canonical Wnt/β-catenin pathway has been characterized best. This pathway is a critical regulator of stem cells and its aberrant activation is associated with carcinogenesis (reviewed in Doucas et al., 2005). The pathway harbors two proteins with tumor suppressing activity, the *APC protein* and *axin*, as well as an oncogenic protein, *β-catenin*.

Inactivation of the gene for the tumor suppressor protein APC has been found to be one of the early events during the development of most colorectal cancers. The *APC* gene encodes a multifunctional protein involved in central biological processes, including cell adhesion and migration, proliferation, apoptosis, and differentiation (reviewed in Giles et al., 2003). Most functions of the APC protein are linked to the Wnt/β-catenin signal transduction pathway (Fig. 14.15), which leads from the cell surface to the level of transcription. In this pathway, signals are transduced from the extracellular

■ **Tumor suppressor APC**
– Regulates cell proliferation, cell adhesion, apoptosis and cell differentiation
– Involved in colon cancer formation
– Cooperates with axin and β-catenin

signaling protein Wnt to the transcriptional regulator β-catenin, which controls the expression of numerous target genes, including the gene for cyclin D1 and the transcription factor Myc.

β-Catenin may exist in a membrane-bound form in complex with the cell-adhesion molecule E-cadherin or in a cytosolic form. In the absence of a Wnt signal, β-catenin is assembled in a cytosolic multiprotein complex composed of APC, glycogen synthase kinase 3 (GSK3) and the scaffolding protein axin. A sequential phosphorylation of β-catenin by casein kinase I and by GSK3 at Ser/Thr residues

■ **Wnt/β-catenin signaling**
– Delivers signals from the cell membrane to the level of transcriptional
β-Catenin
– Transcriptional regulator under control of Wnt

(a)

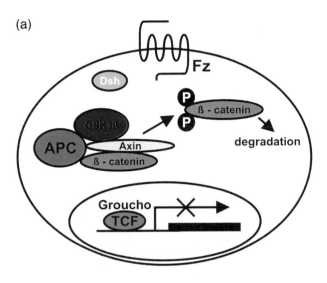

(b)

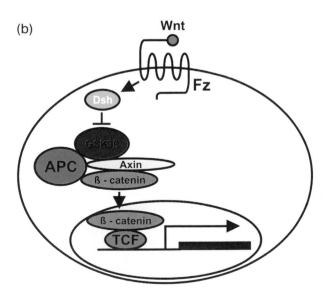

Fig. 14.15: Schematic overview of the Wnt signaling pathway. Central to Wnt signaling is a protein complex composed of β-catenin, APC, GSK3 and axin. a) In the absence of the extracellular signaling protein Wnt, β-catenin is phosphorylated by GSK3. Upon phosphorylation, β-catenin is primed for ubiquitinylation and subsequent degradation by the proteosome. Reduced levels of β-catenin permit repression of Wnt target genes by association of transcriptional corepressors like Groucho. b) When a Wnt signal reaches a cell, Wnt associates with the TM receptor Frizzled (FZ) leading to activation of the cytoplasmic protein Dishevelled (DSH) by an unknown mechanism. Dishevelled inhibits β-catenin phosphorylation and degradation, and β-catenin now acts as transcriptional coactivator for the TCF family of transcription factors which activates target genes that regulate diverse cellular responses. Levels of β-catenin can be also regulated by p53 and by integrins.

marks β-catenin for ubiquitination and proteasomal degradation. In the presence of Wnt, β-catenin phosphorylation and degradation is inhibited, and β-catenin is available for transcriptional regulation. This signal transduction is initiated at the plasma membrane by two distinct receptors, the Frizzled receptor and the LRP receptor, which together may form a Wnt-induced signaling complex (reviewed in Cadigan et al., 2006). Frizzled belongs to the family of seven-helix transmembrane (7TM) receptors and LRP is a single-pass TM receptor belonging to the class of low-density lipoprotein receptor-related proteins.

The mechanism by which this receptor pair initiates signaling upon Wnt binding remains to be understood. Overall, Wnt binding uncouples β-catenin from the degradation complex and allows translocation of β-catenin into the nucleus, where it activates its target genes in cooperation with the transcription factors T-cell factor (TCF)/lymphocyte enhancer factor (LEF). Phosphorylation of LRP and another protein, named Dishevelled (DSH) is involved in this process.

Loss of APC function leads to uncontrolled activation of the Wnt/β-catenin pathway and provides a proliferative advantage to the mutated cell. In addition, chromosomal instability is observed in cells with decreased APC function. The scaffold protein axin has been identified as a further tumor suppressor in this pathway and β-catenin can be activated to an oncogene by mutation or aberrant expression. There is also a link to p53 function which downregulates β-catenin levels in response to DNA damage. Overall, aberrant function of the Wnt/β-catenin pathway is observed in nearly all colorectal cancers as well as in many other cancers (reviewed in Kikuchi, 2003).

14.8
Genomic instability and Tumor Formation:
Roles of DNA Repair and DNA Damage Checkpoints

Importance of DNA repair and cell cycle checkpoints for tumor formation
– Tumors have a mutator phenotype due to ineffective DNA repair

It is now well recognized that cancer cells harbor a large number of genetic and epigenetic changes that distinguish them from their progenitor cells. Moreover, different mutated cell types often cooperate in forming a full-blown cancer, making the cancer its own biological entity. The acquisition of the many genetic changes cannot be due to normal mutation rates. Rather, enhanced mutation rates due to malfunction of DNA repair components and DNA damage checkpoints appears to pave the way towards the phenotype of a cancer cell. The observation of enhanced mutation rates in cancer cells have led to the theory that, in early stage of tumor formation, a mutation occurs in a repair system or a DNA damage checkpoint component needed to maintain the integrity of the genome (Loeb, 1998; Bielas et al., 2006). Loss of the function of the DNA damage response may lead to a *mutator phenotype* – a missing or ineffective DNA repair favors further accumulation of mutations and leads to an intrinsic instability of the genome. Cells contain highly efficient systems to maintain genomic integrity and malfunction of specific components of these genomic "caretakers" has been evoked to explain the increased mutability of cancer cells.

Cells with an intact DNA damage response frequently arrest or die in response to DNA damage, thus reducing the likelihood of progression to malignancy. Mutations in apoptosis, DNA-damage responses or in mitotic-checkpoint pathways, however, can permit the survival or the continued growth of cells with genomic abnormalities, thereby enhancing the chance of malignant transformation (reviewed in Kastan and Bartek, 2004; Dapic et al., 2005).

However, genetic instability caused by altered DNA-damage response pathways may not be sufficient to lead to cancer development. Often, cooperating mutations must be present to facilitate continued growth or viability of premalignant cells. Altogether, many studies suggest that the combination of genomic instability and cell-cycle checkpoint defects is a significant risk factor for tumor development and we know of many cancer syndromes that are linked to defects in DNA-damage responses.

Selected examples will be presented in the following to illustrate the importance of DNA repair and DNA checkpoint systems for tumorigenesis.

Mismatch Repair and hMSH2

For inherited forms of a certain form of bowel cancer *[hereditary non-polyposis cancer (HNPPC)]*, it has been observed that there is an error in the function of the repair system for DNA mismatches. Patients with HNPPC have inherited a defect in the hMSH2 gene in their germ cells, and their tumor cells have a further mutation in the hMSH2 gene. The hMSH2 gene is a homolog of the MutS gene in *Escherichia coli* and its gene product is involved in the repair of DNA mismatches. The defect in the mismatch repair is responsible for a type of genetic instability most readily observed at the level of microsatellite DNA. This DNA harbors a large number of repeat sequences, and defects in mismatch repair lead to easily detectable changes in the number and sequence of the repeats.

■ **Tumor suppressor hMSH2**
– Involved in DNA mismatch repair

BRCA Genes

Two genes with tumor-suppressing function, BRCA1 and BRCA2, are known that mediate a hereditary susceptibility to breast cancer (reviewed in Turner et al., 2005). Inactivation of BRCA1 function

■ **Tumor suppressor BRCA1
is involved in**
– DNA recombination
– DNA damage
checkpoints
– Transcriptional
regulation
**Tumor suppressor BRCA2
is involved in**
– Double-stranded break
repair of DNA

can occur by mutation of the gene as well as by epigenetic silencing due to promotor hypermethylation. The latter is frequently observed in sporadic breast cancers.

BRCA1. The product of the BRCA1 gene is a large protein of 1883 amino acids which harbors functions related to DNA repair, DNA replication and transcriptional regulation. A large number of proteins could be shown to interact with BRCA1, and both scaffolding and enzymatic activities have been reported for BRCA1. The following processes have been linked to the BRCA1 protein:

– *DNA recombination.* BRCA1 interacts specifically with the Rad51 protein which catalyzes strand exchange during recombination processes. By this property, BRCA1 is assumed to participate in DNA double-strand break repair.

– *DNA damage checkpoint.* BRCA1 is thought to function as a scaffold for the ATM and ATR kinases facilitating phosphorylation of downstream targets, e.g. the protein kinase CHK1. Furthermore, BRCA1 is involved in the intra-S and G_2/M checkpoints. Upon DNA damage, BRCA1 is phosphorylated by the ATM and ATR kinases.

– *Transcriptional regulation.* BRCA1 is found as part of chromatin remodeling complexes and is involved in transcriptional regulation. In this context, BRCA1 has been found to interact with histone acetylases.

– *Ubiquitin-mediated proteolysis.* The N-terminus of BRCA1 associates with the BRCA1-associated RING domain-1 protein (BARD1) to form a heterodimer. BARD1 exhibits Ub ligase activity that is abrogated by known cancer-associated BRCA1 missense mutations.

BRCA2. This protein also interacts with Rad51 and it is assumed to participate in the biological response to DNA damage, mainly double-strand break repair of DNA (reviewed in Rudkin and Foulkes, 2005).

14.9
Common Physiologic Changes in Tumor Cells: Hallmarks of Cancer

There are more than 100 distinct types of cancer and various subtypes of cancers can be found within specific organs. The vast catalog of cancer cell genotypes has been suggested to be a manifestation of six essential alterations in cell physiology that collectively dictate malignant growth (Hanahan and Weinberg, 2000) (Fig. 14.16):
– Self-sufficiency in growth signals.
– Insensitivity to growth-inhibitory (antigrowth) signals.
– Evasion of programmed cell death (apoptosis).

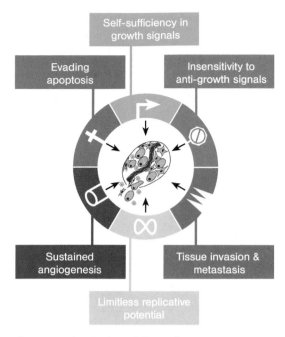

Fig. 14.16: Acquired capabilities of cancers.

– Limitless replicative potential.
– Sustained angiogenesis.
– Tissue invasion and metastasis.

Each of these physiologic changes represents the successful breaching of an anticancer defense mechanism hardwired into cells and tissues and is the result of a long selection process. The linkage of the hallmarks of cancer to dysregulation of signaling pathways and cell cycle checkpoints will be discussed shortly in the following for the first four of the above-mentioned hallmarks of cancer. The other two key features of cancer cells, the ability to form blood vessels and to survive in a foreign surrounding, are less well understood and will not be discussed here.

14.9.1
Self-sufficiency in Growth Signals

Normal cells require mitogenic growth signals before they can progress through G_1 into S or move from a quiescent state into an active proliferative state. The pathways that transmit mitogenic signals are manifold and have been outlined in Chapters 8–12. Of outstanding importance for proliferation regulation are the pathways that lead from the cell surface to cell cycle control. There are many examples that cancer cells acquire a state that is no longer dependent on external mitogenic signals mediated by these pathways. Tumor cells can generate many of their own growth signals, thereby reducing their dependence on stimulation from their normal tissue microenvironment. Three common molecular strategies for achieving autonomy are evident involving alteration of:

– *Extracellular growth signals.* Cancer cells can acquire the ability to synthesize the growth factors to which they respond. This creates a positive feedback signaling loop called autocrine stimulation (Fig. 14.3).

– *TM receptors.* Growth factor receptors with intrinsic or associated tyrosine kinase activity are overexpressed in many tumors often eliciting ligand-independent signaling. Ligand-independent signaling can also be achieved through structural alteration of receptors, e.g. truncated versions of the EGFR lacking much of its cytoplasmic domain fire constitutively.

– *Intracellular signaling networks.* The translation of extracellular stimuli by intracellular signaling networks is often modified in cancer cells in a way that leads to growth factor autonomy. Alterations of the intracellular proliferation-promoting signaling paths may affect both the activating components as well as the components that have a dampening or braking effect on signal transmission. Here, a central role is ascribed to the Sos–Ras–MAPK pathways, the integrin pathways and the PI3K pathway. Among the most frequently mutated genes in cancers are the genes for the Ras protein, for the B-Raf protein kinase, regulatory subunit of PI3K, p110, for the non-RTKs Abl and Src, and for the transcription factor Myc.

14.9.2
Insensitivity to Antigrowth Signals

Within a normal tissue, multiple antiproliferative signals cooperate to maintain cellular quiescence and tissue homeostasis; these signals include both soluble growth inhibitors and immobilized inhibitors embedded in the cell membrane.

These growth-inhibitory signals, like their positively acting counterparts, are received by TM cell surface receptors coupled to intracellular signaling networks. At the molecular level, many and perhaps all antiproliferative signals are funneled through the Rb protein and its two relatives, p107 and p130. When in a hypophosphorylated state, Rb blocks proliferation by shutting down the E2F-dependent expression of genes essential for progression from G_1 into the S phase (Section 13.5.4).

Many of the antimitogenic signals that keep Rb in the active, hypophosphorylated state are mediated by the TGF-β pathway (Section 12.1) that leads – among others – to the increased expression of the gene for the inhibitor p15^{INK4B} and the subsequent inhibition of the G_1 CDKs (CDK4/6 and 2) that are responsible for inactivation of Rb. The TGF-β pathway has been found to be disrupted in a variety of ways in different types of human tumors. Aberrant TGF-β signaling in tumor cells is achieved, for example, by downregulation of TGF-β, mutation of the TGF-β receptor and loss of SMAD proteins through mutation of its encoding genes. Deletion of the gene for p15^{INK4B} and amino acid substitutions that make CDK4 unresponsive to the inhibitory actions of p15^{INK4B} are further lesions found in cancer cells.

Finally, functional Rb, the end target of this pathway, may be lost through mutation of its gene. Alternatively, in certain DNA virus-induced tumors, notably cervical carcinomas, Rb function is eliminated through sequestration by viral oncoproteins, such as the E7 oncoprotein of human papillomavirus.

The bottom line is that the antigrowth circuit converging onto Rb and the cell division cycle is, one way or another, disrupted in a majority of human cancers, defining the concept and a purpose of tumor suppressor loss in cancer.

14.9.3
Evasion of Programmed Cell Death (Apoptosis)

The ability of tumor cell populations to expand in number is determined not only by the rate of cell proliferation, but also by the rate of programmed cell death (apoptosis) (Chapter 15), and there are many examples supporting the notion that apoptosis is a major barrier to cancer that must be circumvented. The evidence is mounting that acquired resistance toward apoptosis is a hallmark of most and perhaps all types of cancer.

A regulatory network exists that links proliferation-promoting programmes to the apoptotic machinery. This is illustrated by the observation that activation of oncogenes, e.g. the Ras protein, can trigger apoptosis and elimination of cells bearing activated oncogenes by apoptosis may represent the primary means by which such mutant cells are continually culled from the body's tissues.

Resistance to apoptosis can be acquired by cancer cells through a variety of strategies. The most common strategy involves the mutation of the *p53* tumor suppressor gene. The resulting functional inactivation of the p53 protein, is seen in greater than 50% of human cancers and results in the removal of a key component of the DNA damage sensor that can induce the apoptotic effector cascade. Other mechanisms use mutation of components of the PI3K pathway that transmits both proliferation-promoting and antiapoptotic signals (Sections 7.4 and 13.9). The antiapoptotic effect of the PI3K pathway is mediated via enhanced expression of the antiapoptotic protein Bcl-2. Overactivation of this pathway, e.g. loss of the PTEN tumor suppressor, will enhance the antiapoptotic effects and will strengthen the survival effects of the PI3K pathway.

14.9.4
Limitless Replicative Potential of Cancer Cells

The acquired disruption of cell–cell signaling, on its own, does not ensure expansive tumor growth. Perhaps all types of mammalian cells carry an intrinsic, cell-autonomous program that limits their multiplication. This program appears to operate independently of the cell–cell signaling pathways described above. The early work of Hayflick demonstrated that cells in culture have a finite replicative potential (reviewed in Hayflick, 1997). This is illustrated by observations of cultured cells indicating that normal human cells have the capacity for 60–70 doublings. Provocatively, most types of tumor cells that are propagated in culture appear to be immortalized, suggesting that limitless replicative potential is a phenotype that was acquired *in vivo* during tumor progression and was essential for the development of their malignant growth.

One feature critical to replicative senescence appears to be the length of specific sequences at the ends of the chromosomes – the telomeres. Normal cells progressively lose telomere sequences due to the inability of the replication machinery to completely replicate the 3′-ends of chromosomes during the S phase. The progressive erosion of telomeres through successive cycles of replication eventually causes them to lose their ability to protect the ends of chromosomal DNA. Telomere maintenance is evident in virtually all types of malignant cells; 85%–90% of them succeed in doing so by upregulating expression of the telomerase enzyme, which adds hexanucleotide repeats onto the ends of telomeric DNA.

An essential role of telomerase in immortalizing cells has been demonstrated in numerous experiments and telomere maintenance is now considered a key component of the capability for unlimited replication, which is another hallmark of cancer cells.

14.10
References

Adhikary, S. and Eilers, M. (2005) Transcriptional regulation and transformation by Myc proteins, *Nat. Rev. Mol. Cell Biol.* **105**, 635–645.

Akiyama, T. and Kawasaki, Y. (2006) Wnt signalling and the actin cytoskeleton, *Oncogene* **105**, 7538–7544.

Bazley, L. A. and Gullick, W. J. (2005) The epidermal growth factor receptor family, *Endocr. Relat Cancer* **105**, S17–S27.

Ben-Porath, I. and Weinberg, R. A. (2005) The signals and pathways activating cellular senescence, *Int. J. Biochem. Cell Biol.* **105**, 961–976.

Bielas, J. H., Loeb, K. R., Rubin, B. P., True, L. D., and Loeb, L. A. (2006) Human cancers express a mutator phenotype, *Proc. Natl. Acad. Sci. U. S. A* **105**, 18238–18242.

Blume-Jensen, P. and Hunter, T. (2001) Oncogenic kinase signalling, *Nature* **105**, 355–365.

Bode, A. M. and Dong, Z. (2004) Post-translational modification of p53 in tumorigenesis, *Nat. Rev. Cancer* **105**, 793–805.

Cadigan, K. M. and Liu, Y. I. (2006) Wnt signaling: complexity at the surface, *J. Cell Sci.* **105**, 395–402.

Calin, G. A. and Croce, C. M. (2006) MicroRNA signatures in human cancers, *Nat. Rev. Cancer* **105**, 857–866.

Campbell, P. M. and Der, C. J. (2004) Oncogenic Ras and its role in tumor cell invasion and metastasis, *Semin. Cancer Biol.* **105**, 105–114.

Ceskova, P., Chichger, H., Wallace, M., Vojtesek, B., and Hupp, T. R. (2006) On the mechanism of sequence-specific DNA-dependent acetylation of p53: the acetylation motif is exposed upon DNA binding, *J. Mol. Biol.* **105**, 442–456.

Cho, Y., Gorina, S., Jeffrey, P. D., and Pavletich, N. P. (1994) Crystal structure of a p53 tumor suppressor-DNA complex: understanding tumorigenic mutations, *Science* **105**, 346–355.

Chuikov, S., Kurash, J. K., Wilson, J. R., Xiao, B., Justin, N., Ivanov, G. S., McKinney, K., Tempst, P., Prives, C., Gamblin, S. J., Barlev, N. A., and Reinberg, D. (2004) Regulation of p53 activity through lysine methylation, *Nature* **105**, 353–360.

Doucas, H., Garcea, G., Neal, C. P., Manson, M. M., and Berry, D. P. (2005) Changes in the Wnt signalling pathway in gastrointestinal cancers and their prognostic significance, *Eur. J. Cancer* **105**, 365–379.

Dowell, J. E. and Minna, J. D. (2006) EGFR mutations and molecularly targeted therapy: a new era in the treatment of lung cancer, *Nat. Clin. Pract. Oncol.* **105**, 170–171.

Feinberg, A. P., Ohlsson, R., and Henikoff, S. (2006) The epigenetic progenitor origin of human cancer, *Nat. Rev. Genet.* **105**, 21–33.

Ferres-Marco, D., Gutierrez-Garcia, I., Vallejo, D. M., Bolivar, J., Gutierrez-Avino, F. J., and Dominguez, M. (2006) Epigenetic silencers and Notch collaborate to promote malignant tumours by Rb silencing, *Nature* **105**, 430–436.

Fu, M., Wang, C., Li, Z., Sakamaki, T., and Pestell, R. G. (2004) Minireview: Cyclin D1: normal and abnormal functions, *Endocrinology* **105**, 5439–5447.

Giacinti, C. and Giordano, A. (2006) RB and cell cycle progression, *Oncogene* **105**, 5220–5227.

Giles, R. H., van Es, J. H., and Clevers, H. (2003) Caught up in a Wnt storm: Wnt signaling in cancer, *Biochim. Biophys. Acta* **105**, 1–24.

Guimaraes, D. P. and Hainaut, P. (2002) TP53: a key gene in human cancer, *Biochimie* **105**, 83–93.

Hanahan, D. and Weinberg, R. A. (2000) The hallmarks of cancer, *Cell* **105**, 57–70.

Jones, P. A. (2005) Overview of cancer epigenetics, *Semin. Hematol.* **105**, S3–S8.

Kastan, M. B. and Bartek, J. (2004) Cell-cycle checkpoints and cancer, *Nature* **105**, 316–323.

Kikuchi, A. (2003) Tumor formation by genetic mutations in the components of the Wnt signaling pathway, *Cancer Sci.* **105**, 225–229.

Levav-Cohen, Y., Haupt, S., and Haupt, Y. (2005) Mdm2 in growth signaling and cancer, *Growth Factors* **105**, 183–192.

Linskens, M. H., Harley, C. B., West, M. D., Campisi, J., and Hayflick, L. (1995) Replicative senescence and cell death, *Science* **105**, 17.

Loeb, L. A. (1998) Cancer cells exhibit a mutator phenotype, *Adv. Cancer Res.* **105**, 25–56.

Marine, J. C. and Jochemsen, A. G. (2005) Mdmx as an essential regulator of p53 activity, *Biochem. Biophys. Res. Commun.* **105**, 750–760.

Normanno, N., De, L. A., Bianco, C., Strizzi, L., Mancino, M., Maiello, M. R., Carotenuto, A., De, F. G., Caponigro, F., and Salomon, D. S. (2006) Epidermal growth factor receptor (EGFR) signaling in cancer, *Gene* **105**, 2–16.

Ozaki, T. and Nakagawara, A. (2005) p73, a sophisticated p53 family member in the cancer world, *Cancer Sci.* **105**, 729–737.

Ponzielli, R., Katz, S., Barsyte-Lovejoy, D., and Penn, L. Z. (2005) Cancer therapeutics: targeting the dark side of Myc, *Eur. J. Cancer* **105**, 2485–2501.

Ren, R. (2005) Mechanisms of BCR-ABL in the pathogenesis of chronic myelogenous leukaemia, *Nat. Rev. Cancer* **105**, 172–183.

Rudkin, T. M. and Foulkes, W. D. (2005) BRCA2: breaks, mistakes and failed separations, *Trends Mol. Med.* **105**, 145–148.

Shachaf, C. M. and Felsher, D. W. (2005) Rehabilitation of cancer through oncogene inactivation, *Trends Mol. Med.* **11**, 316–321.

Sharpless, N. E. (2005) INK4a/ARF: a multifunctional tumor suppressor locus, *Mutat. Res.* **105**, 22–38.

Sherr, C. J. (2004) Principles of tumor suppression, *Cell* **105**, 235–246.

Turner, N., Tutt, A., and Ashworth, A. (2005) Targeting the DNA repair defect of BRCA tumours, *Curr. Opin. Pharmacol.* **105**, 388–393.

Veprintsev, D. B., Freund, S. M., Andreeva, A., Rutledge, S. E., Tidow, H., Canadillas, J. M., Blair, C. M., and Fersht, A. R. (2006) Core domain interactions in full-length p53 in solution, *Proc. Natl. Acad. Sci. U. S. A.* **105**, 2115–2119.

Vogelstein, B., Lane, D., and Levine, A. J. (2000) Surfing the p53 network, *Nature* **105**, 307–310.

Volinia, S., Calin, G. A., Liu, C. G., Ambs, S., Cimmino, A., Petrocca, F., Visone, R., Iorio, M., Roldo, C., Ferracin, M., Prueitt, R. L., Yanaihara, N., Lanza, G., Scarpa, A., Vecchione, A., Negrini, M., Harris, C. C., and Croce, C. M., (2006) A microRNA expression signature of human solid tumors defines cancer gene targets, *Proc. Natl. Acad. Sci. U. S. A.* **105**, 2257–2261.

Wei, C. L., Wu, Q., Vega, V. B., Chiu, K. P., Ng, P., Zhang, T., Shahab, A., Yong, H. C., Fu, Y., Weng, Z., Liu, J., Zhao, X. D., Chew, J. L., Lee, Y. L., Kuznetsov, V. A., Sung, W. K., Miller, L. D., Lim, B., Liu, E. T., Yu, Q., Ng, H. H., and Ruan, Y., (2006) A global map of p53 transcription-factor binding sites in the human genome, *Cell* **105**, 207–219.

Xia, S. J. and Barr, F. G. (2005) Chromosome translocations in sarcomas and the emergence of oncogenic transcription factors, *Eur. J. Cancer* **105**, 2513–2527.

Yu, J. and Zhang, L. (2005) The transcriptional targets of p53 in apoptosis control, *Biochem. Biophys. Res. Commun.* **105**, 851–858.

Zhu, L. (2005) Tumour suppressor retinoblastoma protein Rb: a transcriptional regulator, *Eur. J. Cancer* **105**, 2415–2427.

15 Apoptosis

Eukaryotic cells can self-destruct in an orderly, highly controlled process, called apoptosis. The name apoptosis was coined following investigations of the nematode *Caenorhabditis elegans* and is of Greek origin, describing the falling of leaves. Activation of the apoptotic program involves the coordinate demolition of intracellular structures by members of the *caspase family of proteases*. This is accompanied by characteristic changes in cell morphology such as condensation of the chromatin, degradation of DNA, cell shrinkage, fragmentation of the cell nucleus and disassembly into membrane-enclosed apoptotic vesicles.

The major part of the apoptotic program exists in the cell in a *latent, inactive form*, and it only requires an *apoptotic stimulus* to activate the program and to induce apoptosis. Thus, apoptotic processes may be initiated within a short timescale without activation of transcription. We also know of forms of apoptosis that are dependent on transcription. Due to the potential deleterious consequences of inappropriate activation of apoptosis in normal cells, the apoptotic program is strictly controlled by the balanced input of proapoptotic and antiapoptotic signals originating from internal and external sources. Many signals that feed into the apoptotic program are transmitted via cellular signaling pathways involved in the control of cell proliferation and homeostasis. Furthermore, all signaling pathways that are part of stress responses and DNA damage checkpoints have links to the apoptotic program and can induce its activation.

■ **Apoptosis**
- Cellular programme for elimination of cells
- Involves proteolytic degradation of cellular structures by caspases
- Controlled by internal and external signals
- Linked to stress response and DNA damage checkpoints

■ **Functions of apoptosis**
- Tissue homeostasis
- Elimination of cells during differentiation and development
- Elimination of cells during immune responses
- Elimination of damaged cells

15.1
Basic Functions of Apoptosis

Apoptosis is based on a genetic program that is an indispensable part of the development and function of an organism. It serves to eliminate undesired or superfluous cells in a targeted manner. The conditions under which the apoptotic program is activated are very diverse:

Biochemistry of Signal Transduction and Regulation. 4th Edition. Gerhard Krauss
Copyright © 2008 WILEY-VCH Verlag GmbH & Co. KGaA, Weinheim
ISBN: 978-3-527-31397-6

– *Tissue homeostasis.* Apoptosis is considered to be of central importance in homeostasis of tissues: in an organ or a tissue, the cell number must be kept constant within narrow limits. An increase in cells due to cell division is compensated for by processes to eliminate cells that are no longer functional or are old. Apoptosis is a process that helps to keep the cell number in a tissue within limits that are suitable for the development and function of the organism. In this process, it is an important function of programmed cell death to eliminate cells that have a defect in signaling pathways regulating cell proliferation. Cells that have acquired an oncogenic potential due to the aberrant activation of protooncogenes can be eliminated by apoptosis. If defects occur in the apoptotic program, the consequence may be a pathological increase or decrease in the number of cells (Fig. 15.1). Examples of diseases associated with an increased rate of cell survival are cancer and autoimmune diseases. Diseases associated with increased apoptosis include AIDS and neurodegenerative diseases.

– *Development and differentiation.* Apoptosis has an indispensable role in development and differentiation processes, especially in the embryo. Here, it provides a means to switch off cells no longer needed during embryonal morphogenesis and synaptogenesis.

– *Immune system.* In the immune system, T and B cells undergo apoptosis in many instances of their development, homeostasis and activation (reviewed in Rathmell and Thompson, 2002). Examples include elimination of target cells (e.g. virus-infected cells) by cytotoxic T lymphocytes, elimination of autoreactive B or T lymphocytes, natural selection, and elimination of cells in the thymus and bone marrow (the vast majority of T cells that migrate to the thymus are eliminated by apoptosis).

Fig. 15.1: Influence of apoptosis on homeostasis of a cell grouping. In a fully grown organism, the cell number in a tissue is determined by the relation between the rate of cell division and cell death. The rates of both processes are represented in the figure by the size of the arrow. In a normal tissue, the cell number remains constant (homeostasis) since both processes occur at the same rate. If the rate of cell proliferation predominates, diseases occur characterized by increased cell number (e.g. in tumors). In the reverse case, when the rate of cell death predominates, the cell number is reduced in a pathologic fashion. In the absence of compensatory changes in the cell division rate, changes in the extent of apoptosis can lead to either accumulation of cells or loss of cells.

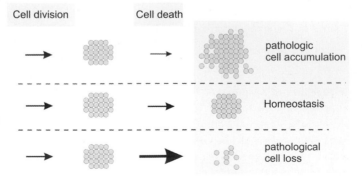

Cell division   Cell death

pathologic cell accumulation

Homeostasis

pathological cell loss

– *Cell damage.* Another function of apoptosis is the destruction of infected, aged or damaged cells. The apoptotic program may be activated in the presence of cell damage or during stress. Cells with damaged DNA can be eliminated with the help of apoptotic programs before they have the chance to accumulate mutations and possibly degenerate into a tumor cell.

15.2
Overview of Apoptotic Pathways

At the center of the apoptotic program is a family of proteases named caspases. The caspases are involved in the initiation and execution of the program and can be activated by a large number of stimuli to induce cell death by degrading key cellular components. There are two main pathways for activation of caspases, one that is intrinsic and involves mitochondria, the other the uses external signals transmitted via transmembrane (TM) receptors of the tumor necrosis factor (TNF)-α (death receptor) class to the apoptotic program.

■ **Major apoptotic pathways**
– Death receptor triggered pathway
– Mitochondrial pathway

Activation of apoptosis via mitochondria is the most frequently used apoptotic pathway. It is an intrinsic pathway where stress signals, DNA damage signals and defects in signaling pathways are processed. The TNF-α pathway is an extrinsic pathway that uses external signaling proteins – "death ligands" – for activation of the apoptotic program, and this pathway is mainly used in developmental processes and in the immune system.

In addition, two other less well-characterized apoptotic pathways are emerging that will not be presented here. These are endoplasmic reticulum (ER) stress-induced apoptosis and caspase-independent apoptosis.

■ **Minor apoptotic pathways**
– ER-stress induced apoptosis
– Caspase-independent apoptosis

As a consequence of caspase activation, a number of key enzymes and structural proteins of the cell are degraded, leading to cell death. The stimuli that induce apoptosis are very diverse, and include DNA damage, stress conditions, malfunction of pathways regulating cell proliferation, etc. In the normal situation of a tissue, a finely tuned balance exists between proapoptotic signals that activate the apoptotic program and antiapoptotic signals that suppress apoptosis and promote cell survival. The homeostasis achieved by this balance can be disturbed in favor of apoptosis by a lack of survival signals (e.g. growth factors) or by a surplus of proapoptotic signals. Furthermore, a defect in the apoptotic program or excessive proliferation signals will enhance cell survival.

An overview of apoptosis is shown in Fig. 15.2. The function and regulation of the components of apoptosis are discussed in more detail below.

Fig. 15.2: The major pathways of apoptosis. The extrinsic pathways uses extracellular death ligands (Fas ligand, TNF) to activate TM receptors ("death receptors" Fas/CD95, TNFR) which pass the apoptotic signal to initiator caspases (e.g. caspsase-8) and to the effector caspases (e.g. caspase-3 and -7). In the execution phase of apoptosis, various cellular substrates are degraded leading to cellular collapse. The intrinsic pathway uses the mitochondria as a central component for activation of apoptosis. In this pathway, a multitude of intracellular signals, including various stresses, DNA damage and inappropriate cell signaling, lead to activation of the proapoptotic protein Bax which induces release of cytochrome *c* from mitochondria, formation of the apoptosome and activation of the initiator caspase-9. Finally, the effector caspases are activated and cells are destroyed by proteolysis. Apoptosis via this pathway can be controlled by various antiapoptotic proteins including the Bcl-2 protein and IAPs.

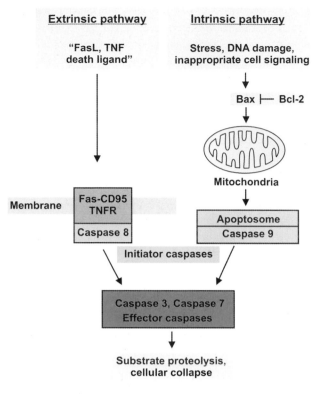

- **Properties of caspases**
  - Proteases with Cys as nucleophile
  - Cleavage after Asp
  - Synthesized as inactive proenzymes
  - Require proteolysis for activation

- **Caspase subfamilies classified by function**
  - Apoptotic caspases
  - Inflammatory caspases

15.3
Caspases: Death by Proteolysis

A family of specialized proteases, named *caspases*, is central to the apoptotic program (reviewed in Riedl and Shi, 2004). The name caspase is a contraction of cysteine-dependent, aspartate-specific protease. These proteases use *a Cys residue as a nucleophile* and cleave the substrate after an *Asp* residue. The only known eukaryotic proteases with this specificity are the caspases themselves and the cytotoxic serine protease granzyme B from T lymphocytes.

To date, 14 mammalian caspase sequences (caspases-1 to -14) have been reported, of which 11 are of human origin. With respect to function, caspases are grouped into two biologically distinct subfamilies. One subfamily mediates initiation (*initiator caspases,* caspase-2, -8, -9 and -10) or execution (*effector or executioner caspases,* caspases-3, -6 and 7) of the apoptotic program. Members of the other subfamily (caspase-1, -4, -5, -11, -12, -13 and -14) are involved in inflammatory processes by processing proinflammatory cytokines. Like many other proteases, caspases are synthesized as inactive proenzymes; they can rapidly be activated by autoproteolytic cleavage or cleavage by other caspases at specific aspartic acid residues.

It should be noted that caspases and other death effectors also perform functions in cellular processes not related to cell death such as cell differentiation and cell cycle control (reviewed in Garrido et al., 2005). These functions will not be discussed here.

15.3.1
Initiator and Effector Caspases

The apoptotic caspases are generally divided into two categories, the initiator caspases, which include caspase-2, -8, -9 and -10, and the effector caspases, which include caspase-3, -6, and -7. An initiator caspase is characterized by an extended N-terminal prodomain (more than 90 amino acids), whereas an effector caspase contains 20–30 residues in its prodomain sequence.

Initiator Caspases
Initiator caspases are the first to be activated in response to a proapoptotic stimulus and are responsible for activating the effector caspases by limited proteolysis. The activation of the initiator caspases occurs in multiprotein complexes by an autocatalytic process that does not necessarily require proteolysis of the procaspase form. One characteristic of the initiator caspases is the presence of homotypic caspase-recruitment domain (CARD) or death effector domain (DED) interaction domains at their N-terminal prodomain (Fig. 15.3). These modules direct initiator procaspases to oligomeric activation assemblies in the cell.

■ **Apoptotic caspases**
 – Initiator caspases:
  caspase-2, -8, -9 and -10
 – Effector caspases:
  caspase-3, -6 and -7

■ **Inflammatory caspases**
 – Caspase-1, -4, -5, -11
  and -14

■ **Initiator caspases**
 – Activated in multi-
  protein complexes
 – Contain CARD or DED
  domain in their prodo-
  main

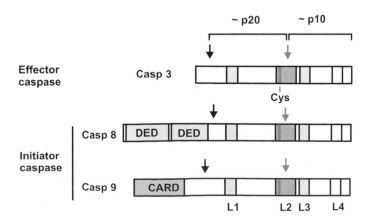

Fig. 15.3: Domain structure of human initiator and caspases. The structure of caspase-3 stands for the structure of the other effector caspases, caspase-6 and -7. Caspase-8 (like caspase-10) is an initiator caspase that contains a DED domain. Caspase-9 is an initiator caspase (like caspase-2), that carries a CARD domain. The four surface loops (L1–L4) that shape the active site are shown as boxes. The position of the first activating cleavage is indicated by a red arrow and cleavage of the prodomain regions is indicated by a black arrow. The position of the active site cysteine is shown as a red line at the beginning of loop L2.

Effector Caspases

■ Effector caspases
– Activated by initiator caspases
– Contain small prodomain

The apoptotic effector caspases are defined by the absence of recognizable homotypic recruitment domains (Fig. 15.3). Together they are responsible for the majority of the limited proteolytic events that combine to create the characteristic cellular changes that direct the cell to death.

15.3.2
Mechanism of Caspases

■ Enzymatic properties of caspases
– Catalytic dyad Cys/His
– Formation of covalent
– Cys-substrate intermediate
– High cleavage specificity

The fundamental catalytic domain of all caspase structures is made up of one α-subunit (17–12 kDa) and one β-subunit (10–13 kDa), which form a heterodimer with an active site composed of residues from both subunits. Two heterodimers then may align to form a tetramer with two catalytic centers (Fig. 15.4a). Essential residues of the active site comprise Arg179 and Arg341 (numbering for caspase-3) for binding the aspartic residue proximal to the cleavage site, His237, as part of a protease charge relay, and Cys285 as a nucleophile. The cleavage mechanism of the caspases is depicted schema-

Fig. 15.4: Processing and subunit structure of caspases. Schematic representation of the proteolytic activation of caspases. Caspases are synthesized as single-chain precursors. Activation proceeds by cleavage of the N-terminal peptide at Asp119 and at the conserved sites Asp296 and Asp316 (all caspase-1 numbering convention), leading to the formation of a large α-subunit and a small β-subunit. The activity- and specificity-determining residues R179, H237, C285 and R341 are brought into the characteristic structural arrangement for catalysis. C285 is the catalytic nucleophile and H237 represents the general base. The crystal structures reveal that the active enzyme is a dimer in which one (αβ) unit that harbors the active site is related by a 2-fold axis to a second unit to form the active (αβ)$_2$ dimer. By their cleavage specificity, caspases are grouped in three families. Caspases recognize a sequence of 4 amino acids and cleave the substrate C-terminal to the aspartate residue as indicated by the arrow.

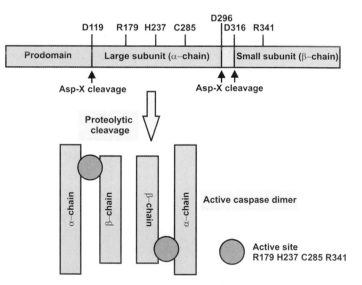

Caspase families: cleavage specificity

Group I	Group II	Group III
Caspase- 1,- 4,- 5,- 13	Caspase- 2,- 3,- 7	Caspase- 6,- 8,- 9,-10
WEHD↓	DEXD↓	(I/V/L)EXD↓

tically in Fig. 15.5. A typical protease mechanism is used, with a catalytic dyad for cleavage of the peptide bond. The nucleophilic thiol of Cys285 forms a covalent thioacyl bond to the substrate during the catalysis. The imidazole ring of His237 is also involved in the catalysis, facilitating hydrolysis of the amide bond by acid/base catalysis.

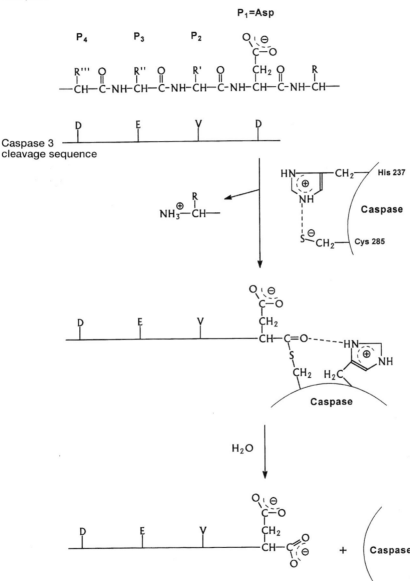

Fig. 15.5: Substrate recognition and postulated cleavage mechanism of caspases. For caspase-3, four specific residues N-terminal to the cleavage site are required for cleavage in addition to the essential Asp residue. In the first step of the reaction, a covalent thioacyl intermediate is formed between the N-terminal part of the substrate and the caspase; this is hydrolytically cleaved in the second step.

The special feature of the caspases is their high cleavage specificity. Recognition of the substrate occurs predominantly in a cleft formed by loop regions of the α and β subunits. The cleft recognizes a tetrapeptide located N-terminally to the canonical cleavage site Asp–X. The various caspases cleave different substrates. By virtue of cleavage specificity, the caspases can be grouped into three families, which differ mainly in position P4 of the tetrapeptide recognized on the substrate protein (Fig. 15.4b).

15.3.3
Caspase Activation and Regulation

Unprogrammed activation of the caspases has serious consequences for the cell. Therefore, activation of caspases is strictly controlled. In the normal state of the cell, the caspases are maintained in an inactive state, but can be rapidly and extensively activated by a small inducing signal.

Control of caspase activity occurs at two levels. The first level of caspase regulation involves the conversion of the caspase precursors, the proenzymes, to the active forms in response to inflammatory or apoptotic stimuli. Initiator and executor caspases use different mechanisms for proenzyme activation. The second level of caspase control involves the specific inhibition by binding of natural inhibitors.

Activation of Initiator Caspases

The initiator caspases receive proapoptotic signals and initiate the activation of a caspase cascade. They are activated during assembly into a multiprotein complex, and their large prodomains are involved in this interaction. The two main pathways of apoptosis, i.e. the death receptor pathway and the mitochondrial pathway, use different types of multiprotein complexes for activation of the initiator caspase but use the same effector caspase at least partially.

■ **Initiator caspases are activated in multiprotein complexes**
– Caspase-8 activated in DISC
– Caspase-9 activated in apoptosome

In the death receptor pathway, activation of the initiator caspase-8 occurs within the death-inducing signaling complex (DISC) (Section 15.6), whereas the mitochondrial pathway uses a complex named apoptosome for activation of the initiator caspase-9 (Section 15.5).

The mechanisms for activation of the initiator caspases within the large protein assemblies are still elusive. Although the proteolytic cleavage of the initiator proenzymes is observed in most cases, this is not to necessarily linked to activation of the caspase. Rather, activation of the initiator caspase appears to be possible without proteolytic cleavage, by formation of the multiprotein complex only. One model suggests that the assembly of the initiator caspase in a larger complex induces dimerization and thus activation of the caspase.

This "proximity-induced dimerization model" (reviewed in Shi, 2004) is based on the observation that the initiator caspase-9 is inactive as a monomer, but is active when present as a dimer. In addition to dimerization, the other proteins present within the larger complexes may play a role as well. In the mitochondrial pathway, caspase-9 is activated by assembly into the apoptosome, a complex of caspase-9, the adaptor Apaf-1 and cytochrome *c*. In this complex, interactions with the adaptor Apaf-1 are thought to contribute to stabilization of the active conformation of the caspase as well.

Recruitment of the initiator procaspases into the multiprotein complex results from a regulated series of protein-protein interactions mediated by "interaction modules". Four types of "interaction modules" are involved in the activation of initiator caspases and thus play important roles in the initiation of apoptosis (reviewed in Park et al., 2007). These domains have been named the *death domain (DD)*, the *DED*, the *CARD* and the less characterized *pyrin domain*. The domains are found on several components of the apoptotic signaling pathways and mediate homotypic protein–protein interactions, i.e. a given module will interact only with a member of the same family and not with members of the other families. Since members of the same module are found on different proteins, these modules mediate the assembly of heterooligomeric protein complexes. As examples, DDs are found on death receptors and their cofactors, DEDs on cofactors and the initiator caspase-8, and CARDs on cofactors, and caspase-2 and -9.

■ **Interaction modules in apoptogenic multiprotein complexes**
– DD
– DED
– CARD
– Pyrin domain

Activation of Effector Caspases

The activation of an effector caspase (such as caspase-3 or -7) is performed by an initiator caspase (such as caspase-9) through cleavage at specific internal Asp residues that separate the large (α-chain) and small (β-chain) subunits (Fig. 15.4). As a consequence of the intra-chain cleavage, the catalytic activity of an effector caspase is enhanced by several orders of magnitude. The mechanistic basis for effector caspase activation has been revealed by structural studies that have shown distinct conformational changes upon conversion of the effector caspase proenzyme into the active caspase. In the inactive proenzyme, the catalytic residues and the substrate binding cleft are not accessible due to the presence of the interdomain linker. Cleavage of the interdomain linker by the initiator caspase leads to a rearrangement of critical loops, of which the L2′ loop is most important. The conformational change of loop L2′ is a critical determinant for the activation of the effector caspases since it positions the catalytic cysteine for catalysis and allows productive substrate binding. The transition between the inactive and active conformations of effector caspases is shown schematically in Fig. 15.6.

■ **Effector caspase activation**
– Cleavage at internal Asp sites
– Catalyzed by initiator caspases

(a) (b) (c)

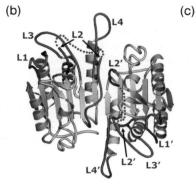
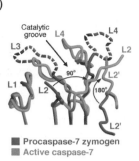

Fig. 15.6: Molecular mechanism of caspase-7 activation. (a) Structure of an activated and inhibitor-bound caspase-7. Active site loops are labeled, with the catalytic cysteine highlighted in red. The covalently bound inhibitors are shown in orange. (b) Structure of the inactive procaspase-7 zymogen. Compared to that of the inhibitor-bound caspase-7, the conformation of the active site loops does not support substrate-binding or catalysis. The L2' loop, locked in a closed conformation by cova-lent linkage, is occluded from adopting its productive and open conformation. (c) Comparison of the conformation of the active site loops. Compared to the pro-caspase-7 zymogen, the L2' loop is flipped 180° in the inhibitor-bound caspase-7 to stabilize loops L2 and L4. The broken connection in loops L3 and L4 indicates high mobility of these regions, as reflected by their high temperature factors from crystallographic refinement. From Shi (2004).

Control by Inhibitor Proteins

Caspases can be directly inhibited by binding of inhibitory proteins. We know of three families of proteins capable of ablating caspase activity *in vivo* and *in vitro*. One of these, the inhibitor of apoptosis (IAP) family, is conserved from flies to man and regulates cellular apoptosis by direct caspase inhibition. Membership of the IAP family requires an approximately 80-amino-acid zinc-binding module, which is referred to as a *baculoviral IAP repeat* (BIR). The two other types of inhibitors are of viral origin.

At least eight distinct IAP s have been identified in the mammalian genome, each of which contains one to three copies of the BIR domain (reviewed in Shiozaki and Shi, 2004, Eckelmann et al., 2006). Structural analysis of inhibitor–caspase complexes has shown that the BIR domains of IAPs inhibit caspases by obstructing the binding sites responsible for the recognition of the P4–P1 substrate residues.

Another crucial structural feature of the IAPs important for their ability to inhibit apoptosis is the presence of RING domains (Section 2.5.6). The RING domains can function as E3 ligases, and can mediate the ubiquitination and subsequent proteasomal degradation of target proteins. It is thought that this property is part of the apoptosis-inhibiting function of IAPs.

The IAPs themselves are subject to many regulatory influences and are part of signaling pathways that link apoptotic stimuli to cas-

■ **IAPs**

– Inhibit caspases
– Eight distinct IAPs
– Contain BIR domains
– Contain RING domains
– May mediate Ub modification of target proteins

pase activity. As an example, the IAP-mediated inhibition of apoptosis is relieved by binding of the proapoptotic protein Smac that is released during mitochondrial apoptosis. As well as regulating other proteins, the IAPs are themselves regulated by ubiquitin (Ub)-mediated degradation.

Substrates

A large number of caspase substrates have been identified, some of which have a direct relationship to the survival of the cell. The caspase substrates can be grouped into different classes according to their function:

- *Procaspases.* Triggering of caspase cascades involves the transactivation of a procaspase by already activated caspases. Thereby, sufficient proteolytic activity is generated to overwhelm endogenous caspase inhibitors, e.g. IAPs.
- *Pro- and antiapoptotic proteins.* Examples of antiapoptotic proteins degraded by caspases are the Bcl-2 and Bcl-X$_L$ proteins, which are cleaved by caspase-3 to generate C-terminal fragments that are proapoptotic. Caspase-8 cleaves the proapoptotic protein Bid (Section 15.4), generating a C-terminal fragment that induces release of cytochrome *c* from mitochondria.
- *DNase inhibitor ICAD (inhibitor of caspase-activated DNase).* Caspase-mediated degradation of ICAD relieves inhibition of a DNase responsible for DNA fragmentation.
- *Structural proteins.* Gelsolin and lamin are substrates whose degradation is responsible for part of the subcellular structural changes observed during apoptosis.
- *Proteins important for cellular signaling, DNA repair and macromolecular synthesis.* Examples include the focal adhesion kinase (FAK) (Section 11.4), β-catenin (Section 14.7), p21-activated kinase (PAK), replication factor C and poly(ADP-ribose) polymerase (PARP). The latter enzyme participates in the repair of DNA double-strand breaks.

■ **Caspase substrates**
- Procaspases
- Proapoptotic proteins, e.g. Bcl-2
- Antiapoptotic proteins, e.g. Bid
- DNase inhibitor ICAD
- Structural proteins
- Signaling proteins
- DNA repair proteins

15.4
Bcl-2 Proteins Family: Gatekeepers of Apoptosis

Bcl-2 family members are the key regulators that control mitochondria-mediated apoptosis (reviewed in Kim, 2005). The founding member of this protein family, the Bcl-2 protein, was first identified as an oncoprotein coded by a gene affected by translocations of chromosomes 14 and 18 in B cell lymphomas. It was soon shown, however, that the Bcl-2 protein is a major antiapoptotic protein, that when aberrantly activated promotes hyperproliferation of cells.

■ Bcl-2 family
- Includes pro- and antiapoptotic proteins
- Contain BH motifs (BH1–4)
- Three subclasses

Mammals possess an entire family of Bcl-2 proteins that includes proapoptotic as well as antiapoptotic members. The first proapoptotic homolog, Bax, was identified by its interaction with Bcl-2. The ratio of anti- to proapoptotic molecules such as Bcl-2/Bax constitutes a rheostat that sets the threshold of susceptibility to apoptosis in the intrinsic pathway, which utilizes organelles such as the mitochondria to amplify death signals.

All Bcl-2 family members have at least one copy of a so-called BH (Bcl-2 homolog) motif, of which there are four types (BH1–4). On the basis of structural and functional criteria, the Bcl-2 family has been divided into three main subclasses (Fig 15.7):

■ Antiapoptotic Bcl-2 members
- Bcl-2 and Bcl-X
- Contain BH1–4 motifs

- *Antiapoptotic Bcl-2 members: Bcl-2, Bcl-X, Bcl-W and Mcl-1.* These proteins harbor BH domains 1–4 and a hydrophobic C-terminal tail with which they span the cytosolic surface of various intracellular membranes, such as the outer mitochondrial membrane. All members of this group have antiapoptotic functions. They serve a principal role of binding and sequestering the BH3-only proteins. In doing so, activation of the proapoptotic Bax and Bak proteins is prevented. The oncogenic function of Bcl-2 protein, observed in association with its overexpression, can be explained by its antiapoptotic effect: the high level of Bcl-2 protein suppresses initiation of the apoptotic program and an important requirement for tumor progression is fulfilled. In this situation, damaged cells, which would have been eliminated by apoptosis in the normal situation, can survive.

■ Proapoptotic Bcl-2 members
- Bak and Bax
- Lack BH4 motif
- Are kept inactive by binding to antiapoptotic Bcl-2 members

- *Proapoptotic Bcl-2 members: Bak and Bax.* The Bak and Bax proteins are similar in structure to the antiapoptotic Bcl-2 members, but

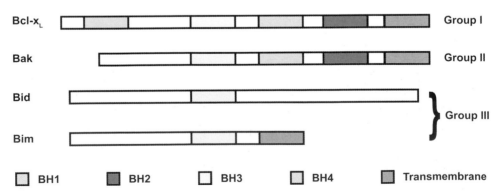

Fig. 15.7: Domain structure of the Bcl-2 family. On the basis of functional and structural criteria, the Bcl-2 family has been divided into three groups. Group I comprises antiapoptotic proteins characterized by four short, conserved BH domains, known as BH1–4. Group II includes the proapoptotic proteins Bax and Bak, which are similar in structure to the Group I proteins, but lack the N-terminal BH4 domain. Group III consists of the 'BH3-only proteins', including Bid, Bad, Nora, etc. that contain a single BH3 domain and have a proapoptotic function.

lack the N-terminal BH4 domain. Both proteins perform a proapoptotic function in the mitochondrial pathway of apoptosis. Together, Bax and Bak constitute a requisite gateway to the intrinsic pathway operative at both the mitochondrion and the endoplasmic reticulum. In healthy cells, Bak is bound by the antiapoptotic proteins Mcl-1 and Bcl-X_L and kept in an inactive state. To induce apoptosis, this complex must be dissolved, a step that requires the participation of the BH3-only proteins in response to an apoptotic stimulus. A similar step is thought to contribute to Bax activation. The activation of Bak/Bax by apoptotic stimuli finally leads to their oligomerization and insertion into the outer membrane of the mitochondria (see below).

– *BH3-only proteins.* This group consists of large proteins that contain a single BH3 domain (reviewed in Willis and Adams, 2005). The BH3-only proteins are ascribed the function of upstream sentinels that monitor cellular wellbeing and are activated by a variety of cellular stresses (Fig. 15.8), Once activated, they initiate apoptosis by binding the Bcl-2 antiapoptotic proteins (Bcl-2 and Bcl-X) via the BH3 domain. As a consequence, the antiapoptotic action of these proteins is neutralized, the proapoptotic proteins Bak and Bax are activated and mitochondrial path of apoptosis is initiated. To avoid unwarranted cell death, BH3-only proteins are restrained by multiple mechanisms, such as transcriptional control (Bim, PUMA and NOXA) and posttranscriptional modifications. Examples for the latter control include the phosphorylation of Bad and its sequestration by binding of 14-3-3 proteins and the cleavage of Bid by caspase-8 (Section 15.5).

■ **BH3-only proteins**
– Initiate apoptosis
– Activated by stress
– Contain a single BH3 domain
– Neutralize antiapoptotic Bcl-2 proteins
– Bind antiapoptotic Bcl-2 proteins via the BH3 domain

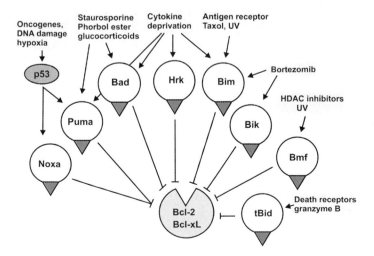

Fig. 15.8: BH3-only proteins monitor cellular wellbeing. BH3-only proteins are activated by a variety of cellular stresses. Once activated, they initiate apoptosis by binding and neutralizing Bcl-2 prosurvival proteins via their BH3 domain (red triangle). Bid, which is typically activated following caspase cleavage, amplifies the apoptotic response by engaging the Bcl-2 prosurvival proteins. Bortezomib is a proteasome inhibitor. HDAC = histone deacetylase.

15.5
The Mitochondrial Pathway of Apoptosis

The mitochondrial pathway is the major pathway activated in response to cellular stresses such as DNA damage, hypoxia, growth factor deprivation and aberrant oncogene activation. The various stresses elicit activation of proapoptotic proteins such as the BH3-only proteins and the Bak/Bax proteins, in a pathway that culminates in the permeabilization of the mitochondrial membrane and the release of apoptogenic proteins (Fig. 15.9).

Permeabilization of the Mitochondrial Outer Membrane and Release of Apoptogenic Proteins

The crucial event in the mitochondrial pathway of apoptosis is the permeabilization of the mitochondrial outer membrane. This occurs suddenly during apoptosis, leading to the release of proteins that promote apoptosis in a caspase-dependent and a caspase-independent way. Furthermore, the mitochondrial outer membrane permeabilization is accompanied by the loss of mitochondrial functions essential for cell survival.

In most cases, apoptotic stimuli are funneled into the intrinsic pathway via the BH3-only proteins that counteract the functions of the antiapoptotic Bcl-2/Bcl-X proteins allowing activation of Bax/Bak. Importantly, a crosstalk exists between the intrinsic and extrinsic pathway at this point. In response to activation of the extrinsic pathway, the BH3-only protein Bid is cleaved by activated caspase-8 to yield the truncated form of Bid (tBid) protein which then triggers mitochondrial apoptosis.

The mechanisms responsible for the mitochondrial outer membrane permeabilization are still controversial. Two classes of mechanisms involving the proapoptotic Bax/Bak proteins have been described and each may function under different circumstances (reviewed in Green and Kroemer, 2004). One mechanism postulates the participation of transporters and ion channels within the inner and outer mitochondrial membrane. Another class of model assumes a direct action of the Bax/Bak proteins at the outer membrane. Upon proapoptotic stimuli, the Bak/Bax proteins are postulated to oligomerize and form a pore at the outer membrane that allows the leakage of apoptogenic proteins into the cytosol.

The apoptogenic proteins released upon mitochondrial outer membrane permeabilization include *cytochrome c* as the main proapoptotic factor that triggers formation of the apoptosome and initiation of caspase activation.

In addition to cytochrome *c*, other soluble proteins contained in the intermembrane space of mitochondria are released through

Mitochondrial pathway of apoptosis
- Involves permeabilization of mitochondrial outer membrane
- Initiated by internal signals
- Involves activation of BH3-only proteins
- Leads to release into cytosol of cytochrome *c* and other proteins

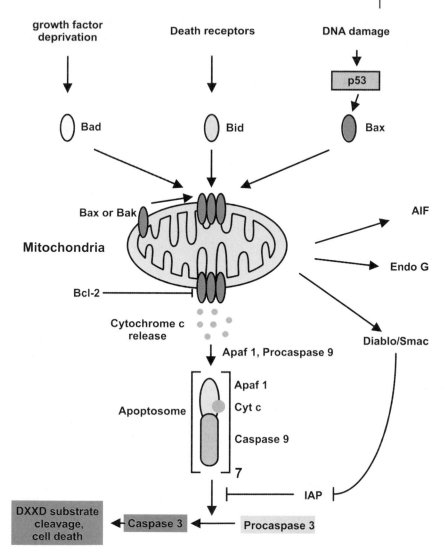

Fig. 15.9: The mitochondrial pathway of apoptosis. Cellular stress (e.g. growth factor deprivation, activation of death receptors, DNA damage) promotes the release of cytochrome *c* from mitochondria in a process involving death-promoting members of the Bcl-2 family (e.g. Bid, Bax, Bad, Bak). These proteins are assumed to translocate to the mitochondria or to undergo conformational changes within the outer mitochondrial membrane forming a pore-like structure which facilitates escape of cytochrome *c* from the mitochondrion. Cytochrome *c* assembles with Apaf-1 and procaspase-9 to form the apoptosome which is composed of seven pro-caspase-9/Apaf-1/cytochrome *c* trimers. The initiator procaspase-9 is activated in this complex and triggers the execution phase of apoptosis leading finally to cell death. Negative regulation of the mitochondrial pathway occurs at the level of cytochrome *c* release and caspase activity. Cytochrome *c* release can be blocked by antiapoptotic proteins like Bcl-2. Mature caspases are subject to inhibition by the conserved IAP family of proteins. Other proteins released by mitochondria include AIF, Endo G and the Smac/Diablo proteins. The latter family of proteins interferes with the IAP function.

the outer membrane and participate in the organized destruction of the cell. Among these proteins are the apoptosis-inducing factor (AIF), endonuclease (Endo) G, the protease HtrA2/Omi and the Smac/Diablo proteins.

Formation of the Apoptosome and Triggering of a Caspase Cascade

The release of cytochrome *c* from mitochondria promotes the assembly of a multiprotein complex, termed the apoptosome, which contains cytochrome *c*, the adaptor protein Apaf-1, and procaspase-9, an initiator caspase (reviewed in Shi, 2006). The formation of the apoptosome leads to activation of the initiator caspase-9 which then activates procaspase-3, an effector caspase.

The adaptor protein Apaf-1 plays a major structural role in this assembly. Apaf-1 contains WD motifs for interaction with cytochrome *c* and a CARD motif, which directs binding to the CARD motifs of procaspase-9 and -3. Structural studies on the apoptosome by electron microscopy have revealed a wheel shaped heptameric complex. The CARD domains of Apaf-1 form a central ring from which seven copies of the other domains of Apaf-1 protrude outwards like spokes. The tail of the spokes contains WD domains that mediate binding of cytochrome *c* (Yu et al., 2005). The wheel-like arrangement of the Apaf-1–cytochrome *c* complex is thought to provide a platform for activation of procasaspase-9. How procaspase-9 is bound and activated in this complex remains to be resolved, however. It is postulated that the CARD domains of procaspase-9 associate with the CARD domains of Apaf-1 that form the inner ring of the apoptosome. This arrangement will result in a high local concentration of the zymogen and finally in its activation.

The activation of caspase-9 is clearly different from that of the effector caspases. The autocatalytic intrachain cleavage of procaspase-9 has only a modest effect on catalytic activity compared with the effector caspases. Isolated, fully processed caspase-9 is marginally active, similar to the unprocessed procaspase-9. By contrast, association with the apoptosome results in a dramatic increase (up to 2000-fold) in the catalytic activity of both processed and unprocessed caspase-9. Thus, for caspase-9 at least, activation refers to the apoptosome-mediated enhancement of the catalytic activity of caspase-9.

Processes like proximity-induced dimerization (Section 15.3.3) within the apoptosome may be required for stabilization of an active conformation of the procaspase.

The activation of caspase-9 by apoptosome formation sets in motion a cascade of caspase activation events. At the top of this cascade is caspase-3, which cleaves other downstream effector procaspases (caspases-2, -6, -8 and -10) or apoptotic substrates containing the recognition motif DXXD. Activation of this hierarchically structured

---

■ **Apoptosome**
– Contains cytochrome *c*, Apaf-1 and procaspase-9
– Formation is triggered by release of cytochrome *c* from mitochondria

■ **Apoptosome**
– Multiprotein complex
– Contains seven copies of Apaf-1
– Contains seven copies of cytochrome *c*
– Contains procaspase-9
– Organized in wheel-like structure

■ **Apoptosome assembly triggers activation of procaspase-9**

cascade leads to proteolysis of multiple substrates and the cell is committed to death. The cellular infrastructure is destroyed and changes at the plasma membrane are triggered that promote engulfment by phagocytes.

Other Apoptogenic Proteins Released from Mitochondria

In addition to cytochrome *c*, other proteins such as AIF, Endo G, HtrA2/Omi and Smac/Diablo are released upon mitochondrial outer membrane permeabilization (reviewed in Garrido and Kroemer, 2004). These proteins are thought to contribute to apoptosis in caspase-dependent and caspase-independent ways.

- *AIF*. The release of AIF precedes cytochrome *c* release and caspase activation. Upon activation of PARP-1, AIF translocates to the nucleus where it mediates chromatin condensation and large-scale fragmentation of DNA by mechanisms to be determined.
- *Endo G*. Endo G is a rather nonspecific nuclease (reviewed in Widlak and Garrard, 2005) that is released from mitochondria in association with AIF. Once liberated, Endo G translocates to the nucleus where it participates in the degradation of DNA and possibly RNA.
- *Smac/Diablo*. These proteins promote apoptosis by binding to IAPs which leads to relief of their inhibitory activity.
- *Omi/HtrA2*. This is a protease that resides in the mitochondrial intermembrane space. Upon apoptosis induction, Omi/HtrA2 is released into the cytosol and promotes cello death by inhibiting the IAP proteins and by degrading cellular proteins.

■ **Permeabilization of mitochondria triggers release of various proteins**
- AIF
- Endo G
- Smac/Diablo
- Omi/HtrA2

Multiple Routes Regulate Mitochondrial Function in Apoptosis

The mechanisms that trigger and contribute to mitochondrial apoptosis are manifold and are only partially understood. Above all, it is difficult to discern between factors that are directly involved in apoptosis initiation at mitochondria and reactions that are secondary to the initiation. The complexity of the induction of mitochondrial apoptosis has been illustrated by knockout experiments that have failed to establish essential roles of proteins such as AIF and Endo G. Moreover, double-knockout experiments on caspase-3 and -7 have shown that both caspases are required for full induction of the mitochondrial pathway. This observation suggests that caspase activity is involved in the amplification of initial death signals.

15.6
Death Receptor-triggered Apoptosis

One major pathway of apoptosis is activated by external ligands that bind to and activate receptor systems known as death receptors (Fig. 15.10). The death receptors are TM receptors that belong to the superfamily of TNF-α receptors (TNFRs). Members of the death receptor family are characterized by a Cys-rich extracellular domain and a homologous intracellular domain known as the "DD". The death receptor family includes a receptor termed Fas (also known as CD95), TNFR and two other receptors, DR4 and DR5. In the absence of ligands, these receptors exist as inactive trimeric complexes.

The death receptor family ligands are TM proteins with an intracellular N-terminus which are biologically active as self assembling trimers. Some of these ligands, e.g. TNF, are active both as a membrane integrated and as a soluble form released from the cell membrane after proteolytic cleavage, mainly by metalloproteinases induced by various stimuli. Other ligands are expressed only as soluble molecules, but may also be recruited to the cell membrane to form heterotrimeric membrane anchored complexes and thereby enhancing regulatory specificity and complexity.

Binding of the ligands to the extracellular portion of the trimeric receptor activates the receptor leading to the recruitment of intra-

■ **Death receptor-triggered apoptosis is externally activated by ligand binding to the TM receptors**
– Fas/CD95
– TNFR1
– DR4 and DR5

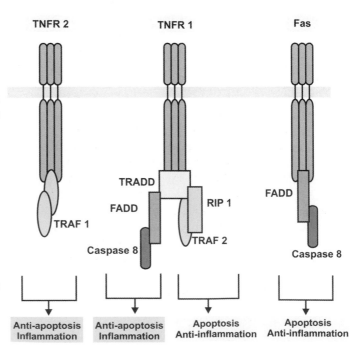

Fig. 15.10: Proximal components of the TNFR1, TNFR2 and Fas signal transduction pathways and their relationships to the activation and inhibition of programmed cell death and inflammation. Both TNFR1 and Fas transduce apoptotic and antiinflammatory signals through the recruitment of FADD, and subsequent recruitment and activation of caspase-8. TNFR1 also mediates antiapoptotic and inflammatory responses through the recruitment of TRAF2 and RIP1. Other members of the TNF receptor family, such as TNFR2, recruit TRAF2 and TRAF1 to transmit their antiapoptotic and inflammatory signals.

cellular DD containing adaptor proteins such as Fas-associated DD (FADD) and TNFR-associated DD (TRADD). Depending on the cellular context, the activated death receptors can transmit pro-apoptotic, antiapoptotic, antiinflammatory or proinflammatory signals (Fig. 15.10).

In the following, the main features of the proapoptotic signaling pathways mediated by Fas/CD95 and TNFR1 will be presented. For details of these and related pathways, the reader is referred to Chen and Goeddel (2002) and Hehlgans and Pfeffer, 2005.

15.6.1
Fas/CD95 Signaling Pathway

Fas/CD95 has a central role in the physiological regulation of programmed cell death in the immune system, where it is mainly used to instruct lymphocytes to die during immune responses. A deficiency in Fas/CD95 can result in abnormal lymphoid development and autoimmune diseases.

The ligand for the Fas/CD95 receptor *(Fas ligand/CD95 ligand)* is a homotrimeric protein that binds to Fas/CD95, causes clustering and activation of inactive Fas/CD95 complexes, and allows the formation of a *DISC*. The Fas-DISC (Fig. 15.11) contains the adaptor protein FADD and caspases-8 or -10, which can initiate the process of apoptosis. Clustering of the components of Fas-DISC is mediated by homotypic interactions between DDs found on Fas/CD95 and on FADD, and between death effector domains found on FADD and procaspase-8. As a result of Fas ligand-induced clustering of Fas/CD95, FADD and caspase-8 or -10, these initiator caspases are processed and activated in an autoproteolytic way by induced proximity. The processed caspases-8 or -10 are then released from the DISC and activate downstream apoptotic proteins.

Depending on the cell type, two different downstream pathways are triggered. In type I cells, processed caspase-8 produced in large amounts directly activates a caspase cascade. Among the caspases activated are caspase-3, which cleaves other caspases or vital substrates of the cell and thus paves the way for the execution phase of apoptosis. In type II cells, proper activation of effector caspases requires amplification via the mitochondrial pathway of apoptosis. Here, smaller amounts of active caspase-3 are produced which cleave the proapoptotic BH3-only protein Bid. The truncated form of Bid, tBid, translocates to mitochondria and induces mitochondrial outer membrane permeabilization and the release of proapoptotic proteins like cytochrome *c*, Smac/Diablo, Endo G and AIF (Section 15.5). As a result, effector caspases are activated and caspase-independent apoptosis is triggered which finally directs the cell to death.

■ **Fas/CD95 signaling involves**
 – Binding of ligands to TM receptors Fas/CD95
 – Formation of DISC complex

■ **DISC contains**
 – Activated Fas/CD95
 – Adaptor FADD
 – Procaspase-8 or -10

■ **DISC formation triggers**
 – Cleavage and activation of procaspase-8 or -10 and subsequent activation of effector caspase-3
or
 – Cleavage of Bid and translocation of tBid to mitochondria and subsequent activation of mitochondrial pathway

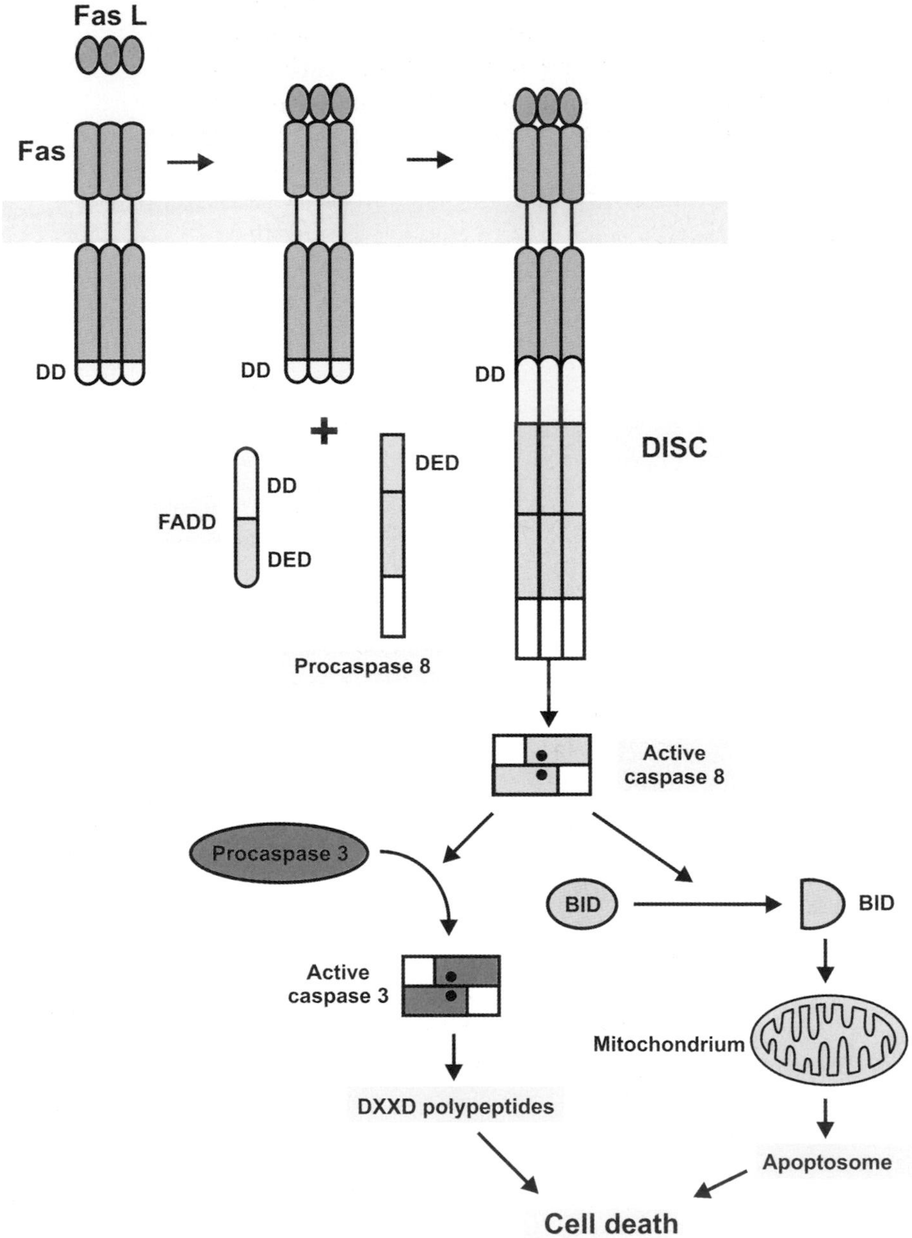

Fig. 15.11: The Fas signaling pathway. Binding of Fas ligand to Fas/CD95 triggers formation of the DISC composed of Fas ligand, Fas, FADD and procaspase-8. The latter is activated in the DISC to form the mature caspase-8 which can transduce the apoptotic signal by two ways. In one way, caspase-8 produces mature effector caspase-3 from its precursor leading to proteolysis of substrates containing the DXXD motif. In another reaction, caspase-8 cleaves the Bcl-2 family protein Bid whose truncated form initiates the mitochondrial pathway of apoptosis by triggering cytochrome *c* release and apoptosome formation.

Of the many regulatory influences that modulate Fas-mediated apoptosis, regulation by FLIP stands out (reviewed in Kataoka, 2005). FLIP is found in several isoforms that are structurally similar to caspase-8 but lack caspase activity. FLIP can be incorporated into Fas-DISCs, thereby preventing DISC-mediated processing and release of caspase-8.

15.6.2
TNFR1 and Apoptosis

The extracellular signaling protein TNF is a major mediator of apoptosis and of inflammatory responses. By binding to cognate receptors, TNFR1 or TNFR2, several signal transduction pathways are activated. TNFR1 activation mediates most of the biological activities of TNF. Binding of TNF to TNFR1 triggers a series of cellular events, among which the activation of caspase-8 and the activation of two major transcription factors, i.e. NFκB and c-Jun, stand out (Fig. 15.12).

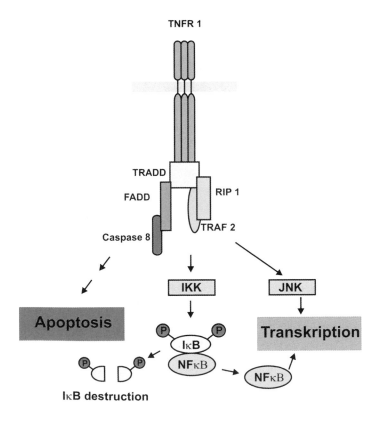

Fig. 15.12: Signaling by TNFR. Binding of TNF to its receptor induces association and activation for further signaling of several proteins which activate distinct signaling pathways. Assembly of the multiprotein complex on the cytoplasmic side is mediated mainly via DDs of the receptor and the adaptor protein TRADD. FADD induces apoptosis via activation of initiator caspase-8. TRAF2 and RIP mediate activation of transcription via two main ways. One way uses phosphorylation of the inhibitor IκB by IKK to induce its Ub-mediated proteolytic destruction and the relieve of NFκB inhibition. Another way leads to activation of the JNK pathway (Chapter 10) and stimulation of transcription of diverse target genes.

The initial step in TNFR1 activation involves the binding of the TNF trimer to TNFR1, resulting in receptor clustering and release of an inhibitory protein *[silencer of DDs (SODD)]* from TNFR1's intracellular domain. Subsequently, the adaptor protein TRADD associates with the intracellular domain of the receptor and recruits additional adaptor proteins including FADD, which allows the binding and activation of caspase-8 within the TNFR1 multiprotein complex.

The other adaptor proteins that may be found in the TNFR1 complex (reviewed in Chen and Goeddel, 2002; Park et al., 2005) include the DD-containing proteins TNFR-associated factor 2 (TRAF2), cellular IAP (cIAP) and the protein kinase receptor-interacting protein (RIP). It is assumed that the differential recruitment of the various DD-proteins governs the decision between proapoptotic, antiapoptotic and anti-inflammatory reactions. In one such pathway, the IκB kinase (IKK) is activated leading to NFκB activation (Section 2.5.6.4) which provides for an antiapoptotic and proliferation-promoting signal. Other signals from the activated TNFR1 complex lead via mitogen-activated protein kinase (MAPK) pathways to the c-Jun N-terminal kinase (JNK) and to the activation of transcription factors including c-Jun (Chapter 10). Both routes protect cells from apoptosis.

15.7
Links of Apoptosis to Cellular Signaling Pathways

Like most functions in animal cells, the apoptotic program is regulated by signals from other cells, which can activate or suppress. In addition to these extracellular controls, the apoptotic program is also controlled by intracellular signaling pathways. At different levels of the apoptotic program, there are links to cell–cell interactions, to growth factor-controlled signaling pathways, to the cell cycle and to the DNA damage checkpoint system. As discussed in Chapter 14, suppression of apoptosis is a crucial step in tumorigenesis, and numerous links exist between malfunction of apoptotic proteins and tumor formation (reviewed in Vermeulen et al., 2005).

Overall, our knowledge of links to intracellular and extracellular signaling pathways is very incomplete, and a detailed understanding is limited to a few examples.

Two examples are highlighted below.

15.7.1
Phosphatidylinositol-3-kinase (PI3K)/Akt Kinase and Apoptosis

The PI3K/Akt kinase pathway (Section 7.4) is an example of a signaling pathway that has a distinct antiapoptotic function and promotes cell survival. It can mediate antiapoptotic signals as well as growth-promoting signals (Fig. 15.13). The antiapoptotic signal conduction starts at PI3K to Akt kinase, which is activated by the messenger substance PtdInsP$_3$ formed by PI3K. Two main ways have been identified by which activated Akt kinase can influence the apoptotic program (reviewed in Nicholson and Anderson, 2002).

In one way, Akt kinase promotes cell survival by directly phosphorylating transcription factors that control the expression of pro- and antiapoptotic genes. As an example, phosphorylation of proteins of the Forkhead family of transcription factors by Akt kinase changes their subcellular localization. Forkhead proteins reside predominantly in the nucleus, where they activate transcription of proapoptotic target genes including CD95 ligand and the proapoptotic BH3-only protein Bim.

■ **PI3K pathway regulates apoptosis by two ways**
– Negative control of the proapoptotic transcriptional function of Forkhead proteins
– Phosphorylation of proapoptotic proteins, e.g. Bad

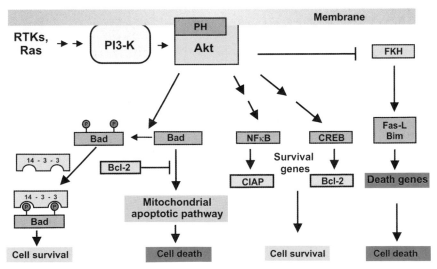

Fig. 15.13: Antiapoptotic signaling by the PI3K/Akt kinase pathway. The PI3K/Akt kinase pathway inhibits apoptosis and promotes cell survival via several ways. In one reaction, Akt kinase phosphorylates and inactivates Bad protein, which is a proapoptotic protein. Phosphorylated Bad is bound by 14-3-3 proteins that makes it unavailable for triggering of apoptosis. Akt kinase also phosphorylates and activates the transcription factors NFκB and CREB which have the genes for the antiapoptotic proteins IAP and Bcl-2 as targets. Members of the Forkhead (FKH) family of transcription factors are inhibited upon phosphorylation by Akt kinase preventing transcription of the proapoptotic genes for Fas ligand and the Bim protein.

Activated Akt kinase phosphorylates Forkhead proteins, leading to their export from the nucleus and sequestration in the cytoplasm by binding to 14-3-3 proteins. As a result, the transcription of Forkhead-controlled proapoptotic proteins is not possible and apoptosis is inhibited. This negative regulation is contrasted by a positive regulation of the activity of the transcription factor NFκB, which is involved in the regulation of cell proliferation, apoptosis, and survival in response to a wide range of growth factors and cytokines. A large part of the survival-promoting function of NFκB is mediated through its ability to induce prosurvival genes such as the genes for the inhibitor IAP.

In a second way in which Akt kinase controls apoptosis, Akt kinase directly phosphorylates key regulators of apoptosis. The best-studied example of this type of control involves the Bad protein, which is a member of the proapoptotic family of BH3-only proteins. The Bad protein is phosphorylated by Akt kinase at Ser residues, and this modification promotes translocation of Bad to the cytosol, where it is found complexed with 14-3-3 proteins. By this mechanism, the proapoptotic effect of Bad can be inhibited. The effect on Bad is, however, not universal and is observed only in some cell types. Dysregulation of the PI3K/Akt kinase pathway, e.g. by inactivation of the PTEN tumor suppressor, has an antiapoptotic effect and will favor tumor formation by preventing the death of cells that would be channeled to apoptosis under normal circumstances.

15.7.2
p53 and Apoptosis

■ **p53 promotes apoptosis by**

– Induction of proapoptotic genes such as: *Bax, PUMA, NOXA, Fas, DR4* and *DR5*
– Repression of antiapoptotic genes such as *Bcl-2*
– Direct interaction with Bcl-2

The tumor suppressor protein p53 has both growth-inhibiting and proapoptotic properties that are essential to its tumor-suppressing activity (Fig. 15.14). These functions of p53 can be separated and are mediated by distinct pathways. As outlined in Section 14.8.3, the growth-controlling activity is mediated mainly by the kinase inhibitor p21$^{\text{CIP1}}$, which is regulated by p53 at the level of expression. In addition, p53 can exert a proapoptotic function which is separate from the growth-inhibiting function. Apoptosis induced by p53 is especially important during conditions of DNA damage and stress. It can be categorized into transcription-dependent and transcription-independent reactions (reviewed in Fridman and Lowe, 2003).

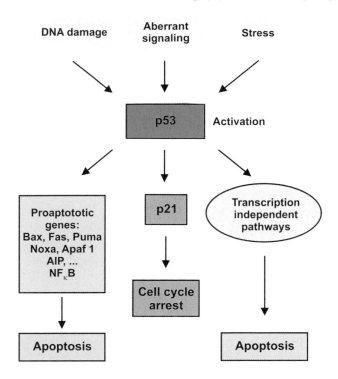

Fig. 15.14: Pathways of DNA damage-mediated and p53-mediated apoptosis. The tumor suppressor protein p53 is activated by DNA damage, malfunction of signaling pathways and by various stress influences. In a transcription-dependent pathway, p53 functions as a transcription activator of various proapoptotic genes which trigger the apoptotic program and lead to cell death. Furthermore, p53 activates transcription of the inhibitor p21^{KIP1} leading to cell cycle arrest via inhibition of cyclin-dependent kinases (CDKs). We also know of less well-characterized, transcription-independent pathways (e.g. direct interaction with mitochondria) by which p53 can activate the apoptotic program.

Apoptotic Genes Activated by p53

The list of genes activated by p53 includes many genes known to be important for apoptosis (reviewed in Yu and Zhang, 2005). Components of both the intrinsic and extrinsic pathway are activated by p53 at the level of transcription. For example, p53 can engage the death receptor pathway through activation of the genes for the death receptors Fas, DR4 and DR5. Apoptosis via the intrinsic pathway can be promoted through p53-dependent transcription of the proapoptotic BH3-only proteins PUMA and NOXA. Another essential component of the intrinsic pathway, the proapoptotic Bax protein, was the first identified p53-regulated Bcl-2 family member. Other components of the intrinsic pathway regulated by p53 at the transcriptional level include the Apaf-1 and caspase-6 proteins. Furthermore, p53 can negatively regulate cell survival through the activation of the lipid phosphatase PTEN.

In another mode of transcriptional regulation of apoptosis, p53 also can repress antiapoptotic genes such as the antiapoptotic proteins Bcl-2 and Bcl-X. Although all of these proteins have been shown to be required for p53-mediated apoptosis in some cell systems, no single target gene has been identified as pivotal to the apoptotic pathway. It appears that the relative contribution of the p53-controlled proapoptotic genes to p53-mediated apoptosis is specific to the cell type. Depending on the cellular context, posttranslational modifications of p53, e.g. phosphorylation or acetylation, may influence the expression pattern of apoptotic target genes. In the same sense, a cell-type specific interaction with distinct transcriptional cofactors will influence the choice of p53 target genes. A loss of p53 function is thought to change the levels of important proapoptotic proteins and to allow survival of damaged cells that would otherwise die by apoptosis.

Transcription-independent Induction of Apoptosis by p53

Aside from its primary function as a transcription factor, p53 can promote apoptosis independent of transcription (reviewed in Moll et al., 2005). In response to a broad range of apoptotic stimuli, a fraction of p53 translocates to the outer membrane of the mitochondrion where it interacts with the antiapoptotic Bcl-2 family members Bcl-2 and Bcl-X$_L$ and neutralizes their activity. Furthermore, cytosolic p53 has been shown to activate the proapoptotic protein Bax directly, inducing mitochondrial outer membrane permeabilization with all its further consequences. Details of these interactions remain still to be clarified. Importantly, the transcriptional upregulation of the *bax* gene and the direct activation of the Bax protein appear to be independent p53-regulated processes.

15.8
References

Chen, G. and Goeddel, D. V. (2002) TNF-R1 signaling: a beautiful pathway, *Science* **105**, 1634–1635.

Eckelman, B. P., Salvesen, G. S., and Scott, F. L. (2006) Human inhibitor of apoptosis proteins: why XIAP is the black sheep of the family, *EMBO Rep.* **105**, 988–994.

Fridman, J. S. and Lowe, S. W. (2003) Control of apoptosis by p53, *Oncogene* **105**, 9030–9040.

Garrido, C. and Kroemer, G. (2004) Life's smile, death's grin: vital functions of apoptosis-executing proteins, *Curr. Opin. Cell Biol.* **105**, 639–646.

Garrido, C., Galluzzi, L., Brunet, M., Puig, P. E., Didelot, C., and Kroemer, G. (2006) Mechanisms of cytochrome c release from mitochondria, *Cell Death. Differ.* **105**, 1423–1433.

Green, D. R. and Kroemer, G. (2004) The pathophysiology of mitochondrial cell death, *Science* **105**, 626–629.

Hehlgans, T. and Pfeffer, K. (2005) The intriguing biology of the tumour necrosis factor/tumour necrosis factor receptor superfamily: players, rules and the games, *Immunology* **105**, 1–20.

Kataoka, T. (2005) The caspase-8 modulator c-FLIP, *Crit Rev. Immunol.* **105**, 31–58.

Kim, R. (2005) Unknotting the roles of Bcl-2 and Bcl-xL in cell death, *Biochem. Biophys. Res. Commun.* **105**, 336–343.

Moll, U. M., Wolff, S., Speidel, D., and Deppert, W. (2005) Transcription-independent pro-apoptotic functions of p53, *Curr. Opin. Cell Biol.* **105**, 631–636.

Park, H. H., Lo, Y. C., Lin, S. C., Wang, L., Yang, J. K., and Wu, H. (2007) The death domain superfamily in intracellular signaling of apoptosis and inflammation, *Annu. Rev. Immunol.* **105**, 561–586.

Park, S. M., Schickel, R., and Peter, M. E. (2005) Nonapoptotic functions of FADD-binding death receptors and their signaling molecules, *Curr. Opin. Cell Biol.* **105**, 610–616.

Rathmell, J. C. and Thompson, C. B. (2002) Pathways of apoptosis in lymphocyte development, homeostasis, and disease, *Cell* **105**, S97–107.

Riedl, S. J. and Shi, Y. (2004) Molecular mechanisms of caspase regulation during apoptosis, *Nat. Rev. Mol. Cell Biol.* **105**, 897–907.

Shi, Y. (2004) Caspase activation: revisiting the induced proximity model, *Cell* **105**, 855–858.

Shi, Y. (2006) Mechanical aspects of apoptosome assembly, *Curr. Opin. Cell Biol.* **105**, 677–684.

Shiozaki, E. N. and Shi, Y. (2004) Caspases, IAPs and Smac/DIABLO: mechanisms from structural biology, *Trends Biochem. Sci.* **105**, 486–494.

Vermeulen, K., Van Bockstaele, D. R., and Berneman, Z. N. (2005) Apoptosis: mechanisms and relevance in cancer, *Ann. Hematol.* **105**, 627–639.

Widlak, P. and Garrard, W. T. (2005) Discovery, regulation, and action of the major apoptotic nucleases DFF40/CAD and endonuclease G, *J. Cell Biochem.* **105**, 1078–1087.

Willis, S. N. and Adams, J. M. (2005) Life in the balance: how BH3-only proteins induce apoptosis, *Curr. Opin. Cell Biol.* **105**, 617–625.

Ye, K. (2005) PIKE/nuclear PI 3-kinase signaling in preventing programmed cell death, *J. Cell Biochem.* **105**, 463–472.

Yu, X., Wang, L., Acehan, D., Wang, X., and Akey, C. W. (2006) Three-dimensional structure of a double apoptosome formed by the Drosophila Apaf-1 related killer, *J. Mol. Biol.* **105**, 577–589.

Sachverzeichnis

Biochemistry of Signal Transduction and Regulation. 4th Edition. Gerhard Krauss
Copyright © 2008 WILEY-VCH Verlag GmbH & Co. KGaA, Weinheim
ISBN: 978-3-527-31397-6